Nuclear Magnetic Resonance Spectroscopy

Nuclear Magnetic Resonance Spectroscopy

Special Issue Editor

Teresa Lehmann

MDPI • Basel • Beijing • Wuhan • Barcelona • Belgrade

MDPI

Special Issue Editor
Teresa Lehmann
University of Wyoming
USA

Editorial Office
MDPI
St. Alban-Anlage 66
Basel, Switzerland

This is a reprint of articles from the Special Issue published online in the open access journal *Magnetochemistry* (ISSN 2312-7481) from 2017 to 2018 (available at: http://www.mdpi.com/journal/magnetochemistry/special_issues/NMR)

For citation purposes, cite each article independently as indicated on the article page online and as indicated below:

LastName, A.A.; LastName, B.B.; LastName, C.C. Article Title. *Journal Name* **Year**, *Article Number*, Page Range.

ISBN 978-3-03842-993-7 (Pbk)
ISBN 978-3-03842-994-4 (PDF)

Contents

About the Special Issue Editor

Teresa Lehmann is an Associate Professor of Chemistry at the University of Wyoming in Laramie, WY. Her research focuses on the understanding of the structure-function relationship for biologically relevant biopolymers. These biopolymers include anti-plasmodial peptides, metal complexes of antitumor antibiotics, and diabetes- and obesity-related natural products. Primary research techniques include nuclear magnetic resonance (NMR) spectroscopy and molecular modeling for the study of structure and dynamics. Prof. Lehmann holds close collaborations with other faculty members at the University of Wyoming using NMR and molecular dynamics to investigate drug-target interactions and crude-oil/water interfacial phenomena.

magnetochemistry

MDPI

Editorial

Nuclear Magnetic Resonance Spectroscopy

Teresa E. Lehmann

Department of Chemistry, University of Wyoming, Laramie, WY 82071, USA; tlehmann@uwyo.edu;
Tel.: +1-307-766-2772

Received: 13 April 2018; Accepted: 16 April 2018; Published: 20 April 2018

Over the past fifty years, nuclear magnetic resonance spectroscopy (NMR) has gained popularity in a wide variety of research areas. Its non-invasive character makes it ideal for the study of biomolecules and tissue samples. The possibility of determining the solution structure of molecules with the use of NMR has led to important advances in many areas including pharmacy, biology, botany, medicine, sensor design, study of polymers and more. The technology used in spectrometer and probe design have advanced to allow the study of molecules in solids, liquids, and gels. Additionally, data in the form of images, multidimensional spectra, or relaxation profiles can be obtained to study systems from different fronts. The possibility of detecting multiple nuclei allows intimate examination and characterization of a wide variety of materials.

This Special Issue devoted to NMR, and the associated Special Issue reprint, contain papers covering some of the most innovative and exiting NMR applications to different systems.

(1) Aiken, K. et al. Nuclear Magnetic Resonance Spectroscopy Investigations of Naphthalene-Based 1,2,3-Triazole Systems for Anion Sensing. Deals with the use of NMR to characterize and study the fluoride binding of 2-(4-(naphthalen-2-yl)-1H-1,2,3-triazol-1-yl)phenol (NpTP). The capability of NpTP to recognize anions makes it an asset in logic gate systems as well as molecular switches, dual detection systems, and in biological environments [1].

(2) Klemm, R. Towards a Microscopic Theory of the Knight Shift in an Anisotropic, Multiband Type-II Superconductor. In this contribution, a method is proposed to extend the zero-temperature Hall-Klemm microscopic theory of the Knight shift K in an anisotropic and correlated, multi-band metal to calculate K(T) at finite temperatures T both above and into its superconducting state. A procedure to obtain a microscopic theory of the Knight shift in an anisotropic Type-II superconductor is outlined [2].

(3) Daigle, J.-C. et al. Solid-State NMR Study of New Copolymers as Solid Polymer Electrolytes. Deals with the use of solid-state NMR to characterize comb-like copolymers, which could be used in the future in the batteries of electrical vehicles [3].

(4) Makulski, W. The Radiofrequency NMR Spectra of Lithium Salts in Water; Reevaluation of Nuclear Magnetic Moments for ^{6}Li and ^{7}Li Nuclei. In this contribution, the dipole moments of ^{6}Li and ^{7}Li were determined through NMR in the presence of dissolved ^{3}He atoms, used as a shielding reference in salt water solutions. The values obtained for the ^{6}Li and ^{7}Li magnetic moments are more reliable than those determined in water solvents [4].

(5) Brass, M. et al. Spatially Resolved Measurements of Crosslinking in UV-Curable Coatings Using Single-Sided NMR. This paper describes the extent of crosslinking in UV-curable coatings quantified through single-sided NMR. This method can be applied in evaluating systems whose crosslinking properties are intentionally varied throughout its thickness [5].

(6) Follett, S. et al. Structural Changes of Zn(II)bleomycin Complexes When Bound to DNA Hairpins Containing the 5′-GT-3′ and 5′-GC-3′ Binding Sites, Studied through NMR Spectroscopy. Describes the use of one- and two-dimensional NMR spectroscopy to determine the conformations adopted by Zn(II)bleomycin-A_2, -A_5, -B_2, and Zn(II)peplomycin in the presence of two different

DNA fragments. The results of the investigation indicate that the bleomycin C-termini and the DNA base sequence have an impact on the conformation of the different drugs in the target-drug complexes [6].

(7) Furihata, K. et al. Application of NMR Screening Methods with [19]F Detection to Fluorinated Compounds Bound to Proteins. The NMR-based screening methods with ^{19}F-detection were applied to the human serum albumin (HSA)-diflunisal complex [7].

(8) Krishnarjuna, B. et al. Accelerating NMR-Based Structural Studies of Proteins by Combining Amino Acid Selective Unlabeling and Fast NMR Methods. Involves the application of fast NMR methods, G-matrix Fourier transform (GFT) and non-uniform sampling (NUS), to proteins samples prepared using a specific labeling scheme. The spectral simplification obtained combined with rapid data collection can help in reducing the time required for data analysis [8].

(9) Springer, T. et al. Calcium-Dependent Interaction Occurs between Slow Skeletal Myosin Binding Protein C and Calmodulin. The conformation of a recombinant N-terminal fragment of slow skeletal myosin binding protein (ssC1C2) is characterized using differential scanning fluorimetry, nuclear magnetic resonance, and molecular modeling. The differential molecular regulation of contractility that exists between skeletal and cardiac muscle are outlined [9].

(10) Price, L. et al. Local and Average Structural Changes in Zeolite A upon Ion Exchange. Solid state NMR was used to study the changes to the local structure of the Linde Type A framework of Zeolite A. This study confirms that using a local probe such as solid state NMR alongside powder diffraction and other long-range methods to study zeolites can reveal an extra level of information about the structure of those useful minerals, which will further their use as potential catalysts and ion exchange materials [10].

(11) Cantarutti, C. et al. Short-Chain Alkanethiol Coating for Small-Size Gold Nanoparticles Supporting Protein Stability. The interaction of the 3-mercaptopropionic acid coating of AuNPs (MPA-AuNPs) with β2-microglobulin (β2m) was investigated to probe protein structure perturbations. NMR and fluorescence spectroscopies were useful in determining that β2m interacts with MPA-AuNPs through a highly localized patch maintaining its overall native structure with minor conformational changes [11].

(12) Vyalikh, A. et al. Early Stages of Biomineral Formation—A Solid-State NMR Investigation of the Mandibles of Minipigs. This contribution describes the changes in the mineral phase upon new tissue formation and maturation in the bone blocks surrounding dental implants in minipigs. Solid-state NMR spectroscopy allowed for the identification of inorganic species during the biomineral formation at very early stages, when crystallite particles visible in direct imaging techniques have not yet been formed [12].

(13) Yamanoi, T. et al. Separation of the α- and β-Anomers of Carbohydrates by Diffusion-Ordered NMR Spectroscopy. This article describes the successful use diffusion-ordered spectroscopy (DOSY) for the separation and analysis of the α- and β-anomers of carbohydrates with different diffusion coefficients. The individual diffusion coefficients were determined. Additionally, DOSY was also used to separate two kinds of glucopyranosides having similar aglycon structures from a mixture [13].

(14) Poirier, D. et al. NMR-Assisted Structure Elucidation of an Anticancer Steroid-β-Enaminone Derivative. The unknown product of the fortuitous modification of a quinoline-proline-piperazine side chain linked to a steroid in the presence of lithium (trimethylsilyl) acetylidethe was identified to be the β-enaminone steroid derivative through two-dimensional NMR spectroscopy [14].

(15) Nam, A.-M. et al. Quantification of Squalene in Olive Oil Using [13]C Nuclear Magnetic Resonance Spectroscopy. A method for direct quantification of squalene using ^{13}C-NMR spectroscopy without saponification, extraction, or fractionation of the investigated samples was developed by these authors. The method led to reliable quantitative determination of squalene in olive oil samples from Corsica with an analysis time of less than three hours using a medium field NMR spectrometer (9.4 T) [15].

(16) Zivkovic, A. et al. Low Field NMR Determination of pKa Values for Hydrophilic Drugs for Students in Medicinal Chemistry. A bench top NMR spectrometer was used for the determination of pKa values of different drugs. The pKa values obtained showed to be in agreement with the literature data for the compounds [16].

(17) Aulikki, M. et al. The NMR^2 Method to Determine Rapidly the Structure of the Binding Pocket of a Protein–Ligand Complex with High Accuracy. A NMR Molecular Replacement (NMR^2) method was developed. It is shown how NMR^2 very quickly provides the complex structure of a binding pocket in a protein as measured by solution-state NMR, circumventing the long process of signal assignment [17].

(18) Proietti, N. et al. Nuclear Magnetic Resonance, a Powerful Tool in Cultural Heritage. The use of diverse NMR techniques in the monitoring and diagnosis of artworks, in order to prevent or delay their degradation, is described in this work. The development of portable NMR sensors suitable for non-destructive and non-invasive analysis in situ has made possible the investigation of precious and unmovable artefacts, and their monitoring over time [18].

(19) Porion, P. et al. Multi-Quanta Spin-Locking Nuclear Magnetic Resonance Relaxation Measurements: An Analysis of the Long-Time Dynamical Properties of Ions and Water Molecules Confined within Dense Clay Sediments. Multi-quanta spin-locking NMR relaxometry of quadrupolar nuclei was tested, in order to investigate the dynamical properties of confined fluids. This technique was shown to be a powerful tool to quantify the average residence time of molecular and ionic probes confined within the interlamellar space of clay lamellae inside dense sediments. The developed protocols are expected to be easily extended to study other interfacial systems, including porous silicate, zeolites, and cements [19].

(20) Sullivan, N. et al. Orientational Glasses: NMR and Electric Susceptibility Studies. The results of a wide range of NMR measurements of the local order parameters and the molecular dynamics of solid ortho-para hydrogen mixtures and solid nitrogen–argon mixtures are reviewed in this account. These mixtures form novel molecular orientational glass states at low temperatures. Additionally, studies of the dielectric susceptibilities of the nitrogen-argon mixtures are reviewed, in terms of replica symmetry breaking analogous to that observed for spin glass states. It is shown that this wide set of experimental results is consistent with orientation or quadrupolar glass ordering of the orientational degrees of freedom [20].

(21) Ciancaleoni, G. Characterization of Halogen Bonded Adducts in Solution by Advanced NMR Techniques. This review shows how crucial information about halogen bonding adducts can be obtained by advanced NMR techniques. It is proposed that advanced NMR techniques could be used increasingly in the near future in the consolidated, but still fruitful and rapidly evolving, field of halogen bonding in solution [21].

Conflicts of Interest: The author declares no conflict of interest.

References

1. Aiken, K.; Bunn, J.; Sutton, S.; Christianson, M.; Winder, D.; Freeman, C.; Padgett, C.; McMillen, C.; Ghosh, D.; Landge, S. Nuclear Magnetic Resonance Spectroscopy Investigations of Naphthalene-Based 1,2,3-Triazole Systems for Anion Sensing. *Magnetochemistry* **2018**, *4*, 15. [CrossRef]

2. Klemm, R.A. Towards a Microscopic Theory of the Knight Shift in an Anisotropic, Multiband Type-II Superconductor. *Magnetochemistry* **2018**, *4*, 14. [CrossRef]

3. Daigle, J.-C.; Arnold, A.A.; Vijh, A.; Zaghib, K. Solid-State NMR Study of New Copolymers as Solid Polymer Electrolytes. *Magnetochemistry* **2018**, *4*, 13. [CrossRef]

4. Makulski, W. The Radiofrequency NMR Spectra of Lithium Salts in Water; Reevaluation of Nuclear Magnetic Moments for ^{6}Li and ^{7}Li Nuclei. *Magnetochemistry* **2018**, *4*, 9. [CrossRef]

5. Brass, M.; Morin, F.; Meldrum, T. Spatially Resolved Measurements of Crosslinking in UV-Curable Coatings Using Single-Sided NMR. *Magnetochemistry* **2018**, *4*, 8. [CrossRef]

6. Follett, S.E.; Murray, S.A.; Ingersoll, A.D.; Reilly, T.M.; Lehmann, T.E. Structural Changes of Zn(II)bleomycin Complexes When Bound to DNA Hairpins Containing the 5′-GT-3′ and 5′-GC-3′ Binding Sites, Studied through NMR Spectroscopy. *Magnetochemistry* **2018**, 4, 4. [CrossRef]

7. Furihata, K.; Usui, M.; Tashiro, M. Application of NMR Screening Methods with [19]F Detection to Fluorinated Compounds Bound to Proteins. *Magnetochemistry* **2018**, 4, 3. [CrossRef]

8. Krishnarjuna, B.; Chandra, K.; Atreya, H.S. Accelerating NMR-Based Structural Studies of Proteins by Combining Amino Acid Selective Unlabeling and Fast NMR Methods. *Magnetochemistry* **2018**, 4, 2. [CrossRef]

9. Springer, T.I.; Johns, C.W.; Cable, J.; Lin, B.L.; Sadayappan, S.; Finley, N.L. Calcium-Dependent Interaction Occurs between Slow Skeletal Myosin Binding Protein C and Calmodulin. *Magnetochemistry* **2018**, 4, 1. [CrossRef]

10. Price, L.; Leung, K.M.; Sartbaeva, A. Local and Average Structural Changes in Zeolite A upon Ion Exchange. *Magnetochemistry* **2017**, 3, 42. [CrossRef]

11. Cantarutti, C.; Bertoncin, P.; Corazza, A.; Giorgetti, S.; Mangione, P.P.; Bellotti, V.; Fogolari, F.; Esposito, G. Short-Chain Alkanethiol Coating for Small-Size Gold Nanoparticles Supporting Protein Stability. *Magnetochemistry* **2017**, 3, 40. [CrossRef]

12. Vyalikh, A.; Elschner, C.; Schulz, M.C.; Mai, R.; Scheler, U. Early Stages of Biomineral Formation—A Solid-State NMR Investigation of the Mandibles of Minipigs. *Magnetochemistry* **2017**, 3, 39. [CrossRef]

13. Yamanoi, T.; Oda, Y.; Katsuraya, K. Separation of the α- and β-Anomers of Carbohydrates by Diffusion-Ordered NMR Spectroscopy. *Magnetochemistry* **2017**, 3, 38. [CrossRef]

14. Poirier, D.; Maltais, R. NMR-Assisted Structure Elucidation of an Anticancer Steroid-β-Enaminone Derivative. *Magnetochemistry* **2017**, 3, 37. [CrossRef]

15. Nam, A.-M.; Bighelli, A.; Tomi, F.; Casanova, J.; Paoli, M. Quantification of Squalene in Olive Oil Using [13]C Nuclear Magnetic Resonance Spectroscopy. *Magnetochemistry* **2017**, 3, 34. [CrossRef]

16. Zivkovic, A.; Bandolik, J.J.; Skerhut, A.J.; Coesfeld, C.; Raos, M.; Zivkovic, N.; Nikolic, V.; Stark, H. Low Field NMR Determination of p*Ka* Values for Hydrophilic Drugs for Students in Medicinal Chemistry. *Magnetochemistry* **2017**, 3, 29. [CrossRef]

17. Wälti, M.A.; Orts, J. The NMR[2] Method to Determine Rapidly the Structure of the Binding Pocket of a Protein–Ligand Complex with High Accuracy. *Magnetochemistry* **2018**, 4, 12. [CrossRef]

18. Proietti, N.; Capitani, D.; Di Tullio, V. Nuclear Magnetic Resonance, a Powerful Tool in Cultural Heritage. *Magnetochemistry* **2018**, 4, 11. [CrossRef]

19. Porion, P.; Delville, A. Multi-Quanta Spin-Locking Nuclear Magnetic Resonance Relaxation Measurements: An Analysis of the Long-Time Dynamical Properties of Ions and Water Molecules Confined within Dense Clay Sediments. *Magnetochemistry* **2017**, 3, 35. [CrossRef]

20. Sullivan, N.; Hamida, J.; Muttalib, K.; Pilla, S.; Genio, E. Orientational Glasses: NMR and Electric Susceptibility Studies. *Magnetochemistry* **2017**, 3, 33. [CrossRef]

21. Ciancaleoni, G. Characterization of Halogen Bonded Adducts in Solution by Advanced NMR Techniques. *Magnetochemistry* **2017**, 3, 30. [CrossRef]

magnetochemistry

MDPI

Article

Structural Changes of Zn(II)bleomycin Complexes When Bound to DNA Hairpins Containing the 5′-GT-3′ and 5′-GC-3′ Binding Sites, Studied through NMR Spectroscopy

Shelby E. Follett [1], Sally A. Murray [1], Azure D. Ingersoll [1], Teresa M. Reilly [2] and Teresa E. Lehmann [1,*]

[1]   Department of Chemistry, University of Wyoming, Laramie, WY 82071, USA;
      Shelby.follett@tolmar.com (S.E.F.); smurra15@uwyo.edu (S.A.M.); azureingersoll@gmail.com (A.D.I.)
[2]   Department of Chemical Engineering, University of Wyoming, Laramie, WY 82017, USA; treilly2@uwyo.edu
*   Correspondence: tlehmann@uwyo.edu; Tel.: +1-307-766-2772

Received: 9 November 2017; Accepted: 12 December 2017; Published: 27 December 2017

Abstract: We have previously investigated the diverse levels of disruption caused by Zn(II)BLMs with different C-termini to DNA hairpins containing 5′-GC-3′ and 5′-GT-3′ binding sites. The results of this investigation indicated that both the DNA-binding site and the bleomycin C-termini have an impact on the final conformation of the aforementioned hairpins in the drug-target complexes, as suggested by the different sets of intramolecular NOEs displayed by both oligonucleotides when bound to each Zn(II)BLM. The NMR signals elicited by ^{1}H nuclei in the oligonucleotide bases and sugar moieties were also affected differently (shifted upfield or downfield in various patterns) depending on the BLM C-termini and the binding site in the oligonucleotides. The overall conclusion derived from the precedent research is that the spatial conformation of target DNA segments in DNA-Zn(II)BLM complexes could be forged by interactions between drug and DNA that are guided by the DNA binding site and the BLM C-termini. The present study focuses on the structural alterations exhibited by Zn(II)bleomycin-A$_2$, -B$_2$, -A$_5$ and Zn(II)peplomycin molecules upon binding to the previously studied hairpins. Our main goal is to determine if different spatial conformations of the drugs in their DNA-bound forms are found in drug-DNA complexes that differ in the oligonucleotide binding site and BLM C-termini. Evidence that suggest that each Zn(II)bleomycin is structurally affected depending these two factors, as indicated by different sets of intramolecular NOE connectivities between drug protons and diverse patterns of shifting of their ^{1}H-NMR signals, is provided.

Keywords: DNA; NMR; pulmonary fibrosis; anticancer drug; structure-function

1. Introduction

Bleomycins (BLMs) compose a family of glycopeptide-derived antibiotics produced by *Streptomyces verticillus* [1]. BLMs have been used as chemotherapeutic agents in the clinical treatment of a wide spectrum of cancers, and their antitumor activity is generally proposed to be related to cleaving single-stranded or double-stranded DNA in carcinoma cells [2–4]. The overall structure of these agents can be thought of as containing four distinct regions (Figure 1): the metal binding domain (D1), which is responsible for metal binding [5,6], oxygen activation [5,7–9], and site-selective DNA cleavage [6,10]; the peptide linker (D2); the DNA binding domain (D3), containing a bithiazole moiety and the C-terminus, which provides the majority of the DNA binding affinity [11,12]; and the disaccharide moiety (D4), which influences metal ion binding [6,13–24] and is proposed to be a tumor-targeting unit [25].

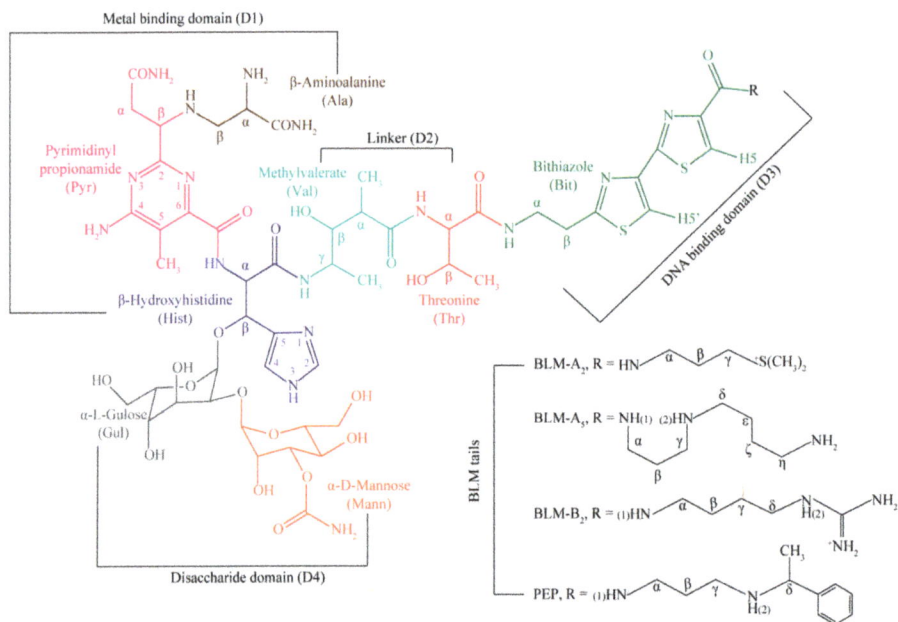

Figure 1. Structures BLM-A$_2$, -A$_5$, -B$_2$, and PEP showing the breakdown of the different domains, residues, and C-termini.

Although successful in the treatment of certain cancers, BLMs are associated with pulmonary toxicity, and extensive research is required to lower this risk to patients [16]. Biological studies performed by Raisfeld et al. [26–31] have linked the cause of pulmonary toxicity to the BLM C-termini (tails). Blenoxane, introduced in 1972, is the clinically used combination of BLMs, with the major components being BLM-A$_2$ and -B$_2$. Over 300 BLM analogs have since been developed with the hope of lowering the risk of pulmonary toxicity and achieving high levels of antitumor activity [32–34].

Research work on the interactions of various metallo-BLMs (MBLMs) with DNA fragments have generated abundant evidence indicating that the mode of binding of MBLMs to DNA is sensitive to various factors. These factors could include the DNA binding site, the DNA base sequence, the metal center bound to BLM, and the C-terminus of the drug. Different specific binding interactions between drug and target have been reported [35–44]; and various modes of binding of the bithiazole unit to different DNA fragments have been described including minor groove binding [35,37,45], and partial [40,42,43] or total [46] intercalation.

Most of the available research on MBLMs bound to DNA fragments focus on the structure of the full drug-target complex, and briefly mention how the MBLM molecule is affected upon binding to DNA. Manderville et al. showed that there are differences between the structural changes of Zn(II)BLM-A$_2$ and Zn(II)BLM-A$_5$ upon complexation with a DNA fragment with a 5′-GC-3′ binding site, and that the structure of Zn(II)BLM-A$_5$ was more disrupted than that of Zn(II)BLM-A$_2$ after binding [37]. Vanderwall et al. showed that when HOO-Co(III)BLM-A$_2$ was bound to DNA fragments with either the 5′-GC-3′ and 5′-GT-3′ binding sites there were little differences in the structural changes to the Zn(II)BLM [38]. Some of the available studies have briefly examined the influence DNA has on the MBLM structure, focusing almost exclusively on the bithiazole (Bit) and the β-hydroxyhistidine (Hist) moieties in BLM [35–39,47,48]. Additionally, the BLMs used in these studies are limited, with the majority using MBLM-A$_2$ [35–43,47], and some investigating MBLM-A$_5$ [37], MBLM-B$_2$ [46] and metallo-PEP [49]. Although the precedent work has provided the scientific community with important information regarding DNA-MBLM interactions, the multiplicity of DNA fragments, metal centers, and BLM C-termini used makes it difficult to generalize the findings.

We have previously performed two studies to determine the significance of various factors that could influence the final conformation of target DNA segments in Zn(II)BLM-DNA triads, using DNA hairpins of sequences 5′-AGCCTTTTGGCCT-3′ (OL$_1$) containing a 5′-GC-3′ binding site [50] and 5′-CCAGTATTTTTACTGG-3′ (OL$_2$) containing 5′-GT-3′ binding site [51]. The first study entailed investigating the effect that the BLM tails have on the binding of Zn(II)BLMs to a DNA hairpin of containing the 5′-GC-3′ binding site. The second study tested the role of the DNA binding site (5′-GC-3′ or 5′-GT-3′) in the relative spatial arrangement of the Zn(II)BLM-target complexes. The results of this work indicated that both the DNA-binding site and the bleomycin C-termini have an impact on the final conformation of the aforementioned hairpins in the drug-target complex, as suggested by the different sets of intramolecular NOEs displayed by both oligonucleotide (OLs) when bound to each Zn(II)BLM. The NMR signals elicited by ^{1}H nuclei in the OL bases and sugar moieties were also affected differently (shifted upfield or downfield in various patterns) depending on the BLM C-termini and the binding site in the OLs. The work presented herein has the goal of determining if the BLM chemical structure and the DNA binding site affect the conformation of Zn(II)BLM molecules in their OL1- and OL2-bound states. These studies all involve a Zn^{2+} metal center. Zn(II)BLMs maintain the same ligands that participate in chelation as Fe(II)BLMs [14,15,20–23] which, in the presence of oxygen, becomes HOO-Fe(III)BLM, the activated form of the MBLM proposed to cleave DNA in vivo. Due to the potential for DNA cleavage and paramagnetic nature of MBLMs containing the Fe(III) ion, Zn(II)BLMs are, in our opinion, the best diamagnetic inactive models for Fe(II)BLM, which is the next MBLM to be studied in our laboratory to determine the relevance of the metal center in MBLM-DNA interactions. The DNA base sequences were selected to be very similar to those used on studies of MBLMs bound to various DNA fragments with the aim of comparing our results to those of other researchers on the field. Due to the short lengths of various oligonucleotide previously reported in the available literature, and their self-complementarity, we decided to use DNA hairpins in order to guarantee that the BLM-binding site was located in a double-stranded region of the DNA segment, while keeping the OL at minimum complexity for the sake of NMR data analysis. The selected hairpins still contain the important inter-strand interactions found in double-stranded DNA, and therefore are a valid test models. In our previous work on the conformational changes exhibited by OL1 [50] and OL2 [51] in the presence of various Zn(II)BLMs we confirmed that the free OLs display sets of inter- and intra-strand NOEs that indicate normal double-stranded structures crowned by loops. We present here the structural changes the Zn(II)BLM molecules suffer upon binding to these hairpins.

2. Results

The ^{1}H-NMR signals elicited by free Zn(II)BLM-A$_2$, -A$_5$, -B$_2$, and Zn(II)PEP were assigned using COSY, TOCSY, and NOESY spectra acquired in H$_2$O at 5 °C. These assignments are collected in Supplementary Table S1. The NOESY spectra of these Zn(II)BLMs bound to OL$_1$ and OL$_2$ (Supplementary Figures S1–S4) acquired in H$_2$O at 5 °C previously examined [50,51] were analyzed this time to identify the signals generated by the bound Zn(II)BLMs, and investigate the effects that OL complexation has on the conformation of each Zn(II)BLMs. The spectra acquired for all Zn(II)BLM-DNA triads in D$_2$O at both 5 °C and 25 °C and in H$_2$O at 25 °C [50,51] were also used for confirmation of some of the peak assignments. The work described herein looks at the structural changes of the entire Zn(II)BLM molecule upon binding to the OLs, and allows for comparability between the different Zn(II)BLMs studied and the preferential binding sites in DNA.

Our previous studies on the Zn(II)BLM-DNA triads [50,51] show that Zn(II)BLMs are bound to both OLs through the analysis of one-dimensional (1D) ^{1}H-NMR spectra at both 5 °C in H$_2$O and 25 °C in D$_2$O. The proton signals in the imino region for the OLs are significantly affected in both studies, exhibiting downfield shifting and broadening for both OLs. The bithiazole (Bit) and β-hydroxyhistidine (Hist) ring protons of the Zn(II)BLMs also shift and broaden upon binding, and are essential when investigating the potential binding mode of the drugs to DNA [50,51]. As previously reported and shown in Table 1, the Bit aromatic signals in each Zn(II)BLM exhibit changes upon complexation to the OL1. In all cases, there is broadening and shifting of these signals for each Zn(II)BLM-OL1 triad indicating binding to OL1. However, there are differences in their behavior depending on the Zn(II)BLM bound. The Zn(II)BLM-A$_5$-OL1 triad exhibits large downfield shifting for the CH5 and CH5$'$ protons, with the -A$_2$-OL1, -B$_2$-OL1, and PEP-OL1 triads displaying upfield shifts in decreasing order, respectively [50]. Upon complexation of Zn(II)BLM-A$_2$ and Z n(II)PEP to OL2, the Bit ring protons also experience broadening and shifting. Examination of the $\Delta\delta$ values shown in Table 1 indicates that the Bit signals have a greater upfield shift in OL$_2$-bound Zn(II)PEP than in the Zn(II)BLM-A$_2$-OL2 triad [51]. This trend is opposite to that exhibited by the same signals in the OL1 triads.

Table 1. Chemical shift differences ($\Delta\delta$) between free and OL-bound Zn(II)BLM for spectra acquired in H$_2$O at 5 °C.

Residue	OL$_1$ Zn(II)BLM-A$_2$	OL$_2$ Zn(II)BLM-A$_2$	OL$_1$ Zn(II)PEP	OL$_2$ Zn(II)PEP	OL$_1$ Zn(II)BLM-B$_2$	OL$_1$ Zn(II)BLM-A$_5$
Val C$^\alpha$H	0.17 ^{a}	−0.07	0.21	−0.06	0.21	0.33
Val C$^\beta$H	−0.03	−0.06	−0.03	−0.29	−0.03	−0.07
Val C$^\gamma$H	−0.02	−0.06	−0.02	−0.08	−0.02	−0.03
Val C$^\alpha$CH$_3$	0.03	0.00	0.04	−0.04	0.03	0.02
Val C$^\gamma$CH$_3$	−0.01	−0.02	−0.01	0.19	−0.01	−0.01
Val NH	0.05	−0.10	0.05	^{b} -	0.08	0.09
Thr C$^\alpha$H	−0.02	−0.05	−0.03	−0.04	−0.04	−0.05
Thr C$^\beta$H	−0.03	−0.05	−0.04	−0.05	−0.04	−0.09
Thr CH$_3$	−0.01	−0.02	−0.02	−0.01	−0.02	−0.04
Thr NH	0.15	−0.02	0.19	-	0.18	0.28
Bit C$^\alpha$H$_2$	0.05	0.08	0.06	0.12	0.07	0.10
Bit C$^\beta$H$_2$	0.02	0.05	0.03	0.09	0.02	0.05
^{c} Bit C5$'$H	0.19	0.19	0.37	0.07	0.09	−0.19
^{c} Bit C5H	-	-	0.36	-	-	−0.37
Bit NH	−0.02	−0.03	−0.03	0.01	−0.04	−0.05
Ala C$^\alpha$H	−0.01	0.00	−0.01	0.04	−0.01	0.00
Ala C$^\beta$H$_{2b}$	0.00	−0.01	−0.01	−0.05	−0.01	0.00
Ala NH	-	-	0.05	-	-	-
Ala NH$_{2a}$	0.00	0.01	−0.01	−0.04	−0.01	−0.01
Ala CONH$_{2a}$	−0.01	−0.01	−0.02	−0.14	−0.01	0.00
Ala CONH$_{2b}$	0.00	0.02	0.00	0.04	0.00	0.02
Pyr NH$_2$	−0.03	−0.04	−0.04	−0.05	−0.03	−0.02
Hist C2H	−0.05	−0.23	−0.01	−0.11	−0.03	−0.03
Hist C4H	−0.02	−0.04	−0.01	−0.03	−0.03	−0.04
Hist C$^\alpha$H	0.00	0.06	−0.03	0.01	0.00	−0.01
Hist C$^\beta$H	0.00	0.05	−0.01	−0.01	−0.01	0.00
Mann C1H	0.03	0.07	0.01	-	0.01	-
Mann C6H	-	-	-	−0.06	-	-
Mann C6H$'$	-	-	-	−0.04	-	-
Mann NH$_{2a}$	−0.01	−0.07	−0.01	−0.07	−0.01	−0.01
Gul C2H	0.00	−0.01	0.01	−0.04	0.01	0.00
^{d} A$_2$ C$^\beta$H$_2$	0.01	0.06				
A$_2$ C$^\gamma$H$_2$	0.05	0.16				
A$_2$ NH	0.09	0.23				
PEP C$^\alpha$H$_2$			0.03	0.13		
PEP C$^\gamma$H$_{2a}$			−0.01	0.04		
PEP C$^\delta$H			0.01	0.04		
PEP NH (2)			0.10	-		
B$_2$ C$^\alpha$H$_{2a}$					0.05	
B$_2$ C$^\gamma$H$_{2a}$					−0.04	
B$_2$ NH (1)					0.13	
A$_5$ C$^\alpha$H$_2$						0.08
A$_5$ NH$_{2a}$						0.24

^{a} Calculated as [(Free Zn(II)BLM)–(Bound Zn(II)BLM)]; ^{b} Unassignable; ^{c} Reported in [50,51]; ^{d} Red labels indicate tail protons for each Zn(II)BLM.

Table 1 also displays the $\Delta\delta$ values between the free and OL-bound forms of Zn(II)BLM for protons in other moieties of the BLM molecules that experience significant shifts ($\Delta\delta \geq 0.04$ ppm) upon complexation to OL1 and OL2. It is clear from this table that, when bound to the OLs, each Zn(II)BLM experiences a wide range of significant shifts from their original positions in the free forms. Additionally both upfield and downfield shifts are observed. For some protons, the preferential binding site (5′-GC-3′ (OL$_1$) vs. 5′-GT-3′ (OL$_2$)) influences the direction and magnitude of the shift (Table 1, columns 2–5). Another interesting result is that some of the signals experience a significant shift for only one of the triads. Additionally, the Methylvalerate (Val) C$^\alpha$H and the C$^\alpha$H$_2$ aliphatic protons in Bit exhibit significant shifts for all triads. Most of the available structural work on MBLM triads formed with DNA fragments focus on the Bit and Hist moieties. Our results show that most of the Bit moiety protons shift upfield, with the exception of the Bit NH proton, for all complexes. However, the degree and direction of this shifting is dependent upon the triad. The Hist ring protons, C2H and C4H, move downfield for all complexes with different degrees of shifting. These shifts show that although all Zn(II)BLMs studied share the same D1, D2, and D4, each of them is impacted differently, in terms of Δδ, based on the chemical structure of the C-terminus and the binding site in the OLs.

Using schematic diagrams, we illustrate which Zn(II)BLM protons experience significant shifts upon binding to OL$_1$ and OL$_2$, based on the results shown in Table 1. Figure 2 shows the Zn(II)BLM protons that experience significant shifts (red circles) in the OL$_1$ triads. Figure 3 shows the Zn(II)BLM protons that experience significant shifts (green circles) in the OL$_2$ triads. The protons circled in purple were assigned in the free form of the Zn(II)BLMs, but could not be assigned in their OL-bound forms. Figures 2 and 3 schematically show that both, the BLM tail and the binding site in DNA have an effect on the magnetic and/or chemical environment experienced by some protons in the Zn(II)BLMs. It is important to indicate that the NMR signals of most of the Zn(II)BLM sugar protons could not be assigned due to significant overlap with the signals coming from the sugar moieties in OL$_1$ and OL$_2$.

5′-GC-3′ DNA Binding Site

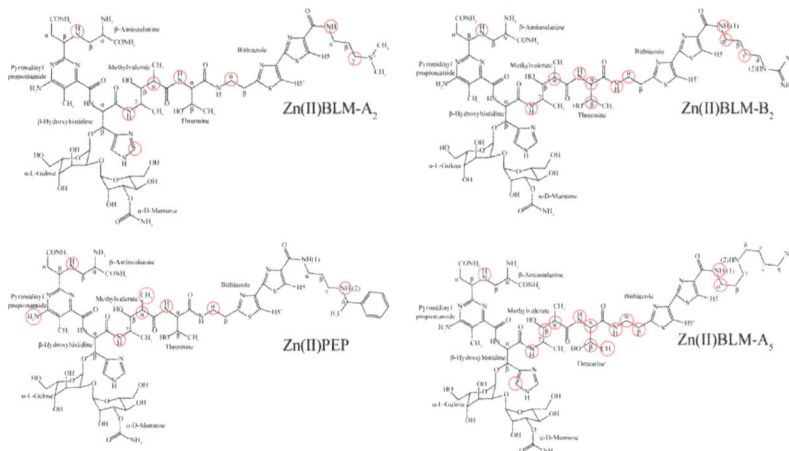

Figure 2. Protons exhibiting significant shifts in their OL$_1$-bound forms. (red circles) and protons that could not be assigned in the bound forms (purple circles) for spectra acquired in H$_2$O at 5 °C.

5'-GT-3' DNA Binding Site

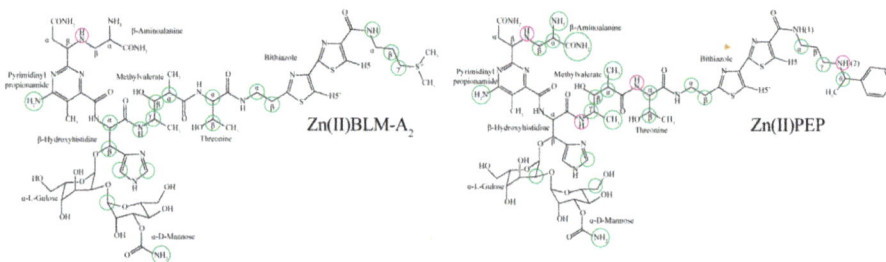

Figure 3. Protons exhibiting significant shifts in their OL$_2$-bound forms (green circles) and protons that could not be assigned in the bound forms (purple circles) for spectra acquired in H$_2$O at 5 °C.

Analysis of Figure 2 shows that in the OL$_1$ triads, Zn(II)BLM-A$_2$ and Zn(II)PEP are the least affected Zn(II)BLMs after binding, compared to Zn(II)BLM-B$_2$ and -A$_5$. In general terms, the significant shifts are contained within D2 and D3. The OL$_1$ triads with Zn(II)BLM-A$_2$ and -A$_5$ experience significant downfield shifts of the Hist C2H and C4H protons, respectively. The Ala NH protons is affected in all OL$_1$ triads, however, this signal in the bound Zn(II)BLMs could only be assigned in the Zn(II)PEP-OL$_1$ complex. Additionally, one of the Pyr NH$_2$ protons experiences significant downfield shifts in the Zn(II)PEP-OL$_1$ triad and in both OL$_2$ triads (Figure 3).

Comparison of Figures 2 and 3 shows that when the Zn(II)BLMs are complexed with the same OL, the chemical and/or magnetic environment of some protons vary based on the C-termini in BLM. However, when the same Zn(II)BLM is bound to both OLs, OL$_2$ has a greater effect on the new environments of the protons in that Zn(II)BLMs than OL$_1$. Specifically, for Zn(II)BLM-A$_2$ and Zn(II)PEP complexed with OL$_2$, many additional protons experience significant shifts, and the affected protons are no longer limited to D2 and D3. For both OL$_2$ triads, many of the D1 protons experience significant shifts. Additionally, the Hist and Ala moieties are greatly shifted in Zn(II)BLM-A$_2$ and Zn(II)PEP, respectively in the OL$_2$ triads. These results indicate that the metal binding domain experiences different environments upon complexation dependent upon the chemical structure of the BLM tail and the BLM-binding site in the OLs.

The spectra acquired for the Zn(II)BLM-DNA triads exhibit signals for the three ligands to the metal center containing protons: Ala NH, Ala NH$_2$, and Mann NH$_2$. The Ala NH proton signal could not be identified for the OL-bound forms of the Zn(II)BLM, except in the Zn(II)PEP-OL$_1$ complex. This result indicates that this ligand is somehow affected by the binding to the OLs, and exhibits a significant $\Delta\delta$ in Zn(II)PEP upon binding to OL$_1$. Furthermore for both of the OL$_2$ triads under study, one of the Mann NH$_2$ protons shows a significant $\Delta\delta$ after binding to OL$_2$, in addition to both protons in Ala NH$_2$ for the Zn(II)PEP-OL$_2$ triad. The Hist N1 and NH (deprotonated upon metal coordination) ligands do not generate ^{1}H-NMR signals, however, the Hist C$^\alpha$H and Hist C2H protons are in close proximity to these ligands. The Hist C2H proton displays significant $\Delta\delta$s for both OL$_2$ triads, and the Zn(II)BLM-A$_2$-OL$_1$ complex. Additionally, the Hist C$^\alpha$H proton is significantly shifted for the Zn(II)BLM-A$_2$-OL$_2$ triad. These results suggest that there are diverse possibilities for the magnetic and/or chemical environments experienced by the metal coordination cage dependent upon the BLM C-terminus and the DNA binding site.

We continued our analysis of DNA-bound Zn(II)BLMs by looking at how the intra-residue and inter-residue intramolecular NOEs for each of them are affected upon binding to the OLs. Figure 4 shows the inter-residue intramolecular NOEs for each Zn(II)BLM before complexation with the OLs for samples in H$_2$O at 5 °C (these NOEs are collected in: Zn(II)BLM-A$_2$, Supplementary Table S2; Zn(II)PEP, Supplementary Table S4; Zn(II)BLM-B$_2$, Supplementary Table S6; and Zn(II)BLM-A$_5$, Supplementary Table S8). This figure shows how each Zn(II)BLM is folded with complex through-space connections, however, each of them is different regarding their native conformations. Both Zn(II)PEP and Zn(II)BLM-A$_2$ have inter-residue intramolecular NOEs connecting the BLM tails to residues in D1, D2, and D4, indicating a more compact folded structure than Zn(II)BLM-B$_2$ and -A$_5$. Also indicative of a folded structure are the inter-residue intramolecular NOEs between D1 and D2, as well as between D4 and D2 observed for all Zn(II)BLMs.

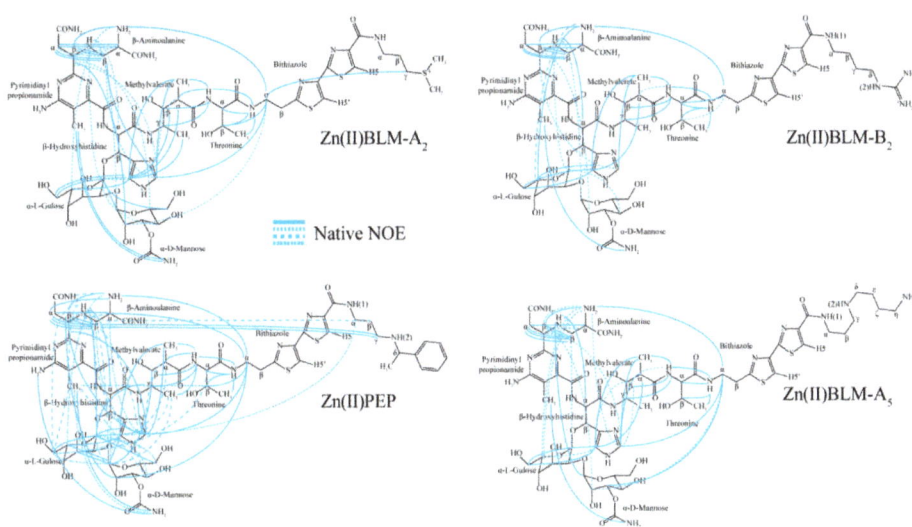

Figure 4. Inter-residue intramolecular NOEs for each Zn(II)BLM under study in their native forms. Data taken from spectra acquired in H$_2$O at 5 °C for Zn^{2+}:BLM samples in a 1:1 molar ratio. Dashed and continuous lines all represent NOE connectivities, they are used to avoid confusion in busy sectors of the figure.

Some of the inter-residue intramolecular NOEs present in the free Zn(II)BLMs (native NOEs) are also detected in their OL-bound forms. These NOEs, together with inter-residue intramolecular NOEs that arise in the Zn(II)BLMs upon binding to the OLs (new NOEs), are presented in Figures 5 and 6 for Zn(II)BLMs bound to OL$_1$ and OL$_2$, respectively. Supplementary Tables S2, S4, S6, and S8 show these NOEs for each OL-bound Zn(II)BLM. Supplementary Tables S3, S5, S7, and S9 display the intra-residue intramolecular NOEs for each free and OL-bound Zn(II)BLM.

Examination of these tables and Figures 4–6 leads to some interesting facts regarding the differences in conformation of the Zn(II)BLMs between their native and OL-bound forms. Comparison of Figures 4 and 5 shows that Zn(II)BLM-A$_5$ and Zn(II)BLM-B$_2$ lose more of their native NOEs than Zn(II)BLM-A$_2$ and Zn(II)PEP after complexation with OL$_1$. This result suggest that there is a greater degree of unfolding of these Zn(II)BLMs in their OL$_1$-bound forms. A multitude of new NOEs are connecting D1 and D2 protons in the OL$_1$-bound Zn(II)BLMs, and suggest refolding of the Zn(II)BLMs around D2. One of these NOEs connects Hist C2H to Threonine (Thr) CH$_3$, which interestingly is present for all OL$_1$ triads, but neither of the OL$_2$ triads (Figures 5 and 6). Comparison of Figures 5 and 6 shows many substantial differences in the conformations of the Zn(II)BLMs when bound to the two OLs. This fact based on the number and differences in the native and new NOEs that each bound Zn(II)BLM displays. Some of the new NOEs in the two OL$_2$ triads are connecting D1 and D3, which is also observed in the Zn(II)BLM-A$_5$-OL$_1$ complex. Examination of Zn(II)PEP bound to both OLs shows that there is a greater reduction in the number of native NOEs when this Zn(II)BLM is complexed with OL$_2$, which is significant around D1. In the free Zn(II)BLMs there is a multitude of NOEs connecting Ala protons with D4 and Hist protons. These NOEs suggest that D1 and D4 are involved in metal ion coordination. The limited number of these connections that remain after the Zn(II)BLMs bind to the OLs provide evidence suggesting that this segment has rearranged upon complexation to the OLs and each Zn(II)BLM-OL triad displays different conformations for it.

5'-GC-3' DNA Binding Site

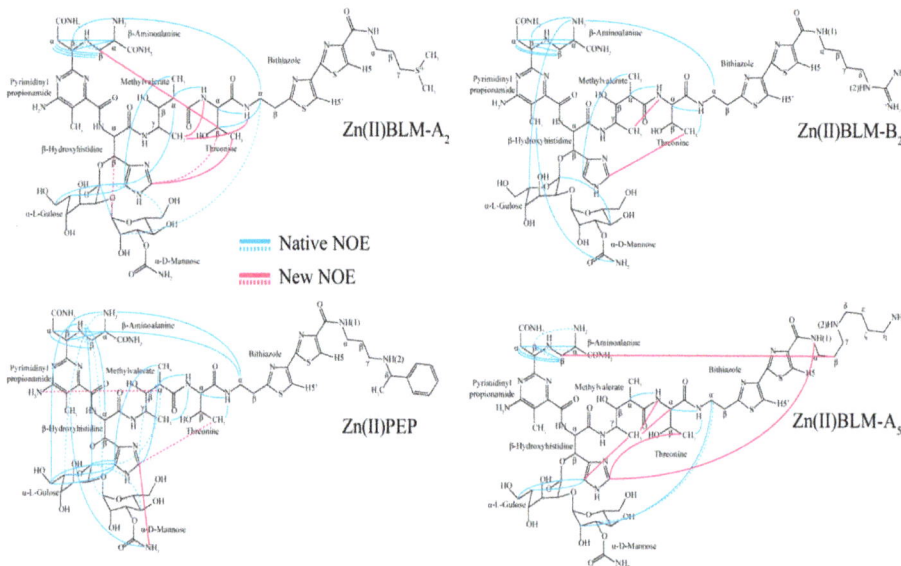

Figure 5. Inter-residue intramolecular NOEs for each OL_1-bound Zn(II)BLMs. Samples of Zn(II)BLMs:OL_1 are in a 1:1 molar ratio in H_2O at 5 °C. Dashed and continuous lines all represent NOE connectivities, they are used to avoid confusion in busy sectors of the figure.

5'-GT-3' DNA Binding Site

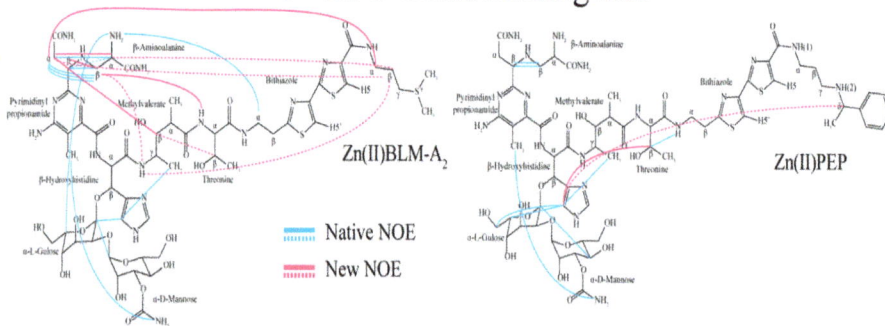

Figure 6. Inter-residue intramolecular NOEs for each OL_2-bound Zn(II)BLM for samples of Zn(II)BLM:OL_2 in a 1:1 molar ratio in H_2O at 5 °C. Dashed and continuous lines all represent NOE connectivities, they are used to avoid confusion in busy sectors of the figure.

The Pyr and Hist moieties show NOEs with the D4 in the free Zn(II)BLMs. Upon complexation to the OLs, changes to these NOEs appear that are diverse for each triad. This fact indicates that there is possible rearrangement of the metal coordination cage upon triad formation. The aforementioned results can be summarized to argue in favor of molecular rearrangement of the Zn(II)BLMs after triad formation. The NOE connections (native and new) present in the triads indicate that OL complexation impact the folding of the Zn(II)BLMs diversely depending on both the preferential binding site in the DNA fragment and the chemical structure of the BLM tail.

Table 2 displays a summary of the changes exhibited by the Zn(II)BLMs upon binding to the OLs. Examination of this table indicates that there is a significant difference between the conformations of Zn(II)PEP when complexed with OL_1 and OL_2. Only 44% of the native inter-residue intramolecular NOEs remain when bound to OL_1, but that percentage decreases to 13% when bound to OL_2. This trend is not displayed by bound Zn(II)BLM-A_2, the percentages being 43% for OL_1 and 38% for OL_2. The finding that there is a difference between the parameters shown in Table 1 when Zn(II)BLMs are bound to different OLs are compared indicates that the DNA-binding site has an impact on the final conformation of the drug. When comparing OL_2 triads, it can be seen that Zn(II)PEP conserves a significantly smaller amount of connections than Zn(II)BLM-A_2, with the percentages being 13% and 38%, respectively. This result could be interpreted to suggest that the chemical structure of the BLM-tail has an effect on the final conformation of the bound drug.

Table 2. Comparison of the overall changes in significant signal shifts and native and new NOEs displayed by the Zn(II)BLMs after binding to the indicated OL.

	OL_1 Triads				OL_2 Triads	
	A_2	PEP	B_2	A_5	A_2	PEP
% Overall native NOEs detected	51	48	56	50	43	26
Number of new intra-residue NOEs	3	2	2	1	0	0
% Native inter-residue NOEs	43	44	42	38	38	13
Number of new inter-residue NOES	6	3	2	6	10	2
Number of intermolecular NOEs	1	3	0	0	4	8
Number of significantly shifted BLM residues	8	8	11	14	19	25

3. Discussion

We have previously investigated the diverse levels of disruption caused by Zn(II)BLMs with different C-termini to DNA hairpins containing 5'-GC-3' (OL1) [50] and 5'-GT-3' (OL2) [51] binding sites. The results of this investigation indicated that, in the presence of different Zn(II)BLMs, both OLs display different patterns of intramolecular NOE connectivities and ^{1}H-NMR signal shifting suggesting that they exhibit different solution conformations in their Zn(II)BLM-bound forms. The overall conclusion derived from the precedent research is that the spatial conformation of target DNA segments in DNA-Zn(II)BLM complexes could be forged by interactions between drug and DNA, that are guided by the DNA binding site and the BLM C-termini.

The information presented herein is focused on the structural disturbances displayed by the same four Zn(II)BLMs that take place after these molecules bind to the aforementioned OLs. We have found that globally, OL_2 causes a greater degree of disturbance to the Zn(II)BLM structures than OL_1, just as the structure of OL_2 was more disturbed upon binding to the Zn(II)BLMs than that of OL_1 [50,51]. Additionally when complexed with OL_1, the shifts of the protons and the inter-residue NOE network of Zn(II)PEP were affected the least followed by Zn(II)BLM-A_2 and -B_2, with Zn(II)BLM-A_5 being the most affected; which is in direct correlation to the effects of these Zn(II)BLMs on the OL_1 structure [50]. We separate our discussion of the structural changes observed in the different Zn(II)BLMs based on the BLM domains indicated in Figure 1.

3.1. Bithiazole (D3)

The DNA-binding domain (D3) in MBLMs has been the focal point of the investigation of many MBLM-DNA triads, due to its ability to closely interact with the DNA bases. The interactions between the Bit moiety and various DNA fragments have been interpreted to highlight a particular binding mode of different MBLMs to DNA [36]. Examination of Figure 4 shows that all Zn(II)BLMs described herein exhibit NOEs connecting the Bit $C^{\alpha}H_2$ and NH protons to other residues in the metal complex before DNA binding occurs. After triad formation, the number of inter-residue intramolecular NOEs displayed by these protons is notably reduced, with the greatest effect occurring in the Zn(II)BLM-OL$_2$ triads, suggesting that the DNA-binding site has an effect on the location of D3 relative to the rest of the BLM molecule. The NOE networks for the Zn(II)BLM-A$_2$, -B$_2$, and Zn(II)PEP in the OL$_1$ triads show that the Bit $C^{\alpha}H_2$ protons remain in close contact with the Ala NH$_2$ protons, one of the ligands involved in metal ion coordination. Meanwhile, Bit $C^{\alpha}H_2$ protons are in close contact with a couple of the sugar protons in the Mann moiety for the Zn(II)BLM-A$_5$-OL$_1$ complex, suggesting a different conformation of the linker. The Mann NH$_2$ protons are involved in metal ion coordination as well, and thus the Bit $C^{\alpha}H_2$ protons are in a different location with respect to the coordinated metal center in this triad. These results suggest that the relative locations of D1 and D3 change upon triad formation, depending on the C-terminus in BLM.

In its OL$_1$ triad, the Bit $C^{\alpha}H_2$ in Zn(II)PEP exhibit NOEs with Ala NH$_2$ and Pyr $C^{\alpha}H_2$, while the Bit NH shows NOEs with two Thr protons. The aforementioned Bit $C^{\alpha}H_2$ NOEs have disappeared in the OL$_2$ triad for this Zn(II)BLM, and the Bit NH to Thr NOEs are different. Previously we have shown that the binding affinity of Zn(II)PEP is greater for OL$_2$ than for OL$_1$ [51] and thus the binding interaction of the Bit moiety is likely different when involving the two preferential binding sites. For both Zn(II)BLM-A$_2$ triads, the Bit $C^{\alpha}H_2$ protons are in close contact with the Ala NH$_2$ protons. On the other hand the NOEs between Bit NH and Thr $C^{\alpha}H$ and $C^{\beta}H$, and Val $C^{\gamma}CH_3$ in the OL$_1$ triad are missing in the OL$_2$ triad for this Zn(II)BLM. Based on these results we can propose that the conformation of the linker in BLM is also affected by the DNA-binding site.

3.2. C-termini (D3)

Before complexation to the OLs (Figure 4), Zn(II)BLM-A$_2$ displays one inter-residue intramolecular NOE between Bit $C^{\alpha}H_2$ and the $C^{\gamma}H_2$ protons of the tail, and Zn(II)PEP has multiple NOE connections connecting the metal binding domain and the tail. On the other hand, Zn(II)BLM-B$_2$ and -A$_5$ do not exhibit any inter-residue intramolecular NOEs involving tail protons. These results seem to indicate a high level of flexibility in solution for this region of the free Zn(II)BLM molecules. Upon complexation with OL$_1$, the inter-residue intramolecular connections of the tails in Zn(II)PEP and Zn(II)BLM-A$_2$ are no longer detected. However, there are two new NOEs for Zn(II)BLM-A$_5$ with the metal binding domain (Figure 5). Complexation of Zn(II)BLM-A$_2$ and Zn(II)PEP to OL$_2$ (Figure 6) leads to multiple NOEs involving the tail protons in the A$_2$ triad, and one of these NOEs for the PEP triad. These results suggest that after the Zn(II)BLMs bind OL$_1$ and OL$_2$, the BLM tails are positioned differently with respect to the rest of the BLM moieties depending on the OL available and possibly their chemical structures. In our previous studies involving the conformational changes of OL$_1$ and OL$_2$ in the presence of the Zn(II)BLMs discussed herein [50,51], we proposed that the conformation of each Zn(II)BLM-bound OL will be affected depending on the C-terminus of each Zn(II)BLM and the binding site present in each OL (5′-GC-3′ vs. 5′-GT-3′). If the tail location in bound Zn(II)BLMs is a consequence of the final conformation of the corresponding OL, or the interactions of D3 with the DNA helix remains to be demonstrated.

3.3. Linker (D2)

The linker region has previously been identified to contribute to the efficiency of DNA cleavage by bleomycin, and is necessary for promoting a compact structure [52]. In the present study, we have shown that the linker region is greatly affected upon binding to both DNA hairpins. For all OL-bound Zn(II)BLMs, the protons in the linker region exhibit some of the most significant $\Delta\delta$s calculated (Table 1), with Zn(II)BLMs bound to OL$_2$ showing the greater effect on their chemical shifts. As it can be seen in Table 1, the Val $C^{\alpha}H$ proton displays upfield shifts for Zn(II)BLMs bound to OL$_1$. On the other hand the $\Delta\delta$s calculated for this proton for Zn(II)BLMs bound to OL$_2$ show downfield shifting. The Val $C^\beta H$ and $C^\gamma H$ protons all shift downfield upon OL complexation, with the most significant shifts observed for drugs complexed with OL_2. Significant changes in the chemical shifts of the linker protons are expected upon OL complexation, due to the different roles attributed to D1 (metal binding) and D3 (DNA binding) in the presence of DNA. However, it is clear from the results presented herein that the chemical and/or magnetic environment experienced by these protons depends on the C-terminus of each drug and the binding site available. Previous studies have reported that the Val $C^\alpha H$ has a significant upfield shift for MBLMs when complexed with DNA, and it is most likely indicative of a structural change in the BLM molecule rather than being involved in DNA base pair stacking [36,53]. Although, studies involving HOO-Co(III)BLM-A_2 complexed to DNA fragments containing 5'-GC-3' and 5'-GT-3' binding sites report a significant downfield shift for this proton [38,41]. It is possible that the metal center (Co(III) vs. Zn(II)) also has an effect on the chemical and/or magnetic environment this proton is exposed to after DNA binding.

The network of inter-residue intramolecular NOEs displayed by the free Zn(II)BLMs for the linker region is also modified upon binding. Only a few of the native NOEs are detected in the OL-bound forms of most Zn(II)BLMs, with OL_1-bound Zn(II)PEP exhibiting the highest number of these NOEs. Simplification of the NOE network of the linker is expected if the Zn(II)BLM molecules refold upon DNA complexation. Our results suggest that the Zn(II)BLM molecule adopts a more open conformation as a consequence of DNA binding. As shown in Figures 5 and 6, new NOEs are detected for the OL-bound forms of the Zn(II)BLMs. For the linker region, the number of native NOEs conserved and the new NOEs detected for each OL_1-bound Zn(II)BLM are different. The same conclusion regarding these factors can be drawn from a comparison of Zn(II)BLM-A_2 and Zn(II)PEP bound to OL_1 and OL_2. NOEs connecting the linker protons to protons in other BLM residues have been detected in previous studies of MBLM-DNA triads, and were interpreted to indicate that the MBLM molecule is folded compactly [35,37,54]. Our results show that the folding of OL-bound Zn(II)BLMs seem to depend on the C-termini in BLM and the DNA-binding site.

3.4. Metal Binding Domain (D1)

The metal binding domain is of great interest due to its chemical interaction with DNA during DNA cleavage by MBLMs [9]. Examination of Table 1 shows interesting differences between OL_1- and OL_2- bound Zn(II)BLMs. OL_1-bound Zn(II)BLM-A_2, Zn(II)PEP, and Zn(II)BLM-A_5 display significant downfield shifts of the Hist C2H, Pyr NH_2 and Ala NH, and Hist C4H, respectively. When the same Zn(II)BLMs are bound to OL_2, the Hist C2H downfield shift increases, and other protons in the Hist, Pyr, and Mann moieties show significant shifts for Zn(II)BLM-A_2. For Zn(II)PEP, additional protons in the Ala and Hist units are significantly shifted, together with Mann and Gul protons. Comparison of the significant shifts generated when all four Zn(II)BLMs bind to OL_1 indicated that just a few protons change their shifts among OL_1-bound Zn(II)BLMs. On the other hand, comparison of the $\Delta\delta$s calculated for OL_2-bound Zn(II)BLM-A_2 and Zn(II)PEP present a different picture for these two Zn(II)BLMs in terms of the chemical and/or magnetic environment their D1 protons experience. Based on these results we can propose that OL_2 has a stronger influence on the environment of D1 than OL_1. Additionally, we can see that binding to OL_2 significantly affects the shifts of the Hist moiety in Zn(II)PEP, and the disaccharide unit in Zn(II)BLM-A_2, which could be interpreted to indicate that each BLM anchor itself differently to the same DNA-binding sites.

Before binding to the OLs, D1 in the free Zn(II)BLMs displays a multitude of NOEs connecting it to the disaccharide, linker, and, in some cases, the BLM tail. This fact indicates that the Zn(II)BLMs are folded in solution. Comparison of Figures 4–6 shows that the network of NOEs displayed by D1 is greatly simplified (only a few native NOEs remain) after the Zn(II)BLMs bind to the OLs, with OL_2 causing more extensive simplification than OL_1. The remaining native and the new NOEs that arise after OL binding are different for each Zn(II)BLM (Figures 5 and 6). Additionally, comparison of the OL_1 and OL_2-bound Zn(II)BLMs indicates that the binding site in OL also has an effect on the folding of the Zn(II)BLM molecule. The Pyr and Ala moieties remain connected to each other in the

OL-bound forms of the drug, although through less NOEs than in the free forms, possibly due to their closeness in the chemical structure of the BLM molecule. On the other hand, the NOE connectivities of these moieties to the disaccharide and Hist units are more tenuous, hinting slight distortions of the metal-coordination cage that are different depending on the Zn(II)BLM and the DNA-binding site. Connections between D1 and the BLM tails are scarce in OL_1-bound forms of the drug (only observed for Zn(II)BLM-A_5), and are found in both forms of the OL_2-bound drugs. The extensive simplification of the NOE network in the OL_2-bound Zn(II)BLMs is consistent with more protons in Zn(II)BLM triads exhibiting significant shifts.

Based on the results of the investigation discussed herein, we are prone to propose that the different anchors (DNA-binding domains) used by each BLM to bind DNA, and the available DNA-binding site can produce different folding of the rest of the BLM molecule around the OLs. It is possible that the interactions of each BLM with specific DNA-binding sites could change upon binding, to arrange the MBLM molecule to achieve the best conformation for optimal DNA binding and cleavage.

Previous studies of Zn(II)BLM-A_2 and -A_5 bound to a DNA fragment of sequence d(CGCTAGCG)$_2$ [37] reported to observe more structural disturbance of the Zn(II)BLM-A_5 structure than that of Zn(II)BLM-A_2. These findings are corroborated by the results described herein. On the other hand, Vanderwall et al. investigated the deviations to the HOO-Co(III)BLM-A_2 structure when complexed with both 5'-GC-3' and 5'-GT-3' binding sites, and concluded that the binding site did not significantly affect the MBLM structure [38,43]. The evidence provided here shows dramatic differences of the Zn(II)BLM structure upon complexation with both binding sites. It is possible that the metal ion coordinated to the BLM could be causing the differences in the results of both investigations, and the influence of the metal center on the structure of the DNA-bound MBLM is a task worth taking.

In our series of studies on the conformational changes exhibited by OLs [50,51] and Zn(II)BLMs upon the formation of Zn(II)BLM-OL triads, we have provided molecular information on the deviations of the DNA and Zn(II)BLM structures upon triad formation with consistency and comparability. We have found that the C-termini and the DNA-binding site have an effect on the conformations of both the OL and the BLM molecule, with the 5'-GT-3' binding site showing the most dramatic changes. At this point in our investigation, we cannot directly correlate the degree of disturbance in the Zn(II)BLM and DNA structures to the level of pulmonary toxicity produced by each of the BLMs considered. However, it is interesting that when comparing the effect of the C-substituents on the conformations of OL_1 (5'-CG-3' binding site), Zn(II)PEP and Zn(II)BLM-A_5 produced the lowest and highest levels of disturbance to this OL, respectively. Additionally when complexed with OL_1, the shifts of the protons and the inter-residue NOE network of Zn(II)PEP were affected the least followed by Zn(II)BLM-A_2 and -B_2, with Zn(II)BLM-A_5 in order of increasing disturbance. PEP and BLM-A_5 are in the opposite ends of the toxicity spectrum of BLMs, with PEP reported to have a lower degree of pulmonary toxicity [32,34,55,56], and BLM-A_5 with a high level of toxicity [26–31]. Based on these results, it is tempting to propose a possible connection between the level of disturbance of both target and drug upon triad formation, and that of pulmonary toxicity resulting from the use of different BLMs in cancer chemotherapy. A better understanding of the molecular mechanism of MBLM-DNA complexes is necessary to advance the development of analogs of bleomycin with lower pulmonary toxicity levels and higher therapeutic activity.

4. Materials and Methods

BLM-A_2 and -B_2 were purchased from TOKU-E (Bellingham, WA, USA). BLM-A_5 was purchased from LKT Laboratories, Inc. (St. Paul, MN, USA). PEP was a generous gift from Nippon Kayaku Co., Ltd. (Tokyo, Japan). Zinc sulfate hexa-hydrate was purchased from VWR (Radnor, PA, USA). Deuterated water (99.9%, d), sodium hydroxide, and sodium chloride were purchased from Sigma-Aldrich (St. Louis, MO, USA). The oligonucleotides: 5'-AGCCTTTTGGCCT-3'

(OL$_1$), and 5'-CCAGTATTTTTACTGG-3' (OL$_2$) used for binding to Zn(II)BLMs were purchased from Integrated DNA Technologies, Inc. (Coralville, IA, USA).

4.1. NMR Sample Preparation

BLM samples, 1.95 µmol, were dissolved in 650 µL of D$_2$O. A 0.12 M aqueous solution of ZnSO$_4$·7H$_2$O was mixed with the BLM solution to achieve a 1:1 molar ratio of Zn(II):BLM. The pH (meter reading uncorrected for the deuterium isotope effect) was adjusted to 6.5 with a 0.1 M NaOD solution. DNA, 0.335 µmol, was dissolved in 603 µL D$_2$O, and 67 µL of a 200 mM NaCl solution was added. The pH adjusted to 6.5 for the DNA samples. The Zn(II)BLM solutions were titrated with the DNA samples until a 1:1 molar ratio for the Zn(II)BLM:DNA complex was achieved. 1D ^{1}H-NMR spectra were used to monitor the changes in the complex formation. No additional changes in the 1D spectra were observed once a 1:1 molar ratio was achieved. Zn(II)BLM and DNA samples in 90% H$_2$O/10% D$_2$O (referred to as spectra in H$_2$O) were prepared by analogous procedures.

4.2. NMR Spectra Collection

NMR spectra were acquired at 600 MHz on a Bruker AVANCE III 600 spectrometer (Bruker BioSpin Corp, Billerica, MA, USA) with a 5.0 mm multi-nuclear broad-band observe probe. Spectra were acquired at both 278 K and 298 K for all samples, and were referenced to HDO and H$_2$O as internal standards. Two-dimensional experiments including correlation spectroscopy (COSY), totally correlated spectroscopy (TOCSY) and nuclear Overhauser effect spectroscopy (NOESY) were acquired utilizing solvent suppression achieved by excitation sculpting with gradients. The mixing times for the experiments were as follows: TOCSY 40 ms, and NOESY 200 ms. The spectral width was set to 10 ppm for D$_2$O samples and 20 ppm for H$_2$O samples in both dimensions, and 512 t$_1$ points were acquired with 2048 complex points for each free induction decay (FID). The number of scans for t$_1$ point for the experiments were as follows: 48 for COSY, 32 for TOCSY, and 48 for NOESY. All spectra were Fourier transformed using Lorentzian-to-Gaussian weighting and phase-shifted sine-bell window functions. NMR spectra were processed and analyzed using Topspin3.0 (Bruker BioSpin Corp., Billerica, MA, USA) and NMR ViewJ software (One Moon Scientific, Inc., Westfield, NJ, USA).

5. Conclusions

We have examined the structural changes of Zn(II)BLM-A$_2$, -A$_5$, -B$_2$, and Zn(II)PEP upon complexation with DNA hairpins of sequences 5'-AGG<u>C</u>CTTTTGGCCT-3' and 5'-CCA<u>GT</u>ATTTTT ACTGG-3'. The information here complements the findings we have presented on) how both the BLM C-termini and DNA binding site cause diverse conformational changes to the same DNA hairpins upon complexation with Zn(II)BLMs. These studies provide consistency and comparability missing in the field of BLM research. We have found that after Zn(II)BLM-DNA triad formation, not only is the DNA structure diversely affected, but the BLM structure is also disturbed, possibly to accommodate to that of the corresponding OL. When comparing the effect of different Zn(II)BLMs bound to the same OL, we found that the C-termini has an effect on both the shifting of protons in the OL and Zn(II)BLM, and the network of native NOEs present in each molecule. Additionally, binding of the same Zn(II)BLMs to OL$_2$ (5'-GT-3' binding site) indicates that the binding site in DNA has an effect on the conformations of the OL and BLM molecules.

The work presented herein and that discussed in our studies of the conformation of MBLM-DNA triads containing the 5'-GC-3' and 5'-GT-3' binding sites [50,51] will be used as the diamagnetic analogs in future studies to be performed in our laboratory to investigate the structural changes to both the DNA hairpins and Fe(II)BLM. The mentioned studies have to goal of probing the effect of the metal center in MBLM-DNA interactions. Extensive detailed research on the mode of binding of MBLMs to DNA will hopefully provide direction for designing studies to result in correlations between pulmonary toxicity and the MBLM-DNA interaction.

Supplementary Materials: The following are available online at www.mdpi.com/2312-7481/4/1/0004/s1, Figure S1: NOESY spectra for both free Zn(II)BLM-A$_2$ and Zn(II)BLM-A$_2$ bound to each of the DNA strands under study, Figure S2: NOESY spectra for both free Zn(II)PEP and Zn(II)PEP bound to each of the OLs under study, Figure S3: NOESY spectra for both free Zn(II)BLM-B$_2$ and Zn(II)BLM-B$_2$ bound to each of the OLs under study, Figure S4: NOESY spectra for both free Zn(II)BLM-A$_5$ and Zn(II)BLM-A$_5$ bound to each of the DNA strands under study, Table S1: Chemical shifts for the bleomycin residues for each of the free Zn(II)BLMs under study, Table S2: Inter-residue intramolecular NOEs for free Zn(II)BLM-A$_2$ and Zn(II)BLM-A$_2$ bound to both OLs, Table S3: Intra-residue intramolecular NOEs for free Zn(II)BLM-A$_2$ and Zn(II)BLM-A$_2$ bound to both OLs, Table S4: Inter-residue intramolecular NOEs for free Zn(II)PEP and Zn(II)PEP bound to both OLs, Table S5: Intra-residue intramolecular NOEs for free Zn(II)PEP and Zn(II)PEP bound to both OLs, Table S6: Inter-residue intramolecular NOEs for free Zn(II)BLM-B$_2$ and Zn(II)BLM-B$_2$ bound to OL$_1$, Table S7: Intra-residue intramolecular NOEs for free Zn(II)BLM-B$_2$ and Zn(II)BLM-B$_2$ bound to OL$_1$, Table S8: Inter-residue intramolecular NOEs for free Zn(II)BLM-A$_5$ and Zn(II)BLM-A$_5$ bound to OL$_1$, Table S9: Intra-residue intramolecular NOEs for free Zn(II)BLM-A$_5$ and Zn(II)BLM-A$_5$ bound to OL$_1$.

Acknowledgments: This work was supported in whole by the National Institute of Health [Grant 1R15GM106285-01A1]. Our gratitude also goes to Nippon Kayaku Co., Ltd. (Tokyo, Japan) for the generous gift of peplomycin. We also acknowledge Alexander Goroncy for help collecting the NMR data presented in this work.

Author Contributions: S.E.F. prepared the NMR samples, collected NMR data, analyzed and interpreted NMR spectra and participated on the writing of this manuscript, S.A.M., A.D.I., and T.M.R. participated in sample preparation and NMR data collection, and they also analyzed and interpreted NMR spectra, T.E.L. provided the research idea, supervised and managed the project, and participated in data interpretation and manuscript writing.

Conflicts of Interest: The authors declare no conflict of interest.

References

1. Blum, R.H.; Carter, S.K.; Agre, K. A clinical review of bleomycin—A new antineoplastic agent. *Cancer* **1973**, *31*, 903–914. [CrossRef]

2. Akiyama, Y.; Ma, Q.; Edgar, E.; Laikhter, A.; Hecht, S.M. Identification of Strong DNA Binding Motifs for Bleomycin. *J. Am. Chem. Soc.* **2008**, *130*, 9650–9651. [CrossRef] [PubMed]

3. Giroux, R.A.; Hecht, S.M. Characterization of Bleomycin Cleavage Sites in Strongly Bound Hairpin DNAs. *J. Am. Chem. Soc.* **2010**, *132*, 16987–16996. [CrossRef] [PubMed]

4. Yu, Z.; Schmaltz, R.M.; Bozeman, T.C.; Paul, R.; Rishel, M.J.; Tsosie, K.S.; Hecht, S.M. Selective Tumor Cell Targeting by the Disaccharide Moiety of Bleomycin. *J. Am. Chem. Soc.* **2013**, *135*, 2883–2886. [CrossRef] [PubMed]

5. Boger, D.L.; Cai, H. Bleomycin: Synthetic and Mechanistic Studies. *Angew. Chem. Int. Ed.* **1999**, *38*, 448–476. [CrossRef]

6. Carter, B.J.; Murty, V.S.; Reddy, K.S.; Wang, S.N. A Role for the Metal-Binding Domain in Determining the DNA-Sequence Selectivity of Fe-Bleomycin. *J. Biol. Chem.* **1990**, *265*, 4193–4196. [PubMed]

7. Kane, S.A.; Hecht, S.M. *Progress in Nucleic Acid Research and Molecular Biology*; Waldo, E.C., Kivie, M., Eds.; Academic Press: Walttham, MA, USA, 1994; pp. 313–352.

8. Hecht, S.M. The Chemistry of Activated Bleomycin. *Acc. Chem. Res.* **1986**, *19*, 383–391. [CrossRef]

9. Stubbe, J.; Kozarich, J.W. Mechanisms of Bleomycin-Induced DNA-Degradation. *Chem. Rev.* **1987**, *87*, 1107–1136. [CrossRef]

10. Povirk, L.F.; Hogan, M.; Dattagupta, N. Binding of Bleomycin to DNA—Intercalation of the Bithiazole Rings. *Biochemistry* **1979**, *18*, 96–101. [CrossRef] [PubMed]

11. Hecht, S.M. RNA degradation by bleomycin, a naturally-occurring bioconjugate. *Bioconjug. Chem.* **1994**, *5*, 513–526. [CrossRef] [PubMed]

12. Zuber, G.; Quada, J.C.; Hecht, S.M. Sequence Selective Cleavage of A DNA Octanucleotide by Chlorinated Bithiazoles and Bleomycins. *J. Am. Chem. Soc.* **1998**, *120*, 9368–9369. [CrossRef]

13. Loeb, K.E.; Zaleski, J.M.; Hess, C.D.; Hecht, S.M.; Solomon, E.I. Spectroscopic Investigation of the Metal Ligation and Reactivity of the Ferrous Active Sites of Bleomycin and Bleomycin Derivatives. *J. Am. Chem. Soc.* **1998**, *120*, 1249–1259. [CrossRef]

14. Lehmann, T.E. Molecular Modeling of the Three-Dimensional Structure of Fe(II)-Bleomycin: Are the Co(II) and Fe(II) Adducts Isostructural? *J. Biol. Inorg. Chem.* **2002**, *7*, 305–312. [CrossRef] [PubMed]

15. Lehmann, T.E.; Serrano, M.L.; Que, L., Jr. Coordination Chemistry of Co(II)-Bleomycin: Its Investigation Through NMR and Molecular Dynamics. *Biochemistry* **2000**, *39*, 3886–3898. [CrossRef] [PubMed]

16. Akkerman, M.A.J.; Neijman, E.W.J.F.; Wijmenga, S.S.; Hilbers, C.W.; Bermel, W. Studies of the Solution Structure of the Bleomycin-A2 Iron(II) Carbon-Monoxide Complex by Means of 2-Dimensional NMR-Spectroscopy and Distance Geometry Calculations. *J. Am. Chem. Soc.* **1990**, *112*, 7462–7474. [CrossRef]

17. Akkerman, M.A.J.; Haasnoot, C.A.G.; Pandit, U.K.; Hilbers, C.W. Complete Assignment of the ^{13}C-NMR Spectra of Bleomycin-A2 and its Zinc Complex by Means of Two-Dimensional NMR-Spectroscopy. *Magn. Reson. Chem.* **1988**, *26*, 793–802. [CrossRef]

18. Akkerman, M.A.J.; Haasnoot, C.A.G.; Hilbers, C.W. Studies of the Solution Structure of the Bleomycin-A2 Zinc Complex by Means of Two-Dimensional NMR-Spectroscopy and Distance Geometry Calculations. *Eur. J. Biochem.* **1988**, *173*, 211–225. [CrossRef] [PubMed]

19. Oppenheimer, N.J.; Rodriguez, L.O.; Hecht, S.M. Structural Studies of Active Complex of Bleomycin—Assignment of Ligands to the Ferrous Ion in a Ferrous-Bleomycin Carbon Monoxide Complex. *Proc. Natl. Acad. Sci. USA* **1979**, *76*, 5616–5620. [CrossRef] [PubMed]

20. Lehmann, T.E.; Ming, L.J.; Rosen, M.E.; Que, L., Jr. NMR Studies of the Paramagnetic Complex Fe(II)-Bleomycin. *Biochemistry* **1997**, *36*, 2807–2816. [CrossRef] [PubMed]

21. Lehmann, T.E.; Li, Y. Possible Structural Role of the Disaccharide Unit in Fe-Bleomycin before and after Oxygen Activation. *J. Antibiot.* **2012**, *65*, 25–33. [CrossRef] [PubMed]

22. Lehmann, T.E.; Li, Y. Solution Structure of Fe(II)-Azide-Bleomycin Derived from NMR Data: Transition from Fe(II)-Bleomycin to Fe(II)-Azide-Bleomycin as Derived from NMR Data and Structural Calculations. *J. Biol. Inorg. Chem.* **2012**, *17*, 761–771. [CrossRef] [PubMed]

23. Li, Y.; Lehmann, T.E. Coordination Chemistry and Solution Structure of Fe(II)-Peplomycin. Two Possible Coordination Geometries. *J. Inorg. Biochem.* **2012**, *111*, 50–58. [CrossRef] [PubMed]

24. Lehmann, T.E.; Topchiy, E. Contributions of NMR to the Understanding of the Coordination Chemistry and DNA Interactions of Metallo-Bleomycins. *Molecules* **2013**, *18*, 9253–9277. [CrossRef] [PubMed]

25. Schroeder, B.R.; Ghare, M.I.; Bhattacharya, C.; Paul, R.; Yu, Z.Q.; Zaleski, P.A.; Bozeman, T.C.; Rishel, M.J.; Hecht, S.M. The Disaccharide Moiety of Bleomycin Facilitates Uptake by Cancer Cells. *J. Am. Chem. Soc.* **2014**, *136*, 13641–13656. [CrossRef] [PubMed]

26. Raisfeld, I.H. Pulmonary Toxicity of Bleomycin Analogs. *Toxicol. Appl. Pharmacol.* **1980**, *56*, 326–336. [CrossRef]

27. Raisfeld, I.H.; Chovan, J.P.; Frost, S. Bleomycin Pulmonary Toxicity—Production of Fibrosis by Bithiazole-Terminal Amine and Terminal Amine Moieties of Bleomycin-A2. *Life Sci.* **1982**, *30*, 1391–1398. [CrossRef]

28. Raisfeld, I.H.; Chu, P.; Hart, N.K.; Lane, A. A Comparison of the Pulmonary Toxicity Produced by Metal-Free and Copper-Complexed Analogs of Bleomycin and Phleomycin. *Toxicol. Appl. Pharmacol.* **1982**, *63*, 351–362. [CrossRef]

29. Raisfeld, I.H. Relation between Bleomycin Structure and Pulmonary Fibrosis. *Clin. Pharmacol. Ther.* **1981**, *29*, 274.

30. Raisfeld, I.H. Relation of Bleomycin Structure to Pulmonary Toxicity. *Clin. Res.* **1980**, *28*, A530.

31. Raisfeld, I.H. Bleomycin Terminal Groups Produce Pulmonary Fibrosis. *Clin. Res.* **1979**, *27*, A445.

32. Oka, S. A Review of Clinical-Studies of Pepleomycin. *Recent Results Cancer Res.* **1980**, *74*, 163–171. [PubMed]

33. Umezawa, H. Recent Studies on Bleomycin. *Lloydia* **1977**, *40*, 67–81.

34. Tanaka, W.; Takita, T. Pepleomycin—2nd Generation Bleomycin Chemically Derived From Bleomycin A2. *Heterocycles* **1979**, *13*, 469–476. [CrossRef]

35. Sucheck, S.J.; Ellena, J.F.; Hecht, S.M. Characterization of Zn(II) Deglycobleomycin A2 and Interaction with d(CGCTAGCG)2: Direct Evidence for Minor Groove Binding of the Bithiazole Moiety. *J. Am. Chem. Soc.* **1998**, *120*, 7450–7460. [CrossRef]

36. Keck, M.V.; Manderville, R.A.; Hecht, S.M. Chemical and Structural Characterization of the Interaction of Bleomycin A2 with d(CGCGAATTCGCG)(2). Efficient, Double-Strand DNA Cleavage Accessible without Structural Reorganization. *J. Am. Chem. Soc.* **2001**, *123*, 8690–8700. [CrossRef] [PubMed]

37. Manderville, R.A.; Ellena, J.F.; Hecht, S.M. Interaction of Zn(II)-Bleomycin With d(CGCTAGCG)2—A Binding Model-Based on NMR Experiments and Restrained Molecular-Dynamics Calculations. *J. Am. Chem. Soc.* **1995**, *117*, 7891–7903. [CrossRef]

38. Vanderwall, D.E.; Lui, S.M.; Wu, W.; Turner, C.J.; Kozarich, J.W.; Stubbe, J. A Model of the Structure of HOO-CoBleomycin Bound to d(CCAGTACTGG): Recognition at the d(GpT) Site and Implications for Double-Stranded DNA Cleavage. *Chem. Biol.* **1997**, *4*, 373–387. [CrossRef]

39. Wu, W.; Vanderwall, D.E.; Lui, S.M.; Tang, X.J.; Turner, C.J.; Kozarich, J.W.; Stubbe, J. Studies of CoBleomycin A2 Green: Its Detailed Structural Characterization by NMR and Molecular Modeling and its Sequence-Specific Interaction with DNA Oligonucleotides. *J. Am. Chem. Soc.* **1996**, *118*, 1268–1280. [CrossRef]

40. Wu, W.; Vanderwall, D.E.; Stubbe, J.; Kozarich, J.W.; Turner, C.J. Interaction of CoBleomycin A2 (Green) with d(CCAGGCCTGG)2—Evidence for Intercalation Using 2D NMR. *J. Am. Chem. Soc.* **1994**, *116*, 10843–10844. [CrossRef]

41. Wu, W.; Vanderwall, D.E.; Teramoto, S.; Lui, S.M.; Hoehn, S.T.; Tang, X.J.; Turner, C.J.; Boger, D.L.; Kozarich, J.W.; Stubbe, J. NMR Studies of Codeglycobleomycin A2 Green and its Complex with d(CCAGGCCTGG). *J. Am. Chem. Soc.* **1998**, *120*, 2239–2250. [CrossRef]

42. Wu, W.; Vanderwall, D.E.; Turner, C.J.; Hoehn, S.; Chen, J.Y.; Kozarich, J.W.; Stubbe, J. Solution Structure of the Hydroperoxide of Co(III) Phleomycin Complexed With d(CCAGGCCTGG)2: Evidence for Binding by Partial Intercalation. *Nucleic Acids Res.* **2002**, *30*, 4881–4891. [CrossRef] [PubMed]

43. Wu, W.; Vanderwall, D.E.; Turner, C.J.; Kozarich, J.W.; Stubbe, J. Solution Structure of CoBleomycin A2 Green Complexed With d(CCAGGCCTGG). *J. Am. Chem. Soc.* **1996**, *118*, 1281–1294. [CrossRef]

44. Zhao, C.Q.; Xia, C.W.; Mao, Q.K.; Forsterling, H.; DeRose, E.; Antholine, W.E.; Subczynski, W.K.; Petering, D.H. Structures of HO2-Co(III)Bleomycin A2 Bound to d(GAGCTC)2 and d(GGAAGCTTCC)2): Structure-Reactivity Relationships of Co and Fe Bleomycins. *J. Inorg. Biochem.* **2002**, *91*, 259–268. [CrossRef]

45. Kuwahara, J.; Sugiura, Y. Sequence-Specific Recognition and Cleavage of DNA by Metallobleomycin—Minor Groove Binding and Possible Interaction Mode. *Proc. Natl. Acad. Sci. USA* **1988**, *85*, 2459–2463. [CrossRef] [PubMed]

46. Goodwin, K.D.; Lewis, M.A.; Long, E.C.; Georgiadis, M.M. Crystal Structure of DNA-Bound Co(III)-Bleomycin B-2: Insights on Intercalation and Minor Groove Binding. *Proc. Natl. Acad. Sci. USA* **2008**, *105*, 5052–5056. [CrossRef] [PubMed]

47. Lui, S.M.; Vanderwall, D.E.; Wu, W.; Tang, X.J.; Turner, C.J.; Kozarich, J.W.; Stubbe, J. Structural Characterization of CoBleomycin A2 Brown: Free and Bound to d(CCAGGCCTGG). *J. Am. Chem. Soc.* **1997**, *119*, 9603–9613. [CrossRef]

48. Caceres-Cortes, J.; Sugiyama, H.; Ikudome, K.; Saito, I.; Wang, A.H.J. Interactions of Deglycosylated Cobalt(III)—Pepleomycin (Green Form) with DNA Based on NMR Structural Studies. *Biochemistry* **1997**, *36*, 9995–10005. [CrossRef] [PubMed]

49. Caceres-Cortes, J.; Sugiyama, H.; Ikudome, K.; Saito, I.; Wang, A.H.J. *Structure, Motion, Interaction and Expression of Biological Macromolecules*; Sarma, R.H., Sarma, M.H., Eds.; Adenine Press: New York, NY, USA, 1998; pp. 207–225.

50. Lehmann, T.E.; Murray, S.A.; Ingersoll, A.D.; Reilly, T.M.; Follett, S.E.; Macartney, K.E.; Harpster, M.H. NMR Study of the Effects of Some Bleomycin C-Termini on the Structure af a DNA Hairpin With the 5′-GC-3′ Binding Site. *J. Biol. Inorg. Chem.* **2017**, *22*, 121–136. [CrossRef] [PubMed]

51. Follett, S.E.; Ingersoll, A.D.; Murray, S.A.; Reilly, T.M.; Lehmann, T.E. Interaction of Zn(II)Bleomycin-A2 and Zn(II)Peplomycin With a DNA Hairpin Containing the 5′-GT-3′ Binding Site in Comparison with the 5′-GC-3′ Binding Site Studied by NMR Spectroscopy. *J. Biol. Inorg. Chem.* **2017**, *22*, 1039–1054. [CrossRef] [PubMed]

52. Ohno, M.; Otsuka, M. *Recent Progress in the Chemical Synthesis of Antibiotics*; Lukacs, G., Ohno, M., Eds.; Springer: Berlin, Germany, 1990; pp. 387–414.

53. Glickson, J.D.; Pillai, R.P.; Sakai, T.T. Proton NMR-Studies of the Zn(II)-Bleomycin-A2-Poly(dA-dT) Ternary Complex. *Proc. Natl. Acad. Sci. USA* **1981**, *78*, 2967–2971. [CrossRef] [PubMed]

54. Xu, R.X.; Nettesheim, D.; Otvos, J.D.; Petering, D.H. NMR Determination of the Structures of Peroxycobalt(III) Bleomycin and Cobalt(III) Bleomycin, Products of the Aerobic Oxidation of Cobalt(II) Bleomycin by Dioxygen. *Biochemistry* **1994**, *33*, 907–916. [CrossRef] [PubMed]

55. Raisfeld, I.H.; Kundahl, E.R.; Sawey, M.J.; Chovan, J.P.; Depasquale, J. Selective Toxicity of Specific Lung-Cells to Bleomycin. *Clin. Res.* **1982**, *30*, A437.
56. Takahashi, K.; Aoyagi, H.S.; Koyu, A.; Kuramochi, H.; Yoshioka, O.; Matsuda, A.; Fujii, A.; Umezawa, H. Biological Studies on the Degradation Products of 3-[(S)-1′-Phenylethylamino]Propylaminobleomycin— Novel Analog (Pepleomycin). *J. Antibiot.* **1979**, *32*, 36–42. [CrossRef] [PubMed]

![magnetochemistry logo] *magnetochemistry*

MDPI

Communication

Application of NMR Screening Methods with ^{19}F Detection to Fluorinated Compounds Bound to Proteins

Kazuo Furihata [1], Moe Usui [2] and Mitsuru Tashiro [2,*]

[1] Division of Agriculture and Agricultural Life Sciences, The University of Tokyo, Yayoi, Bunkyo-ku, Tokyo 113-8657, Japan; afuriha@mail.ecc.u-tokyo.ac.jp

[2] Department of Chemistry, College of Science and Technology, Meisei University, Hino, Tokyo 191-8506, Japan; 16m2004@stu.meisei-u.ac.jp

* Correspondence: tashiro@chem.meisei-u.ac.jp; Tel.: +81-42-591-5597

Received: 11 November 2017; Accepted: 21 December 2017; Published: 27 December 2017

Abstract: The combinational use of one-dimensional (1D) NMR-based screening techniques with ^{1}H and ^{19}F detections were applied to a human serum albumin–diflunisal complex. Since most NMR screening methods observe ^{1}H spectra, the overlapped ^{1}H signals were unavailable in the binding epitope mapping. However, the NMR experiments with ^{19}F detection can be used as an effective complementary method. For the purpose of identifying the ^{1}H and ^{19}F binding epitopes of diflunisal, this paper carries out a combinatorial analysis using ^{1}H{^{1}H} and ^{19}F{^{1}H} saturation transfer difference experiments. The differences of the ^{1}H-inversion recovery rates with and without target irradiation are also analyzed for a comprehensive interpretation of binding epitope mapping.

Keywords: NMR-based screening; fluorinated compound; diflunisal; ^{19}F NMR

---

1. Introduction

Protein–ligand interactions can provide useful insights for understanding the molecular recognition system. However, arriving at such understandings requires the developments of useful methods for selectively observing the ligand. Although X-ray analyses can determine such interactions of the complex at the atomic level, difficulties in crystallization often interfere with the process of X-ray studies. In some cases, NMR spectroscopy can be a useful alternative for analyzing macromolecular complexes and screening compounds with an affinity to target proteins. Various NMR-based screening methods to observe the ligand signals have been proposed. It has been shown that NOE-pumping [1], saturation transfer difference (STD) [2], water–ligand observed via gradient spectroscopy (WaterLOGSY) [3,4], and reverse NOE-pumping [5] experiments could directly detect ^{1}H of the bound ligands. Recently, the NMR-based methods have been extended to fluorine detection [6–8]. Since the spectral elucidation in the aforementioned experiments [1–5] depends on the dispersion of ^{1}H signals, its signal degeneracy leads to a lack of information for the target molecules. Considering these difficulties, the NMR-based screening methods with ^{19}F-detection were applied to the human serum albumin (HSA)-diflunisal complex. HSA is an abundant plasma protein that binds to a wide range of drugs. Diflunisal contains two fluorine atoms in a molecule, and is a nonsteroidal anti-inflammatory drug that is effective in treating fever, pain, and inflammation. Since the X-ray crystal structure of a diflunisal-HSA complex has been determined (pdb: 2BXE), this complex could be a suitable model system for studying the molecular interactions of ^{1}H and ^{19}F using NMR spectroscopy. Information of the binding epitopes can be obtained for ^{19}F as well as ^{1}H of the fluorinated compound.

2. Results and Discussion

To investigate the ^{1}H binding epitopes of ligands, two representative methods—the ^{1}H{^{1}H} STD method acquired with various saturation times [9], or the difference of inversion recovery rates with and without target irradiation (DIRECTION) [10] method—were generally used. In the present study, the binding epitopes of diflunisal (Figure 1) were investigated using both methods. In the ^{1}H{^{1}H} STD experiments, the STD build-up curves were obtained at various saturation times. The slope of the STD build-up curve at a saturation time of 0 s was obtained by fitting to the monoexponential equation: $STD = STD_{max}(1 - e^{(-k_{sat} \times t)})$, where STD stands for the STD signal intensity at saturation time t, STD_{max} is the maximal STD intensity at long saturation times, and k_{sat} stands for the observed saturation constant. The values of $k_{sat} \times STD_{max}$ correspond to the slope of the curve at zero saturation time with an elimination of T_1 bias. In the DIRECTION experiments, ^{1}H-T_1 were measured with and without the selective irradiation of protein, and its reciprocals, corresponding to the inversion recovery rates, were calculated for each separated ^{1}H signal.

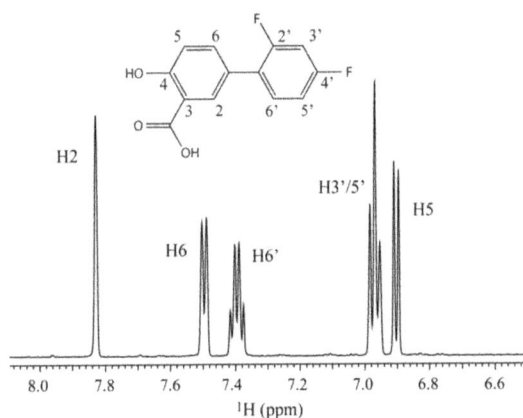

Figure 1. Structure and ^{1}H NMR spectrum of diflunisal.

The values of the STD effect were normalized by referencing the signal of H6 with the largest STD effect. The relative values are shown in Figure 2a. The values of the STD effect were larger in H5 and H6, indicating that these protons contributed as the binding epitopes. The smallest value was obtained in H2, which made less contact with the protein. The binding epitopes were also investigated using the DIRECTION method, evaluating the difference between the ^{1}H-inversion recovery rates with and without the irradiation of protein [10]. The large differences reflect the proximity to the protein surface. H6 and H6$'$ showed relatively large values (Figure 2b). Since the H3$'$ and H5$'$ signals overlapped, H6$'$ was the only signal available for analysis in a 2$'$,4$'$-difluoro ring (Figure 1), indicating that the incomplete information was obtained in the ^{1}H-detection NMR methods. To obtain more detailed information of the binding epitopes for the 2$'$,4$'$-difluoro ring, the ^{19}F{^{1}H} STD spectra were acquired with the arrayed saturation times (Figure 3). The ^{19}F{^{1}H} STD experiment was more insensitive than the ^{1}H{^{1}H} STD experiment. It can be considered that the saturation transfer from ^{1}H to ^{19}F is much less effective than that from ^{1}H to ^{1}H. However, the ^{19}F{^{1}H} STD experiment provided the useful information regarding the ^{19}F binding epitopes. The normalized values of the STD effect of F2$'$ and F4$'$ were 100% and 41.9%, respectively, and the values of ^{19}F-T_1 were 0.82 s and 1.6 s in the aforesaid order. This result indicated that F2$'$ made more close contact to HSA than F4$'$. It can be considered that a portion comprising H6, H6$'$, and F2$'$ could play a key role as the binding portion of diflunisal. The H2 made less close contact, which could be caused by an interruption of the carboxyl group at position 3. In the X-ray crystal structure of HSA complexed with diflunisal (pdb: 2BXE), three molecules of diflunisal were bound with one molecule of HSA, where various close contacts were made in each binding site between two fluorine atoms of diflunisal, and protons of HSA. Since information from an epitope mapping that was obtained by the NMR experiments revealed average contacts in three HSA binding sites, some differences in the close contacts need to be considered between crystal and solution states.

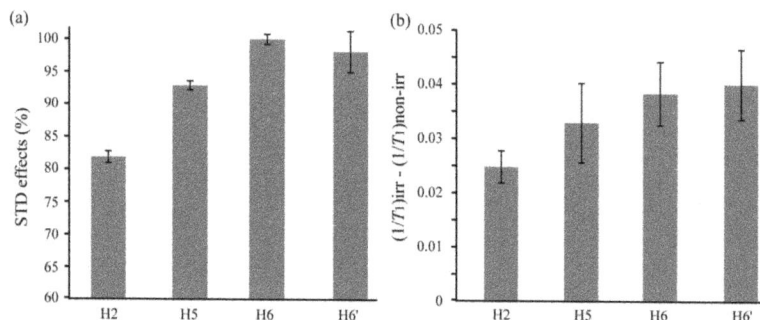

Figure 2. (**a**) The values of the saturation transfer difference (STD) effect of diflunisal. The values were normalized by referencing the signal of H6 with the largest STD effect; (**b**) The difference of inversion recovery rate with and without target irradiation.

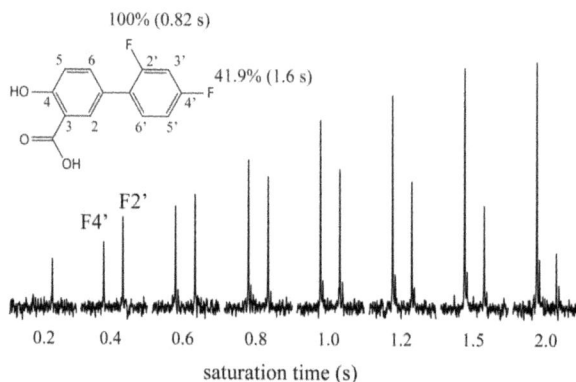

Figure 3. The $^{19}F\{^{1}H\}$ STD spectra acquired with the arrayed saturation times. The normalized values of the STD effect (%) and the values of $^{19}F\text{-}T_1$ (s) are shown.

3. Materials and Methods

3.1. Instrumentation and Chemicals

All of the NMR spectra were recorded at 20 °C on a Varian 600 MHz NMR system (Vaian, Palo Alto, CA, USA) or JEOL ECA-500 MHz spectrometer (JEOL Ltd., Tokyo, Japan). Diflunisal and HSA were purchased from Sigma-Aldrich (Tokyo, Japan). A 600-µL of solution containing 0.05 mM HSA and 5.0 mM diflunisal was prepared in 100% ^2H_2O.

3.2. NMR Spectroscopy

The experimental parameters of the $^1H\{^1H\}$ STD experiment were as follows: data points = 16,384, spectral width of ^1H = 8012 Hz, number of scans = 1024, recycle time = 1.0 s. The saturation times for the selective excitation of proteins were arrayed in the range of 0.2–3.5 s, and the arrayed spectra were acquired five times. The on and off resonance frequencies of ^{1}H were 0.6 and -20 ppm, respectively. Those of the ^{19}F{^{1}H} STD experiment were as follows: data points = 8192, spectral width of ^{19}F = 6012 Hz, number of scans = 10,240, recycle time = 1.0 s. The saturation times for the selective excitation of protein were arrayed in the range of 0.2–2.0 s. The on and off resonance frequencies of ^{1}H were 0.6 and -20 ppm, respectively. The values of the initial slope in the STD build-up curves were obtained by the least-square fitting in both of the STD experiments [9]. The experimental parameters for measuring ^{19}F-T_1 were as follows: data points = 8192, spectral width of ^{19}F = 6012 Hz, number of scans = 128, recycle time = 5.0 s. The inversion recovery pulse sequence was used. In measurements of ^{1}H-T_1 with and without the selective excitation of protein resonance (DIRECTION method) [10], the measurements were repeated five times, and the program in the JEOL Delta software (JEOL Ltd., Tokyo, Japan) was used for calculation of ^{1}H-T_1. The on and off resonance frequencies of ^{1}H were 0.6 and -20 ppm, respectively. The exponential window function was used with zero-filling by a factor of 2. The ^{1}H and ^{19}F chemical shifts were relative to 3-(Trimethylsilyl)-1-propanesulfonic acid sodium salt (DSS) and trichlorofluoromethane, respectively, as external standards.

4. Conclusions

Although the sensitivity of the ^{19}F{^{1}H} STD experiment was lower than that of the ^{1}H{^{1}H} STD experiment, the obtained information was useful for the fluorinated compounds with the degenerated ^{1}H signals. Comprehensive interpretations for the binding epitope mapping are essential, while considering some discrepancies in the results of various NMR experiments. The ^{19}F{^{1}H} STD experiment can be a complimentary method for the ^{1}H detection methods.

Acknowledgments: This study was supported by two Grant-in-Aids for Scientific Research (15K05550 for M.T.) from the Ministry of Education, Culture, Sports, Science, and Technology.

Conflicts of Interest: The authors declare no conflict of interest.

References

1. Chen, A.; Shapiro, M.J. NOE Pumping: A novel NMR technique for identification of compounds with binding affinity to macromolecules. *J. Am. Chem. Soc.* **1998**, *120*, 10258–10259. [CrossRef]

2. Maye, M.; Meyer, B. Group epitope mapping by saturation transfer difference NMR to identify segments of a ligand in direct contact with a protein receptor. *J. Am. Chem. Soc.* **2001**, *123*, 6108–6117.

3. Dalvit, C.; Pevarello, P.; Tatò, M.; Veronesi, M.; Vulpetti, A.; Sundström, M. Identification of compounds with binding affinity to proteins via magnetization transfer from bulk water. *J. Biomol. NMR* **2000**, *18*, 65–68. [CrossRef] [PubMed]

4. Dalvit, C.; Fogliatto, G.P.; Stewart, A.; Veronesi, M.; Stockman, B.J. WaterLOGSY as a method for primary NMR screening: Practical aspects and range of applicability. *J. Biomol. NMR* **2001**, *21*, 349–359. [CrossRef] [PubMed]

5. Chen, A.; Shapiro, M.J. NOE Pumping. 2. A high-throughput method to determine compounds with binding affinity to macromolecules by NMR. *J. Am. Chem. Soc.* **2000**, *122*, 414–415. [CrossRef]

6. Dalvit, C.; Fagerness, P.E.; Hadden, S.T.A.; Sarver, R.W.; Stockman, B.J. Fluorine-NMR experiments for high-throughput screening: Theoretical aspects, practical considerations, and range of applicability. *J. Am. Chem. Soc.* **2003**, *125*, 7696–7703. [CrossRef] [PubMed]

7. Dalvit, C.; Flocco, M.; Stockman, B.J.; Veronesi, M. Competition binding experiments for rapidly ranking lead molecules for their binding affinity to human serum albumin. *Comb. Chem. High Throughput Screen.* **2002**, *5*, 645–650. [CrossRef] [PubMed]

8. Sakuma, C.; Kurita, J.; Furihata, K.; Tashiro, M. Achievement of ^{1}H-^{19}F heteronuclear experiments using the conventional spectrometer with a shared single high band amplifier. *Magn. Reson. Chem.* **2015**, *53*, 327–329. [CrossRef] [PubMed]

9.	Mayer, M.; James, T.L. NMR-based characterization of phenothiazines as a RNA binding scaffold. *J. Am. Chem. Soc.* **2004**, *126*, 4453–4460. [CrossRef] [PubMed]

10.	Mizukoshi, Y.; Abe, A.; Takizawa, T.; Hanzawa, H.; Fukunishi, Y.; Shimada, I.; Takahashi, H. An accurate pharmacophore mapping method by NMR spectroscopy. *Angew. Chem. Int. Ed.* **2012**, *51*, 1362–1365. [CrossRef] [PubMed]

magnetochemistry

MDPI

Article

Accelerating NMR-Based Structural Studies of Proteins by Combining Amino Acid Selective Unlabeling and Fast NMR Methods

Bankala Krishnarjuna [1], Kousik Chandra [1],* and Hanudatta S. Atreya [1,2],*

1   NMR Research Centre, Indian Institute of Science, Bangalore 560012, India; krishnarjuna.bio@gmail.com
2   Solid State and Structural Chemistry Unit, Indian Institute of Science, Bangalore 560012, India
*   Correspondence: kousikc@iisc.ac.in (K.C.); hsatreya@iisc.ac.in (H.S.A.);
    Tel.: +91-80-2293-3302 (K.C. & H.S.A.); Fax: +91-80-2360-1550 (H.S.A.)

Received: 30 October 2017; Accepted: 7 December 2017; Published: 26 December 2017

Abstract: In recent years, there has been a growing interest in fast acquisition and analysis of nuclear magnetic resonance (NMR) spectroscopy data for high throughput protein structure determination. Towards this end, rapid data collection techniques and methods to simplify the NMR spectrum such as amino acid selective unlabeling have been proposed recently. Combining these two approaches can speed up further the structure determination process. Based on this idea, we present three new two-dimensional (2D) NMR experiments, which together provide ^{15}N, ^1HN, $^{13}C^\alpha$, $^{13}C^\beta$, $^{13}C'$ chemical shifts for amino acid residues which are immediate C-terminal neighbors ($i + 1$) of residues that are selectively unlabeled. These experiments have high sensitivity and can be acquired rapidly using the methodology of G-matrix Fourier transform (GFT) NMR spectroscopy combined with non-uniform sampling (NUS). This is a first study involving the application of fast NMR methods to proteins samples prepared using a specific labeling scheme. Taken together, this opens up new avenues to using the method of selective unlabeling for rapid resonance assignment of proteins.

Keywords: selective unlabeling; GFT NMR; non-uniform sampling; fast NMR methods; protein resonance assignments

1. Introduction

During the last decade, there has been a growing emphasis on speeding up the structure determination process of proteins by NMR spectroscopy, especially in the context of structural genomics projects, which require high throughput structure determination [1–4]. This is due to the fact that the conventional approaches for NMR data acquisition and analysis are time consuming. Typically, protein samples enriched uniformly with ^{13}C and/or ^{15}N isotopes are prepared and a suite of 2D and 3D heteronuclear spectra are acquired for sequence specific resonance assignments [5]. With the conventional approach of data collection (e.g., linear sampling) this requires a few days to weeks of measurement time. Acquiring data in two different ways can accelerate this. One involves the rapid collection of multidimensional data using the recently proposed methods for fast data acquisition [1–4,6–9]. These methods reduce the time taken for data collection by an order of magnitude or more. The second approach involves augmenting the data acquired using uniformly ^{13}C, ^{15}N labeled protein sample with those acquired with selectively labeled/unlabeled samples [10–12]. Selective unlabeling or "reverse" labeling involves the $^{13}C/^{15}N$ enrichment of all but specific chosen amino acids in a protein, which are rendered unlabeled ($^{12}C/^{14}N$) [11,13–17]. This helps in simplifying the NMR spectrum by reducing the number of peaks in the spectrum, and thereby, aiding unambiguous resonance assignments.

We propose here a novel approach, which combines different methods for fast data acquisition with amino acid selective unlabeling to accelerate NMR data collection and analysis. Three new NMR experiments are proposed, two of which involve G-matrix Fourier transform (GFT) NMR spectroscopy [18,19] applicable to a selectively unlabeled protein sample. The experiments together provide the chemical shifts of ^{15}N, ^{1}HN, $^{13}C^{\alpha}$, $^{13}C^{\beta}$, $^{13}C'$ nuclei of the C-terminal neighbor ('$i + 1$') of amino acid residue, 'i', which is selectively unlabeled. The experiments, namely, 2D HN(CA)(i + 1), GFT (3,2)D HNCACB(i + 1) and GFT (3,2)D HNCACO(i + 1) are further accelerated by employing non-uniform sampling (NUS) [20–26]. The methodology is demonstrated on a selectively unlabeled protein sample of ubiquitin. Its application to larger proteins is discussed.

2. Results

2.1. Implementation of NMR Experiments

All the experiments proposed are "HNCA" [5] based and involve the transfer of ^{15}N magnetization of residue $i + 1$ ($^{15}N_{i+1}$) to $^{13}C^{\alpha}_{i+1}$ or $^{13}CO_{i+1}$. They, however, detect selectively the resonances ($^{1}H/^{13}C/^{15}N$) of residue $i + 1$ corresponding to the selectively unlabeled residue, i. This selection is achieved by tuning the delay periods in the radio frequency (r.f.) pulse sequence appropriately such that the magnetization on $^{15}N_{i+1}$ is attenuated by coupling to both $^{13}CO_i$ and $^{13}C^{\alpha}_i$ by one-bond and two-bond scalar couplings, $^{1}J_{NCO}$ (~15 Hz) and $^{2}J_{NC\alpha}$ (~7 Hz), respectively. For this purpose, a delay period of $1/2J$ (J is the scalar coupling) is used which converts an in-phase magnetization of a nuclei to anti-phase magnetization with respect to its J-coupled partner. This is depicted schematically in Figure 1a. In the case where both residues i and $i + 1$ are ^{13}C, ^{15}N labeled, $^{15}N_{i+1}$ magnetization is attenuated by the scalar coupling evolution to $^{13}CO_i$ and $^{13}C^{\alpha}_i$ (passive coupling via $^{2}J_{NC\alpha}$). Thus, if residue i is a labeled ($^{13}C/^{15}N$) residue the chemical shift correlations from its neighbor; $i + 1$ are not detected. However, if the residue i among a given pair ($i, i + 1$) is unlabeled (i.e., $^{12}C/^{14}N$), $^{15}N_{i+1}$ is coupled only to $^{13}C^{\alpha}_{i+1}$ and hence the delay period of $1/2J$ corresponding to $^{1}J_{NCO}$ and $^{2}J_{NC\alpha}$ has no effect and $^{15}N_{i+1}$ magnetization gets selected (product operator treatment is discussed below).

The delay periods used in the r.f. pulse schemes are thus optimized to achieve minimal selection of labeled ($i, i + 1$) pair while maximizing the intensity of unlabeled (i)-labeled ($i + 1$) pair. Since the experiments are "HNCA" type, the transfer function for the desired or selected $^{15}N_{i+1}$ magnetization (from unlabeled i-labeled $i + 1$ pair) and undesired or suppressed $^{15}N_{i+1}$ magnetization (in case of labeled $i, i + 1$ pair) during the filter element 'a' shown in Figure 1a can be expressed (ignoring relaxation) as:

$$\Gamma(\text{selected}): \sin(\pi\,^{1}J_{NC\alpha}\,\tau_{NC\alpha}) \tag{1}$$

$$\Gamma(\text{suppressed}): \sin(\pi\,^{1}J_{NC\alpha}\,\tau_{NC\alpha}) \times \cos(\pi\,^{2}J_{NC\alpha}\,\tau_{NC\alpha}) \tag{2}$$

where $\tau_{NC\alpha}$ is the delay period during which $^{15}N_{i+1}$ is coupled passively to $^{13}C^{\alpha}_i$ via $^{2}J_{NC\alpha}$ if residue i is labeled. Next, during this period the $^{15}N_{i+1}$ is also allowed to couple with $^{13}CO_i$ (for labeled i–$i + 1$ pair) for the duration $\tau_{HNC'}$ (~$1/2^{1}J_{NCO}$) and the transfer function above gets modified as:

$$\Gamma(\text{selected}): \sin(\pi\,^{1}J_{NC\alpha}\,\tau_{NC\alpha}) \tag{3}$$

$$\Gamma(\text{supressed}): \sin(\pi\,^{1}J_{NC\alpha}\,\tau_{NC\alpha}) \times \cos(\pi\,^{2}J_{NC\alpha}\,\tau_{NC\alpha}) \times \cos(\pi\,^{1}J_{NCO}\,\tau_{NCO}) \tag{4}$$

where τ_{NCO} is the delay period during which $^{15}N_{i+1}$ is coupled to $^{13}CO_i$ via $^{1}J_{NCO}$. Figure 1b depicts a plot of the transfer function as a function of $\tau_{NC\alpha}$ delays. Also shown in the figure is a plot of ratio of $\Gamma(\text{selected})/\Gamma(\text{suppressed})$, which is maximum for $\tau_{NC\alpha}$ ~43 ms and τ_{NCO} ~33 ms (assuming $^{1}J_{NCO}$ = 15 Hz, $^{1}J_{NC\alpha}$ = 11 Hz and $^{2}J_{NC\alpha}$ = 7 Hz). Variations in $^{1}J_{NCO}$, $^{1}J_{NC\alpha}$ and $^{2}J_{NC\alpha}$ among different secondary structural elements has been ignored.

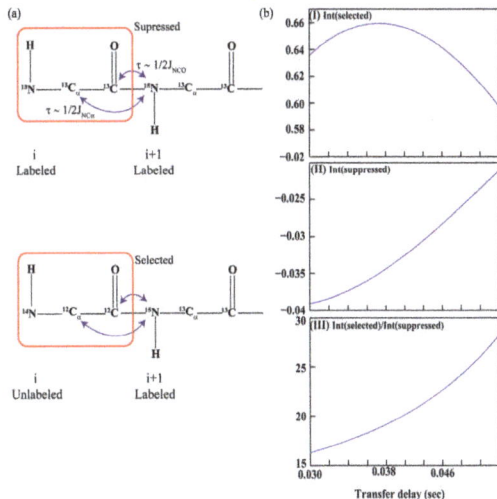

Figure 1. (a) Schematic illustration of the selection of magnetization in all the filter experiments employing one-bond and two-bond scalar couplings, $^1J_{NCO}$ (~15 Hz) and $^2J_{NC\alpha}$ (~7 Hz), respectively; (b) Figure 1b shows a plot of the transfer function as a function of $\tau_{NC\alpha}$ delays. Panels (I), (II) and (III) shows variation of Γ(selected), Γ(suppressed) the ratio of Γ(selected)/Γ(suppressed) as a function of $\tau_{NC\alpha}$ delays respectively.

The above selection scheme was implemented in the 2D HN(CA)(i + 1), GFT (3,2)D HNCACB(i + 1) and GFT (3,2)D HN(CA)CO(i + 1) the r.f. pulse schemes of which are shown in Figure 2. The first experiment, 2D HN(CA)(i + 1) provides a 2D [$^{15}N_{i+1}$, $^1H_{i+1}$] correlation spectrum analogues to a 2D [^{15}N, ^1H] HSQC spectrum. The delay periods $\tau_{NC\alpha}$ and τ_{NCO} in Equations (1)–(4) above correspond to 2 * τ_3 and 2 * ($\tau_3 - \tau_4$) and, respectively. In the (3,2)D GFT experiments, for nuclei shown underlined (e.g., N and CO in (3,2)D HN(CA)CO(i + 1)), chemical shifts are jointly sampled. That is, the chemical shift evolution periods of ^{15}N and ^{13}CO are co-incremented resulting in sums and differences of chemical shifts [1,2,18]. Thus, the following shift correlations are detected in the GFT (ω_1) dimension: (i) $\Omega(^{15}N_{i+1} \pm \kappa * ^{13}C\alpha_{i+1})$, $\Omega(^{15}N_{i+1} \pm \kappa * ^{13}C\beta_{i+1})$ in (3,2)D HNCACB(i + 1) and (ii) $\Omega(^{15}N_{i+1} \pm \kappa * ^{13}CO_{i+1})$ in (3,2)D HN(CA)CO(i + 1). The factor, κ, scales the relative shifts of $^{13}C\alpha_{i+1}$, $^{13}C\beta_{i+1}$ in (3,2)D HNCACB(i + 1) and CO$_{i+1}$ in (3,2)D HN(CA)CO(i + 1) with respect $^{15}N_{i+1}$ (Figure 2). In all spectra acquired in the present study κ = 0.5 was used. The $^{15}N_{i+1}$ serves as the center shift and hence 2D HN(CA)(i + 1) providing [$^1H_{i+1}$, $^{15}N_{i+1}$] shift correlations serves the central peak spectrum.

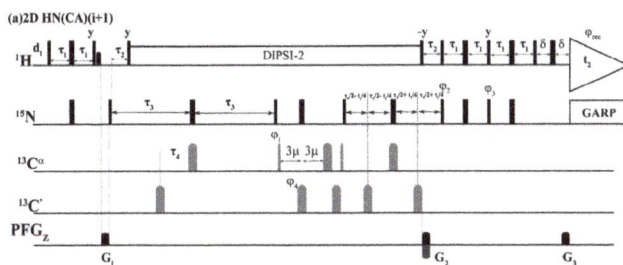

Figure 2. *Cont.*

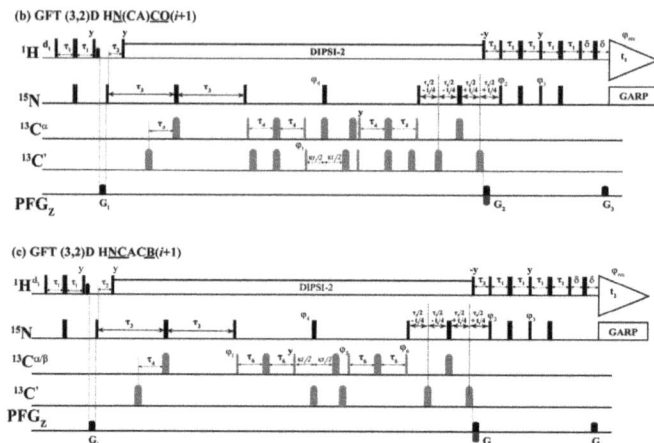

Figure 2. Radio frequency (r.f.) pulse sequences of (**a**) 2D HN(CA)(*i* + 1); (**b**) GFT (3,2)D HN(CA)CO(*i* + 1) and (**c**) GFT (3,2)D HNCACB(*i* + 1). Rectangular 90° and 180° hard pulses on ^{1}H and ^{15}N channel are indicated by thin and thick vertical bars, respectively and the same thin and thick notations are used for 90° and 180° shape pulses in ^{13}C channel. The representative 90° pulse widths are 11.5 μs, 37.5 μs and 9.8 μs for ^{1}H, ^{15}N and ^{13}C channels respectively. The corresponding phases of the applied pulses are indicated above and in places where no r.f. phase is marked, the pulse is applied along *x*. The ^{1}H offset is placed at the position of the solvent line at 4.7 ppm and the ^{15}N carrier was adjusted according to the spectral width observed in ^{15}N dimension which was 119.5 ppm for Ubiquitin. The ^{13}C$^\alpha$ r.f. carrier was placed at 54 ppm throughout the sequence and ^{13}C′ carrier frequency was set at 176.0 ppm. The shaped pulse on ^{13}C$^\alpha$ are of Gaussian cascade type with a pulse width of 240 μs and 196 μs, respectively, for 90° and 180° on resonance. The 180° off resonance pulse (Gaussian cascade) on ^{13}C′ was applied for duration of 192 μs. In (**a**) DIPSI-2 is employed for decoupling ^{1}H during ^{15}N shift evolution periods and in other cases (**b**,**c**), ^{15}N and ^{13}C chemical shifts were jointly incremented. GARP was employed to decouple ^{15}N during acquisition (r.f. strength = 3 kHz) all the sequences. All pulsed z-field gradients (PFGs) are sinc shaped with gradient recovery delay of 200 μs. The duration of gradient was 1.0 ms each and the strengths of the PFGs were G1: 16 G/cm, G2: 43 G/cm, G3: 4.3 G/cm. The delays and the phase cycling employed were as follows: For (**a**) 2D HN(CA)(*i* + 1), delays employed were τ_1 = 2.3 m, τ_2 = 5.5 m, τ_3 = 21.5 m, τ_4 = 4.5 m, τ_5 = 12 m and Phase cycling: φ_2 = 2(*x*), 2(−*x*); φ_3 = 2(−*y*), 2(*y*); φ_4 = 4(*x*), 4(−*x*); φ_5 = 8(*x*), 8(−*x*) and φ_{rec} = 2(*x*,−*x*,−*x*,*x*), 2(−*x*,*x*,*x*,−*x*). For (**b**) GFT (3,2)D HN(CA)CO(*i* + 1), same delays were employed and Phase cycling: φ_1 = *x*, −*x*; φ_2 = 2(*x*), 2(−*x*); φ_3 = 2(−*y*), 2(*y*); φ_4 = 4(*x*), 4(−*x*) and φ_{rec} = 2(*x*,−*x*,−*x*,*x*), 2(−*x*,*x*,*x*,−*x*). For GFT NMR: two data sets with phase cycle φ_1 = (*x*), (−*x*) and (*y*), (−*y*) were acquired, in conjunction with quadrature detection in ^{15}N which were linearly combined later employing a G-matrix transformation. For (**c**) GFT (3,2)D HNCACB(*i* + 1), delays employed were τ_1 = 2.3 m, τ_2 = 5.5 m, τ_3 = 21.5 m, τ_4 = 1 m, τ_5 = 12.5 m and τ_6 = 3.6 m. Phase cycling: φ_1 = *x*; φ_2 = 2(*x*), 2(−*x*); φ_3 = 2(−*y*), 2(*y*); φ_4 = 2(*x*), 2(−*x*) and φ_{rec} = 2(*x*,−*x*,−*x*,*x*), 2(-*x*,*x*,*x*,−*x*). For GFT NMR: two data sets with phase cycle φ_5 = (−*y*), (*y*) along with φ_6 = 2(*x*,−*x*), 2(−*x*,*x*) and φ_5 = (*x*), (−*x*) along with φ_6 = 2(*y*,−*y*), 2(−*y*,*y*) were acquired, in conjunction with quadrature detection in ^{15}N and later were linearly combined employing a G-matrix transformation. Quadrature detection in $t_2(^{15}$N) is accomplished using the sensitivity enhanced scheme by inverting the sign of gradient G$_2$ in concert with phases φ_3. Chemical shift evolution in ^{15}N channel (t_2) is achieved in a constant manner 'κ' is the scaling factor. For GFT experiments, at the same time ^{13}C chemical shift evolution period is co-incremented leading to the linear combination: (**b**) $\Omega(^{15}$N$_{i+1} \pm$ κ * ^{13}CO$_{i+1})$ in (3,2)D HN(CA)CO(*i* + 1) and (**c**) $\Omega(^{15}$N$_{i+1} \pm$ κ * ^{13}C$^\alpha{}_{i+1})$, $\Omega(^{15}$N$_{i+1} \pm$ κ * ^{13}C$^\beta{}_{i+1})$ in (3,2)D HNCACB(*i* + 1) respectively. In all spectra acquired in the present study κ = 0.5 was used.

2.2. Resonance Assignment Strategy

Using the above set of experiments, resonance assignment is carried out as follows. For each $[^{15}N_{i+1}, {}^{1}H_{i+1}]$ chemical shifts identified in 2D HN(CA)(i + 1), the amino acid type corresponding to residue i + 1 (self) is identified using (3,2)D HNCACB(i + 1) based on $^{13}C\alpha_{i+1}, {}^{13}C\beta_{i+1}$ values. Similarly using the (3,2)D HN(CA)CO(i + 1) the CO_{i+1} shifts are identified. Once the identity of the amino acid type corresponding to i + 1 is identified, and given that the amino acid type of i is known which is selectively unlabeled, the dipeptide pair i–i + 1 can be mapped onto the primary sequence for sequence-specific resonance assignment. Note that this is approach, that is different from the conventional strategy where the di-peptide pairs i–i − 1 are identified using 3D HNCACB and 3D CBCA(CO)NH for given residue i. Thus, the GFT experiments using selectively unlabeling can augment the conventional assignment approach resulting in a tri-peptide stretch (i − 1, i, i + 1) around a selectively unlabeled residue i. This is depicted in Figure 3. The experiments presented here thus increase the assignment speed compared to the approach presented earlier which did not yield directly the $^{13}C\alpha_{i+1}, {}^{13}C\beta_{i+1}$ shifts.

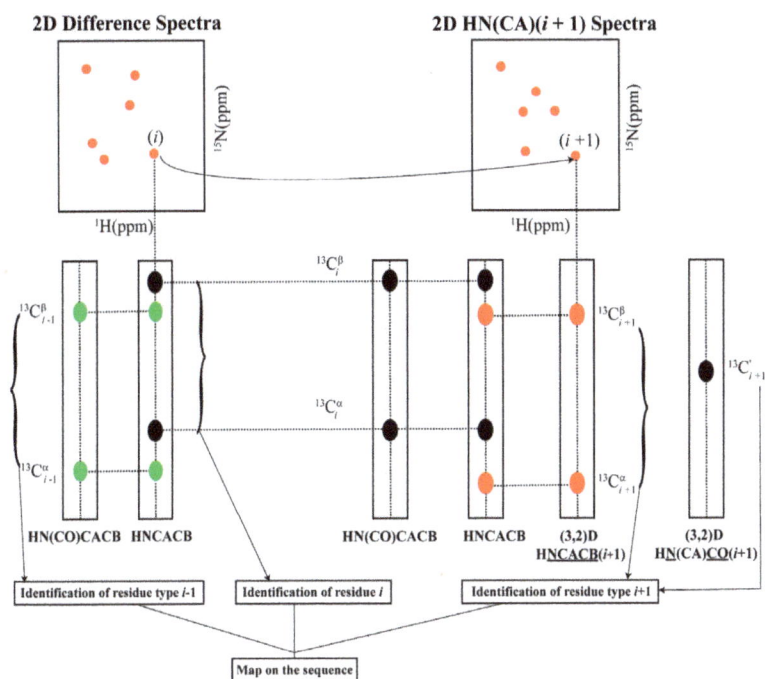

Figure 3. Schematic illustration of the sequential assignment strategy used with selective unlabeling.

In a given sample, more than one amino acid type can be chosen for simultaneous selective unlabeling so that the number of samples to be prepared is minimized. Two or more amino acid types are chosen for selectively unlabeling such that their $^{13}C\beta_{i+1}$ shifts lie in distinct spectral regions enabling their type identification. For instance, in the current study, Arg and Asn were used for selective unlabeling in the same sample due to their distinct $^{13}C\beta_{i+1}$ shifts. However, in such cases the $[^{15}N_{i+1}, {}^{1}H_{i+1}]$ pair of shifts obtained from 2D HN(CA)(i + 1) do not provide any information directly on the type of amino acid residue i. Hence, a 3D CBCA(CO)NH spectrum acquired on the uniformly $^{13}C, {}^{15}N$ labeled sample is needed. In the 3D CBCA(CO)NH spectrum, at a given $[^{15}N_{i+1}, {}^{1}H_{i+1}]$ pair of shifts, the amino acid type corresponding to residue i can be identified based on $^{13}C^{\alpha}{}_{i}$, $^{13}C^{\beta}{}_{i}$ values (Figure 3).

2.3. Data Acquisition Using Non-Uniform Sampling (NUS)

The measurement time required for acquiring the GFT spectra can be reduced further by using the non-uniform sampling (NUS) approach [24–26]. The NUS approach is based on the premise that the conventional method involving linear sampling of interferogram in the inderct dimension requires a lot more number of points although the number of frequencies encoded in the interferogram is much less. Thus, by reducing the number of sampling points in the indirect dimensions the total measurement time can be proportionately reduced. The dataset obtained using NUS is then reconstructed either in the time domain using the multiway decomposition method (MDD) or directly in the frequency domain using the maximum entropy reconstruction (MER) approach. The sampling points are chosen based on the decay of the Free Induction Decay (FID). In case of a constant time experiment, the NUS points can be chosen randomly. In the present study, the MDD approach was used for spectral reconstruction in the time domain. The data were acquired a random sampling of 25% of the points (i.e., the omission of 75% of the time domain points) in the GFT dimension.

Figure 4a shows an overlay of the 2D [^{15}N, ^{1}H] HSQC spectrum of the uniformly ^{13}C, ^{15}N labeled (shown in red) and Arg, Asn selectively unlabeled (in blue) ubiquitin. The resonances, which are absent in the selectively unlabeled sample, corresponding to Arg and Asn residues as indicated. No other residues were observed to be absent indicating minimal mis-incorporation of ^{14}N isotope of Arg and Asn (also referred to as 'isotope scrambling'). Figure 4b shows 2D HN(CA)(i + 1) spectrum acquired on the Arg, Asn selectively unlabeled sample. All expected [$^{15}N_{i+1}$, $^{1}H_{i+1}$] correlations are observed in the spectrum as indicated.

MQIFVKTLTGKTITLEVEPSDTIENVKAKIQDKEGIPPDQQRLIFAGK
QLEDGRTLSDYNIQKESTLHLVLRLRGG

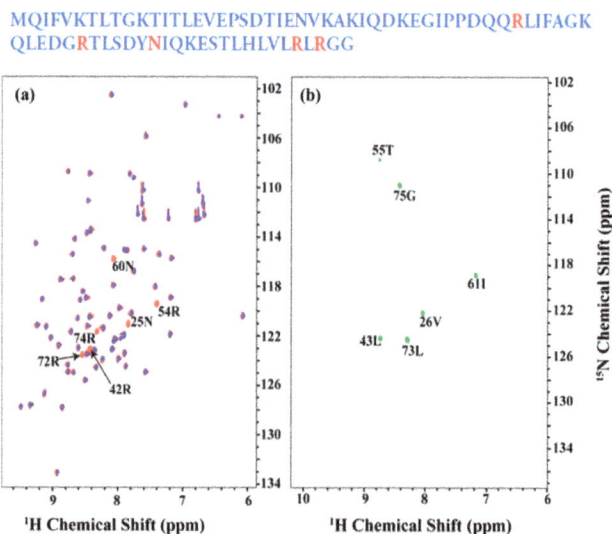

Figure 4. (**a**) overlay of the 2D [^{15}N, ^{1}H] HSQC spectrum of the uniformly ^{13}C, ^{15}N labeled ubiquitin as shown in red and Arg, Asn selectively unlabeled sample as shown in blue; (**b**) 2D HN(CA)(i + 1) spectrum acquired on the Arg, Asn selectively unlabeled sample. Assignments are indicated on the spectra. The sequence is shown above and the unlabeled residues are highlighted in red.

Figures 5 and 6 show the (3,2)D HNCACB(i + 1) and (3,2)D HN(CA)CO(i + 1) of Arg, Asn selectively unlabeled sample of ubiquitin acquired in 54 and 35 min with 8 scans each for both datasets, respectively. All expected $^{13}C^{\alpha}{}_{i+1}$, $^{13}C^{\beta}{}_{i+1}$ and $^{13}CO_{i+1}$ correlations except that of 55T are observed.

The correlation of 55T is presumably absent due to its weak intensity in the HSQC spectrum and shorter transverse relaxation time of its ^{13}C nuclei. The ^{13}Cα_{i+1}, ^{13}Cβ_{i+1} and ^{13}CO$_{i+1}$ chemical shifts values are obtained by linearly combining the sums and differences of the chemical shifts observed in the two GFT sub-spectra of each experiment and taking into account the appropriate scaling factor. Note that a 3D spectrum with an equivalent resolution would have taken more than 3 days of measurement time (2 days 7 h 9 min for 3D HNCACB(i + 1) and 19 h 23 min for 3D HN(CA)CO(i + 1)). Thus, the GFT experiments potentially reduce the measurement time by about an order of magnitude.

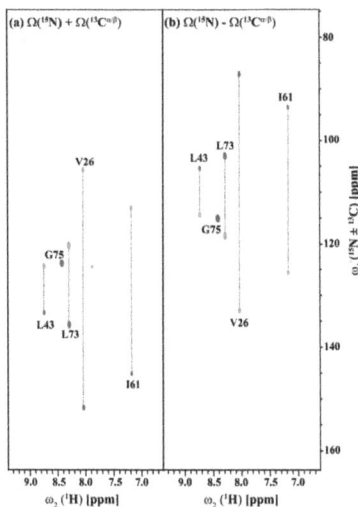

Figure 5. (**a,b**) shows the different linear combinations of (3,2)D HNCACB(i + 1) of Arg, Asn selectively unlabeled sample of ubiquitin. Total measurement time for two spectra was 54 min with 8 scans each. Peaks shown in blue and green correspond to $\Omega(^{15}N_{i+1} \pm \kappa * {}^{13}C\alpha_{i+1})$ and $\Omega(^{15}N_{i+1} \pm \kappa * {}^{13}C\beta_{i+1})$, respectively. The scaling factor κ was set to 0.5.

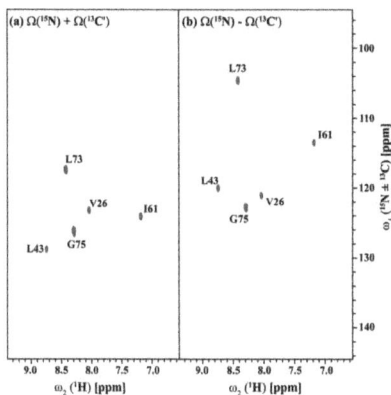

Figure 6. (**a,b**) shows the different linear combinations of (3,2)D HN(CA)CO(i + 1) of Arg, Asn selectively unlabeled sample of ubiquitin. Total measurement time for two spectra was 35 min with 8 scans each. Peaks shown in blue in figure (**a,b**) correspond to $\Omega(^{15}N_{i+1} + \kappa * {}^{13}C'_{i+1})$ and $\Omega(^{15}N_{i+1}-\kappa * {}^{13}C'_{i+1})$, respectively. The scaling factor κ was set to 0.5.

The acquisition of GFT spectra is further accelerated using the NUS approach. Figures 7 and 8 show the 25% NUS spectrum of (3,2)D HNCACB(i + 1) and (3,2)D HN(CA)CO(i + 1) experiments acquired in less than 14 and 9 min respectively. All expected correlations as observed in the non-NUS counterparts are observed without any spectral distortions or artifacts. The signal-to-noise ratio (S/N) by NUS is not compromised as shown in Figure 7c,d and Figure 8c,d which compares the 1D traces from the non-NUS and NUS GFT data.

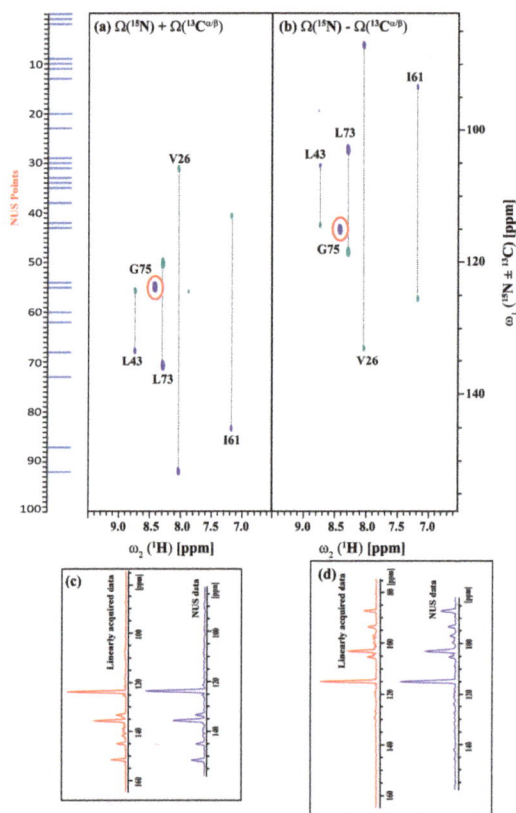

Figure 7. (**a,b**) shows the different linear combinations of non-uniformly sampled (3,2)D HNCACB(i + 1) of Arg, Asn selectively unlabeled sample of ubiquitin. Total measurement time for two spectra was 14 min with 8 scans each. Peaks shown in blue and green correspond to $\Omega(^{15}N_{i+1} \pm \kappa * {}^{13}C^{\alpha}{}_{i+1})$ and $\Omega(^{15}N_{i+1} \pm \kappa * {}^{13}C^{\beta}{}_{i+1})$, respectively. The scaling factor κ was set to 0.5. The NUS time increments in the shared dimension which was actually acquired in the experiment are shown in blue bars; (**c,d**), represents the overlay of 1D projection of linearly acquired data (red) and non-uniformly acquired data (blue) along ω_1 for $\Omega(^{15}N_{i+1} + \kappa * {}^{13}C^{\alpha}{}_{i+1}/{}^{13}C^{\beta}{}_{i+1})$ and $\Omega(^{15}N_{i+1} - \kappa * {}^{13}C^{\alpha}{}_{i+1}/{}^{13}C^{\beta}{}_{i+1})$ respectively. There is 28% reduction in signal to noise (SNR) ratio ongoing from linearly acquired data to NUS data as measured by the SNR of the projections. We also have calculated SNR of G75 (highlighted in red) individually and there also we have seen ~30% decrease in SNR on going from linearly acquired data to NUS data.

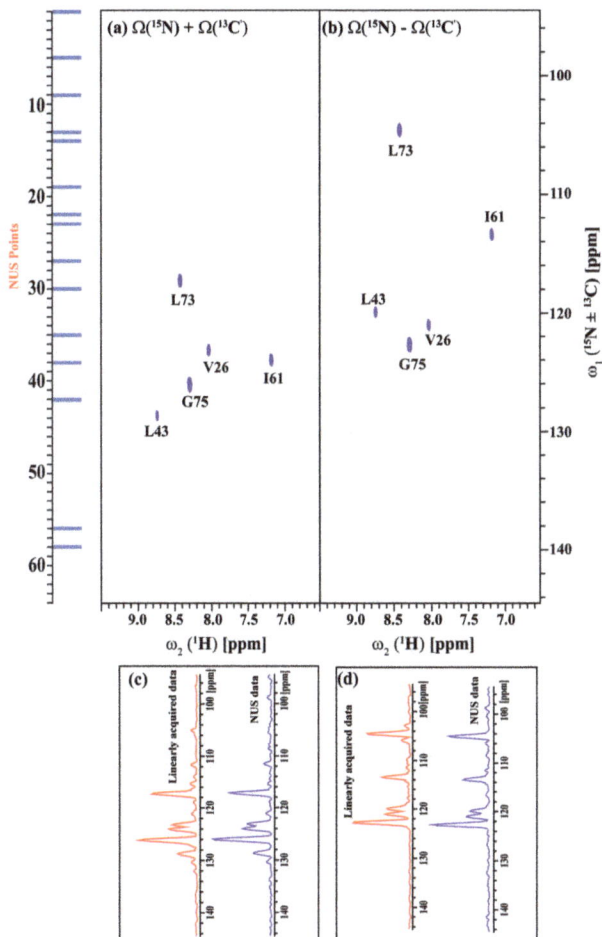

Figure 8. (**a,b**) shows the different linear combinations of non-uniformly sampled (3,2)D HN(CA)CO(*i* + 1) of Arg, Asn selectively unlabeled sample of ubiquitin. Total measurement time for two spectra was ~9 min with 8 scans each. Peaks shown in blue in figure (**a,b**) correspond to $\Omega(^{15}N_{i+1} + \kappa * {}^{13}C'_{i+1})$ and $\Omega(^{15}N_{i+1} - \kappa * {}^{13}C'_{i+1})$, respectively. The scaling factor κ was set to 0.5. The NUS time increments in the shared dimension which was actually acquired in the experiment are shown in blue bars; (**c,d**), represents the overlay of 1D projection of linearly acquired data (cyan) and non-uniformly acquired data (blue) along ω_1 for $\Omega(^{15}N_{i+1} + \kappa * {}^{13}C'_{i+1})$ and $\Omega(^{15}N_{i+1} - \kappa * {}^{13}C'_{i+1})$ respectively. There is 27% reduction in signal to noise ratio ongoing from linearly acquired data to NUS data as measured by the SNR of the projections.

3. Discussion

Amino acid selective unlabeling has been shown to be a robust, easy and cost-effective method for simplifying NMR spectrum and obtaining useful structural information in proteins [11,15,27–30]. On the other hand, GFT NMR is a powerful technique for speeding up NMR data acquisition and has been shown to be useful in various applications [2,18,19,31–46]. In the present study, the application of two fast NMR methods, namely GFT NMR and non-uniform sampling to selectively unlabeled protein samples expands the repertoire of applications possible with selective unlabeling. In our previous study, we reported an experiment for identification of $[^{15}N_{i+1}, {}^1H_{i+1}]$ shifts based on a 2D ^{12}CO-filtered $[^{15}N, {}^1H]$ HSQC. The new experiment proposed herein, namely, 2D HN(CA)(i + 1) improves upon the selection of unlabeled (i)-labeled (i + 1) pair by incorporating two filters simultaneously. That is, attenuating the $^{15}N_{i+1}$ magnetization by allowing its scalar coupling to both $^{13}CO_i$ and $^{13}C^{\alpha}{}_i$. The delay periods used in the r.f. pulse sequences are appropriately tuned to achieve the desired selection. The only limitation of the approach is the additional sample requirement. Additionally, not all the amino acids can be used for unlabeling as cross metabolism is an issue. However, we have extensively studied on the cross metabolism issue and came up with excellent solutions of many different combinations of amino acids which can be used as starting point. These combinations greatly cover a large percentage of residues in the sequences and the resonance assignment strategies can be simplified to a large extent. Considering the cost-effective solution this approach is much better than selective labeling as the cross metabolism is also an issue there.

Having thus achieved a selection of $[^{15}N_{i+1}, {}^1H_{i+1}]$ resonances, the application of GFT NMR and NUS methodology helps to provide high dimensional spectral information, namely $^{13}C^{\alpha}{}_{i+1}$, $^{13}C^{\beta}{}_{i+1}$ and $^{13}CO_{i+1}$ correlations, rapidly. The two (3,2)D GFT experiments proposed thus augment the conventional assignment strategy involving 3D HNCACB and 3D CBCA(CO)NH without severely increasing the overall measurement time. Further, the information obtained can be used for automated resonance assignment strategies to speed up data analysis. Since the experiments are based on the "HNCA" based approach, the methodology can be extended in a straightforward manner to detect $^1H^{\alpha}$ or measure scalar/dipolar couplings, which involve HNCA type editing.

4. Materials and Methods

4.1. Sample Preparation

The plasmid (P^{GLUB}) coding for ubiquitin was transformed into *E. coli* BL21 cells (New England Biolabs, Ipswich, MA, USA). Cells were grown at 37 °C in M9 medium containing 1 g/L of ^{15}NH$_4$Cl (Cambridge Isotope Laboratories, Tewksbury, MA, USA) and 4 g/L of ^{13}C-Glucose (Cambridge Isotope Laboratories, Tewksbury, MA, USA). For selective unlabeling, 1.0 g/L each of desired unlabeled amino acid(s): Arg and Asn (stock solution of 1.0 g in 50 mL of H$_2$O was prepared and filter sterilized) was added to the growth medium. To induce protein expression, 1.0 mM isopropyl β-D-thiogalactoside (IPTG) (Sigma-Aldrich, St. Louis, MO, USA) was added at midlog phase (O.D$_{600}$ ~0.6). Cells were harvested by centrifugation and suspended in acetate buffer (5 mM EDTA (Sigma-Aldrich, St. Louis, MO, USA), 50 mM Na acetate (Sigma-Aldrich, St. Louis, MO, USA), pH 5) and taken up for sonication. Following sonication, the supernatant containing the protein was loaded on to a pre-equilibrated ion exchange column (SP Sepharose fast flow from GE) and the protein eluted with a salt gradient of 0–0.6 M NaCl. The protein sample was further purified using size-exclusion chromatography with Superdex 75 (Sigma-Aldrich, St. Louis, MO, USA). For NMR data acquisition, a sample containing ~1.0 mM of protein in 50 mM phosphate buffer (10% ^{2}H$_2$O, pH 6.0) was prepared.

4.2. NMR Data Collection

All NMR data were acquired at 298 K on a Bruker Avance 800 MHz NMR spectrometer (Billerica, MA, USA) equipped with a room temperature triple resonance probe with a z-axis shielded gradient. Data were processed with NMRPipe [47] and analyzed using XEASY [48].

5. Conclusions

Three new NMR experiments are presented which are applicable to protein samples prepared with amino acid selective unlabeling. The experiments provide ^{15}N, ^{1}HN, $^{13}C^{\alpha}$, $^{13}C^{\beta}$, $^{13}C'$ chemical shifts for amino acid residues which are immediate C-terminal neighbors (i + 1) of selectively unlabeled residues. Two of the experiments provide 3D shift correlations rapidly using the methodology of GFT NMR and non-uniform sampling. The data acquisition time can be further reduced using the method of longitudinal ^1H relaxation optimization [19]. The spectral simplification obtained combined with rapid data collection will help in reducing the time required for data analysis. In addition to resonance assignments, the proposed methodology can also be used for studies such as protein-ligand and protein-protein interactions, which involve monitoring the changes in shifts of certain residues against a background of unshifted resonances. Taken together, the proposed methodology expands the repertoire of applications possible with amino acid selective unlabeling for rapid protein structural studies.

Acknowledgments: The facilities provided by NMR Research Centre at IISc supported by Department of Science and Technology (DST), India is gratefully acknowledged. K.C. acknowledges support from DST Inspire Faculty Scheme (IFA 13-CH 106). H.S.A. acknowledges support from DBT and DAE-BRNS research awards.

Author Contributions: H.S.A. and K.C. conceived and designed the experiments; B.K. and K.C. performed the experiments and analyzed the data; B.K., K.C. and H.S.A. wrote the paper.

Conflicts of Interest: The authors declare no conflict of interest.

References

1. Atreya, H.S.; Szyperski, T. Rapid NMR data collection. *Methods Enzymol.* **2005**, *394*, 78–108. [PubMed]
2. Szyperski, T.; Atreya, H.S. Principles and applications of GFT projection NMR spectroscopy. *Magn. Reson. Chem.* **2006**, *44*, S51–S60. [CrossRef] [PubMed]
3. Schanda, P. Fast-pulsing longitudinal relaxation optimized techniques: Enriching the toolbox of fast biomolecular NMR spectroscopy. *Prog. NMR Spectrosc.* **2009**, *55*, 238–265. [CrossRef]
4. Felli, I.C.; Brutscher, B. Recent Advances in Solution NMR: Fast Methods and Heteronuclear Direct Detection. *ChemPhysChem* **2009**, *10*, 1356–1368. [CrossRef] [PubMed]
5. Cavanagh, J.; Fairbrother, W.J.; Palmer, A.G.; Skelton, N.J. *Protein NMR Spectroscopy*; Academic Press: San Diego, CA, USA, 1996.
6. Pudakalakatti, S.; Chandra, K.; Thirupathi, R.; Atreya, H.S. Rapid Characterization of Molecular Diffusion by NMR. *Chem. Eur. J.* **2014**, *20*, 15719–15722. [CrossRef] [PubMed]
7. Mulleti, S.; Singh, A.; Brahmkhatri, V.P.; Chandra, K.; Raza, T.; Mukherjee, S.P.; Seelamantula, C.S.; Atreya, H.S. Super-resolved nuclear magnetic resonance spectroscopy. *Sci. Rep.* **2017**, *7*, 9651.
8. Khaneja, N.; Dubey, A.; Atreya, H.S. Ultra broadband NMR spectroscopy using multiple rotating frame technique. *J. Magn. Reson.* **2016**, *265* (Suppl. C), 117–128. [CrossRef] [PubMed]
9. Atreya, H.S. Isotope labeling in Biomolecular NMR. In *Advances in Experimental Medicine and Biology*; Springer: Amsterdam, The Netherlands, 2012; pp. 1–219.
10. Krishnarjuna, B.; Jaipuria, G.; Thakur, A.; D'Silva, P.; Atreya, H.S. Amino acid selective unlabeling for sequence specific resonance assignments in proteins. *J. Biomol. NMR* **2011**, *49*, 39–51. [CrossRef] [PubMed]
11. Prasanna, C.; Dubey, A.; Atreya, H.S. Amino acid selective unlabeling in protein nmr spectroscopy. In *Methods in Enzymology*; Kelman, Z., Ed.; Academic Press: Cambridge, MA, USA, 2015; pp. 167–189.
12. Vuister, G.W.; Kim, S.J.; Wu, C.; Bax, A. 2D and 3D NMR-Study of Phenylalanine Residues in Proteins by Reverse Isotopic Labeling. *J. Am. Chem. Soc.* **1994**, *116*, 9206–9210. [CrossRef]
13. Shortle, D. Assignment of amino acid type in ^1H-^{15}N correlation spectra by labeling with ^{14}N-amino acids. *J. Magn. Reson.* **1994**, *105*, 88–90. [CrossRef]
14. Atreya, H.S.; Chary, K.V.R. Amino acid selective 'unlabelling' for residue-specific NMR assignments in proteins. *Curr. Sci.* **2000**, *79*, 504–507.
15. Atreya, H.S.; Chary, K.V.R. Selective 'unlabeling' of amino acids in fractionally ^{13}C labeled proteins: An approach for stereospecific NMR assignments of CH_3 groups in Val and Leu residues. *J. Biomol. NMR* **2001**, *19*, 267–272. [CrossRef] [PubMed]
16. Dubey, A.; Kadumuri, R.V.; Jaipuria, G.; Vadrevu, R.; Atreya, H.S. Rapid NMR assignments of proteins by using optimized combinatorial selective unlabeling. *ChemBioChem* **2016**, *17*, 334–340. [CrossRef] [PubMed]
17. Kim, S.; Szyperski, T. GFT NMR, a new approach to rapidly obtain precise high-dimensional NMR spectral information. *J. Am. Chem. Soc.* **2003**, *125*, 1385–1393. [CrossRef] [PubMed]
18. Atreya, H.S.; Szyperski, T. G-matrix Fourier transform NMR spectroscopy for complete protein resonance assignment. *Proc. Natl. Acad. Sci. USA* **2004**, *101*, 9642–9647. [CrossRef] [PubMed]

19. Orekhov, V.Y.; Ibraghimov, I.V.; Billeter, M. MUNIN: A new approach to multi-dimensional NMR spectra interpretation. *J. Biomol. NMR* **2001**, *20*, 49–60. [CrossRef] [PubMed]

20. Orekhov, V.Y.; Ibraghimov, I.; Billeter, M. Optimizing resolution in multidimensional NMR by three-way decomposition. *J. Biomol. NMR* **2003**, *27*, 165–173. [CrossRef] [PubMed]

21. Rovnyak, D.; Frueh, D.P.; Sastry, M.; Sun, Z.Y.J.; Stern, A.S.; Hoch, J.C.; Wagner, G. Accelerated acquisition of high resolution triple-resonance spectra using non-uniform sampling and maximum entropy reconstruction. *J. Magn. Reson.* **2004**, *170*, 15–21. [CrossRef] [PubMed]

22. Hoch, J.C.; Stern, A.S. Maximum entropy reconstruction, spectrum analysis and deconvolution in multidimensional nuclear magnetic resonance. In *Nuclear Magnetic Resonance of Biologica Macromolecules*; Academic Press: Cambridge, MA, USA, 2001; Part A, pp. 159–178.

23. Hoch, J.C.; Maciejewski, M.W.; Filipovic, B. Randomization improves sparse sampling in multidimensional NMR. *J. Magn. Reson.* **2008**, *193*, 317–320. [CrossRef] [PubMed]

24. Jaravine, V.A.; Orekhov, V.Y. Targeted acquisition for real-time NMR spectroscopy. *J. Am. Chem. Soc.* **2006**, *128*, 13421–13426. [CrossRef] [PubMed]

25. Tugarinov, V.; Kay, L.E.; Ibraghimov, I.; Orekhov, V.Y. High-resolution four-dimensional H-1-C-13 NOE spectroscopy using methyl-TROSY, sparse data acquisition, and multidimensional decomposition. *J. Am. Chem. Soc.* **2005**, *127*, 2767–2775. [CrossRef] [PubMed]

26. Rasia, R.M.; Brutscher, B.; Plevin, M.J. Selective isotopic unlabeling of proteins using metabolic precursors: Application to NMR assignment of intrinsically disordered proteins. *ChemBioChem* **2012**, *13*, 732–739. [CrossRef] [PubMed]

27. Kelly, M.J.S.; Krieger, C.; Ball, L.J.; Yu, Y.; Richter, G.; Schmieder, P.; Bacher, A.; Oschkinat, H. Application of amino acid type-specific 1H and 14N labeling in a 2H-, 15N-labeled background to a 47 kDa homodimer: Potential for NMR structure determination of large proteins. *J. Biomol. NMR* **1999**, *14*, 79–83. [CrossRef] [PubMed]

28. Mohan, P.M.K.; Barve, M.A.; Chatteljee, A.; Ghosh-Roy, R.V. Hosur, NMR comparison of the native energy landscapes of DLC8 dimer and monomer. *Biophys. Chem.* **2008**, *134*, 10–19. [CrossRef] [PubMed]

29. Tugarinov, V.; Kay, L.E. Stereospecific NMR assignments of prochiral methyls, rotameric states and dynamics of valine residues in malate synthase G. *J. Am. Chem. Soc.* **2004**, *126*, 9827–9836. [CrossRef] [PubMed]

30. Mukherjee, S.; Mustafi, S.M.; Atreya, H.S.; Chary, K.V.R. Measurement of $^{1}J(N_i,C^{\alpha}_i)$, $^{1}J(N_i,C'_{i-1})$, $^{2}J(N_i,C^{\alpha}_{i-1})$, $^{2}J(H^N_i,C'_{i-1})$ and $^{2}J(H^N_i,C^{\alpha}_i)$ values in $^{13}C/^{15}N$-labeled proteins. *Magn. Reson. Chem.* **2005**, *43*, 326–329. [CrossRef] [PubMed]

31. Rout, M.; Mishra, P.; Atreya, H.S.; Hosur, R.V. Reduced dimensionality 3D HNCAN for unambiguous HN, CA and N assignments in proteins. *J. Magn. Reson.* **2012**, *216*, 161–168. [CrossRef] [PubMed]

32. Chandra, K.; Jaipuria, G.; Shet, D.; Atreya, H.S. Efficient sequential assignments in proteins with reduced dimensionality 3D HN(CA)NH. *J. Biomol. NMR* **2011**, *52*, 115–126. [CrossRef] [PubMed]

33. Franks, W.T.; Atreya, H.S.; Szyperski, T.; Rienstra, C.M. GFT projection NMR spectroscopy for proteins in the solid state. *J. Biomol. NMR* **2010**, *48*, 213–223. [CrossRef] [PubMed]

34. Jaipuria, G.; Thakur, A.; D'Silva, P.; Atreya, H.S. High-resolution methyl edited GFT NMR experiments for protein resonance assignments and structure determination. *J. Biomol. NMR* **2010**, *48*, 137–145. [CrossRef] [PubMed]

35. Swain, M.; Atreya, H.S. CSSI-PRO: A method for secondary structure type editing, assignment and estimation in proteins using linear combination of backbone chemical shifts. *J. Biomol. NMR* **2009**, *44*, 185–194. [CrossRef] [PubMed]

36. Barnwal, R.P.; Atreya, H.S.; Chary, K.V.R. Chemical shift based editing of CH_3 groups in fractionally C-13-labelled proteins using GFT (3,2)D CT-HCCH-COSY: Stereospecific assignments of CH_3 groups of Val and Leu residues. *J. Biomol. NMR* **2008**, *42*, 149–154. [CrossRef] [PubMed]

37. Barnwal, R.P.; Rout, A.K.; Atreya, H.S.; Chary, K.V.R. Identification of C-terminal neighbours of amino acid residues without an aliphatic C-13(gamma) supercript stop as an aid to NMR assignments in proteins. *J. Biomol. NMR* **2008**, *41*, 191–197. [CrossRef] [PubMed]

38. Barnwal, R.P.; Rout, A.K.; Chary, K.V.R.; Atreya, H.S. Rapid measurement of pseudocontact shifts in paramagnetic proteins by GFT NMR spectroscopy. *Open Magn. Reson. J.* **2008**, *1*, 16–28. [CrossRef]

39. Zhang, Q.; Atreya, H.S.; Kamen, D.E.; Girvin, M.E.; Szyperski, T. GFT projection NMR based resonance assignment of membrane proteins: Application to subunit c of *E. coli* F_1F_0 ATP synthase in LPPG micelles. *J. Biomol. NMR* **2008**, *40*, 157–163. [CrossRef] [PubMed]
40. Atreya, H.S.; Garcia, E.; Shen, Y.; Szyperski, T. J-GFT NMR for precise measurement of mutually correlated nuclear spin-spin couplings. *J. Am. Chem. Soc.* **2007**, *129*, 680–692. [CrossRef] [PubMed]
41. Barnwal, R.P.; Rout, A.K.; Chary, K.V.R.; Atreya, H.S. Rapid measurement of $^3J(H^N\text{-}H^\alpha)$ and $^3J(N\text{-}H^\beta)$ coupling constants in polypeptides. *J. Biomol. NMR* **2007**, *39*, 259–263. [CrossRef] [PubMed]
42. Atreya, H.S.; Eletsky, A.; Szyperski, T. Resonance assignment of proteins with high shift degeneracy based on 5D spectral information encoded in G^2FT NMR experiments. *J. Am. Chem. Soc.* **2005**, *127*, 4554–4555. [CrossRef] [PubMed]
43. Eletsky, A.; Atreya, H.S.; Liu, G.H.; Szyperski, T. Probing structure and functional dynamics of (large) proteins with aromatic rings: L-GFT-TROSY (4,3)D HCCHNMR spectroscopy. *J. Am. Chem. Soc.* **2005**, *127*, 14578–14579. [CrossRef] [PubMed]
44. Liu, G.H.; Aramini, J.; Atreya, H.S.; Eletsky, A.; Xiao, R.; Acton, T.; Ma, L.C.; Montelione, G.T.; Szyperski, T. GFT NMR based resonance assignment for the 21 kDa human protein UFC1. *J. Biomol. NMR* **2005**, *32*, 261. [CrossRef] [PubMed]
45. Liu, G.H.; Shen, Y.; Atreya, H.S.; Parish, D.; Shao, Y.; Sukumaran, D.K.; Xiao, R.; Yee, A.; Lemak, A.; Bhattacharya, A.; et al. NMR data collection and analysis protocol for high-throughput protein structure determination. *Proc. Natl. Acad. Sci. USA* **2005**, *102*, 10487–10492. [CrossRef] [PubMed]
46. Shen, Y.; Atreya, H.S.; Liu, G.H.; Szyperski, T. G-matrix Fourier transform NOESY-based protocol for high-quality protein structure determination. *J. Am. Chem. Soc.* **2005**, *127*, 9085–9099. [CrossRef] [PubMed]
47. Delaglio, F.; Grzesiek, S.; Vuister, G.W.; Zhu, G.; Pfeifer, J.; Bax, A. NMRpipe—A Multidimensional Spectral Processing System Based on Unix Pipes. *J. Biomol. NMR* **1995**, *6*, 277–293. [CrossRef] [PubMed]
48. Bartels, C.; Xia, T.H.; Billeter, M.; Güntert, P.; Wüthrich, K. The Program Xeasy for Computer-Supported NMR Spectral-Analysis of Biological Macromolecules. *J. Biomol. NMR* **1995**, *6*, 1–10. [CrossRef] [PubMed]

magnetochemistry

MDPI

Article

Calcium-Dependent Interaction Occurs between Slow Skeletal Myosin Binding Protein C and Calmodulin

Tzvia I. Springer [1,†], Christian W. Johns [2], Jana Cable [1], Brian Leei Lin [3], Sakthivel Sadayappan [4] and Natosha L. Finley [1,2,*]

[1]   Department of Microbiology, Miami University, Oxford, OH 45056, USA; tspringer@mcw.edu (T.I.S.);
       cablejm@miamioh.edu (J.C.)
[2]   Cell, Molecular, and Structural Biology Program, Miami University, Oxford, OH 45056, USA;
       johnscw@miamioh.edu
[3]   Department of Cardiology, Johns Hopkins University, Baltimore, MD 21205, USA; blin29@jhmi.edu
[4]   Department of Internal Medicine, Heart Branch of the Heart, Lung and Vascular Institute,
       University of Cincinnati College of Medicine, Cincinnati, OH 45267, USA; SADAYASL@ucmail.uc.edu
*   Correspondence: finleynl@miamioh.edu; Tel.: +1-513-529-0950
†   Present affiliation Department of Biophysics, Medical College of Wisconsin, 8701 W. Watertown Plank Road,
       Milwaukee, WI 2042, USA.

Received: 1 November 2017; Accepted: 15 December 2017; Published: 21 December 2017

Abstract: Myosin binding protein C (MyBP-C) is a multi-domain protein that participates in the regulation of muscle contraction through dynamic interactions with actin and myosin. Three primary isoforms of MyBP-C exist: cardiac (cMyBP-C), fast skeletal (fsMyBP-C), and slow skeletal (ssMyBP-C). The N-terminal region of cMyBP-C contains the M-motif, a three-helix bundle that binds Ca^{2+}-loaded calmodulin (CaM), but less is known about N-terminal ssMyBP-C and fsMyBP-C. Here, we characterized the conformation of a recombinant N-terminal fragment of ssMyBP-C (ssC1C2) using differential scanning fluorimetry, nuclear magnetic resonance, and molecular modeling. Our studies revealed that ssC1C2 has altered thermal stability in the presence and absence of CaM. We observed that site-specific interaction between CaM and the M-motif of ssC1C2 occurs in a Ca^{2+}-dependent manner. Molecular modeling supported that the M-motif of ssC1C2 likely adopts a three-helix bundle fold comparable to cMyBP-C. Our study provides evidence that ssMyBP-C has overlapping structural determinants, in common with the cardiac isoform, which are important in controlling protein–protein interactions. We shed light on the differential molecular regulation of contractility that exists between skeletal and cardiac muscle.

Keywords: calcium; calmodulin; molecular model; MyBP-C; NMR; protein

1. Introduction

Myosin binding protein C (MyBP-C) is a modular protein composed of immunoglobulin (Ig) domains and fibronectin type III (FN3) repeats. It is reported to span between thin and thick filaments [1] while participating in the regulation of actin–myosin association. The N-terminal domains of MyBP-C interact with actin [2–8], myosin S2 [9,10], and the regulatory light chains [11], while the C-terminal domains are tethered to titin [12] and the myosin rod [13–15]. There are three different isoforms of MyBP-C expressed in striated muscles: cardiac (cMyBP-C), fast skeletal (fsMyBP-C), and slow skeletal (ssMyBP-C). Primarily, fsMyBP-C expression is localized to fast skeletal muscle, whereas ssMyBP-C is expressed in both slow and fast skeletal muscle [16]. The expression of cMyBP-C is localized to the heart, where its role as a dynamic regulator of cardiac contractility is established. Mutations in the genes encoding for each MyBP-C isoforms are associated with the development diseases such as distal arthrogryposis and hypertrophic cardiomyopathy. Interestingly, expression of mutant cMyBP-C appears to be associated with the progression of skeletal myopathies, but the molecular basis is poorly understood. Ablation of cMyBP-C leads to increased expression of fsMyBP-C in heart muscle, but its presence does not restore cardiac contractility in heart failure (HF) models [17]. While there seems to be a link between spatial and temporal expression of MyBP-C isoforms in normal and diseased muscles, the functional consequences of isoform switching remain unclear. Certainly, the fact that skeletal and cardiac MyBP-C preferentially interact with their isoform-specific variants of actin and myosin [18] suggests that each protein may be structurally and functionally distinct.

Although skeletal and cardiac MyBP-C isoforms have approximately 50–70% sequence homology at the amino acid level, there are global differences in protein architecture that potentially impact control of contractility. Unlike fsMyBP-C and ssMyBP-C, cMyBP-C has a cardiac specific N-terminal domain denoted C0 and an additional loop in the C5 domain [19,20]. The presence of unique phosphorylation sites in the M-motif of cMyBP-C influences the force and frequency of contraction in the heart. In addition to association with myosin S2, the M-motif of cMyBP-C exerts its regulatory effects through actin [3,21–24] and calcium-calmodulin (Ca^{2+}-CaM) interactions [25]. The Ca^{2+}-dependent CaM protein kinase (CaMK) covalently modifies S282, which in turn promotes phosphorylation by protein kinase A (PKA) or protein kinase C (PKC) at sites S273 and S302 [8,10]. PKA phosphorylation sites have been identified in ssMyBP-C [26], but phosphorylation in fsMyBP-C is largely uncharacterized. Considerably less is known about the role of phosphorylation in regulating contraction in fast and slow skeletal muscles. However, there is a critical link between Ca^{2+}-signaling and cardiac function that is mediated in part by the interaction of the M-motif with CaM.

CaM is composed of two globular lobes connected by a flexible tether. Each N- and C-terminal domain has two EF-hand motifs that coordinate Ca^{2+} in the presence of saturating intracellular Mg^{2+} concentrations. In response to ligating Ca^{2+}, the globular domains undergo conformational transitions that expose the hydrophobic pockets necessary for target recognition. CaM exhibits a great degree of conformational plasticity, facilitated by dynamic motion in the linker region. One mode of interaction involves the formation of an extended complex, whereby both lobes of CaM are held apart by binding to protein targets, as is observed in numerous biological systems [27,28]. Alternatively, the lobes of CaM can engage a target protein by collapsing around it in a fashion similar to the complex formed with CaM-dependent kinases [29]. Although there is no single canonical CaM-binding motif, protein targets involved in its interaction typically have hydrophobic and positively charged amino residues involved in CaM recognition. Lu et al. report that CaM mostly associates with the M-motif of cMyBP-C through the insertion of a tryptophan residue into the hydrophobic pockets of Ca^{2+}-loaded CaM. Moreover, basic amino acid residues mapping to the M-motif are conformationally perturbed in the presence of CaM, further supporting the idea that this region is involved in protein–protein association. Comparison of amino acid sequence alignments reveals that M-motif residues are highly conserved in MyBP-C, including the skeletal isoforms, but the structural and functional significance remain to be determined [30].

Based on previous animal model studies [17], skeletal and cardiac MyBP-C are proposed to have functionally disparate roles in regulating contractility, but the molecular and structural bases are unknown. In this study, we use biophysical techniques to structurally characterize slow skeletal MyBP-C (ssMyBP-C). The conformation and protein binding properties of an N-terminal recombinant fragment of ssMyBP-C (ssC1C2) were examined by differential scanning fluorimetry (DSF), nuclear magnetic resonance (NMR) spectroscopy, and molecular modeling. The conformation of ssC1C2 was modulated in the presence of Ca^{2+}-loaded CaM as evidenced by the detection of altered protein thermal stability. NMR binding experiments were performed to demonstrate that site-specific interactions between ssC1C2 and Ca^{2+}-CaM are mediated through hydrophobic surfaces. Using protein homology modeling, we examined the tertiary structure of the M-motif, a region of ssMyBP-C known to be of importance in protein binding. Our findings support that ssC1C2 associates with CaM in a Ca^{2+}-dependent manner through hydrophobic interaction mediated by the M-motif, which may have significance in the differential regulation of skeletal and cardiac muscle contractility.

2. Results

2.1. Analysis of Amino Acid Conservation Provides Clues about Regulatory Functions in MyBP-C Isoforms

Using the CLUSTAL Omega server, we aligned the primary sequences corresponding to the N-terminal fragments of MyBP-C [31]. Analyses of primary structure of amino acid sequences corresponding to the N-terminal domains of MyBP-C revealed that significant amino acid similarities exist between the skeletal and cardiac isoforms (~50%) (Figure 1). Most notably, the regions known to be involved in actin and myosin binding in cMyBP-C domains (C0C2) [2], exhibited the highest degree of amino acid conservation between the isoforms. Overlapping binding determinants for actin and myosin were located in the C1 domain for skeletal and cardiac isoforms, suggesting that ssC1C2 and fsC1C2 associates with cardiac filament proteins in manner similar to that observed for cMyBP-C. Similarly, many amino acid residues in the M-motif of cMyBP-C that are known to experience conformational perturbations in the presence of CaM [25] were also conserved in ssC1C2 and fsC1C2. While other studies report PKA and PKC sites in ssMyBP-C [32], the structural and functional roles of CaMK phosphorylation in skeletal MyBP-C remain to be determined. Using bioinformatics tools, we identified phosphorylation sites for CaMK located at specific regions in the C1 and C2, near the M-motif, of both ssC1C2 and fsC1C2 (Table S1). Interestingly, these predicted phosphorylation sites in ssMyBP-C and fsMyPB-C proteins are proximal to highly conserved actin, CaM, and myosin binding motifs in cMyBP-C. Taken together, these observations suggest that the skeletal proteins might have similar target recognition sites as compared to cMyBP-C and that post-translational modifications within these regions may modulate protein–protein interactions.

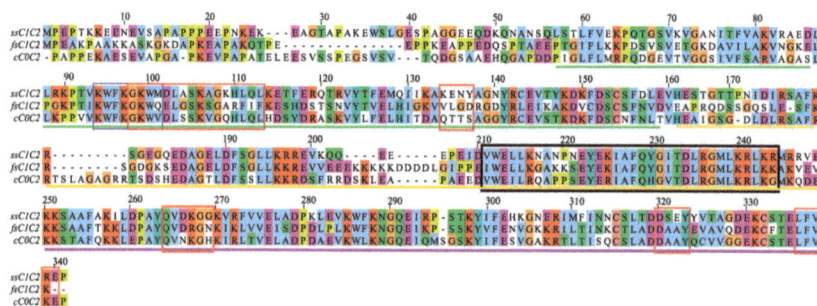

Figure 1. Clustal Omega alignments reveal conservation of amino acid residues in the N-terminal regions containing domains C1-C2 consisting of determinants for myofilament protein interaction in MyBP-C isoforms. The amino acid residues for mouse MyBP-C are aligned using Clustal Omega, demonstrating conservation from Jalview (ClustalX coloring). The corresponding domains are illustrated by a green line (C1), a gold line (M-motif), and a purple line (C2). Known regions of actin binding (red boxes), myosin binding (blue boxes), and overlapping CaM/actin binding sites (black line) for cC0C2 are depicted.

2.2. The Thermal Stability of ssMyBP-C Is Impacted by Ca^{2+}-CaM

Recombinant N-terminal His-tagged ssC1C2 was produced using a T7 expression system and purified with HisTrap columns. CaM was produced and purified as previously described [28]. Following purification, protein samples were visualized by SDS-PAGE and determined to be homogenous. We examined the thermal stability of ssC1C2 in the presence and absence of Ca^{2+}-CaM using DSF experiments. Proteins are subjected to thermal denaturation during DSF, exposing the hydrophobic protein core, which allows for SYPRO Orange dye to interact [33]. Upon binding to hydrophobic environments, the fluorescence of the dye increases relative to the degree of exposed hydrophobic surface area. This permits the protein unfolding to be monitored and the midpoint of this thermal transition is considered to be the apparent melting temperature (T_m). Changes in T_m occur upon ligand or protein–protein binding, which can increase or decrease the thermal stability of proteins depending on the nature of the binding [33,34], are detected as shift changes in peaks. Given that the N- and C-lobes of CaM have exposed hydrophobic patches, high intrinsic fluorescence was observed for Ca^{2+}-CaM free, which precluded determining its T_m (Figure 2a). To monitor complex formation, MyBP-C proteins were combined with CaM in a final molar ratio of [ssMyBP-C:CaM][1.0:0.5]. The ssC1C2 protein exhibited an apparent T_m 52.4 °C in the absence of CaM, but two T_m measurements of 48.7 °C and 58.0 °C were observed in the presence of Ca^{2+}-CaM (Figure 2; see Figure S1 for derivative data), which suggests that interaction with bilobal CaM may impact the structure in multiple ways. Increased thermal stability upon the addition of Ca^{2+}-CaM is indicative of complex formation. However, the decrease in thermal stability suggests that Ca^{2+}-CaM binding might destabilize a region of ssC1C2.

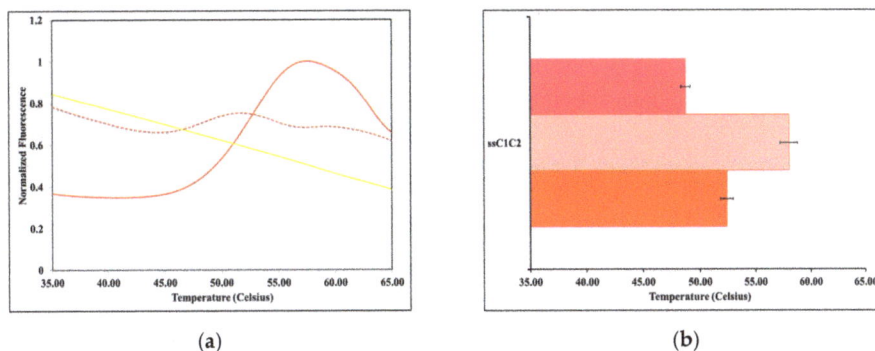

(a)   (b)

Figure 2. Differential scanning fluorimetry experiments show that ssC1C2 has altered thermal stability in the presence of Ca^{2+}-CaM. Normalized fluorescence versus temperature plots are shown for ssC1C2 (red), ssC1C2/CaM (red dash), and free CaM (yellow) (a). The T_m determined for ssC1C2 in the presence (hatched and pink bars) and absence (solid red bar) of Ca^{2+}-CaM are shown (b).

2.3. CaM-Induced Conformational Modulation of ssC1C2 Is Ca^{2+}-Dependent and Localized Primarily to the M-Motif

NMR binding experiments were performed to monitor conformational changes in [^{15}N]ssC1C2 in the presence of unlabeled Ca^{2+}-CaM. NMR chemical shift values for [^{15}N]ssC1C2 were similar to those reported in the literature [25,30,35–37]. Amide proton nitrogen chemical shift assignments, in particular for indole peaks corresponding to W95, W100, W212, and W286 (W191, W196, W318, and W396 in mouse cMyBP-C), were determined by direct comparison to BMRB accession numbers 6015, 5591, and 17,867. Assessment of 2D ^{1}H-^{15}N spectra for [^{15}N]ssC1C2 free and bound to unlabeled CaM revealed that Ca^{2+}-CaM induced conformational perturbations in numerous peaks (Figure 3a). In particular, M-motif residues V211, W212, E213, K216, N217, A218, Y223, E224, R236, G237, K240, L242, and K243 showed reduced intensities upon the addition of CaM suggesting that Ca^{2+}-CaM interacts with this region. Furthermore, the indole peak for W212, also mapping to the M-motif, experienced significant broadening upon saturation with Ca^{2+}-CaM. After the addition of an excess of CaM, W212 in the M-motif broadened beyond detection while the indole peaks mapping to the C1 (W95 and W100) and C2 (W286) domains did not experience conformational changes. Modest conformational perturbations were detected in residues A253 and A254, suggesting a limited role for the C2 domain in Ca^{2+}-CaM binding. The dissociation constant was determined to be 15–30 μM (data not shown), which is similar to that previously reported [25]. When the binding was repeated in the presence of the Ca^{2+} chelator EDTA (Figure 3b), no detectable conformational changes were observed in ssC1C2, which suggests that Ca^{2+} plays a significant role in mediating this interaction.

Taken together, the NMR binding studies support that the M-motif of ssC1C2 associates with CaM in a Ca^{2+}-dependent manner.

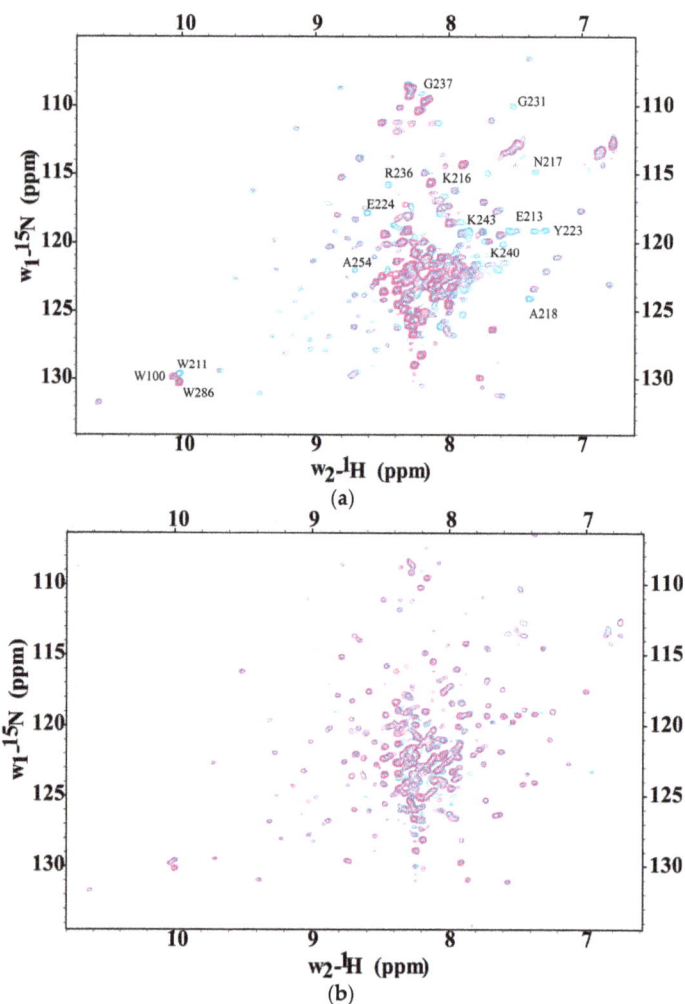

Figure 3. CaM binding is mediated through hydrophobic interaction with primarily the M-motif of ssC1C2. Protein interaction is monitored by comparison of 2D ^1H-^{15}N TROSY-HSQC spectra of [^{15}N]ssC1C2 free and bound to unlabeled CaM. Protein interaction is monitored by comparison of 2D ^1H-^{15}N TROSY-HSQC spectra of [^{15}N]ssC1C2 free and bound to unlabeled CaM in the presence of 10 mM $CaCl_2$ at 600 MHz. TROSY-HSQC spectra of labeled ssC1C2 (cyan) following the binding with an equimolar amount of Ca^{2+}-loaded CaM (magenta) are superposed (**a**). Multiple resonances mapping to the M-motif and a few mapping to the C2 domain experience conformational perturbation upon the addition of Ca^{2+}-CaM. Select resonances that experience broadening in the presence Ca^{2+}-CaM and the indole resonances for W100, W212, and W286 are labeled. Overlay of spectra of [^{15}N]ssC1C2 free and bound to unlabeled CaM collected in the presence of 5 mM EDTA at 850 MHz. TROSY-HSQC spectra of labeled ssC1C2 (cyan) following the binding with an equimolar amount of CaM (magenta) are superposed in the absence of Ca^{2+}-saturation (**b**). The absence of observable chemical shift perturbations suggests that CaM binding to ssC1C2 is Ca^{2+}-dependent.

The reverse binding experiment was performed where unlabeled ssC1C2 was added to [^{15}N, ^{13}C, ^{2}H]Ca^{2+}-CaM and complex formation was monitored by ^{1}H-^{15}N 2D correlation spectra. Amide proton-nitrogen resonances in [^{15}N, ^{13}C, ^{2}H]Ca^{2+}-CaM shifted or decreased in intensity during ssC1C2 binding. The chemical shift differences were calculated and plotted versus CaM amino acid sequence number (Figure 4a). The following resonances mapping to both the N-terminal and C-terminal domains of CaM experienced significant chemical shift differences in the presence of ssC1C2: S17, L18, F19 D22, D24, V35, I52, D56, D58, M73, Y99, D131, N137, T146. When these residues were colored onto a surface representation of Ca^{2+}-CaM (PDB 1CLL), it is clear that ssC1C2-dependent conformational perturbations are localized to the lobes of CaM (Figure 4b). As predicted, the structural changes occur predominantly in the hydrophobic clefts of CaM, indicating that nonpolar residues, such as methionine, are important in binding ssC1C2. Lu et al. report that cMyBP-C preferentially associates with the C-terminal domain of CaM. More recently, Trewhella and colleagues report that both domains of CaM are involved in interactions with cMyBP-C [38].

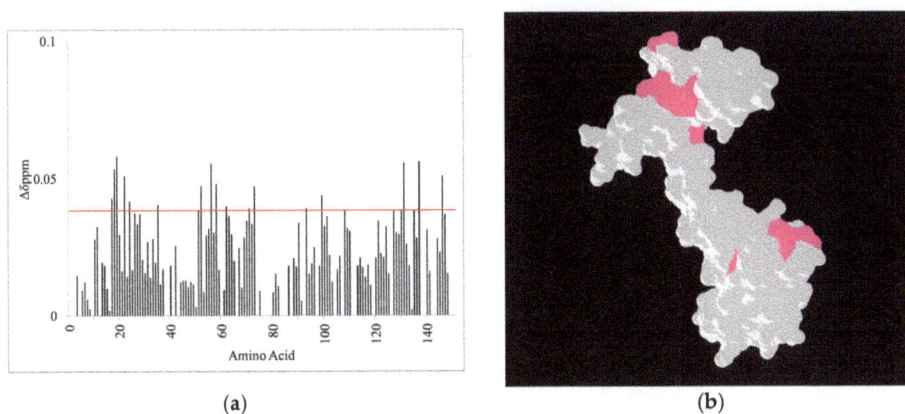

(a)   (b)

Figure 4. The association of ssC1C2 M-motif perturbs both lobes of CaM. Composite amide-proton nitrogen chemical shift differences were calculated between labeled CaM and ssC1C2-bound CaM and plotted versus CaM amino acid residue. The horizontal line is indicative of the average chemical shift difference plus one standard deviation (**a**). The amino acid residues that experience significant chemical shift perturbations in the presence of ssC1C2 are colored pink on the surface representation of Ca^{2+}-loaded CaM (PDB 1CLL) (**b**).

2.4. A Molecular Model Sheds Light on How the M-Motif of MyBP-C Engages Protein Targets

To better understand how the M-motif of ssMyBP-C is folded, we used the SWISS-MODEL to predict 3D structures of this region [39]. The M-motif of mouse cMyBP-C (PDB 2LHU) was used as a template structure. A minimum of 86% sequence coverage was attained and a valid model was chosen based on QMEAN score and PROCHECK analysis [40,41]. Molecular models may be assessed using the QMEAN score, which examines global and local structural qualities. QMEAN scores range from 0 to 1, with a score of 1 being indicative of a high-quality model. The selected ssC1C2 model had a QMEAN score of 0.76, which is considered to be acceptable. In the PROCHECK analysis, the stereochemical quality of a structure was assessed, and we found that the M-motif structure had greater than 90% of residues in the most favored regions, which confirms that the model quality is high. The global fold of M-motif structure for ssC1C2 was similar to the three-helix bundle reported for cMyBP-C (Figure 5a). Both skeletal and cardiac isoforms were comprised of three-helix bundle containing M-motifs and ssC1C2 exhibits similar presentations of exposed charged and nonpolar surface areas (Figure 5b). Furthermore, comparison of M-motif exposed surfaces revealed that positively charged residues such as R236, which have been shown to be involved in CaM and actin binding in cardiac, are conformationally similar in both isoforms. While the M-motif of ssC1C2 has similarities in accessible hydrophobic surface areas, providing further evidence that this region might engage hydrophobic targets such as Ca^{2+}-CaM in a similar way, a slight difference in polar surfaces is noted (Figure 5b). This reflects the presence of N219 in ssC1C2, which is a P residue in C0C2 cMyBP-C (Figure 1 and Table S2). The P is conserved in the human and mouse cardiac isoform, but varies in the skeletal isoforms.

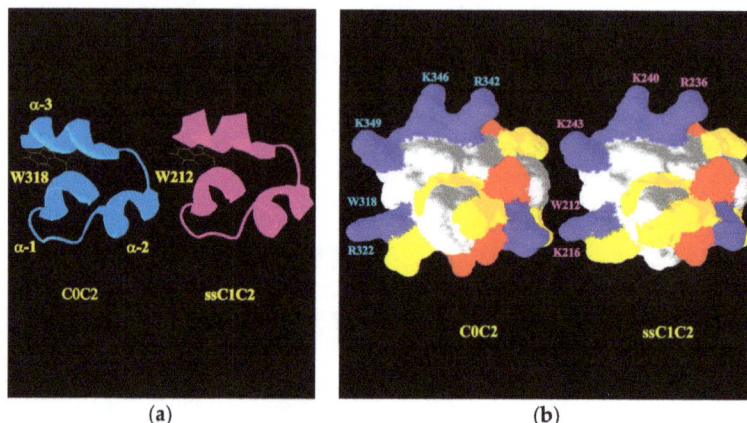

| (a) | (b) |

Figure 5. Comparison of M-motif structural features in mouse MyBP-Cs. Ribbon structures of the average solution NMR structure of the M-motif from cMyBP-C (PDB 2LHU) (C0C2) (blue) and the molecular model of the M-motif of ssC1C2 (fuchsia) show the triple helix bundle reported to be involved in actin and CaM interactions (**a**). The conserved tryptophan residues from the M-motif of each isoform are indicated in yellow and the helices are numbered from α-helix 1(α-1) through α-helix 3(α-3). Electrostatic surface representations of mouse cardiac and skeletal M-motif structures are shown for C0C2 (PDB 2LHU) and ssC1C2 (molecular model) (**b**). Surfaces are colored according to amino acid type: basic (blue), acidic (red), nonpolar (gray), and polar (yellow). Select amino acid residues W318, R322, R342, K346 for mouse cardiac (blue), which were identified by Lu et al. to be involved in CaM binding to M-motif of cMyBP-C, are labeled. In this study, we found that these key conserved residues labeled W212, K216, R236, K240, and K243 (fuchsia), are also perturbed in ssC1C2 in the presence of Ca^{2+}-CaM.

3. Discussion

In this study, we provide the first known evidence demonstrating that Ca^{2+}-CaM interacts with the M-motif of ssC1C2. We performed DSF and NMR experiments to examine the interaction between ssC1C2 and Ca^{2+}-CaM. We found that ssC1C2 experiences conformational changes in the presence of CaM as evidenced by distinct thermal transitions measured using DSF. Our DSF experiments found that in the presence of Ca^{2+}-CaM, there are both stabilizing and destabilizing binding events occurring, as evidenced by the measurement of two T_m values. These findings might be explained by the differential interactions occurring between ssC1C2 and the N- and C-lobes of Ca^{2+}-CaM. For example, it is possible that association with Ca^{2+}-CaM stabilizes a region of ssC1C2 at or near the M-motif, which would increase the thermal stability. However, binding CaM might promote a structural rearrangement in ssC1C2 that favors unfolding, which would result in reduced thermal stability. Interestingly, Michie and Trewhella report on the modular nature of the M-motif, the C2 domain, and the flexible tether connecting these regions of cMyBP-C [38]. In that same study, they indicate that the tri-helix bundle of the M-motif undergoes changes in helical content, including destabilization of helices in the M-motif, and tertiary organization upon Ca^{2+}-CaM association. However, their data support the idea that stabilizing interactions involving insertion of W318 (W322 in Human cMyBP-C) into the hydrophobic cleft of Ca^{2+}-CaM occur. Our findings in the current study of ssC1C2 are in good agreement with the observations that Ca^{2+}-CaM binding induces both stabilization and destabilization within the modular regions of N-terminal cMyBP-C. We propose that, in the presence of Ca^{2+}-CaM, these conformational changes result from either direct intermolecular association with ssC1C2 or indirectly through allosteric mechanisms, common regulatory mechanisms by which modular proteins function.

Our NMR binding studies revealed that the M-motif of ssC1C2 engages in site-specific interactions with the hydrophobic lobes of CaM. In the presence of ssC1C2, we found conformational perturbations in the hydrophobic clefts of Ca^{2+}-CaM, indicating its involvement in multi-domain interactions with CaM. We observed conformational perturbation largely localized to the M-motif of ssC1C2 in the presence of CaM, while residues mapping to the C1 and C2 domains remain unchanged. Most notably, we found that W212, which broadens beyond detection in the presence of Ca^{2+}-CaM, is most likely involved in site-specific hydrophobic interactions occurring between these proteins. The Ca^{2+}-CaM-induced exchange broadening of residues mapping to the M-motif of ssC1C2 is likely the consequence of highly dynamic interactions at the binding interface. This exchange broadening phenomenon and conformational fluctuations have been reported in other muscle proteins [42,43]. The NMR observation of conformational plasticity in the M-motif and the C2 linker in ssC1C2 is also consistent with our DSF data showing partial destabilization in the ssC1C2/CaM complex as evidenced by the lowered T_m. Our structural modeling supports that ssC1C2 has an exposed hydrophobic surface area consisting of W212 which would provide a binding surface for Ca^{2+}-CaM. The M-motif W212 is conserved across isoforms and species (Figure 1 and Table S2), which further indicates its importance in MyBP-C structure and function. Moreover, our NMR studies revealed no detectable conformational perturbations in ssC1C2 in the presence of CaM and EDTA, which is in agreement with reports of Ca^{2+}-dependent interaction between CaM and cMyBP-C [25]. This is likely the result of a reduction in hydrophobic surface area in CaM in the presence of EDTA, which disrupts W212 association, further emphasizing the importance of hydrophobic binding in this system. In ssC1C2, the W212R mutation was identified in distal arthrogryposis patients [44]. While the W212R mutant protein localized to the sarcomere, it was predicted to result in dysfunctional regulation of contractility in these patients [44]. Based on our findings, we predict that the M-motif of ssC1C2 may be capable of switching between CaM, actin, and myosin interactions, which would allow it to be an important mediator in interpreting Ca^{2+} signaling in muscle.

Our work provides compelling evidence that ssC1C2 interacts with CaM in a Ca^{2+}-dependent manner. It is possible that interaction with Ca^{2+}-CaM facilitates recruitment of CaMK to this region. Because Ca^{2+}-CaM has a moderate affinity for MyBP-C, it would be possible to rapidly reverse this interaction in the muscle so that Ca^{2+}-CaM may activate CaMK and promote phosphorylation in this region, which may facilitate its dynamic interaction with other muscle regulatory proteins. This finding is significant because it provides justification for pursuing more detailed studies into the role of CaMK and Ca^{2+}-CaM in molecular regulation of skeletal contractility. Previously, CaMK phosphorylation was shown to modulate cMyBP-C conformation, which plays a role in normal contractile function [8]. Lu et al. demonstrated the M-motif of cMyBP-C contains the molecular determinants for Ca^{2+}-CaM binding and its association is not phosphorylation-dependent [25], suggesting that another mechanism must account for CaM's dissociation. However, the conformation of cMyBP-C has been shown to be affected by its phosphorylation state [45], which is directly linked to functional adaption in muscles.

In fact, phosphorylation of myofibrillar proteins is a well-known mechanism for modulating skeletal and cardiac muscle activity, including MyBP-C. In skeletal muscles, CaMK2 activity is affected by cellular $[Ca^{2+}]$, which provides a conduit for fine-tuning Ca^{2+}-responsiveness through Ca^{2+}-CaM-dependent interactions, but the molecular targets are not completely characterized [46]. Phosphorylation-induced modulation of troponin structure in cardiac muscle provides a molecular switch by which the force and frequency of muscle contraction is controlled [47–52]. In previous studies, we have shown that protein kinase A (PKA) phosphorylation of S273, S282, S302, S307 in mouse MyBP-C are critical to the regulation of normal muscle function [8,53]. Phosphorylation of cMyBP-C was demonstrated to modulate its association with actin [54]. A region of importance in phosphorylation-dependent modulation of motility is the N-terminal region of the M-motif [55]. In addition to PKA, both protein kinase C (PKC) [56] and CaMK2 phosphorylation of cMyBP-C [57] were demonstrated to influence muscle contraction. There is evidence that phosphorylation of the N-terminal region in ssMyBP-C may be critical in skeletal muscle health and disease. In ssMyBP-C, site-specific phosphorylation sites at residues S6 and T84 have been identified as functionally important [26,32], but the role of phosphorylation in fsMyBP-C remains to be determined. Based on the importance of phosphorylation in muscle structure and function, we wanted to look for predicted phosphorylation sites in skeletal MyBP-C, in particular those that might be near the CaM binding site we found in ssC1C2. We used the Group-based Prediction System V3.0 (GPS) software [58] to examine ssC1C2 and fsC1C2 protein sequences for CaMK phosphorylation sites. We found that CaMK phosphorylation sites are predicted at residues S55, T92, T125, T166, S252, S296, S316, and S333 in ssMyBP-C and at positions T81, T110, and S250, and S294 in fsMyBP-C (Table S1). We used GPS to identify known CaMK phosphorylation sites at positions S328, S402, S423, and S440 C0C2. In agreement with other studies, we identified similar CaMK sites that were previously reported in cMyBP-C [53]. These data suggest that functionally relevant CaMK phosphorylation sites are likely to be found in ssMyBP-C and fsMyBP-C. However, more solution structure studies are needed to better understand whether or not phosphorylation of skeletal MyBP-C modulates its target binding.

The current evidence in the field supports that MyBP-C proteins have overlapping structural determinants that modulate target recognition. Even though ssMyBP-C, fsMyBP-C, and cMyBP-C proteins preferentially associate with myofilament proteins in an isoform-dependent manner, these homologous proteins retain the core determinants necessary for myofibrillar association [10,18,59,60]. Conserved amino acid residues in regions mapping to actin and myosin binding sites suggests that MyBP-C proteins share common determinants for interactions with key myofilament proteins (Table S2). Overlapping regions of myosin and actin interaction in N-terminal MyBP-C strongly suggest that it alternates between thin and thick filament binding during the contraction cycle [4]. Dynamic movement between thin and thick filament association would require that MyBP-C's structure be extended in such a way so as to permit it to traverse this conformational space in muscle fibers. Jefferies et al. demonstrated that N-terminal fragments of cMyBP-C, in particular C0C2, are elongated in solution, which would be consistent with the ability to switch between thin and thick interactions. In addition to MyBP-C, other muscle proteins, such as the troponin complex, are known to modulate the force–frequency relationship in muscles by alterations in global conformation propagated through overlapping protein–protein binding motifs. In response to Ca^{2+}-signaling, troponin I alternates between actin and troponin C association, and its interaction with these proteins can be modified by changes cardiac troponin I structure [47,48,52]. Conformational transitions of muscle proteins in response to Ca^{2+}-flux and hormonal stimulation are key regulatory features of fine-tuning muscle performance.

Conformational plasticity has also been reported for MyBP-C. The N-terminal region has been reported to switch between actin and myosin binding, which underscores its regulatory importance in muscles. Hypertrophic mutations in the M-motif (R322Q, E334K, V338D, and L348P in human) are known to impact function and organization of the sarcomere [61]. Corresponding mutations in ssC1C2 (K216Q and E224K) differentially altered sarcomeric localization and regulatory functions of this protein [44]. In a recent report, Michie et al. determined that the C-terminal region of the M-motif and the N-terminal region of the C2 domain are critical in CaM association [38]. In this study we also found that the M-motif is conformationally perturbed in the presence of Ca^{2+}-CaM, with modest changes in residues mapping to the C2 linker region occurring. We found that K216, R236, K240, and K243 in the M-motif and A254 in the C2 linker are perturbed in the presence of Ca^{2+}-CaM. These findings are significant because a number of mutations mapping to M-motif and the region connecting it to C2 in both skeletal and cardiac MyBP-C have been identified as pathogenic in humans, emphasizing that it likely plays a critical role in muscle regulation. Our findings are in agreement with the proposal that this region of MyBP-C is likely involved in dynamic interactions that utilize switching between CaM, actin, and myosin association as a means by which to modulate contractility. Taken together, these data hint that highly conserved structural determinants are present in skeletal and cardiac isoforms of MyBP-C, which are necessary for maintaining core muscle functions, but structural variations outside of these conserved regions may provide the basis for unique isoform-specific features that participate in controlling structure and function is a tissue-dependent manner.

In summary, we have shown that in solution ssC1C2 has distinct conformation and thermal stability in the presence and absence of Ca^{2+}-CaM. We believe that this is the first report indicating that ssC1C2 interacts with Ca^{2+}-CaM by associating with the M-motif, a finding that is relevant given that this region of cMyBP-C is known to contain the core components necessary for actin, myosin, and CaM interactions. Using NMR binding studies, we show that site-specific interaction occurs between the M-motif of ssC1C2 and Ca^{2+}-CaM in a manner that is similar to cMyBP-C, which suggests that Ca^{2+}-CaM-regulated phosphorylation of MyBP-C may also be relevant in skeletal muscle function. Based on these observations, we propose that a critical link between Ca^{2+}-signaling and phosphorylation-induced modulation of contractility may exist for skeletal isoforms of MyBP-C as well. Other studies demonstrate that PKA and PKC phosphorylation of ssMyBP-C impacts normal and diseased states [26,32], but considerably less is known about the role of CaMK-dependent phosphorylation of MyBP-C in skeletal muscle. Current studies in our lab are focused on determining the structural mechanisms of interaction between N-terminal MyBP-C and muscle associated proteins such as actin, myosin, and CaM and what potential role Ca^{2+} has in modulating these interactions. We are using biophysical and biochemical techniques to examine how unphosphorylated and phosphorylated MyBP-C N-terminal fragments associate with myofibrillar proteins. An improved understanding of the molecular mechanisms by which ssMyBP-C, fsMyBP-C, and cMyBP-C regulate muscle contraction offers the potential to develop novel therapeutic and diagnostic approaches to treat muscle diseases.

4. Materials and Methods

4.1. Recombinant Protein Expression and Purification

The mouse gene encoding for ssC1C2 (domains C1C2 for slow skeletal MyBP-C; Uniprot Q6P6L5) was cloned into the pET-28a$^+$ vector (EMD Millipore, Burlington, MA, USA) and recombinant proteins were overproduced in *Escherichia coli* BL21(DE3) cells (Lucigen, Middleton, WI, USA). Briefly, the transformed cells were inoculated in Luria-Bertani (LB) media with the appropriate antibiotic selection and grown at 37 °C with shaking (250 rpm). When the optical density (OD_{600}) reached 1.0–1.5, recombinant protein expression was induced with the addition of isopropyl-1-thiogalactopyroside (IPTG) and growth was continued for 20 h at 15 °C. Stable isotope enrichment of ssC1C2 was achieved using the protocol described above, except the cells were grown in M9 media supplemented with $^{15}NH_4Cl$ (1.0 g/L) or $^{15}NH_4Cl$ and $[^{13}C_6]$-glucose. Recombinant MyBP-C proteins were extracted from cell pellets by sonication in the presence of a lysis buffer containing 500 mM NaCl, 20 mM imidazole, 20 mM Tris-HCl pH 8.0, 5 mM 2-mercaptoethanol (βME), and 1 mM phenylmethanesulfonylfluoride (PMSF). N-terminal fragments of MyBP-C were resolved using HisTrap^{TM} HP Nickel-Sepharose resin (GE Healthcare, Little Chalfont, UK). The isoforms were eluted from columns with increasing concentrations of imidazole. Recombinant proteins were identified as homogenous as analyzed by sodium dodecyl sulfate polyacrylamide gel electrophoresis (SDS-PAGE). Samples containing MyBP-C were dialyzed against 1X phosphate buffered saline (PBS) supplemented with 1.0 mM tris(2-carboxyethyl)phosphine (TCEP). Protein concentrations were calculated by Bradford assay and UV-absorbance at 280 nm based on molecular extinction coefficients for each recombinant MyBP-C.

Stable isotope-labeled and unlabeled CaM were expressed, purified, and quantified as previously described [28]. ApoCaM was prepared by dialyzing purified recombinant proteins against 4 L of buffer containing 250 mM NaCl, 20 mM EDTA, 20 mM EGTA, 20 mM Hepes pH 7.3, and 1 mM PMSF. Following preparation, the metal bound state of CaM was monitored by ^{1}H-NMR.

4.2. Differential Scanning Fluorimetry (DSF) Experiments

For DSF analyses, stock solutions ssC1C2 and CaM were suspended at 20 µM in 1X PBS supplemented with 1.0 mM TCEP. A master mix was prepared by adding 50 µL of protein (5 µM final concentration), 2 µL of SYPRO® Orange (final concentration was 5×), and the volume was adjusted to 200 µL using buffer solution supplemented with tris(2-carboxyethyl)phosphine and 1 mM $CaCl_2$. For protein complexes, master mix samples consisting of 5 µM of ssC1C2 was combined with 2.5 µM of CaM in the presence of 5× SYPRO orange. All samples of free ssC1C2 and CaM-containing MyBP-C mixtures were incubated for 1 h at room temperature. Allotments of 40 µL were loaded into 96-well 0.2 mL thin-wall PCR plates and sealed with iCycler optical quality sealing tape (BioRad, Hercules, CA, USA). Three independent measurements were collected in triplicate for each reaction on a Bio-Rad, Hercules, CA, USA iCycler iQ Real-Time Detection System. Thermal denaturation experiments were performed from 25 °C to 95 °C (0.5 °C/min using 5 s equilibration intervals) with fluorescence excitation and detection at 490 to 575 nm respectively, with the HEX filter. Fluorescence data were normalized over the temperature range with respect to differences in molar ratios. The midpoint for each MyBP-C thermal transition was calculated by taking the first order derivative of the melt curve for each independent measurement (see Figure S1 for derivative data). The average T_m and standard deviation were calculated for ssC1C2 in the presence and absence of Ca^{2+}-CaM.

4.3. Nuclear Magnetic Resonance Spectroscopy (NMR)

NMR experiments were performed on Bruker Avance (Bruker BioSpin, Billerica, MA, USA) III 600 MHz and 850 MHz spectrometers equipped with conventional 5-mm probes. Two-dimensional (2D) Heteronuclear Single Quantum Coherence (^{1}H-^{15}N-HSQC) or Transverse Relaxation Optimized (^{1}H-^{15}N TROSY-HSQC) spectra were collected for all samples at 298 K. For NMR analyses, a sample consisting of [^{15}N]ssC1C2 or [^{15}N, ^{13}C]ssC1C2 was suspended in NMR buffer composed of 1X PBS buffer supplemented with 1.0 mM TCEP and 10% ^2H_2O at a final concentration of 130 µM–500 µM in the presence of 10 mM $CaCl_2$ or 5 mM EDTA. The backbone chemical shift assignments for stable isotope-labeled ssC1C2 were determined in part by direct comparison to the previously known values (BMRB accession numbers 17,867, 11,212 and 5591). The following suite of triple-resonance experiments were used to confirm assignments: ^{15}N edited NOESY-HSQC, CβCαCONH, HNCA, TROSY-HNCO, and TROSY-HNCA [28]. Samples of [^{15}N, ^{13}C, ^{2}H]Ca^{2+}-CaM were also suspended at 130 µM in NMR buffer in the presence of 10 mM $CaCl_2$. Based on our previous NMR studies [28], the chemical shifts for [^{15}N, ^{13}C, ^{2}H]Ca^{2+}-CaM were readily assigned in the current NMR buffer. NMR data were processed using NMRPipe [62] and analyzed using Sparky [63].

For NMR binding experiments, 2D ^{1}H-^{15}N-HSQC or TROSY-HSQC spectra of free [^{15}N, ^{13}C]ssC1C2 and free [^{15}N, ^{13}C, ^{2}H]Ca^{2+}-CaM were separately collected. NMR binding studies were performed by adding allotments of unlabeled CaM to [^{15}N, ^{13}C]ssC1C2 to a final CaM:ssC1C2 molar equivalent of [1.5:1.0] in the presence of 10 mM $CaCl_2$ or 5 mM EDTA. Following addition, samples were mechanically mixed, incubated at ambient temperature, and 2D ^{1}H-^{15}N correlation spectra were collected. In the reverse experiment, samples consisting of [^{15}N, ^{13}C, ^{2}H]Ca^{2+}-CaM were analyzed for ssC1C2 binding by NMR. The interaction of [^{15}N, ^{13}C, ^{2}H]Ca^{2+}-CaM with unlabeled ssC1C2 was monitored after addition ssC1C2 to a final molar ratio of [^{15}N, ^{13}C, ^{2}H]Ca^{2+}-CaM:ssC1C2] at [1:1]. Changes in NMR spectra were monitored by measuring peak intensities and the calculation of amide-proton nitrogen chemical shift differences, which were determined as described [28].

4.4. Homology Modeling and Bioinformatics

Protein homology models were generated for the M-motif of ssMyBP-C using the SWISS-MODEL Workspace [39]. The amino acid sequences corresponding to the M-motif of ssMyBP-C were used as the target, and mouse template structures (PDB 2LHU) were used for tertiary structure prediction. The quality of each model was estimated based on QMEAN scores [40] followed by PROCHECK evaluation using the PBDsum server [41]. The amino acid sequences of mouse C0C2, fsC1C2, and ssC1C2 were aligned using CLUSTAL Omega [31]. To examine protein sequences for putative phosphorylation sites, we used the GPS 3.0 software [58].

Supplementary Materials: The following are available online at www.mdpi.com/2312-7481/4/1/1/s1, Table S1. Predicted CaMK phosphorylation sites of MyBP-C N-terminal fragments identified using GPS 3.0 (medium threshold); Table S2. Amino acid sequences show conservation between mouse and human M-motif regions; Figure S1. The derivatives of DSF curves are shown with minima revealing the T_m: (a) ssC1C2; (b) ssC1C2 in the presence of Ca^{2+}-CaM.

Acknowledgments: We gratefully acknowledge the contribution of Andor Kiss in supervising and maintaining the instrumentation in the Center for Bioinformatics and Functional Genomics (CBFG). Also, we would like to express our gratitude to Xiaoyun Deng for her technical assistance in the CBFG. We thank Theresa Ramelot for maintaining the NMR spectrometers. The authors appreciate the helpful discussion and insightful comments provided by Gary Lorigan during the preparation of this manuscript. T.I.S., C.W.J., J.C., and N.L.F. were supported in part by DUOS and CFR grants from Miami University. B.L.L. and S.S. were supported by National Institutes of Health (NIH) grants R01HL130356, R01HL105826, K0²HL114749, and American Heart Association, Cardiovascular Genome-Phenome Study 15CVGPSD27020012. N.L.F. was supported in part by N.I.H. grant R15GM117478. N.L.F. and C.W.J. were supported in part by United States Department of Agriculture (USDA) Project number 6034-22000-041-24.

Author Contributions: T.I.S., C.W.J., and J.C. carried out experiments and data analyses. J.C. performed bioinformatics analyses. B.L.L. and S.S. established target regions for gene cloning and carried out sub-cloning for plasmid constructs. N.L.F. conceived study design and performed data analyses. T.I.S. and N.L.F. wrote the manuscript.

Conflicts of Interest: The authors declare no conflict of interest.

References

1. Luther, P.K.; Winkler, H.; Taylor, K.; Zoghbi, M.E.; Craig, R.; Padron, R.; Squire, J.M.; Liu, J. Direct visualization of myosin-binding protein C bridging myosin and actin filaments in intact muscle. *Proc. Natl. Acad. Sci. USA* **2011**, *108*, 11423–11428. [CrossRef] [PubMed]

2. Shaffer, J.F.; Kensler, R.W.; Harris, S.P. The myosin-binding protein C motif binds to F-actin in a phosphorylation-sensitive manner. *J. Biol. Chem.* **2009**, *284*, 12318–12327. [CrossRef] [PubMed]

3. Kensler, R.W.; Shaffer, J.F.; Harris, S.P. Binding of the N-terminal fragment C0–C2 of cardiac MyBP-C to cardiac F-actin. *J. Struct. Biol.* **2011**, *174*, 44–51. [CrossRef] [PubMed]

4. Lu, Y.; Kwan, A.H.; Trewhella, J.; Jeffries, C.M. The C0C1 fragment of human cardiac myosin binding protein C has common binding determinants for both actin and myosin. *J. Mol. Biol.* **2011**, *413*, 908–913. [CrossRef] [PubMed]

5. Rybakova, I.N.; Greaser, M.L.; Moss, R.L. Myosin Binding Protein C Interaction with Actin: Characterization and Mapping of the Binding Site. *J. Biol. Chem.* **2011**, *286*, 2008–2016. [CrossRef] [PubMed]

6. Bhuiyan, M.S.; Gulick, J.; Osinska, H.; Gupta, M.; Robbins, J. Determination of the critical residues responsible for cardiac myosin binding protein {C's} interactions. *J. Mol. Cell. Cardiol.* **2012**, *53*, 838–847. [CrossRef] [PubMed]

7. Whitten, A.E.; Jeffries, C.M.; Harris, S.P.; Trewhella, J. Cardiac myosin-binding protein C decorates F-actin: Implications for cardiac function. *Proc. Natl. Acad. Sci. USA* **2008**, *105*, 18360–18365. [CrossRef] [PubMed]

8. Sadayappan, S.; Gulick, J.; Osinska, H.; Barefield, D.; Cuello, F.; Avkiran, M.; Lasko, V.M.; Lorenz, J.N.; Maillet, M.; Martin, J.L.; et al. A critical function for Ser-282 in cardiac myosin binding protein-C phosphorylation and cardiac function. *Circ. Res.* **2011**, *109*, 141–150. [CrossRef] [PubMed]

9. Okagaki, T.; Weber, F.E.; Fischman, D.A.; Vaughan, K.T.; Mikawa, T.; Reinach, F.C. The Major Myosin-Binding Domain of Skeletal Muscle MyBP-C (C-Protein) Resides in the COOH-Terminal, Immunoglobulin-C2 Motif. *J. Cell Biol.* **1993**, *123*, 619–626. [CrossRef] [PubMed]

10. Gruen, M.; Prinz, H.; Gautel, M. cAPK-phosphorylation controls the interaction of the regulatory domain of cardiac myosin binding protein C with myosin-S2 in an on-off fashion. *FEBS Lett.* **1999**, *453*, 254–259. [CrossRef]

11. Ratti, J.; Rostkova, E.; Gautel, M.; Pfuhl, M. Structure and interactions of myosin-binding protein C domain C0: Cardiac-specific regulation of myosin at its neck? *J. Biol. Chem.* **2011**, *286*, 12650–12658. [CrossRef] [PubMed]

12. Freiburg, A.; Gautel, M. A molecular map of the interactions between titin and myosin-binding protein C. Implications for sarcomeric assembly in familial hypertrophic cardiomyopathy. *Eur. J. Biochem.* **1996**, *235*, 317–323. [CrossRef] [PubMed]

13. Miyamoto, C.A.; Fischman, D.A.; Reinach, F.C. The interface between MyBP-C and myosin: Site-directed mutagenesis of the CX myosin-binding domain of MyBP-C. *J. Muscle Res. Cell Motil.* **1999**, *20*, 703–715. [CrossRef] [PubMed]

14. Flashman, E.; Watkins, H.; Redwood, C. Localization of the binding site of the C-terminal domain of cardiac myosin-binding protein-C on the myosin rod. *Biochem. J.* **2007**, *401*, 97–102. [CrossRef] [PubMed]

15. Kuster, D.W.D.; Govindan, S.; Springer, T.I.; Martin, J.L.; Finley, N.L.; Sadayappan, S. A hypertrophic cardiomyopathy-associated MYBPC3 mutation common in populations of South Asian descent causes contractile dysfunction. *J. Biol. Chem.* **2015**, *290*, 5855–5867. [CrossRef] [PubMed]

16. Weber, F.E.; Vaughan, K.T.; Reinach, F.C.; Fischman, D.A. Complete sequence of human fast-type and slow-type muscle myosin-binding-protein C (MyBP-C). Differential expression, conserved domain structure and chromosome assignment. *Eur. J. Biochem.* **1993**, *216*, 661–669. [PubMed]

17. Lin, B.; Govindan, S.; Lee, K.; Zhao, P.; Han, R.; Runte, K.E.; Craig, R.; Palmer, B.M.; Sadayappan, S. Cardiac Myosin Binding Protein-C Plays No Regulatory Role in Skeletal Muscle Structure and Function. *PLoS ONE* **2013**, *8*, e69671. [CrossRef] [PubMed]

18. Alyonycheva, T.N.; Mikawa, T.; Reinach, F.C.; Fischman, D.A. Isoform-specific interaction of the myosin-binding proteins (MyBPs) with skeletal and cardiac myosin is a property of the C-terminal immunoglobulin domain. *J. Biol. Chem.* **1997**, *272*, 20866–20872. [CrossRef] [PubMed]

19. Idowu, S.M.; Gautel, M.; Perkins, S.J.; Pfuhl, M. Structure, stability and dynamics of the central domain of cardiac myosin binding protein C (MyBP-C): Implications for multidomain assembly and causes for cardiomyopathy. *J. Mol. Biol.* **2003**, *329*, 745–761. [CrossRef]

20. Cecconi, F.; Guardiani, C.; Livi, R. Analyzing pathogenic mutations of C5 domain from cardiac myosin binding protein C through MD simulations. *Eur. Biophys. J.* **2008**, *37*, 683–691. [CrossRef] [PubMed]

21. Shaffer, J.F.; Razumova, M.V.; Tu, A.Y.; Regnier, M.; Harris, S.P. Myosin S2 is not required for effects of myosin binding protein-C on motility. *FEBS Lett.* **2007**, *581*, 1501–1504. [CrossRef] [PubMed]

22. Mun, J.Y.; Kensler, R.W.; Harris, S.P.; Craig, R. The cMyBP-C HCM variant L348P enhances thin filament activation through an increased shift in tropomyosin position. *J. Mol. Cell. Cardiol.* **2016**, *91*, 141–147. [CrossRef] [PubMed]

23. Bezold, K.L.; Shaffer, J.F.; Khosa, J.K.; Hoye, E.R.; Harris, S.P. A gain-of-function mutation in the M-domain of cardiac myosin-binding protein-C increases binding to actin. *J. Biol. Chem.* **2013**, *288*, 21496–21505. [CrossRef] [PubMed]

24. Orlova, A.; Galkin, V.E.; Jeffries, C.M.J.; Egelman, E.H.; Trewhella, J. The N-terminal domains of myosin binding protein C can bind polymorphically to F-actin. *J. Mol. Biol.* **2011**, *412*, 379–386. [CrossRef] [PubMed]

25. Lu, Y.; Kwan, A.H.; Jeffries, C.M.; Guss, J.M.; Trewhella, J. The motif of human cardiac myosin-binding protein C is required for its Ca^{2+}-dependent interaction with calmodulin. *J. Biol. Chem.* **2012**, *287*, 31596–31607. [CrossRef] [PubMed]

26. Ackermann, M.A.; Kontrogianni-Konstantopoulos, A. Myosin binding protein-C slow is a novel substrate for protein kinase A (PKA) and C (PKC) in skeletal muscle. *J. Proteome Res.* **2011**, *10*, 4547–4555. [CrossRef] [PubMed]

27. Schumacher, M.A.; Rivard, A.F.; Bachinger, H.P.; Adelman, J.P. Structure of the gating domain of a Ca^{2+}-activated K$^+$ channel complexed with Ca^{2+}/calmodulin. *Nature* **2001**, *410*, 1120–1124. [CrossRef] [PubMed]

28. Springer, T.I.; Goebel, E.; Hariraju, D.; Finley, N.L. Mutation in the beta-hairpin of the Bordetella pertussis adenylate cyclase toxin modulates N-lobe conformation in calmodulin. *Biochem. Biophys. Res. Commun.* **2014**, *453*, 43–48. [CrossRef] [PubMed]

29. Heller, W.T.; Krueger, J.K.; Trewhella, J. Further insights into calmodulin-myosin light chain kinase interaction from solution scattering and shape restoration. *Biochemistry* **2003**, *42*, 10579–10588. [CrossRef] [PubMed]

30. Howarth, J.W.; Ramisetti, S.; Nolan, K.; Sadayappan, S.; Rosevear, P.R. Structural insight into unique cardiac myosin-binding protein-C motif: A partially folded domain. *J. Biol. Chem.* **2012**, *287*, 8254–8262. [CrossRef] [PubMed]

31. Sievers, F.; Wilm, A.; Dineen, D.; Gibson, T.J.; Karplus, K.; Li, W.; Lopez, R.; McWilliam, H.; Remmert, M.; Soding, J.; et al. Fast, scalable generation of high-quality protein multiple sequence alignments using Clustal Omega. *Mol. Syst. Biol.* **2014**, *7*, 539. [CrossRef] [PubMed]

32. Ackermann, M.A.; Ward, C.W.; Gurnett, C.; Kontrogianni-Konstantopoulos, A. Myosin Binding Protein-C Slow Phosphorylation is Altered in Duchenne Dystrophy and Arthrogryposis Myopathy in Fast-Twitch Skeletal Muscles. *Sci. Rep.* **2015**, *5*, 13235. [CrossRef] [PubMed]

33. Lavinder, J.J.; Hari, S.B.; Sullivan, B.J.; Magliery, T.J. High-Throughput Thermal Scanning: A General, Rapid Dye-Binding Thermal Shift Screen for Protein Engineering. *J. Am. Chem. Soc.* **2009**, *131*, 3794–3795. [CrossRef] [PubMed]

34. Cimmperman, P.; Baranauskienė, L.; Jachimovičiūtė, S.; Jachno, J.; Torresan, J.; Michailovienė, V.; Matulienė, J.; Sereikaitė, J.; Bumelis, V.; Matulis, D. A Quantitative Model of Thermal Stabilization and Destabilization of Proteins by Ligands. *Biophys. J.* **2008**, *95*, 3222–3231. [CrossRef] [PubMed]

35. Jeffries, C.M.; Lu, Y.; Hynson, R.M.G.; Taylor, J.E.; Ballesteros, M.; Kwan, A.H.; Trewhella, J. Human cardiac myosin binding protein C: Structural flexibility within an extended modular architecture. *J. Mol. Biol.* **2011**, *414*, 735–748. [CrossRef] [PubMed]

36. Ababou, A.; Zhou, L.; Gautel, M.; Pfuhl, M. Sequence specific assignment of domain C1 of the N-terminal myosin-binding site of human cardiac myosin binding protein C (MyBP-C). *J. Biomol. NMR* **2004**, *29*, 431–432. [CrossRef] [PubMed]

37. Ababou, A.; Gautel, M.; Pfuhl, M. Dissecting the *N*-terminal myosin binding site of human cardiac myosin-binding protein C: Structure and myosin binding of domain C2. *J. Biol. Chem.* **2007**, *282*, 9204–9215. [CrossRef] [PubMed]

38. Michie, K.A.; Kwan, A.H.; Tung, C.S.; Guss, J.M.; Trewhella, J. A Highly Conserved Yet Flexible Linker Is Part of a Polymorphic Protein-Binding Domain in Myosin-Binding Protein C. *Structure* **2016**, *24*, 2000–2007. [CrossRef] [PubMed]

39. Biasini, M.; Bienert, S.; Waterhouse, A.; Arnold, K.; Studer, G.; Schmidt, T.; Kiefer, F.; Cassarino, T.G.; Bertoni, M.; Bordoli, L.; et al. SWISS-MODEL: Modelling protein tertiary and quaternary structure using evolutionary information. *Nucleic Acids Res.* **2014**, *42*, W252–W258. [CrossRef] [PubMed]

40. Benkert, P.; Tosatto, S.C.E.; Schomburg, D. QMEAN: A comprehensive scoring function for model quality assessment. *Proteins Struct. Funct. Bioinform.* **2008**, *71*, 261–277. [CrossRef] [PubMed]

41. De Beer, T.A.P.; Berka, K.; Thornton, J.M.; Laskowski, R.A. PDBsum additions. *Nucleic Acids Res.* **2014**, *42*, D292–D296. [CrossRef] [PubMed]

42. Gasmi-Seabrook, G.M.; Howarth, J.W.; Finley, N.; Abusamhadneh, E.; Gaponenko, V.; Brito, R.M.; Solaro, R.J.; Rosevear, P.R. Solution structures of the *C*-terminal domain of cardiac troponin C free and bound to the *N*-terminal domain of cardiac troponin I. *Biochemistry* **1999**, *38*, 8313–8322. [CrossRef] [PubMed]

43. Hoffman, R.M.B.; Blumenschein, T.M.A.; Sykes, B.D. An interplay between protein disorder and structure confers the Ca^{2+} regulation of striated muscle. *J. Mol. Biol.* **2006**, *361*, 625–633. [CrossRef] [PubMed]

44. Gurnett, C.A.; Desruisseau, D.M.; McCall, K.; Choi, R.; Meyer, Z.I.; Talerico, M.; Miller, S.E.; Ju, J.S.; Pestronk, A.; Connolly, A.M.; et al. Myosin binding protein C1: A novel gene for autosomal dominant distal arthrogryposis type 1. *Hum. Mol. Genet.* **2010**, *19*, 1165–1173. [CrossRef] [PubMed]

45. Michalek, A.J.; Howarth, J.W.; Gulick, J.; Previs, M.J.; Robbins, J.; Rosevear, P.R.; Warshaw, D.M. Phosphorylation modulates the mechanical stability of the cardiac myosin-binding protein C motif. *Biophys. J.* **2013**, *104*, 442–452. [CrossRef] [PubMed]

46. Tavi, P.; Westerblad, H. The role of in vivo Ca^{2+} signals acting on Ca^{2+}-calmodulin-dependent proteins for skeletal muscle plasticity. *J. Physiol.* **2011**, *589*, 5021–5031. [CrossRef] [PubMed]

47. Abbott, M.B.; Dong, W.J.; Dvoretsky, A.; DaGue, B.; Caprioli, R.M.; Cheung, H.C.; Rosevear, P.R. Modulation of cardiac troponin c-cardiac troponin I regulatory interactions by the amino-terminus of cardiac troponin I. *Biochemistry* **2001**, *40*, 5992–6001. [CrossRef] [PubMed]

48. Finley, N.; Dvoretsky, A.; Rosevear, P.R. Magnesium—Calcium Exchange in Cardiac Troponin C Bound to Cardiac Troponin I. *J. Mol. Cell. Cardiol.* **2000**, *1446*, 1439–1446. [CrossRef] [PubMed]

49. Dong, W.J.; Chandra, M.; Xing, J.; She, M.; Solaro, R.J.; Cheung, H.C. Phosphorylation-induced distance change in a cardiac muscle troponin I mutant. *Biochemistry* **1997**, *36*, 6754–6761. [CrossRef] [PubMed]

50. Sakthivel, S.; Finley, N.L.; Rosevear, P.R.; Lorenz, J.N.; Gulick, J.; Kim, S.; VanBuren, P.; Martin, L.A.; Robbins, J. In Vivo and in Vitro Analysis of Cardiac Troponin I Phosphorylation. *J. Biol. Chem.* **2005**, *280*, 703–714. [CrossRef] [PubMed]

51. Baryshnikova, O.K.; Li, M.X.; Sykes, B.D. Modulation of cardiac troponin C function by the cardiac-specific N-terminus of troponin I: Influence of PKA phosphorylation and involvement in cardiomyopathies. *J. Mol. Biol.* **2008**, *375*, 735–751. [CrossRef] [PubMed]

52. Finley, N.; Abbott, M.B.; Abusamhadneh, E.; Gaponenko, V.; Dong, W.; Gasmi-Seabrook, G.; Howarth, J.W.; Rance, M.; Solaro, R.J.; Cheung, H.C.; et al. NMR analysis of cardiac troponin C-troponin I complexes: Effects of phosphorylation. *FEBS Lett.* **1999**, *453*, 107–112. [CrossRef]

53. Sadayappan, S.; Osinska, H.; Klevitsky, R.; Lorenz, J.N.; Sargent, M.; Molkentin, J.D.; Seidman, C.E.; Seidman, J.G.; Robbins, J. Cardiac myosin binding protein C phosphorylation is cardioprotective. *Proc. Natl. Acad. Sci. USA* **2006**, *103*, 16918–16923. [CrossRef] [PubMed]

54. Colson, B.A.; Rybakova, I.N.; Prochniewicz, E.; Moss, R.L.; Thomas, D.D. Cardiac myosin binding protein-C restricts intrafilament torsional dynamics of actin in a phosphorylation-dependent manner. *Proc. Natl. Acad. Sci. USA* **2012**, *109*, 20437–20442. [CrossRef] [PubMed]

55. Weith, A.; Sadayappan, S.; Gulick, J.; Previs, M.J.; VanBuren, P.; Robbins, J.; Warshaw, D.M. Unique single molecule binding of cardiac myosin binding protein-C to actin and phosphorylation-dependent inhibition of actomyosin motility requires 17 amino acids of the motif domain. *J. Mol. Cell. Cardiol.* **2012**, *52*, 219–227. [CrossRef] [PubMed]

56. Kooij, V.; Boontje, N.; Zaremba, R.; Jaquet, K.; dos Remedios, C.; Stienen, G.J.M.; van der Velden, J. Protein kinase C alpha and epsilon phosphorylation of troponin and myosin binding protein C reduce Ca^{2+} sensitivity in human myocardium. *Basic Res. Cardiol.* **2010**, *105*, 289–300. [CrossRef] [PubMed]

57. Tong, C.W.; Gaffin, R.D.; Zawieja, D.C.; Muthuchamy, M. Roles of phosphorylation of myosin binding protein-C and troponin I in mouse cardiac muscle twitch dynamics. *J. Physiol.* **2004**, *558*, 927–941. [CrossRef] [PubMed]

58. Xue, Y.; Liu, Z.; Cao, J.; Ma, Q.; Gao, X.; Wang, Q.; Jin, C.; Zhou, Y.; Wen, L.; Ren, J. GPS 2.1: Enhanced prediction of kinase-specific phosphorylation sites with an algorithm of motif length selection. *Protein Eng. Des. Sel.* **2011**, *24*, 255–260. [CrossRef] [PubMed]

59. Offer, G.; Moos, C.; Starr, R. A new protein of the thick filaments of vertebrate skeletal myofibrils. *J. Mol. Biol.* **1973**, *74*, 653–676. [CrossRef]

60. Squire, J.M.; Luther, P.K.; Knupp, C. Structural evidence for the interaction of C-protein (MyBP-C) with actin and sequence identification of a possible actin-binding domain. *J. Mol. Biol.* **2003**, *331*, 713–724. [CrossRef]

61. Harris, S.P.; Lyons, R.G.; Bezold, K.L. In the thick of it: HCM-causing mutations in myosin binding proteins of the thick filament. *Circ. Res.* **2011**, *108*, 751–764. [CrossRef] [PubMed]

62. Delaglio, F.; Grzesiek, S.; Vuister, G.W.; Zhu, G.; Pfeifer, J.; Bax, A. NMRPipe: A multidimensional spectral processing system based on UNIX pipes. *J. Biomol. NMR* **1995**, *6*, 277–293. [CrossRef] [PubMed]

63. Goddard, T.D.; Kneller, D.G. Sparky—NMR Assignment and Integration Software. Available online: https://www.cgl.ucsf.edu/home/sparky/ (accessed on 16 December 2017).

magnetochemistry

MDPI

Article

Short-Chain Alkanethiol Coating for Small-Size Gold Nanoparticles Supporting Protein Stability

Cristina Cantarutti [1], Paolo Bertoncin [2], Alessandra Corazza [1,3], Sofia Giorgetti [4],
P. Patrizia Mangione [4,5], Vittorio Bellotti [4,5], Federico Fogolari [3,6] and Gennaro Esposito [3,6,7,*]

[1] DAME, Università di Udine, P.le Kolbe 4, 33100 Udine, Italy; cantarutti.cristina@spes.uniud.it (C.C.);
    alessandra.corazza@uniud.it (A.C.)
[2] Dipartimento di Scienze della Vita, Università di Trieste, 34128 Trieste, Italy; pbertoncin@units.it
[3] INBB—Viale Medaglie d'Oro 305, 00136 Roma, Italy; federico.fogolari@uniud.it
[4] Dipartimento di Medicina Molecolare, Università di Pavia, Via Taramelli 3, 27100 Pavia, Italy;
    sofia.giorgetti@unipv.it (S.G.); p.mangione@ucl.ac.uk (P.P.M.); v.bellotti@ucl.ac.uk (V.B.)
[5] Division of Medicine, University College of London, London NW3 2PF, UK
[6] DMIF, Università di Udine, Viale delle Scienze, 33100 Udine, Italy
[7] Science and Math Division, New York University Abu Dhabi, P.O. Box 129188, Abu Dhabi, UAE
*   Correspondence: rino.esposito@uniud.it; Tel.: +39-0432-494321

Received: 30 October 2017; Accepted: 21 November 2017; Published: 27 November 2017

Abstract: The application of gold nanoparticles (AuNPs) is emerging in many fields, raising the need for a systematic investigation on their safety. In particular, for biomedical purposes, a relevant issue are certainly AuNP interactions with biomolecules, among which proteins are the most abundant ones. Elucidating the effects of those interactions on protein structure and on nanoparticle stability is a major task towards understanding their mechanisms at a molecular level. We investigated the interaction of the 3-mercaptopropionic acid coating of AuNPs (MPA-AuNPs) with β2-microglobulin (β2m), which is a paradigmatic amyloidogenic protein. To this aim, we prepared and characterized MPA-AuNPs with an average diameter of 3.6 nm and we employed NMR spectroscopy and fluorescence spectroscopy to probe protein structure perturbations. We found that β2m interacts with MPA-AuNPs through a highly localized patch maintaining its overall native structure with minor conformational changes. The interaction causes the reversible precipitation of clusters that can be easily re-dispersed through brief sonication.

Keywords: amyloidogenic protein-nanoparticle systems; nanoparticle stability; protein unfolding

---

1. Introduction

Proteins play a fundamental role in biological processes. Their activity, indeed, on one hand supports the correct operation of an organism, but on the other, could be responsible for disease onset. Many protein functions are affected by their interaction with other molecules e.g., other proteins, oligonucleotides, hormones, and so on. The fact that the interaction profile can highlight the functions a protein performs implies that understanding the behaviour of a protein at the molecular level is a valuable strategy to get deep insights into the functional role and possibly into the design of new tools to master the protein activity. The spreading application of nanomaterials in different fields such as biomedicine, food, environmental, and material sciences [1,2], has stressed the relevance of understanding at a molecular level protein-nanoparticle interaction, because any contact of nanomaterials with a biological fluid is suddenly followed by the adsorption of proteins [3]. However, the challenging investigation of protein-nanomaterial interface also proved so intriguing that no general trends could be drawn so far. For example, it has been reported that some enzymes, e.g., lysozyme, chymotrypsin, and fibrinogen [4,5], lose their catalytic activity upon interacting with gold nanoparticles (AuNPs), other ones, e.g., pepsin [6], retain their functionality and some other ones, e.g., bovine catalase [7], show a higher stability in harsh conditions in presence of AuNPs. On the other hand, when Aβ peptide, the amyloidogenic fragment from Amyloid Precursor Protein (APP), whose fibril deposition has been related to Alzheimer's disease onset [8], was incubated with different nanoparticles (NPs), various effects on fibrillogenesis were observed. While titanium oxide NPs promote Aβ aggregation [9], silica NPs leave it unaffected [9], and fullerene even inhibited it [10]. Thus, the interaction between proteins and nanomaterials and the ensuing effects appear to be highly dependent both on the specific protein and on the nanomaterial physico-chemical properties.

Here, we present a NMR and fluorescence-based study of the interaction between 3-mercaptopropionic acid-coated AuNPs (MPA-AuNPs) and β2-microglobulin (β2m).

Gold nanoparticles have been widely used in biomedical research because of their optical features, large surface to volume ratio, gold inertness, ease of production, and surface functionalization [11]. MPA-AuNPs were synthesized through a one-phase direct synthesis that proceeds through three steps [12].

β2m is an amyloidogenic protein responsible for dialysis related amyloidosis (DRA) [13], and it is considered a model for amyloidogenic proteins since it recapitulates the typical features of this class of proteins. β2m is the light chain of class I major histocompatibility complex and in healthy organisms, after its detachment from the heavy chain, it is removed from the blood through the kidneys. In patients that are affected by chronic renal failure, it accumulates in the blood and precipitates into amyloid deposits in correspondence of the joints. In literature, the interaction between β2m and citrate-stabilized AuNPs (Cit-AuNPs) was reported for both the wild-type and D76N variants [14,15]. In both of the cases, a labile interaction was observed that does not affect the overall folding and is mainly located at the N-terminal apical part of the protein structure. Furthermore, it was shown that Cit-AuNPs are able to partially hinder the fibrillogenesis of the most amyloidogenic variant, namely D76N β2m [15].

2. Results

2.1. AuNP Synthesis and Characterization

To synthesize MPA-AuNPs, a reported procedure was employed [12] (Figure 1). Briefly, at the beginning, Au^{III} is reduced to Au^I by the thiols, and then Au^I forms with the thiolate polymeric structures ($[Au^ISR]_n$), and finally gold is further reduced to Au^0 by $NaBH_4$, leading to gold clusters that are stabilized by the covalently bound thiolate organic monolayer. When compared to the reference recipe [12], the ligand/Au ratio was changed from 3 to 9 in order to obtain smaller AuNPs.

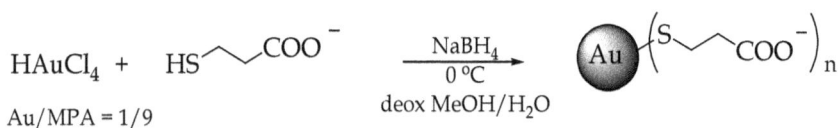

Figure 1. Scheme representing 3-mercaptopropionic acid-coated AuNPs (MPA-AuNP) synthesis.

From transmission electron microscopy (TEM) images, an average diameter of 3.6 nm was determined (Figure 2a). The small dimensions of these AuNPs are consistent with the weak SPB that is recorded in the UV-Vis spectrum (Figure 2b). MPA-AuNPs can be centrifuged, dried, and dispersed again without any aggregation. To evaluate the NP organic percentage content originating from MPA, thermogravimetric analysis (TGA) analysis was performed (Figure 2c). From TGA results and the NP diameter obtained from TEM, it was possible to estimate the average composition and the molecular weight of MPA-AuNPs (Table 1).

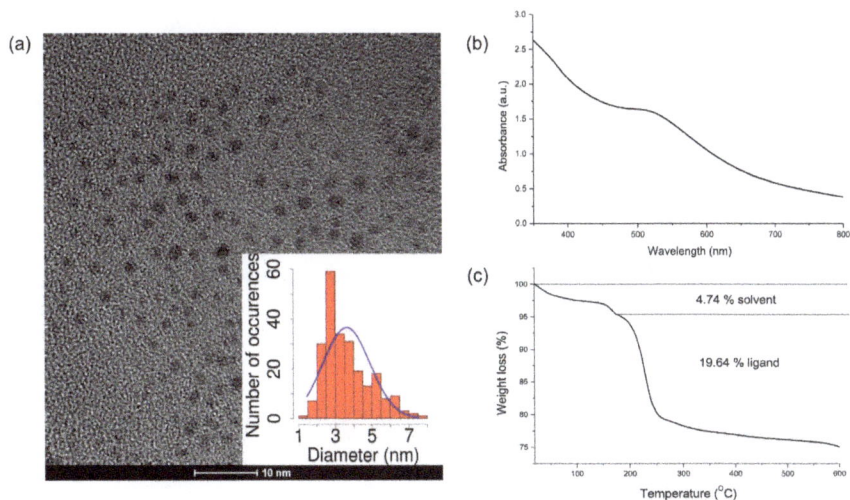

Figure 2. Characterization of synthesized MPA-AuNPs: (**a**) transmission electron microscopy (TEM) micrograph of MPA-AuNPs along with the corresponding size histogram; (**b**) UV-Vis spectrum of MPA-AuNPs; and, (**c**) thermogravimetric analysis (TGA) analysis of MPA-AuNPs.

Table 1. Composition of MPA-AuNPs calculated from TEM and TGA.

Core Diameter (nm)	Organic Percentage (%)	Average Composition	Molecular Weight (g/mol)
3.6	19.64	$Au_{1441}(SCH_2CH_2COO^-)_{661}$	353,315.33

2.2. Protein-AuNP Interaction

When MPA-AuNPs and β2m were mixed together, and within a few hours a brown precipitate developed on the bottom of the flask. This precipitate could be easily re-dispersed by sonication and dropped onto a TEM grid for imaging. TEM micrographs (Figure 3) showed a well dispersed nanoparticle sample in which the average size of the NP cores was the same as the control (Figure 2a).

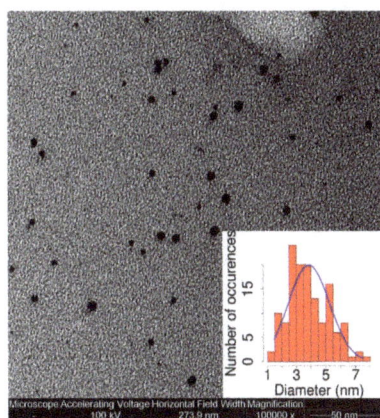

Figure 3. TEM micrograph of 2.5 μM MPA-AuNPs and 25 μM β2m. The average NP diameter measured in presence of the protein was 3.8 ± 1.3 nm.

By doing simple geometrical considerations, it is possible to estimate the protein adsorption capacity of a single nanoparticle. If the ratio between the volumes is considered, as proposed by Calzolai et al. [16], it is possible to calculate the number of proteins (N) per nanoparticle applying the following equation:

$$N = 0.65 \times \frac{(R^3_{complex} - R^3_{NP})}{R^3_{protein}}, \qquad (1)$$

where $R_{complex}$ corresponds to the sum of the NP radius and the diameter of the protein, R_{NP} is the radius of the nanoparticle, and $R_{protein}$ is the protein radius. However, this equation is suited for spherical proteins, while β2m has an oblate three-dimensional structure with longitudinal and transverse axes of 4.3–3.8 and 2.5–2.0 nm, respectively [17]. When considering the crystallographic cylindrical shape of β2m, the Equation (1) becomes

$$N = 0.65 \times \frac{\frac{4}{3}(R^3_{complex} - R^3_{NP})}{h_{cyl} \times r^2_{cyl}}, \qquad (2)$$

where $R_{complex}$ is given by the sum of the NP radius and the height of the β2m cylindroid, R_{NP} is the radius of the nanoparticle, and h_{cyl} and r_{cyl} are the height and the base radius of β2m cylindroid. The NP radius is given by the addition of the alkanethiolate monolayer thickness to the radius of the gold core obtained from TEM. When considering that MPA is 0.55 nm long, approximately, we can assume that the monolayer protected cluster has a diameter of 4.7 nm. Following Equation (2), on the surface of a spherical MPA-AuNP with average diameter of 4.7 nm 36–50 β2m monomers can be accommodated. The maximum packing density factor that is used in this model is referred to spherical proteins. Since the base of β2m cylindroid is a half with respect to its height, substituting symmetric tetramers to cylindroid monomers can improve the geometrical model. This adjustment leads to 9–12 tetramers per NP that means 36–48 monomers. If the ratio between the NP surface area and the cylindroid base is considered, then the number of protein monomers per NP is reduced to 14–22. From all of these considerations, beyond any critical evaluation, the number of β2m molecules that can be adsorbed on a MPA-AuNP goes from 14 to 50, approximately.

Two-dimensional ^{1}H ^{15}N NMR experiments were acquired to ascertain the state of the protein in presence of MPA-AuNPs. Five different protein/NP ratios were examined going from 100 to 10. SOFAST HMQC [18] spectra of β2m alone and in the presence of MPA-AuNPs at protein/NP ratios of 40 and 15 are shown in Figure 4.

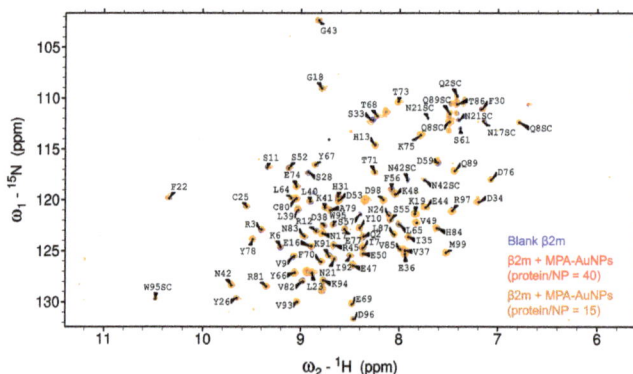

Figure 4. Superimposition of β2m ^{1}H-^{15}N SOFAST-HMQC spectra in absence of MPA-AuNPs (blue) and in presence of MPA-AuNPs in protein/NP ratio of 40 (red) and of 15 (orange). The label SC indicates side chain NH's.

By analyzing the spectra intensity, in addition to an overall attenuation reflecting the increase of protein recruitment by the increasing number of available interaction sites on NPs, there was also a preferential intensity decrease (Figure 5a). Moreover, the gradual intensity attenuation was associated with progressive chemical shift variation (Figure 5b). By plotting for each residue, the relative intensity (RI) against the chemical shift perturbation ($\Delta\delta$), recorded at protein/NP = 25, i.e., the lowest ratio at which most of the peaks were still visible (Figure 5c), it can be seen that there is a quite good correlation between the two variations, i.e., the amino acids that showed the largest chemical shift deviation were usually also characterized by the lowest relative intensity. This suggests that both of the perturbations arose from the same process, namely the exchange interaction with the nanoparticles.

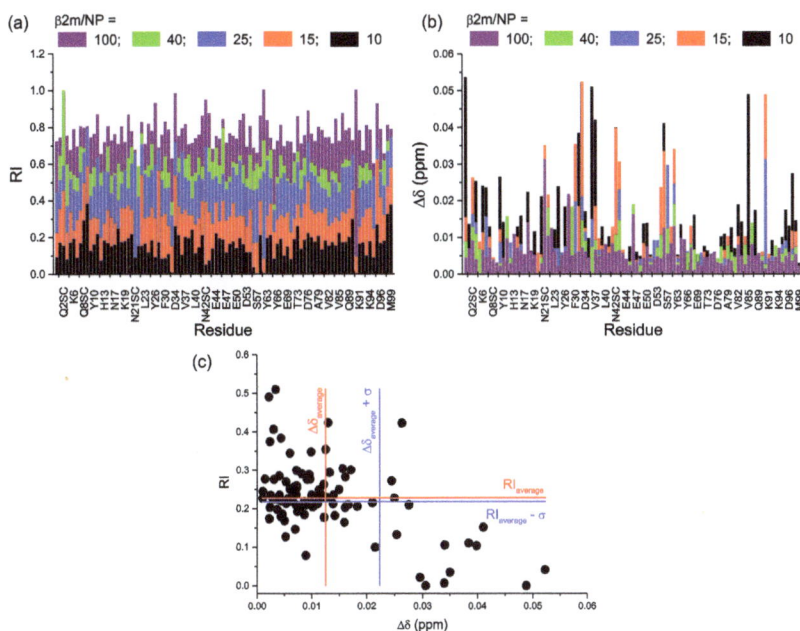

Figure 5. (a,b) Bar plots of amide resonance chemical shift perturbations ($\Delta\delta$) and cross-peak attenuations (RI), respectively. Refer to color legend for the protein/NP ratios; (c) Scatter plot indicating the correlation between chemical shift deviation ($\Delta\delta$) and relative intensity (RI) attenuation, each point corresponds to an amide resonance. Tick marks and labels are reported every 3 residues, except for the missing and undetected ones, i.e., I1, P5, P14, G29, P32, L54, K58, D59, W60, F62, P72, S88, and P90. The observed side chain (SC) NHs are also included, i.e., two for Q2, two for Q8, one for N17, two for N21, two for N24, two for Q89, and one for W95.

The variation of the resonance position accounts for a change in the chemical environment around a specific amide group and the accompanying signal intensity decrease reflects the line broadening, i.e., transverse relaxation rate increase, from additional exchange and possibly cross-relaxation contributions. These two features are both related to the exchange process between free and NP-contacting states. The occurrence of single resonances whose position reflects the population-weighted chemical shift average of the free and the NP-close forms of the protein implies that the exchange regime is not slow, nor intermediate with respect to the NMR chemical shift scale. The observed pattern is consistent with a nearly fast exchange regime with residual line-broadening contributions that, besides the dissociation rate from the NP adduct, may also reflect more or less transient changes in the local dynamics and the overall rotational tumbling rate that affect both dipolar (DD) and chemical shift anisotropy (CSA) contributions to relaxation, and therefore, the linewidth.

The possibility of a progressively slowing exchange rate on increasing the NP concentration that would approach the intermediate exchange regime and stress further the signal attenuation could account for the substantial intensity reductions that were observed at the lowest protein/NP ratios.

To find the residues that proved more affected by the presence of MPA-AuNPs, the amide signals whose RI and $\Delta\delta$ was displaced more than one standard deviation from the average values were identified (Figure 6a). By mapping their positions on the protein three-dimensional structure (Figure 6b), a highly specific and localized region was found to be involved in the interaction. The patch includes two loops, namely BC loop and DE loop, and the spatially close N-terminal tail.

(a)

Structure region	Δδ outliers	RI outliers
N-term, A strand	R3	R3
AB loop		
B strand		S28
BC loop	F30, H31, S33	F30, H31, S33
CC', C'D loops	E36, V37	
D strand		
DE loop	S55, F56, S57, S61	F56, S57, S61
E strand		L65
EF loop		
F strand		
FG loop		
G strand, C-term		

(b)

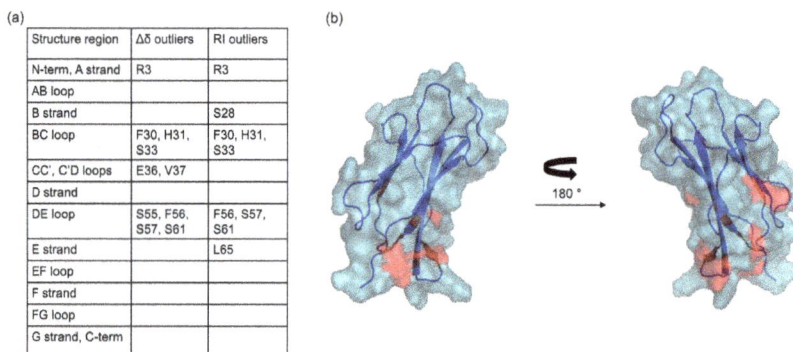

180°

Figure 6. (**a**) Synopsis of β2m positions that proved most affected, i.e., displaced more than one standard deviation from the average, by the presence of MPA-AuNP and their secondary structure element location; (**b**) β2m cartoon highlighting in red the locations of the outlier residues.

This result is consistent with a strong localization of the electrostatic interaction, probably being due to the compactness of the NP electron plasma [14,15]. The character of the interaction surface, as described in Figure 6, can be appreciated when compared to the β2m regions that are involved in the contacts with the heavy chain of type I major histocompatibility complex (MHC-I) [19]. In this complex, β2m establishes the closest contacts with the partner species through a hydrophobic area that extends over strands B (fragment 23–27) and E (fragment 62–66). The interface is also comprised of contacts with hydrophilic stretches at strand A (fragment 8–12), strand D (fragment 51–55) and loop DE (fragment 57–61) that appear less tightly packed in the complex, and are thus more accessible to the solvent. The non-uniform distribution of hydrophilic and hydrophobic patches on β2m surface that is recognized in the quaternary organization of MHC-I is probably relevant for the amyloidogenic propensity of the protein [20]. The regions that are highlighted in Figure 6b, however, appear distinctly different from those involved in the typical hydrophobic-driven interaction that β2m engages in MHC-I. It was not possible to fit the chemical shift perturbation data with a binding isotherm because the signals that showed the highest deviations were also the ones undergoing extensive attenuation up to the complete cancellation in the first titration points.

To further investigate the NP effect on protein conformation, fluorescence experiments were performed. β2m intrinsic fluorescence is mainly due to a buried Trp residue, namely Trp95 [21]. The second Trp residue of the molecule, Trp60, contributes only marginally (20% approximatively) to the overall fluorescence because of its exposure on the protein surface and the consequent quenching effect of the solvent. The β2m intrinsic fluorescence was recorded upon the progressive addition of MPA-AuNPs. After an initial decrease (around 20%), the fluorescence intensity increased and the emission peak shifted (Figure 7a). The initial intensity decrease is likely to be due to the quenching of the limited Trp60 emission contribution to the whole fluorescence. Since Trp60 is exposed on the surface, it is accessible to nanoparticle direct contact. The fluorescence of W60G variant is, indeed, approximately 20% lower than the value of wild-type protein.

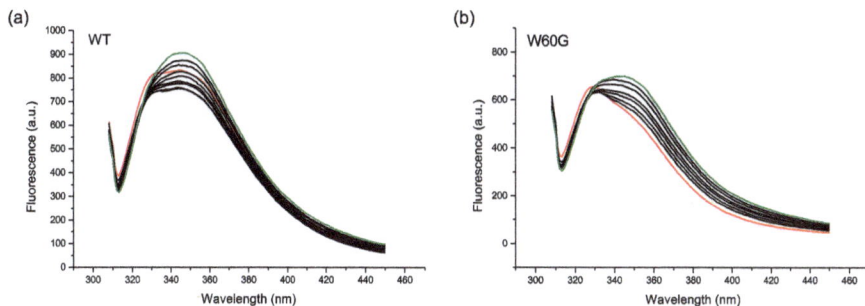

Figure 7. (a,b) Fluorescence quenching of 0.5 μM wild-type (WT) and W60G β2m, respectively, with MPA-AuNP concentrations ranging from 0 to 50 nM. The control sample spectrum is shown in red, while the last titration point in green. Each titration point corresponds to a NP increase of 5 or 7 nM, approximately, for the experiments done with WT or W60G, respectively. For WT, the first two additions caused an intensity decrease and the following ones a progressive increase. For W60G, all of the consecutive additions led to progressive intensity increase.

When considering only the data prior to the shift of the fluorescence emission frequency that certainly pertain to the natively folded β2m species, the bimolecular quenching constant (k_q) values, calculated from the apparent Stern-Volmer constant (Table 2), assuming a fluorescence lifetime of 1–10 ns for the indole chromophore [22], were higher than the collisional rate limit, i.e., 2×10^{10} M^{-1}·s^{-1}. Hence, it can be concluded that the quenching of the external tryptophan is not purely collisional.

Table 2. Parameters obtained from fluorescence quenching data fitted with Stern-Volmer equation [23]. Only the initial points of the titration in which the fluorescence decreases increasing the titrant concentration were used for the Stern-Volmer fitting.

K_{sv} (M^{-1})	R^2	k_q (M^{-1}·s^{-1})
1×10^7	0.89	1×10^{16}–1×10^{15}

After Trp60 complete quenching, the effect of the actual NP interaction becomes evident. Accordingly, the protein conformation perturbation induced by MPA-AuNPs leads to a Trp95 fluorescence intensity increase and an emission spectrum shift. This interpretation is supported by the result observed with W60G β2m variant that exhibited only the intensity increase and the shift of the emission band (Figure 7b).

3. Discussion

The interaction of β2m and its variants with AuNPs has been thoroughly investigated by our group [14,15]. We found that the actual interaction between the citrate-coated surface of AuNPs and β2m is essentially of electrostatic nature, although the overall protein charge should be around zero or slightly negative. The protein regions that are involved in the preferential contacts with AuNPs possess a local dipole or polarity distribution that must favor pairing with the NP surface, irrespective of the overall net charge and dipole moment. An overall weak interaction was observed that involved the N-terminal apical part of the protein, in particular, Q2 and R3 in the N-terminus, and K58 and D59 in the DE loop, in good agreement the simulations results [14,15]. In addition, we could also identify other close interaction sites, namely S55 and F56 in DE loop, and residues Y26, S28, G29, F30, and S33 of the adjacent BC loop. Additional residues appeared selectively perturbed when the very amyloidogenic variant D76N β2m was assayed [15]. These additional involvements did not map the protein-NP interface, but rather the protein-protein association equilibria. The citrate coated AuNPs that were employed had average diameters of 7.5 nm at most and we could even test the in vitro amyloidogenesis inhibition of those NP preparations [15]. Larger AuNPs, as checked with thiol-coated ones (unpublished results), do not support β2m stability towards unfolding as the protein precipitates out of solution shortly after preparation. In general, large NPs had been previously shown to rather accelerate β2m fibrillogenesis [24]. For AuNPs, this effect stems from the well-established affinity increase of proteins for gold surfaces with small curvature, i.e., large sphere diameters, that enhances the contacts, thereby destabilizing the protein folding [25].

Now, we have shown that the same regions of β2m, which establish contacts with citrate-coated AuNPs, which support the conformational stability of the protein and even inhibit fibrillogenesis in vitro, remain involved in the interaction also when the citrate is replaced by 3-mercaptopropionate. With this coating, however, the NP dimensions are sensibly reduced. With an average diameter of 3.6 nm, MPA-AuNPs become much more easy to handle for any preparation, keep the protein stable in solution while engaging an efficient fast-exchange mild interaction, and can be exploited to reach larger NP/protein concentration ratios that may be necessary to exploit the efficiency and the mildness of the interaction in vivo.

4. Materials and Methods

4.1. MPA-AuNP Synthesis and Characterization

All the reagents used in the synthesis were purchased from Sigma Aldrich (St. Louis, MO, USA) To a solution of $HAuCl_4 \cdot 3H_2O$ (49.2 mg, 0.125 mmol) in deoxygenated methanol (5 mL) cooled at 0 °C and purged with nitrogen, three equivalents (65.3 μL) of 3-mercaptopropionic acid, dissolved in deoxygenated water (5 mL), were added under stirring. Upon the addition of the alkanethiol, the solution colour changed from yellow to cloudy white. After two hours of stirring, a freshly prepared cooled aqueous solution of $NaBH_4$ (1.25 mmol, 47.3 mg, in 2 mL of water) was dripped inside the gold/thiol solution. One hour later, the brown solution that was obtained was purified by centrifugation (5000 rpm, 15 min at 25 °C), repeating the removal of the supernatant and the dispersion in methanol five times.

For the UV-Vis characterization, a V-750 spectrophotometer (Jasco, Oklahoma City, OK, USA) was used and the spectra were recorded at 20 °C, from 400 nm to 800 nm with a data pitch of 0.2 nm, a scan rate of 200 nm/min and a bandwidth of 2 nm. To prepare the samples for the TEM imaging, a small amount of the nanoparticle solution was dropped on a TEM grid and left for 5 min. Filter paper was used to remove the excess of the solution. TEM images were acquired with EM 208 microscope (Philips, Amsterdam, The Netherlands). The size distribution was calculated by measuring a minimum of 200 particles using ImageJ software (National Institutes of Health, Bethesda, MD, USA). The average number of gold atoms per NP (N_{Au}) was calculated from the following equation:

$$N_{Au} = \pi \rho d^3 / 6M_{Au} = 30.89602d^3,$$ (3)

where d is the nanoparticle diameter expressed in nm, ρ is the density for face-centered cubic gold (19.3 g/cm^3) and M_{Au} stands for the atomic weight of gold (197 g/mol).

The number of ligands that are bound to the gold core was estimated from thermogravimetric analysis (TGA, collected with TGA 8000, Perkin Elmer, Waltham, MA, USA). This experiment gives the percentage of weight loss during a temperature ramp that is addressed exclusively to the burning of the organic component, following solvent removal. Applying the following equation, it is possible to calculate the monolayer composition:

$$N_L = (N_{Au} \cdot M_{Au} \cdot W\%)/((1 - W\%) \, M_{thiolate}), \tag{4}$$

where N_L is the number of ligands, $W\%$ is the percentage of weight loss due to organic ligands burning and $M_{thiolate}$ is the molecular weight of the thiolate molecule. TGA analysis was performed with a SDT Q600 instrument (TA instruments, New Castle, DE, USA) under N_2 at a heating rate of 10 °C/min going from 0 to 600 °C.

4.2. NMR Experiments

NMR experiments were performed on uniformly ^{15}N-labeled β2m wild-type dissolved in HEPES 50 mM pH 7, and diluted to 25 μM with different concentrations of MPA-AuNPs or 2 mM MPA for the control sample. Proteins were expressed and purified, as previously described [14,15]. D_2O (5%) was added to each sample for lock purposes. [15]N-^{1}H SOFAST-HMQC experiments were collected on the Bruker Avance spectrometer (Bruker, Billerica, MA, USA) (-) at the Udine University Biophysics Laboratory, operating at 500 MHz (^{1}H). Experiments were run at 298 K over spectral widths of 30 ppm (^{15}N, t1) and 14 ppm (^{1}H, t2) with 128 and 1024 points, respectively. For each t1 dimension point, 800 or 1600 scans were accumulated. The data were processed with Topspin 2.1 and were analyzed with Sparky. The β2m assignment was based on the file that was deposited on the Biological Magnetic Resonance Data Bank (Accession Code: 17165). Chemical shift perturbations were calculated as $\Delta\delta$ (ppm) $= [(\Delta\delta_H)^2 + (\Delta\delta_N/6.5)^2]^{1/2}$, where $\Delta\delta_H$ and $\Delta\delta_N$ are the chemical shift variations for ^{1}H and ^{15}N, respectively [26,27]. The relative intensities (RI) correspond to the ratio between the signal intensity in presence of NPs and in absence of NPs.

4.3. TEM Imaging of Stained Samples

To prepare samples for the imaging, a small amount of the nanoparticle-β2m solution was dropped on a TEM grid and left for 5 min. Filter paper was used to remove the excess of the solution. The solution was stained with 2% uranyl acetate solution in water for 2 min.

4.4. Fluorescence Experiments

β2m intrinsic fluorescence was recorded in absence of AuNPs and after the progressive addition of small amounts of nanoparticles using a Cary Eclipse Fluorescence Spectrophotometer (Agilent, Santa Clara, CA, USA). For the measurement, fluorescence semi-micro cuvettes were used (5 mm × 5 mm). The samples were excited at 295 nm and the emission was recorded from 300 to 450 nm, using 5 nm slit for both excitation and emission. Each spectrum was the average of 5 consecutive measurements, and three individual experiments were repeated for each sample. The initial quenching data were fitted with the linear Stern-Volmer equation [23]:

$$\frac{F_0}{F} = 1 + K_{SV}[NP] = 1 + k_q\tau_0[NP], \tag{5}$$

where F_0 and F are the fluorescence intensities of the protein in absence and in presence of Cit-AuNPs, respectively, and the Stern-Volmer constant, K_{SV}, is the product of the diffusion-limited bimolecular quenching constant, k_q, and the fluorophore fluorescence lifetime, τ_0.

Acknowledgments: This work received financial support from PRIN project No. 2012A7LMS3. We acknowledge the New York University Abu Dhabi for access to the Core Technology Platform. We also thank Makek A. for the assistance.

Author Contributions: C.C. and G.E. conceived and designed the experiments; C.C. and P.B. performed the experiments; C.C. and G.E. analyzed the data; S.G., P.P.M. and V.B. expressed and purified the proteins; C.C. and G.E. wrote the draft. All authors read, corrected and discussed the manuscript.

Conflicts of Interest: The authors declare no conflict of interest.

References

1. De, M.; Ghosh, P.S.; Rotello, V.M. Applications of nanoparticles in biology. *Adv. Mater.* **2008**, *20*, 4225–4241. [CrossRef]
2. Liu, W.-T. Nanoparticles and their biological and environmental applications. *J. Biosci. Bioeng.* **2006**, *102*, 1–7. [CrossRef] [PubMed]
3. Lynch, I.; Dawson, K.A. Protein-nanoparticle interactions. *Nano Today* **2008**, *3*, 40–47. [CrossRef]
4. Gagner, J.E.; Lopez, M.D.; Dordick, J.S.; Siegel, R.W. Effect of gold nanoparticle morphology on adsorbed protein structure and function. *Biomaterials* **2011**, *32*, 7241–7252. [CrossRef] [PubMed]
5. Deng, Z.J.; Liang, M.; Monteiro, M.; Toth, I.; Minchin, R.F. Nanoparticle-induced unfolding of fibrinogen promotes Mac-1 receptor activation and inflammation. *Nat. Nanotechnol.* **2011**, *6*, 39–44. [CrossRef] [PubMed]
6. Gole, A.; Dash, C.; Ramakrishnan, V.; Sainkar, S.R.; Mandale, A.B.; Rao, M.; Sastry, M. Pepsin-gold colloid conjugates: Preparation, characterization, and enzymatic activity. *Langmuir* **2001**, *17*, 1674–1679. [CrossRef]
7. Bailes, J.; Gazi, S.; Ivanova, R.; Soloviev, M. Effect of gold nanoparticle conjugation on the activity and stability of functional proteins. In *Nanoparticles in Biology and Medicine*; Methods in Molecular Biology; Humana Press: Totowa, NJ, USA, 2012; pp. 89–99, ISBN 978-1-61779-952-5.
8. Murphy, M.P.; LeVine, H. Alzheimer's disease and the β-amyloid peptide. *J. Alzheimers Dis.* **2010**, *19*, 311. [CrossRef] [PubMed]
9. Wu, W.; Sun, X.; Yu, Y.; Hu, J.; Zhao, L.; Liu, Q.; Zhao, Y.; Li, Y. TiO_2 nanoparticles promote β-amyloid fibrillation in vitro. *Biochem. Biophys. Res. Commun.* **2008**, *373*, 315–318. [CrossRef] [PubMed]
10. Kim, J.E.; Lee, M. Fullerene inhibits β-amyloid peptide aggregation. *Biochem. Biophys. Res. Commun.* **2003**, *303*, 576–579. [CrossRef]
11. Giljohann, D.A.; Seferos, D.S.; Daniel, W.L.; Massich, M.D.; Patel, P.C.; Mirkin, C.A. Gold nanoparticles for biology and medicine. *Angew. Chem. Int. Ed.* **2010**, *49*, 3280–3294. [CrossRef] [PubMed]
12. Wang, Z.; Wu, L.; Cai, W. Size-tunable synthesis of monodisperse water-soluble gold nanoparticles with high X-ray attenuation. *Chem. Eur. J.* **2010**, *16*, 1459–1463. [CrossRef] [PubMed]
13. Gejyo, F.; Yamada, T.; Odani, S.; Nakagawa, Y.; Arakawa, M.; Kunitomo, T.; Kataoka, H.; Suzuki, M.; Hirasawa, Y.; Shirahama, T.; et al. A new form of amyloid protein associated with chronic hemodialysis was identified as β2-microglobulin. *Biochem. Biophys. Res. Commun.* **1985**, *129*, 701–706. [CrossRef]
14. Brancolini, G.; Corazza, A.; Vuano, M.; Fogolari, F.; Mimmi, M.C.; Bellotti, V.; Stoppini, M.; Corni, S.; Esposito, G. Probing the influence of citrate-capped gold nanoparticles on an amyloidogenic protein. *ACS Nano* **2015**, *9*, 2600–2613. [CrossRef] [PubMed]
15. Cantarutti, C.; Raimondi, S.; Brancolini, G.; Corazza, A.; Giorgetti, S.; Ballico, M.; Zanini, S.; Palmisano, G.; Bertoncin, P.; Marchese, L.; et al. Citrate-stabilized gold nanoparticles hinder fibrillogenesis of a pathological variant of β2-microglobulin. *Nanoscale* **2017**, *9*, 3941–3951. [CrossRef] [PubMed]
16. Calzolai, L.; Franchini, F.; Gilliland, D.; Rossi, F. Protein−Nanoparticle interaction: Identification of the Ubiquitin−Gold nanoparticle interaction site. *Nano Lett.* **2010**, *10*, 3101–3105. [CrossRef] [PubMed]
17. Becker, J.W.; Reeke, G.N. Three-dimensional structure of β2-microglobulin. *Proc. Natl. Acad. Sci. USA* **1985**, *82*, 4225–4229. [CrossRef] [PubMed]
18. Schanda, P.; Kupče, Ē.; Brutscher, B. SOFAST-HMQC experiments for recording two-dimensional deteronuclear correlation spectra of proteins within a few seconds. *J. Biomol. NMR* **2005**, *33*, 199–211. [CrossRef] [PubMed]
19. Bjorkman, P.J.; Saper, M.A.; Samaroui, B.; Bennet, W.S.; Strominger, J.L.; Wiley, D.C. Structure of the human class I histocompatibility antigen, HLA-A2. *Nature* **1987**, *329*, 506–512. [CrossRef] [PubMed]
20. Esposito, G.; Michelutti, R.; Verdone, G.; Viglino, P.; Hernandez, H.; Robinson, C.V.; Amoresano, A.; Dal Piaz, F.; Monti, M.; Pucci, P.; et al. Removal of the N-terminal hexapeptide from human β2-microglobulin facilitates protein aggregation and fibril formation. *Protein Sci.* **2000**, *9*, 831–845. [CrossRef] [PubMed]
21. Kihara, M.; Chatani, E.; Iwata, K.; Yamamoto, K.; Matsuura, T.; Nakagawa, A.; Naiki, H.; Goto, Y. Conformation of amyloid fibrils of β2-microglobulin probed by tryptophan mutagenesis. *J. Biol. Chem.* **2006**, *281*, 31061–31069. [CrossRef] [PubMed]
22. Van de Weert, M.; Stella, L. Fluorescence quenching and ligand binding: A critical discussion of a popular methodology. *J. Mol. Struct.* **2011**, *998*, 144–150. [CrossRef]
23. Lakowicz, J.R. *Principles of Fluorescence Spectroscopy*; Springer: New York, NY, USA, 2006.

24. Linse, S.; Cabaleiro-Lago, C.; Xue, W.-F.; Lynch, I.; Lindman, S.; Thulin, E.; Radford, S.E.; Dawson, K.A. Nucleation of protein fibrillation by nanoparticles. *Proc. Natl. Acad. Sci. USA* **2007**, *104*, 8691–8696. [CrossRef] [PubMed]
25. Lacerda, S.H.D.P.; Park, J.J.; Meuse, C.; Pristinski, D.; Becker, M.L.; Karim, A.; Douglas, J.F. Interaction of gold nanoparticles with common human blood proteins. *ACS Nano* **2010**, *4*, 365–379. [CrossRef] [PubMed]
26. Mulder, F.A.A.; Schipper, D.; Bott, R.; Boelens, R. Altered flexibility in the substrate-binding site of related native and engineered high-alkaline Bacillus subtilisins. *J. Mol. Biol.* **1999**, *292*, 111–123. [CrossRef] [PubMed]
27. Williamson, M.P. Using chemical shift perturbation to characterise ligand binding. *Prog. Nucl. Magn. Reson. Spectrosc.* **2013**, *73*, 1–16. [CrossRef] [PubMed]

magnetochemistry

MDPI

Review

The NMR² Method to Determine Rapidly the Structure of the Binding Pocket of a Protein–Ligand Complex with High Accuracy

Marielle Aulikki Wälti and Julien Orts *

Laboratorium für Physikalische Chemie, ETH Zürich, Vladimir-Prelog-Weg 2, 8093 Zürich, Switzerland; marielle.walti@nih.gov
* Correspondence: julien.orts@phys.chem.ethz.ch; Tel.: +41-(0)-44-632-28-64

Received: 22 November 2017; Accepted: 4 January 2018; Published: 22 January 2018

Abstract: Structural characterization of complexes is crucial for a better understanding of biological processes and structure-based drug design. However, many protein–ligand structures are not solvable by X-ray crystallography, for example those with low affinity binders or dynamic binding sites. Such complexes are usually targeted by solution-state NMR spectroscopy. Unfortunately, structure calculation by NMR is very time consuming since all atoms in the complex need to be assigned to their respective chemical shifts. To circumvent this problem, we recently developed the Nuclear Magnetic Resonance Molecular Replacement (NMR²) method. NMR² very quickly provides the complex structure of a binding pocket as measured by solution-state NMR. NMR² circumvents the assignment of the protein by using previously determined structures and therefore speeds up the whole process from a couple of months to a couple of days. Here, we recall the main aspects of the method, show how to apply it, discuss its advantages over other methods and outline its limitations and future directions.

Keywords: complex structure; drug design; NMR spectroscopy; NMR²; structure elucidation

---

1. Introduction

1.1. Structure-Based Drug Design

Most biological processes rely on highly specific protein–protein or protein–ligand inter-molecular interactions. Understanding and manipulating these interactions is the ultimate goal of drug design. Drug research, as we know it today, dates back nearly 100 years with advances in chemistry, including Avogadro's atomic hypothesis, the benzene theory, and the ability to isolate and purify active ingredients from pharmaceutical plants [1]. The finding of active components was initially a serendipitous accident; a famous example is the discovery of penicillin by Alexander Fleming. However, while the ligand was found to have a specific effect, the target receptor(s) remained unknown. The search for its target was very time consuming, and the idea of rational design emerged as a possible solution to speed up the process. Drug research based on the structure–activity relationship, where a molecule is designed to specifically inhibit or promote an interaction, was augmented when X-ray crystallography started to be used to derive protein structures; the protein whose structure was first determined by X-ray was myoglobin and led to a Nobel Prize in 1962. Nowadays, drug discovery commonly starts by screening large libraries of molecules or fragments against a carefully selected drug target, with identified binders further optimized by molecular refinement or fragment-based design approaches. This approach was enabled by advances in biology (e.g., biochemistry, molecular biology and genomics) that drive the search for better drug targets. Further, progress in chemistry and bioinformatics allowed for the synthesis and screening of enormous compound libraries. These methods, however, are very error prone and require validation, preferentially by a complex structure at atomic resolution. To obtain atomic-level structures, X-ray crystallography is still the most widely used method, followed by NMR spectroscopy and cryo-electron microscopy. The latter method is quickly developing (Nobel Prize in chemistry 2017) and shows great potential in drug discovery for large systems. Of particular interest are methods that combine several approaches, including the Nuclear Magnetic Resonance Molecular Replacement (NMR^2) method (Figure 1).

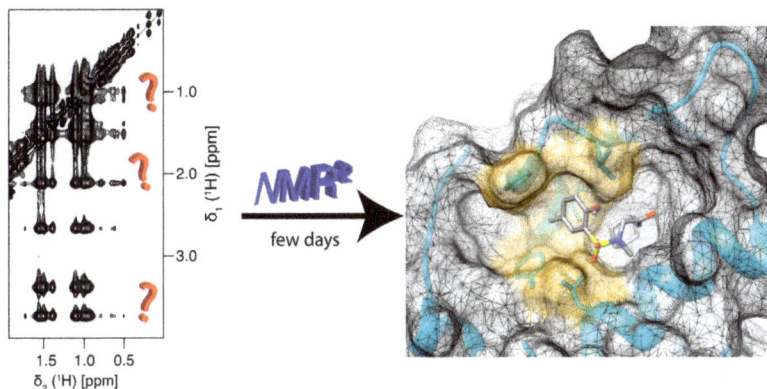

Figure 1. Nuclear Magnetic Resonance Molecular Replacement (NMR^2) derives the complex structure of the binding site within a few days without protein resonance assignment and using only standard 2D NMR experiments.

1.2. The NMR^2 Method

NMR spectroscopy is often the only method able to determine complex structures with ligands, which are typically either part of very dynamic interactions or in fast exchange. Unfortunately, NMR is rather slow in structure determination, since all atoms must be assigned to their respective chemical shifts, which requires long measurements and intensive analysis. However, most often the information in the binding site instead of the whole protein is of interest. In those cases, NMR^2 represents a good alternative. NMR^2 utilizes exact spatial information provided by solution-state NMR to locate and refine the binding pocket of the complex structure using an independent starting model of the receptor (e.g., X-ray structure of a homolog), and performs this analysis without the need for protein resonance assignment. NMR^2 has successfully determined several structures of complexes very accurately (within 1 Å) with only a few days of measurement and calculation time.

1.3. The NMR^2 Protocol

To successfully use NMR^2, the following steps are required (Figure 2):

(i)   sample preparation for NMR measurements: uniformly $^{13}C,^{15}N$ labeled, or selective labeling schemes (e.g., isoleucine, leucine, and valine methyl labeling) can be used for the protein [2]. This can be achieved by recombinant expression, e.g., in *E. coli* [3]. Only one of the two molecules in the complex should be isotopically labeled. For strong binders, i.e., low μM and higher affinity ($k_{off} < \Delta CS$ and $k_{off} < \sigma$, where k_{off} represents the dissociation rate, ΔCS the chemical shift difference of the bound and free states, and σ the cross-relaxation rate), an equimolar ligand to protein ratio is optimal; whereas for weak binders, i.e., high μM and lower affinity ($k_{off} > \Delta CS$), an excess of ligand is required to saturate the receptor as much as possible. This can be monitored by so-called chemical shift mapping experiments, where the ligand is titrated to the protein and binding is detected through perturbation of the backbone NH chemical shifts of the receptor in ^{1}H,^{15}N-HSQC or TROSY experiments [4–7]. Knowing the affinity of the small molecule for its receptor, the protein saturation can be calculated with the following formula:

$$\frac{[PL]}{[P]_{tot}} = \frac{[L]_{tot} + [P]_{tot} + K_D - \sqrt{([L]_{tot} + [P]_{tot} + K_D)^2 - 4\,[P]_{tot}\,[L]_{tot}}}{2\,[P]_{tot}}$$

where PL, L, P and K_D are the concentration of the complex, the concentration of the ligand, the concentration of the protein, and the affinity of the ligand for the protein. The subscript 'tot' stands for total concentration.

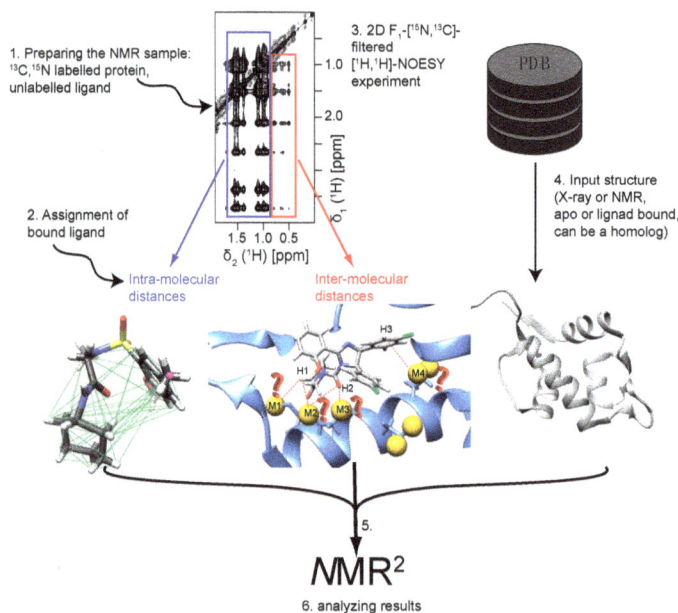

Figure 2. Overview of the NMR^2 method. The following steps are required for NMR^2 to determine the complex structure of the binding pocket: (i) Sample preparation for NMR measurements; (ii) Recording experiments to assign the ligand; (iii) Measurement of the ligand intra- and ligand–protein inter-molecular distances; (iv) Choosing the input structure; (v) Running NMR^2; (vi) Analyzing the results.

(ii)   Recording experiments to assign the ligand. Usually standard NMR spectra are sufficient to assign the compound in the bound state, e.g., any combination of ^{13}C 1D, 1D DEPT-90, and 1D DEPT-135 spectra [8], 2D ^{13}C,^{1}H-HMQC [9], 2D ^{13}C,^{1}H-HMBC, 2D ^{1}H,^{1}H-DQF COSY [10,11], F_1,F_2-^{15}N,^{13}C-filtered ^{1}H,^{1}H-TOCSY, or 2D F_1,F_2-^{15}N,^{13}C-filtered ^{1}H,^{1}H-NOESY spectra [12–21].

(iii)   Measurement of the ligand intra- and ligand–protein inter-molecular distances. All distance restraints for NMR^2 are derived from NOE (nuclear Overhauser enhancement) cross-peaks of F_1-^{15}N,^{13}C-filtered ^{1}H,^{1}H-NOESY spectra [16–21]. These experiments suppress the intra-molecular NOEs peaks from the receptor and render the spectra easier to interpret. In theory, any moiety of the receptor can be analyzed, but to reduce the ambiguity of possible options, the NOEs should be assigned to methyls, amides, or aromatics with respect to their chemical shifts. Focusing only on distinct groups of resonances in the receptor helps to minimize the computational time of the structure calculation. Using methyl groups was so far successful for all complexes. In addition, the NOESY mixing times have to be chosen carefully. The optimal mixing times for the NOE build-ups depend on the correlation time of the complex. Too short of a mixing time would not allow for enough transfer of magnetization and inter-molecular NOE peaks will stay weak or below the noise level. Too long of a mixing time would increase spin diffusion and lead to large signal intensities, but these would require heavy calculations to translate into meaningful distances. In general, NOESY mixing times between 40 and 150 ms are reasonable for a 15–20 kDa protein, exhibiting a correlation time of approximately 10 ns.

The slope of the linear growth of the NOE build-up curve contains the information about inter-protons distances. Under the assumption of an isolated spin-pair system, the inter-molecular NOE cross-peak intensity, $\Delta M_{ij}(t)$, is

$$\frac{\Delta M_{ij}(t)}{\Delta M_{ii}(0)} = -\frac{\sigma_{ij}}{\lambda_+ - \lambda_-}\left(e^{-\lambda_- t} - e^{-\lambda_+ t}\right) \tag{1}$$

$$\frac{\Delta M_{ii}(t)}{\Delta M_{ii}(0)} = e^{-\rho_i t} \tag{2}$$

$$\lambda_\pm = \frac{\rho_i + \rho_j}{2} \pm \sqrt{\left(\frac{\rho_i - \rho_j}{2}\right)^2 + \sigma_{ij}^2} \tag{3}$$

$$\sigma_{ij} = \frac{b^2}{r_{ij}^6}\left(6J(2\omega) - J(0)\right) \tag{4}$$

$$J(\omega) = \frac{2}{5}\left(\frac{\tau_c}{1 + (\omega\tau_c)^2}\right) \tag{5}$$

$$b = \frac{1}{2}\frac{\mu_0}{4\pi}\hbar\gamma_H^2 \tag{6}$$

where ρ_i is the auto-relaxation rate of the proton i, $\Delta M_{ii}(0)$ the initial magnetization, σ_{ij} the cross-relaxation rate, r_{ij} the proton(i)–proton(j) distance, μ_0 the permeability of free space, $\hbar$ the reduced Planck constant, γ_H the gyromagnetic ratio of the proton and τ_c the rotational correlation time of the protein–ligand complex [22,23].

From Equations (1) and (2), we can derive σ_{ij} given that fitting the decays of the ligand diagonal peaks provides the auto-relaxation rates and the initial magnetization. If the auto-relaxation rates of the protein groups are missing, because the protein diagonal peaks are suppressed from the F_1-$^{15}N,^{13}C$-filtered $^1H,^1H$-NOESY, the median of other groups is a good estimate. The fits can be made using general software such as matlab, python, and R or using the previously published eNORA software that contains an applet for fitting NOE build-up curves [24,25]. The influence of a slightly incorrect auto-relaxation rate on the inter-proton distance, r_{ij}, is negligible. However, the initial magnetization is crucial because it is directly multiplied with the cross-relaxation rate. After the fitting of all build-up and decay curves, we can derive a set of intra-ligand and inter-protein–ligand cross-relaxation rates that need to be converted into distances.

To convert cross-relaxation rates into distances, the following has to be kept in mind: in the case of a strong binder, slow exchange regime on the NMR time scale, the correlation time of the complex is the same as the one of the protein, since the influence of the small molecule on the tumbling of the protein can be neglected. In this case, Equations (1)–(6) can be readily used.

In the case of a weak binder, fast exchange regime on the NMR time scale, the effective cross-relaxation rate is the population average between the free and bound states of the ligand [26,27]:

$$\sigma_{eff} = p_{free}\,\sigma_{free} + p_{bound}\,\sigma_{bound} \tag{7}$$

Since the correlation time of the ligand is on the order of picoseconds, the first term can be neglected and, as mentioned above, the correlation time of the complex can be displayed as the correlation time of the protein. Consequently, the effective cross-relaxation rate is defined by the bound population of the ligand and the correlation time of the protein:

$$\sigma_{eff} \approx p_{bound}\,\sigma_{protein} \tag{8}$$

Finally, the correlation time of the protein can be determined by standard ^{15}N relaxation experiments and used to convert the cross-relaxation rates to distances using Equations (4–6) [28].

A second way to derive distances from cross-relaxation rates is by using the known fixed intra-molecular distances within the ligand (e.g., protons in an aromatic ring) to calibrate the cross-relaxation rates and derive all other distances using Equation (4):

$$r_{ij} = \left(\frac{Constant}{\sigma_{i,j}}\right)^{\frac{1}{6}} \tag{9}$$

$$Constant = \sigma_{fixed}\,r_{fixed}^{6} \tag{10}$$

As a rule of thumb, the intra-ligand distances should be slightly shorter than the inter-molecular distances and the median value of all distance should be around ~4.0–4.2 Å, while the median distance of the intermolecular distances is around of 4.4 Å.

(iv) Choosing the input structure. As an input structure, the protein in its apo form, with another bound ligand, or a homolog can be used to derive a starting model of the receptor. Either X-ray or NMR structures can be provided. In the current state of the program, the user should prepare the following input files: a CYANA-regularized protein PDB file, a ligand CYANA library file that can be generated with the program cylib [29], a sequence file containing the amino acid residues of the protein followed by sufficient linker residues (long enough so that the ligand can access all the protein surface) and the ligand residue name as defined in the ligand library file. All these files are needed to produce the starting structure of the complex where the protein structure is identical to the chosen receptor and the ligand is randomly positioned in space but attached to the protein by the linker. Further details can be found in the CYANA manual.

(v) Running NMR^2. The NMR^2 program screens all possible assignment moieties (usually methyl groups) of the protein and calculates the complex structures for all options. However, it is crucial to diminish the number of options in order to complete the calculations in a reasonable amount of time. This is achieved primarily by using only a fraction of the inter-molecular distances in the first calculation cycle, where only around 3–4 methyl groups of the protein are taken into account. The use of an input structure, the previously derived network of inter-molecular distances, and the use of triangle or tetra angle smoothing to rule out most of the false assignment possibilities are equally important for a manageable calculation time. As of now, NMR^2 is a CYANA-based program and calculates all structures using the standard simulated annealing protocol [30]. The results are scored with respect to the target function, which represents a measure of how well the calculated structure fulfills the data. CYANA is the most widely used NMR structure calculation program, which is solely based on experimental data and the repulsive part of the van der Waals potential modeling the atom radii. No other force field is used and therefore the electrostatic potential of the molecules is not modelled. Nonetheless, if specific interactions are known or determined by experiments they can be added following the program syntax [30]. Only the best structures are kept for the next calculation cycle where more methyl groups with their respective inter-molecular distances are included. The calculation is finished when all experimental data have been used.

(vi) Analyzing the results. The final complex structures have to be analyzed carefully to detect potential errors. NMR^2 requires a definition of the receptor flexibility; however, if there are no restraints on backbone and side chain atoms, the protein will freely move to fulfill the distance restraints, which could potentially yield false positives. Another source of false positives is when the ligand finds its binding site at the N- or C-terminus of the protein or where the protein atom density is the lowest. There, the ligand can freely adopt its position and orientation to fulfill the distance restraints because little or no steric inter-molecular interactions are present. One should keep in mind that this is happening only if the protein contains methyl groups at these sites.

Finally, the quantity and quality of inter-molecular distances are critical. While theoretical considerations indicate that a maximum of six distances should, in principle, be sufficient, practically we observe that ~12–15 distances are the minimum needed to calculate a NMR^2 structure (vide infra).

The NMR spectra should have a high signal to noise ratio as well as good resolution. They should also be free from water suppression artifacts, e.g., so-called pulse trains water suppression 'w5' or excitation sculpting that strongly modify nearby peak intensities [31].

1.4. Current Applications of NMR^2

NMR^2 has been successfully applied to calculate complex structures containing ligands in fast and slow exchange (Figure 3) [32,33]. Structures containing strong binders where the ligand is a peptidomimetic (MDMX-comp2, Figure 3a) or a small compound (HDM2-pip, HDM2-nutlin complex, Figure 3b,c) have been determined with an accuracy relative to reference structure of 0.9–1.5 Å. Presently, the receptors were up to 32.1 kDa in size, exemplified by ABL kinase-destatinib (in silico data) where the NMR^2-derived structure has a root-mean-square deviation (RMSD) of 1.1 Å to the previously published complex (Figure 3d). For ligands in fast exchange, so far two structures have been determined (MDMX-SJ212 and HDM2-#845, Figure 3e,f). The NMR^2-derived SJ212-MDMX structure is consistent with the previously published complex structure with an RMSD of 1.35 Å. HDM2-#845 represents a new complex, where no previous structure existed. A thorough validation of the NMR^2 structure was performed showing the correctness of the structure with 3D $^{15}N,^{13}C$-resolved $^1H,^1H$-NOESY, and F_1- $^{13}C,^{15}N$-filtered 3D N-resolved $^1H,^1H$-NOESY-HSQC, Saturation Transfer Difference (STD) experiments and chemical shift perturbations [33].

Strong binding ligands

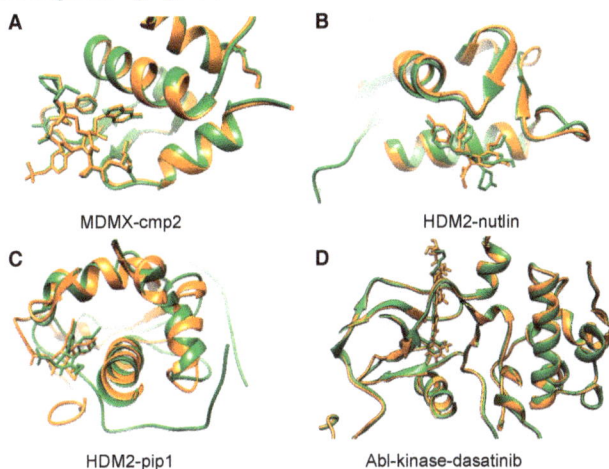

Figure 3. *Cont.*

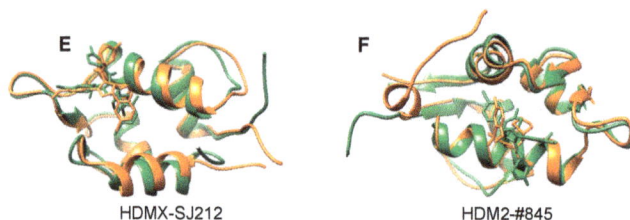

Figure 3. All complex structures so far solved by NMR^2. They consist of four high-affinity ligands (**a**–**d**) and two low-affinity ones (**e**,**f**), all of which are consistent with previously published structures (**a**) 3fea [34] with an RMSD of 1.1 Å, (**b**) 5c5a [32] with an RMSD of 0.9 Å, (**c**) 2lzg [35] with an RMSD of 1.5 Å, (**d**) 2gqg [36] with an RMSD of 1.1 Å (in silico data), and (**e**) 2n0w [37] with an RMSD of 1.35; (**f**) represents a complex with a ligand having a new scaffold, where no other structural data were known, and therefore it is compared to the complex structure with nutlin (5c5a). In orange are the NMR^2-derived structures and in green the reference structures.

How many distances are needed to successfully run NMR^2 depends on the complex structure: How large and well defined is the binding pocket, how flexible is the ligand, etc. For the previously published complexes, the ligands in slow exchange contained 16–23 inter-molecular restraints or 29 in silico restraints (for the ABL kinase-destatinib), all of which comprise distances between methyl groups of the receptor and the ligand protons. For the weakly binding ligands, 14 and 21 inter-molecular distances between the ligand and the receptor were collected, with most distances involving methyl groups of the receptor. However, in the case of HDM2-#845, one distance was included in either an amide or aromatic group of the protein.

Choosing the right input structure for NMR^2 is not very critical. In the example of HDMX in complex with cmpd2, it was shown that the input structure can be either the apo-protein or a structure with another ligand, or from a homolog. The input structures can also be determined by NMR or X-ray [32]. Remarkably, NMR^2 also succeeded in finding the right complex structure of the binding site using an apo-protein as the input structure, wherein the ligand binding site was closed by one receptor helix. This case was very challenging, since the receptor undergoes an allosteric conformational change upon ligand binding, which moves the helix away from the binding site. During the NMR^2 calculations, enough flexibility was given in the loops, with the helices and β-sheets being constrained by hydrogen bonds, and finally yielded to a NMR^2 structure with an RMSD of 1.8 Å to the previously published structure.

1.5. NMR^2 versus Other Methods for Rapid Structure Calculations of Protein–Ligand Complexes

Most complex structures are analyzed by X-ray crystallography due to its speed and high degree of automation. However, weak binders often do not crystalize well. Furthermore, X-ray does not contain information on dynamics, and crystal packing can lead to artifacts. The latter is demonstrated in the case of HDM2-nutlin where the NMR^2 structure is different compared to previously published structures (PDB: 4hg7, 4e3j) that contain crystal packing artifacts, but matches perfectly the artifact-free structure, 5c5a [32]. In cases involving weak binders, NMR spectroscopy is currently the best method to provide high resolution structural data. Recently, attempts to derive structures and/or dynamics of protein–ligand complexes by NMR more efficiently, when compared to the traditional structure calculation protocol, have been proposed including the use of ambiguous restraints [38,39] (such as ambiguous NOEs), chemical shift perturbations [40–44], or saturation transfer experiments [45,46] in combination with computational methods such as docking and scoring [47–49]. Here, we describe the advantages and disadvantages of NMR^2 over some of the most commonly used techniques to quickly determine complex structures by NMR. The methods can be divided into two main classes: the data are derived from chemical shift perturbation (CSP) or NOEs.

Methods using CSP usually record an ^{1}H,^{15}N-HSQC spectrum of the apo-receptor, where each peak corresponds to one amino acid of the backbone. The ligand is then titrated and residues in close proximity to the interaction site are perturbed. These shifts are remarkably large when caused by ring currents produced by aromatic moieties in the ligand. While CSP is difficult to quantitatively interpret, progress in simulations and correlating shifts with secondary and tertiary structure has made it possible to transfer chemical shifts into structural restraints [50–52]. This made a more quantitative interpretation of CSP possible [6,44,53–58]. One example of quantitative CSP is the J-surface-based method: it uses the finding that most of the drugs have aromatic rings involved in the binding (95% in one major drug design database [43]) and that the chemical shift difference due to ring current shift can be converted into a distance [59]. This information is used to construct a so-called J-surface, designed from spheres of the distances, where the ligand could be located. The intersection of the spheres from all of the shifted protons represents the ring location. Because of the complexity of the chemical shifts' dependence with respect to the structure of the complex, the structure prediction initially requires a spatial sampling and scoring step to define the ligand binding site (high density region of the J-surface). This is subsequently followed by an experimentally restraint-based optimization of the ligand binding mode.

An advantage of CSP-based methods over NMR^2 is that, in many cases, CSP is detectable even when no inter-nuclear NOEs are observable [53]. Poor solubility, low affinity, conformational variation of the ligand or few protons in the ligand are the most common difficulties that limit the detection of NOEs.

The disadvantages of CSP are that the protein backbone resonances have to be known for the free and the bound state, which can be very time consuming or sometimes not possible. The latter can occur when the protein undergoes chemical exchange in the intermediate regime, which leads to severe intensity loss of the amide resonances, like in the case of the apo-HDMX. Furthermore, chemical shifts are generally measured for the protein backbone atoms, but usually side chains (such as methyl groups) are primarily involved in binding of the ligand. Note, the CSP method works also on shifts on side chain atoms; however, this would require resonance assignment of the whole protein and is therefore usually not performed. Additionally, CSP works best for weak binders in fast exchange with the receptor (usually K_D weaker than 1 μM) so that the resonances can be followed during a titration. Additionally, CSP will not be treated differently for ligands with slight chemical modifications. This is a clear drawback since often already small chemical modification in the ligand can induce a change in its orientation. Finally, the CSP-based methods also use a docking scoring protocol that relies on force fields or scoring function. The most popular program used in NMR is CSP-HADDOCK [40,47,60], which can make use of a large set of additional experimental restraints such as residual dipolar couplings or pseudocontact shifts [41,61,62]. Other docking programs are BiGGER [63], AutoDockFilter [64], SAMPLEX [65], and LIGDOCK [47].

The second class of protein–ligand structure determination methods involves the usage of NOEs or spin diffusion as experimental restraints. Example methods include SOS-NMR [45,47], NOE matching [49], INPHARMA [48,66,67], CORCEMA [46,68,69], or NMR^2 [32,33]. Except for NMR^2, these methods require a docking step prior to the experimentally based scoring of the found poses and eventually perform an experimentally based refinement step. For example, the NOE matching method generates trial ligand binding poses (e.g., from docking), uses them to back predict the 3D ^{13}C-edited-^{13}C,^{15}N-filtered HSQC-NOESY spectrum and scores each complex with respect to how well its back predicted spectrum matches the measured data. This method has the same advantages as NMR^2: there is no need for protein resonance assignment and one sample is sufficient for these studies. Similarly, as is the case for NMR^2, NOE matching is applicable for ligands in fast and slow exchange. One limiting factor is the strong dependence on the input binding poses. The true binding poses have to be sampled in the first place in order to be found by the program.

SOS-NMR (structural information using Overhauser effects and selective labeling) utilizes STD NMR on many ligand–protein complexes where the receptor is labeled specifically on certain amino acid types while the rest of the receptor is deuterated. With this approach, STD shows the contacts to the specific amino acid types in the receptor and the NOEs derive the respective distances. SOS-NMR gives the amino acid composition of the ligand binding site and, if an input structure of the receptor is available, leads to the 3D structure of the complex. The advantages are that no protein resonance assignment is necessary, only a very little amount of protein (less than 1 mg) is needed, and it is applicable for high molecular weight targets since only the free ligand is detected. The disadvantages are that many samples are required using specific labeling schemes, which may be tedious and it needs a prior docking step of the ligand into the binding site, such as DOCK [70,71].

CORCEMA [46,68,69] and INPHARMA [48,67,72,73] are methods that back predict intra-ligand, intra-protein and protein–ligand NOEs or spin diffusion using the full relaxation matrix formalism. They are powerful tools that can also handle systems undergoing multistate conformational exchange and chemical exchange between the free and bound states. Protein resonance assignment is not required but input structures of the complex should be provided as well as the exchange rates and the correlation time of the complex. The INPHARMA method additionally requires two ligands that compete for the same binding site. As for the other methods, the back predicted data are compared to the experimental data to assess the quality of the docking poses.

To summarize, NMR^2 is currently a purely NOE-based method and requires at least ~12–15 inter-molecular NOEs. This is a limiting factor, especially for low-affinity binders, which may lack enough or sufficiently strong inter-molecular NOEs. Furthermore, NMR^2 is not applicable for completely unknown complexes or protein families, since it requires an input structure. NMR^2 is applicable to most exchange regimes, the only limit being the case of severe exchange-broadening. The main advantages are that it does not need any protein resonance assignment, relies on simple and interpretable NMR experiments, requires only one sample, and performs standard NMR structure calculations instead of relying on docking poses. It provides the full structure of the complex of the binding site with high accuracy, since the distance restraints are based on accurate NOEs [24,74,75]. Additionally, it is applicable to weak and strong binders in fast or slow exchange and the method is fast.

2. Conclusions and Outlook

X-ray crystallography molecular replacement [76] is the prime method used to establish structure–activity relationships of relevant small molecules [77]. Such an approach was not feasible by NMR, as NMR structure determination relies on the assignment of the protein resonances, which can be extremely long and tedious [28]. In recent decades, various methods have been developed in order to derive protein–ligand complex structures faster than with the classical NMR structure calculation protocol, but these methods mostly rely on a preliminary docking step rather than on experimentally driven calculations. Moreover, sometimes partial resonance assignments of the receptor are required [39–45,47,49,72]. A complex structure calculation method that is based on defined and accurate NOEs [78–80] but also bypasses the long and tedious protein assignment step was missing. Therefore, the NMR molecular replacement method (NMR^2), a new molecular replacement-like approach in NMR, allows for the fast determination of protein–ligand complex structures and fills an important gap in structural biology. NMR^2 yielded the structures of ligand (peptide and small molecule)/protein complexes with an accuracy of 1 Å. It requires the measurement of few accurate inter-molecular distances and only a model of the protein receptor. It is a highly efficient way to determine protein/ligand complex structures, without the need to perform the tedious protein resonance assignment, and structures can be calculated within a couple of days. The method was demonstrated on several different complexes with strong or weak binders and will potentially compete with X-ray crystallography for rapid complex structure determination. Furthermore, the development of specific methyl labelling schemes and automatic methyl resonance assignment methods have opened an avenue toward the study of large molecular complexes [81–83]. Since our method strongly relies on sharp methyl NMR signals, the path to structure-based drug design on a large system, where classical NMR methods are limited, is wide open. We foresee great potential for our NMR Molecular Replacement method in drug discovery research where structural information is the gold standard for rational design of new active molecules.

Acknowledgments: We would like to thank Reid Alderson for proofreading this review. This work was supported by ETH Zürich. M.A.W. was supported by an Early Postdoc. Mobility Fellowship from the Swiss National Science Foundation.

Conflicts of Interest: The authors declare no conflict of interest.

References

1. Drews, J. Drug discovery: A historical perspective. *Science* **2000**, *287*, 1960–1964. [CrossRef] [PubMed]
2. Tugarinov, V.; Kay, L.E. Methyl groups as probes of structure and dynamics in nmr studies of high-molecular-weight proteins. *Chembiochem* **2005**, *6*, 1567–1577. [CrossRef] [PubMed]
3. Mcintosh, L.P.; Dahlquist, F.W. Biosynthetic incorporation of n-15 and c-13 for assignment and interpretation of nuclear-magnetic-resonance spectra of proteins. *Q. Rev. Biophys.* **1990**, *23*, 1–38. [CrossRef] [PubMed]
4. Pervushin, K.; Riek, R.; Wider, G.; Wuthrich, K. Attenuated t-2 relaxation by mutual cancellation of dipole-dipole coupling and chemical shift anisotropy indicates an avenue to nmr structures of very large biological macromolecules in solution. *Proc. Natl. Acad. Sci. USA* **1997**, *94*, 12366–12371. [CrossRef] [PubMed]
5. Ziarek, J.J.; Peterson, F.C.; Lytle, B.L.; Volkman, B.F. Binding site identification and structure determination of protein-ligand complexes by nmr: A semiautomated approach. *Methods Enzymol.* **2011**, *493*, 241–275. [PubMed]
6. Williamson, M.P. Using chemical shift perturbation to characterise ligand binding. *Prog. Nucl. Mag. Reson. Spectrosc.* **2013**, *73*, 1–16. [CrossRef] [PubMed]
7. Zuiderweg, E.R.P. Mapping protein-protein interactions in solution by nmr spectroscopy. *Biochemistry* **2002**, *41*, 1–7. [CrossRef] [PubMed]
8. Bendall, M.R.; Doddrell, D.M.; Pegg, D.T. Editing of c-13 nmr-spectra—A pulse sequence for the generation of subspectra. *J. Am. Chem. Soc.* **1981**, *103*, 4603–4605. [CrossRef]
9. Bax, A.; Griffey, R.H.; Hawkins, B.L. Correlation of proton and n-15 chemical-shifts by multiple quantum nmr. *J. Magn. Reson.* **1983**, *55*, 301–315. [CrossRef]
10. Aue, W.P.; Bartholdi, E.; Ernst, R.R. 2-dimensional spectroscopy—Application to nuclear magnetic-resonance. *J. Chem. Phys.* **1976**, *64*, 2229–2246. [CrossRef]
11. Piantini, U.; Sorensen, O.W.; Ernst, R.R. Multiple quantum filters for elucidating nmr coupling networks. *J. Am. Chem. Soc.* **1982**, *104*, 6800–6801. [CrossRef]
12. Ni, F. Complete relaxation matrix analysis of transferred nuclear overhauser effects. *J. Magn. Reson.* **1992**, *96*, 651–656. [CrossRef]
13. Ni, F. Recent developments in transferred noe methods. *Prog. Nucl. Magn. Reson. Spectrosc.* **1994**, *26*, 517–606. [CrossRef]
14. Ni, F.; Zhu, Y. Accounting for ligand-protein interactions in the relaxation-matrix analysis of transferred nuclear overhauser effects. *J. Magn. Reson. Ser. B* **1994**, *103*, 180–184. [CrossRef]
15. Berger, S.; Braun, S. *200 and More Nmr Experiments: A Practical Course*; Wiley: Weinheim, Germany, 2004.
16. Breeze, A.L. Isotope-filtered nmr methods for the study of biomolecular structure and interactions. *Prog. Nucl. Magn. Reson. Spectrosc.* **2000**, *36*, 323–372. [CrossRef]
17. Iwahara, J.; Wojciak, J.M.; Clubb, R.T. Improved nmr spectra of a protein-DNA complex through rational mutagenesis and the application of a sensitivity optimized isotope-filtered noesy experiment. *J. Biomol. NMR* **2001**, *19*, 231–241. [CrossRef] [PubMed]
18. Kogler, H.; Sorensen, O.W.; Bodenhausen, G.; Ernst, R.R. Low-pass j-filters—Suppression of neighbor peaks in heteronuclear relayed correlation spectra. *J. Magn. Reson.* **1983**, *55*, 157–163. [CrossRef]
19. Ogura, K.; Terasawa, H.; Inagaki, F. An improved double-tuned and isotope-filtered pulse scheme based on a pulsed field gradient and a wide-band inversion shaped pulse. *J. Biomol. NMR* **1996**, *8*, 492–498. [CrossRef] [PubMed]

20. Otting, G.; Wuthrich, K. Heteronuclear filters in 2-dimensional [h-1, h-1] nmr-spectroscopy—Combined use with isotope labeling for studies of macromolecular conformation and intermolecular interactions. *Q. Rev. Biophys.* **1990**, *23*, 39–96. [CrossRef] [PubMed]

21. Zwahlen, C.; Legault, P.; Vincent, S.J.F.; Greenblatt, J.; Konrat, R.; Kay, L.E. Methods for measurement of intermolecular noes by multinuclear nmr spectroscopy: Application to a bacteriophage lambda n-peptide/boxb rna complex. *J. Am. Chem. Soc.* **1997**, *119*, 6711–6721. [CrossRef]

22. Macura, S.; Ernst, R.R. Elucidation of cross relaxation in liquids by two-dimensional nmr-spectroscopy. *Mol. Phys.* **1980**, *41*, 95–117. [CrossRef]

23. Solomon, I. Relaxation processes in a system of 2 spins. *Phys. Rev.* **1955**, *99*, 559–565. [CrossRef]

24. Orts, J.; Vogeli, B.; Riek, R. Relaxation matrix analysis of spin diffusion for the nmr structure calculation with enoes. *J. Chem. Theory Comput.* **2012**, *8*, 3483–3492. [CrossRef] [PubMed]

25. Strotz, D.; Orts, J.; Chi, C.N.; Riek, R.; Vogeli, B. Enora2 exact noe analysis program. *J. Chem. Theory Comput.* **2017**, *13*, 4336–4346. [CrossRef] [PubMed]

26. Reuben, J.; Fiat, D. Nuclear magnetic resonance studies of solutions of rare-earth ions and their complexes. 4. Concentration and temperature dependence of oxygen-14 transverse relaxation in aqueous solutions. *J. Chem. Phys.* **1969**, *51*, 4918–4927. [CrossRef]

27. Lippens, G.M.; Cerf, C.; Hallenga, K. Theory and experimental results of transfer-noe experiments. 1. The influence of the off rate versus cross-relaxation rates. *J. Magn. Reson.* **1992**, *99*, 268–281. [CrossRef]

28. Cavanagh, J.; Fairbrother, W.J.; Palmer, A.G.; Rance, M.; Skelton, N.J. Protein nmr spectroscopy principles and practice second edition preface. In *Protein Nmr Spectroscopy: Principles and Practice*, 2nd ed.; Academic Press: Burlington, MA, USA, 2007; pp. V–VI.

29. Yilmaz, E.M.; Guntert, P. Nmr structure calculation for all small molecule ligands and non-standard residues from the pdb chemical component dictionary. *J. Biomol. NMR* **2015**, *63*, 21–37. [CrossRef] [PubMed]

30. Guntert, P.; Mumenthaler, C.; Wuthrich, K. Torsion angle dynamics for nmr structure calculation with the new program dyana. *J. Mol. Biol.* **1997**, *273*, 283–298. [CrossRef] [PubMed]

31. Liu, M.L.; Mao, X.A.; Ye, C.H.; Huang, H.; Nicholson, J.K.; Lindon, J.C. Improved watergate pulse sequences for solvent suppression in nmr spectroscopy. *J. Magn. Reson.* **1998**, *132*, 125–129. [CrossRef]

32. Orts, J.; Walti, M.A.; Marsh, M.; Vera, L.; Gossert, A.D.; Guntert, P.; Riek, R. Nmr-based determination of the 3d structure of the ligand-protein interaction site without protein resonance assignment. *J. Am. Chem. Soc.* **2016**, *138*, 4393–4400. [CrossRef] [PubMed]

33. Walti, M.A.; Riek, R.; Orts, J. Fast nmr-based determination of the 3d structure of the binding site of protein-ligand complexes with weak affinity binders. *Angew. Chem. Int. Ed.* **2017**, *56*, 5208–5211. [CrossRef] [PubMed]

34. Kallen, J.; Goepfert, A.; Blechschmidt, A.; Izaac, A.; Geiser, M.; Tavares, G.; Ramage, P.; Furet, P.; Masuya, K.; Lisztwan, J. Crystal structures of human mdmx (hdmx) in complex with p53 peptide analogues reveal surprising conformational changes. *J. Biol. Chem.* **2009**, *284*, 8803–8812. [CrossRef] [PubMed]

35. Michelsen, K.; Jordan, J.B.; Lewis, J.; Long, A.M.; Yang, E.; Rew, Y.; Zhou, J.; Yakowec, P.; Schnier, P.D.; Huang, X.; et al. Ordering of the n-terminus of human mdm2 by small molecule inhibitors. *J. Am. Chem. Soc.* **2012**, *134*, 17059–17067. [CrossRef] [PubMed]

36. Tokarski, J.S.; Newitt, J.A.; Chang, C.Y.J.; Cheng, J.D.; Wittekind, M.; Kiefer, S.E.; Kish, K.; Lee, F.Y.F.; Borzillerri, R.; Lombardo, L.J.; et al. The structure of dasatinib (bms-354825) bound to activated abl kinase domain elucidates its inhibitory activity against imatinib-resistant abl mutants. *Cancer Res.* **2006**, *66*, 5790–5797. [CrossRef] [PubMed]

37. Grace, C.R.; Ban, D.; Min, J.; Mayasundari, A.; Min, L.; Finch, K.E.; Griffiths, L.; Bharatham, N.; Bashford, D.; Guy, R.K.; et al. Monitoring ligand-induced protein ordering in drug discovery. *J. Mol. Biol.* **2016**, *428*, 1290–1303. [CrossRef] [PubMed]

38. Nilges, M.; Macias, M.J.; ODonoghue, S.I.; Oschkinat, H. Automated noesy interpretation with ambiguous distance restraints: The refined nmr solution structure of the pleckstrin homology domain from beta-spectrin. *J. Mol. Biol.* **1997**, *269*, 408–422. [CrossRef] [PubMed]

39. Nilges, M.; O'Donoghue, S.I. Ambiguous noes and automated noe assignment. *Prog. Nucl. Magn. Reson. Spectrosc.* **1998**, *32*, 107–139. [CrossRef]

40. Dominguez, C.; Boelens, R.; Bonvin, A.M.J.J. Haddock: A protein-protein docking approach based on biochemical or biophysical information. *J. Am. Chem. Soc.* **2003**, *125*, 1731–1737. [CrossRef] [PubMed]

41. Clore, G.M.; Schwieters, C.D. Docking of protein-protein complexes on the basis of highly ambiguous intermolecular distance restraints derived from h-1(n)/n-15 chemical shift mapping and backbone n-15-h-1 residual dipolar couplings using conjoined rigid body/torsion angle dynamics. *J. Am. Chem. Soc.* **2003**, *125*, 2902–2912. [CrossRef] [PubMed]

42. Wang, B.; Westerhoff, L.M.; Merz, K.M. A critical assessment of the performance of protein-ligand scoring functions based on nmr chemical shift perturbations. *J. Med. Chem.* **2007**, *50*, 5128–5134. [CrossRef] [PubMed]

43. McCoy, M.A.; Wyss, D.F. Spatial localization of ligand binding sites from electron current density surfaces calculated from nmr chemical shift perturbations. *J. Am. Chem. Soc.* **2002**, *124*, 11758–11763. [CrossRef] [PubMed]

44. Cioffi, M.; Hunter, C.A.; Packer, M.J.; Spitaleri, A. Determination of protein-ligand binding modes using complexation-induced changes in h-1 nmr chemical shift. *J. Med. Chem.* **2008**, *51*, 2512–2517. [CrossRef] [PubMed]

45. Hajduk, P.J.; Mack, J.C.; Olejniczak, E.T.; Park, C.; Dandliker, P.J.; Beutel, B.A. Sos-nmr: A saturation transfer nmr-based method for determining the structures of protein-ligand complexes. *J. Am. Chem. Soc.* **2004**, *126*, 2390–2398. [CrossRef] [PubMed]

46. Jayalakshmi, V.; Krishna, N.R. Corcema refinement of the bound ligand conformation within the protein binding pocket in reversibly forming weak complexes using std-nmr intensities. *J. Magn. Reson.* **2004**, *168*, 36–45. [CrossRef] [PubMed]

47. Schieborr, U.; Vogtherr, M.; Elshorst, B.; Betz, M.; Grimme, S.; Pescatore, B.; Langer, T.; Saxena, K.; Schwalbe, H. How much nmr data is required to determine a protein-ligand complex structure? *Chembiochem* **2005**, *6*, 1891–1898. [CrossRef] [PubMed]

48. Orts, J.; Bartoschek, S.; Griesinger, C.; Monecke, P.; Carlomagno, T. An nmr-based scoring function improves the accuracy of binding pose predictions by docking by two orders of magnitude. *J. Biomol. NMR* **2012**, *52*, 23–30. [CrossRef] [PubMed]

49. Constantine, K.L.; Davis, M.E.; Metzler, W.J.; Mueller, L.; Claus, B.L. Protein-ligand noe matching: A high-throughput method for binding pose evaluation that does not require protein nmr resonance assignments. *J. Am. Chem. Soc.* **2006**, *128*, 7252–7263. [CrossRef] [PubMed]

50. Wishart, D.S.; Sykes, B.D. The c-13 chemical-shift index—A simple method for the identification of protein secondary structure using c-13 chemical-shift data. *J. Biomol. NMR* **1994**, *4*, 171–180. [CrossRef] [PubMed]

51. Williamson, M.P.; Kikuchi, J.; Asakura, T. Application of h-1-nmr chemical-shifts to measure the quality of protein structures. *J. Mol. Biol.* **1995**, *247*, 541–546. [CrossRef]

52. Wishart, D.S.; Watson, M.S.; Boyko, R.F.; Sykes, B.D. Automated h-1 and c-13 chemical shift prediction using the biomagresbank. *J. Biomol. NMR* **1997**, *10*, 329–336. [CrossRef] [PubMed]

53. McCoy, M.A.; Wyss, D.F. Alignment of weakly interacting molecules to protein surfaces using simulations of chemical shift perturbations. *J. Biomol. NMR* **2000**, *18*, 189–198. [CrossRef] [PubMed]

54. Cioffi, M.; Hunter, C.A.; Packer, M.J.; Pandya, M.J.; Williamson, M.P. Use of quantitative (1)h nmr chemical shift changes for ligand docking into barnase. *J. Biomol. NMR* **2009**, *43*, 11–19. [CrossRef] [PubMed]

55. Shuker, S.B.; Hajduk, P.J.; Meadows, R.P.; Fesik, S.W. Discovering high-affinity ligands for proteins: Sar by nmr. *Science* **1996**, *274*, 1531–1534. [CrossRef] [PubMed]

56. Farmer, B.T. Localizing the nadp(+) binding site on the murb enzyme by nmr. *Nat. Struct. Biol.* **1996**, *3*, 995–997. [CrossRef] [PubMed]

57. Rajagopal, P.; Waygood, E.B.; Reizer, J.; Saier, M.H.; Klevit, R.E. Demonstration of protein-protein interaction specificity by nmr chemical shift mapping. *Protein Sci.* **1997**, *6*, 2624–2627. [CrossRef] [PubMed]

58. Schmiedeskamp, M.; Rajagopal, P.; Klevit, R.E. Nmr chemical shift perturbation mapping of DNA binding by a zinc-finger domain from the yeast transcription factor adr1. *Protein Sci.* **1997**, *6*, 1835–1848. [CrossRef] [PubMed]

59. McCoy, M.A.; Wyss, D.F. Structures of protein-protein complexes are docked using only nmr restraints from residual dipolar coupling and chemical shift perturbations. *J. Am. Chem. Soc.* **2002**, *124*, 2104–2105. [CrossRef] [PubMed]

60. Trellet, M.; Melquiond, A.S.J.; Bonvin, A.M.J.J. A unified conformational selection and induced fit approach to protein-peptide docking. *PLoS ONE* **2013**, *8*, e58769. [CrossRef] [PubMed]

61. Arnesano, F.; Banci, L.; Piccioli, M. Nmr structures of paramagnetic metalloproteins. *Q. Rev. Biophys.* **2005**, *38*, 167–219. [CrossRef] [PubMed]

62. Otting, G. Prospects for lanthanides in structural biology by nmr. *J. Biomol. NMR* **2008**, *42*, 1–9. [CrossRef] [PubMed]

63. Palma, P.N.; Krippahl, L.; Wampler, J.E.; Moura, J.J.G. Bigger: A new (soft) docking algorithm for predicting protein interactions. *Proteins* **2000**, *39*, 372–384. [CrossRef]

64. Stark, J.; Powers, R. Rapid protein-ligand costructures using chemical shift perturbations. *J. Am. Chem. Soc.* **2008**, *130*, 535–545. [CrossRef] [PubMed]

65. Krzeminski, M.; Loth, K.; Boelens, R.; Bonvin, A.M.J.J. Samplex: Automatic mapping of perturbed and unperturbed regions of proteins and complexes. *BMC Bioinform.* **2010**, *11*, 51. [CrossRef] [PubMed]

66. Sanchez-Pedregal, V.M.; Reese, M.; Meiler, J.; Blommers, M.J.J.; Griesinger, C.; Carlomagno, T. The inpharma method: Protein-mediated interligand noes for pharmacophore mapping. *Angew. Chem. Int. Ed.* **2005**, *44*, 4172–4175. [CrossRef] [PubMed]

67. Stauch, B.; Orts, J.; Carlomagno, T. The description of protein internal motions aids selection of ligand binding poses by the inpharma method. *J. Biomol. NMR* **2012**, *54*, 245–256. [CrossRef] [PubMed]

68. Curto, E.V.; Moseley, H.N.B.; Krishna, N.R. Corcema evaluation of the potential role of intermolecular transferred noesy in the characterization of ligand-receptor complexes. *J. Comput. Aided Mol. Des.* **1996**, *10*, 361–371. [CrossRef] [PubMed]

69. Moseley, H.N.B.; Curto, E.V.; Krishna, N.R. Complete relaxation and conformational exchange matrix (corcema) analysis of noesy spectra of interacting systems—2-dimensional transferred noesy. *J. Magn. Reson. Ser. B* **1995**, *108*, 243–261. [CrossRef]

70. Kuntz, I.D.; Blaney, J.M.; Oatley, S.J.; Langridge, R.; Ferrin, T.E. A geometric approach to macromolecule-ligand interactions. *J. Mol. Biol.* **1982**, *161*, 269–288. [CrossRef]

71. Kuntz, I.D.; Meng, E.C.; Shoichet, B.K. Structure-based molecular design. *Accounts Chem. Res.* **1994**, *27*, 117–123. [CrossRef]

72. Orts, J.; Grimm, S.K.; Griesinger, C.; Wendt, K.U.; Bartoschek, S.; Carlomagno, T. Specific methyl group protonation for the measurement of pharmacophore-specific interligand noe interactions. *Chem. Eur. J.* **2008**, *14*, 7517–7520. [CrossRef] [PubMed]

73. Orts, J.; Tuma, J.; Reese, M.; Grimm, S.K.; Monecke, P.; Bartoschek, S.; Schiffer, A.; Wendt, K.U.; Griesinger, C.; Carlomagno, T. Crystallography-independent determination of ligand binding modes. *Angew. Chem. Int. Ed.* **2008**, *47*, 7736–7740. [CrossRef] [PubMed]

74. Vogeli, B.; Segawa, T.F.; Leitz, D.; Sobol, A.; Choutko, A.; Trzesniak, D.; van Gunsteren, W.; Riek, R. Exact distances and internal dynamics of perdeuterated ubiquitin from noe buildups. *J. Am. Chem. Soc.* **2009**, *131*, 17215–17225. [CrossRef] [PubMed]

75. Vogeli, B.; Orts, J.; Strotz, D.; Chi, C.; Minges, M.; Walti, M.A.; Guntert, P.; Riek, R. Towards a true protein movie: A perspective on the potential impact of the ensemble-based structure determination using exact noes. *J. Magn. Reson.* **2014**, *241*, 53–59. [CrossRef] [PubMed]

76. Rossmann, M.G.; Blow, D.M. Detection of sub-units within crystallographic asymmetric unit. *Acta Crystallogr.* **1962**, *15*, 24–31. [CrossRef]

77. Hillisch, A.; Pineda, L.F.; Hilgenfeld, R. Utility of homology models in the drug discovery process. *Drug Discov. Today* **2004**, *9*, 659–669. [CrossRef]

78. Schirmer, R.E.; Noggle, J.H. Quantitative application of nuclear overhauser effect to determination of molecular structure. *J. Am. Chem. Soc.* **1972**, *94*, 2947–2952. [CrossRef]

79. Balaram, P.; Bothnerb, A.; Breslow, E. Localization of tyrosine at binding-site of neurophysin ii by negative nuclear overhauser effects. *J. Am. Chem. Soc.* **1972**, *94*, 4017–4018. [CrossRef]

80. Kaiser, R. Intermolecular nuclear overhauser effect in liquid solutions. *J. Chem. Phys.* **1965**, *42*, 1838–1839. [CrossRef]

81. Pritisanac, I.; Degiacomi, M.T.; Alderson, T.R.; Carneiro, M.G.; Eiso, A.B.; Siegal, G.; Baldwin, A.J. Automatic assignment of methyl-nmr spectra of supramolecular machines using graph theory. *J. Am. Chem. Soc.* **2017**, *139*, 9523–9533. [CrossRef] [PubMed]

82. Chao, F.A.; Kim, J.; Xia, Y.; Milligan, M.; Rowe, N.; Veglia, G. Flamengo 2.0: An enhanced fuzzy logic algorithm for structure-based assignment of methyl group resonances. *J. Magn. Reson.* **2014**, *245*, 17–23. [CrossRef] [PubMed]

83. Xu, Y.; Matthews, S. Map-xsii: An improved program for the automatic assignment of methyl resonances in large proteins. *J. Biomol. NMR* **2013**, *55*, 179–187. [CrossRef] [PubMed]

magnetochemistry

MDPI

Article

Early Stages of Biomineral Formation—A Solid-State NMR Investigation of the Mandibles of Minipigs

Anastasia Vyalikh [1,2,*]**, Cindy Elschner** [2]**, Matthias C. Schulz** [3]**, Ronald Mai** [3,4] **and Ulrich Scheler** [2,*]

[1]   Institute of Experimental Physics, Technische Universität Bergakademie Freiberg,
     D-09599 Freiberg, Germany
[2]   Leibniz-Institut für Polymerforschung Dresden, e.V., D-01069 Dresden, Germany; elschner@ipfdd.de
[3]   Department of Oral and Maxillofacial Surgery, University Hospital "Carl Gustav Carus",
     Technische Universität Dresden, D-01307 Dresden, Germany;
     Matthias.Schulz@uniklinikum-dresden.de (M.C.S.); ronald.mai@gmx.de (R.M.)
[4]   Mund-Kiefer- und Gesichtschirurgie Dr. Dr. Ronald Mai, SN D-01561 Großenhain, Germany
[*]   Correspondence: anastasia.vyalikh@physik.tu-freiberg.de (A.V.); scheler@ipfdd.de (U.S.);
     Tel.: +49-373-1393-341 (A.V.); Tel.: +49-351-4658-275 (U.S.)

Received: 27 October 2017; Accepted: 17 November 2017; Published: 22 November 2017

Abstract: Solid-state nuclear magnetic resonance (NMR) spectroscopy allows for the identification of inorganic species during the biomineral formation, when crystallite particles visible in direct imaging techniques have not yet been formed. The bone blocks surrounding dental implants in minipigs were dissected after the healing periods of two, four, and eight weeks, and newly formed tissues formed around the implants were investigated ex vivo. Two-dimensional ^{31}P-^1H heteronuclear correlation (HETCOR) spectroscopy is based on the distance-dependent heteronuclear dipolar coupling between phosphate- and hydrogen-containing species and provides sufficient spectral resolution for the identification of different phosphate minerals. The nature of inorganic species present at different mineralization stages has been determined based on the ^{31}P chemical shift information. After a healing time of two weeks, pre-stages of mineralization with a rather unstructured distribution of structural motives were found. After four weeks, different structures, which can be described as nanocrystals exhibiting a high surface-to-volume ratio were detected. They grew and, after eight weeks, showed chemical structures similar to those of matured bone. In addition to hydroxyapatite, amorphous calcium phosphate, and octacalcium phosphate, observed in a reference sample of mature bone, signatures of ß-tricalcium phosphate and brushite-like structures were determined at the earlier stages of bone healing.

Keywords: solid-state NMR; heteronuclear correlation spectroscopy; biomineralization; pre-mineralization stage

1. Introduction

As the main component of vertebrates' hard tissues in bones and teeth, calcium phosphate is the most common biomineral. Unravelling the process of calcium phosphate formation in the biological environment during bone healing or remodeling is a prerequisite for an understanding of the mechanisms of biomineralization and pathological mineralization. This knowledge might lead to novel bioinspired strategies for developing advanced materials as well as preventing and treatment of calcified tissue diseases.

The mechanisms of biomineralization and pathological mineralization still remain a subject of debate. The classical physico-chemical view on nucleation and crystallization based on the association of ions from a supersaturated solution has been revised. In mineralized tissues, a key factor for the control of the nucleation, growth, and organization of hierarchical structures at different length scales is the presence of the organic matrix, which can either localize crystallization or stabilize the otherwise metastable or unstable amorphous phases [1]. It has been demonstrated that calcium-based biominerals can be formed through a complex multistage process that involves stable pre-nucleation clusters with aggregation and densification into an amorphous precursor phase and subsequent transition into a crystal [2,3]. This formation mechanism for biomineralization has been supported by in vitro studies using a combination of cryo-transmission electron microscopy and molecular modeling [4,5]. Dey et al. demonstrated heterogeneous surface-induced crystallization of hydroxyapatite (HAp) in the presence of a nucleating surface using a model system of simulated body fluid as a mineral source [5]. In the model system, the crystals were never observed together with aggregates of clusters. The in vitro mineralization studies are in agreement with in vivo experiments that have also suggested the involvement of an initial amorphous calcium phosphate phase (ACP) followed by transformation to the final HAp in forming fin bone of zebra fish [6] and in tooth enamel [7]. The degree and the time sequence of mineralization have been studied in human [8] and animal [9] bone tissues. It has been reported that the formation of new bone is a multistep process, within which the new matrix begins to mineralize after about 5–10 days from the time of deposition. The process is followed by increases in the crystal size and the number of crystals, as well as remodeling, and then stops after a period of 30 months, having reached the physiological limit of the mineral content at the tissue level. However, for a detailed understanding of the biomineralization processes, characterization on the molecular scale at all stages, including initial, intermediate, and mature mineralization phases, is required. It has been postulated long ago that ACP is formed from nanometer-sized clusters—so-called Posner's CaP clusters—with a chemical composition $Ca_9(PO_4)_6$, which are the basic units of the final apatite crystals [10]. The identification of the initial mineral phase proved to be a challenging task due to its poor crystallinity, its highly substituted nature, and a very small size of mineral clusters for visualization by microscopy techniques. In contrast to the systems studied in vitro, in organisms, other transient calcium phosphate phases such as brushite (DCPD) and octacalcium phosphate (OCP) have been implicated as phases that are intermediate to the formation of hydroxyapatite (HAp). Indeed, evidence for brushite and OCP has been reported in bone using Raman spectroscopy [11]. This demonstrates that in vivo bone formation may show different behavior compared to biomimic and in vitro grown model systems. Therefore, a detailed study on a molecular level on how biominerals are formed in organisms can shed light on the natural mineralization mechanisms. However, implementation of in vivo experiments tracing various stages of mineralization is difficult to achieve, because any non-destructive method based on real-space imaging such as magnetic resonance imaging (MRI) and computed tomography (CT) will only find structures exceeding a minimal size in space due to their inherent detection limits caused by the limits of the spatial resolution of the particular method.

Solid-state nuclear magnetic resonance (NMR) spectroscopy is ideally suited for the characterization of disordered or nanoscale materials and has been shown to provide important structural information in natural dentine [12,13], cartilage [14], and bone minerals [15–18] and their model compounds [19–23]. The presence of certain moieties known to exist in calcified tissues such as calcium phosphates, hydroxide groups, and water molecules as well as their arrangement and order, can be examined using ^{31}P and ^{1}H solid-state NMR. Two-dimensional ^{1}H-^{31}P heteronuclear correlation (HETCOR) spectroscopy utilizes the distance-dependent heteronuclear dipolar coupling and thus emphasizes spatial proximity between phosphate- and hydrogen-containing species. Therefore it allows for the determination of the nature of inorganic species present at different mineralization stages based on the ^{31}P chemical shift information.

In the present study, we present a molecular-level ex vivo NMR characterization of in vivo grown tissues. Therefore, screw-type dental implants were inserted into the mandibles of minipigs. For ex vivo analysis by solid-state NMR, the bone blocks surrounding the dental implants were dissected after healing periods of two, four, and eight weeks in order to study how biominerals develop in organisms. The NMR spectra of newly formed tissues were compared to mature bone obtained from the same organism.

2. Results

Figure 1 shows a magnetic resonance (MR) proton density slice image of the bone block of a $20 \times 10 \times 10$ mm^3 size extracted after the healing period of two weeks. The dental implant inserted into the mandibular block and dense mineralized tissue appears dark due to the lack of signal. The adjacent connective tissue is depicted in different gray scales. The red rectangle in Figure 1 indicates the area in a cavity beneath the implant apex, where tissue has grown after the healing period without direct contact to the implant. The proton density and the tissue structure in this region deviate from the surrounding tissues and thus reveal the formation of a new tissue. The morphology and quality of the newly formed tissue has been studied by MRI in the same time steps and published elsewhere [24]. Using solid-state NMR, the composition of newly formed tissue and its promotion with increasing healing time has been investigated in this work.

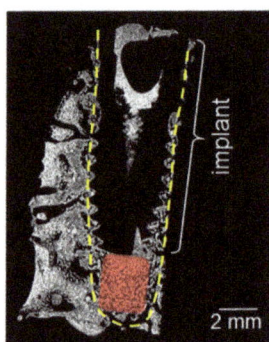

Figure 1. Magnetic resonance (MR) proton density slice image of the bone block containing the implant after a healing time of two weeks. The yellow dotted lines indicate the implant cavity, and the red rectangle shows the region with newly formed tissues, extracted for NMR measurements. The highest proton density is indicated by light gray, the lowest one by black.

Figure 2 shows the ^{31}P magic angle spinning (MAS) NMR spectra of a series of the samples after two, four, and eight weeks of healing, measured by direct polarization (DP) (Figure 2a) and cross-polarization (CP) from ^{1}H (Figure 2b). All ^{31}P MAS NMR spectra from newly formed tissues are very similar irrespective of the healing time, and reproduce the spectrum of the mature bone presented for comparison in Figure 2c (bottom). The DP spectra are characterized by a single, featureless, and almost symmetric line centered at 3.0 ppm with a rather large linewidth of 3.5 ppm. In the literature, the peak at 3 ppm is attributed to ACP, which is known to compose the bone mineral structure [25,26]. The reference spectrum of crystalline HAp shown in Figure 2c is significantly narrower and demonstrates an intrinsic peak at 2.7 ppm.

For selectivity, CP experiments in which signals of the sites in the vicinity of protons were enhanced were subsequently performed. The ^{31}P{^{1}H} CP MAS NMR spectra based on the magnetization transfer from ^{1}H to ^{31}P sites strongly indicated that the ^{31}P species strongly coupled to protons. The CP MAS NMR spectra of newly formed tissues (Figure 2b) demonstrated similar featureless signals centered at ca. 3 ppm. They were characterized by even larger linewidths when compared to the DP spectra. Inhomogeneous broadening of the latter resulted from the distribution in the local environment of ^{31}P sites in close proximity to ^{1}H. The CP spectrum of the 2w sample demonstrated a much weaker intensity compared to other CP spectra, as shown in Figure S3. This is explained by a very low concentration of the phosphorous-containing species due to a nearly complete absence of mineralized tissue. This is supported by the fact that the newly formed tissue after two weeks appeared as soft rather than hard tissue.

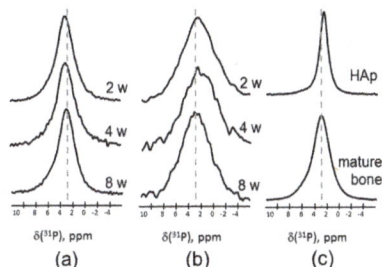

Figure 2. ^{31}P magic angle spinning nuclear magnetic resonance (MAS NMR) spectra of newly formed tissues after two, four, and eight weeks healing time, measured by (**a**) direct polarization (DP), and (**b**) using cross-polarization (CP) from ^{1}H at a contact time of 0.1 ms. (**c**) The ^{31}P MAS NMR spectra of the mature bone (bottom) and hydroxyapatite (top). The spectra are normalized to maximal intensity.

Two-dimensional ^{1}H-^{31}P heteronuclear correlation (HETCOR) spectra of tissues formed after different healing periods as well as the spectrum of mature mandibular bone of a minipig are shown in Figure 3. For assignment, we summarized the data on ^{31}P and ^{1}H chemical shifts previously reported in the literature for various calcium phosphate minerals suggested to be present in bone mineral tissues. This data is shown in Table 1. In the spectrum of bone, two characteristic correlation signals at ca. 0 ppm and 5 ppm in the ^{1}H dimension are assigned, respectively, to PO_4^{3-}/OH^- groups in the apatite structure and to structural/surface water protons spatially related to the mineral phase, according to Yesinowski and Eckert [27]. Similar correlation signals have been observed in synthetic biomimetic nanocomposites [22,28] as well as in biominerals such as animal bone [17,29], joint mineralized cartilage [14], and rat dentine [12].

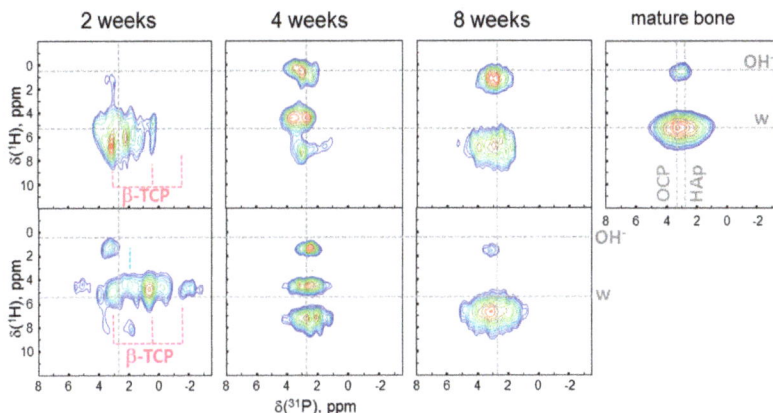

Figure 3. ^{1}H-^{31}P heteronuclear correlation (HETCOR) spectra of the newly formed tissues after two, four, and eight weeks of healing time and of mature bone. For each healing period samples from two animals (top and bottom) were measured. Guidelines at the ^{1}H chemical shifts of 0.2 ppm and 5.4 ppm characteristic of apatitic hydroxides and water, respectively, are shown. In the ^{31}P direction, the guideline at 2.7 ppm characteristic of hydroxyapatite is indicated in all samples. In addition, in the spectrum of mature bone, the guideline at 3.3 ppm, characteristic of octacalcium phosphate (OCP), is shown.

Table 1. ^{31}P and ^{1}H chemical shift parameters for calcium phosphate structures reported in the literature.

Mineral	Chemical Shift		References
	^{31}P, ppm	^{1}H, ±0.05 ppm	
HAp, hydroxyapatite	2.8 ± 0.2	0.2 5.5	[25,30,31]
Bru, brushite (dicalcium phosphate, dehydrate)	1.7 ± 0.3	4.1 6.4 10.4	[25,30–33]
Monetite (dicalcium phosphate, anhydrous)	0.0 ± 0.4 −1.5 ± 0.4		[25,30,31,33]
ACP, amorphous calcium phosphate	3.0	5.5	[25,26]
OCP, octacalcium phosphate	−0.2 ± 0.4 1.9	0.18 5.5	[30,31,34]
	3.3 ± 0.3 3.7	13.6	
β-TCP, tricalcium phosphate	−1.6 0.2 2.9		[31,35]

Based on the correlation signals in the ^{31}P dimension, the dominating contribution at 3.3 ppm in mature bone is attributed to octacalcium phosphate (OCP). The presence of HAp at 2.7 ppm and ACP at 3.0 ppm has also been found. The spectra of newly formed tissues show distinct differences to that of mature bone, particularly, at the earlier stages. The two-week (2w) spectra from two different animals represent a complex superposition of low intensity signals, whose maxima are spread from 4 to −2 ppm. After four weeks, the lines become narrower in the ^{31}P direction and split into three distinct contributions in the ^{1}H direction. After eight weeks, the spectra of both samples become more similar to the spectrum of the reference bone yielding only two correlation peaks, well separated in the ^{1}H dimension. The mineral components identified in the 2D spectra at different healing stages and in the mature bone in our work are summarized in Table 2 and discussed in the section below. In the work of Wu et al. [36], the presence of phosphorylated proteins at the very early stages of bone mineralization in chick embryos aged 8–14 days was shown using solid-state NMR. However, by Day 19 of embryonic development, no phosphoproteins could be detected because their signals were overwhelmed by large amounts of inorganic mineral phases. In our study, we performed the ^{31}P slow MAS NMR measurements at −20 °C (see Figure S4), as the phospholipids were expected to show lower shift anisotropy and specific chemical shifts. These additional measurements did not reveal phospholipid signals.

Table 2. The presence of calcium phosphate minerals in the samples under study at different healing stages (ACP = amorphous calcium phosphate; OCP = octacalcium phosphate; Bru = brushite; TCP = tricalcium phosphate; Hap = hydroxiapaptite).

Sample	Mineral Components				
	ACP	OCP	Bru	β-TCP	HAp
2 w	x	x	x		
	x	x	x	x	
4 w	x	x	x		x
	x	x	x		x
8 w	x	x			x
	x	x	x		x
Mature bone (reference)	x	x			x

3. Discussion

Our results show that solid mineral phosphate species were present in the defect zone around the dental implant in the mandibular bone of minipigs two weeks after implant insertion, although hard tissue had not yet been formed at this stage. The quantitative information available from the DP NMR spectra allowed us to estimate the fractions of mineral and organic components in the tissues after different healing times and to compare it to mature bone. Therefore, we integrated ^{1}H and ^{31}P MAS NMR spectra and calculated the ratio of hydrogen and phosphorous atoms. A spectrum of hydroxyapatite was used for calibration of the respective signal ratios from proton and phosphorous spectra. The contribution from water in the ^{1}H spectra identified by the chemical shift was ignored. The results show a 10-fold increase in the ratio of phosphorous atoms to hydrogen atoms in the organic phase in the four-week (4w) and eight-week (8w) samples when compared to the 2w sample. Moreover, the mineral/organic component ratio remained unchanged after four weeks of healing and was similar to that of the mature bone. This observation is in agreement with the work of Bala et al., who reported that the majority of newly formed tissue within two weeks must be organic soft tissue [9].

In order to follow changes in the mineral phase upon new tissue formation and maturation, 2D HETCOR spectroscopy, which provides spatial correlations between ^{1}H spins (from organic component and water) and ^{31}P spins (primarily phosphate species in the mineral component), was applied. The signal separation available in the 2D spectra provided a higher resolution in the ^{31}P dimension as compared to the 1D ^{31}P spectra and thus allowed local mineral structures to be analyzed based on their chemical shifts. To note, the decreased linewidth in the 2D spectra as compared to the 1D CP spectra is related to the longer mixing time in the former, which results in a narrower distribution of the emphasized components.

The HETCOR spectra in Figure 3 demonstrate the development of the newly formed tissue with healing time. All the 2D correlation peaks in our work were centered within the range from −3 ppm to ca. 5 ppm, corresponding to phosphate groups in various calcium phosphate minerals. The HETCOR spectra of both 2w samples were slightly different. This can be explained by the variation in the concentration and location of the surrounding calcium atoms, because the ^{31}P chemical shift of a phosphate unit is very sensitive to calcium content [37]. Moreover, protonation of the phosphate is also known to lead to an upfield ^{31}P shift [30]. Nevertheless, signatures of ACP, OCP, and brushite (Bru) were found in both 2w samples, and the presence of β-TCP was revealed in only one of them. Such differences might be associated with in vivo interindividual variability and might be explained by the way different animals respond to implant insertion in various ways. Whereas ACP and OCP are known to constitute biological minerals, brushite has been only proposed as an intermediate in biomineralization processes such as bone formation [38] and the dissolution of enamel in acids (dental caries) [39–41]. Hence, we term it here as a brushite-like component. It is worth noting that, although it is a major component of most vertebrate hard tissues, HAp was not found in either 2w sample. Thus, we can conclude that, rather than the formation of biominerals, the pre-stages of biomineralization were observed here. After four weeks, the presence of ACP, OCP, the brushite-like component, and HAp was found in both 4w samples. Finally, at the last stage of healing, ACP, HAp, and OCP were visible in the 8w spectra. This composition is close to that of the mature bone serving as reference, except for the presence of Bru phase in one of two 8w samples. However, this phase could be stabilized by an environment before conversion to hydroxyapatite. To sum up, the general trend in the formation of biomineral tissue with time evolution is clearly observed.

The ^{1}H signals from newly formed tissues strongly deviated from those of mature bone, where the protons related to apatitic OH^{-} groups and water molecules were found. An absence or undetectable amount of apatitic OH^{-} groups expected at ca. 0 ppm was observed at all early stages in all samples, except for one 4w sample. This supports reports of a strong deficiency of hydroxide groups in mature bone and dental tissues [12,42,43]. In our previous studies on water absorbance on silica surfaces [44] and on hydroxyapatite-gelatin nanocomposites [22,28], we demonstrated the variation in the ^{1}H chemical shift depending on (1) the dimensionality of the water structures formed on the surface and related to the strength of hydrogen bonds, and (2) the chemical groups donating hydrogen atoms for hydrogen bond formation. As a fast exchange of hydrogen bonded protons is known to occur at ambient temperatures, the exchanging protons appear in the ^{1}H spectrum at the chemical shift, which is the weighted average of the chemical shifts from the individual structures. Thus, the ^{1}H values discussed here can result from OH groups participating in hydrogen bonding between hydrogenated phosphates, bulk water, isolated water molecules, and organic molecules present in bone (e.g., collagen and non-collagen proteins), giving rise to variability in the ^{1}H shifts of natural bone tissues. Actually, three ^{1}H signals, similar to the 4w samples, have been observed in the biomimetic mesocrystal system of fluoroapatite and the organic matrix at the early growth stage [28]. In the latter, the signals have been attributed to three different water states associated with (i) isolated water molecules, being structural defects in the crystalline mineral domains, (ii) bound water on the mineral surface, and (iii) mobile water included in the amorphous organic layer, where it strongly interacts with organic molecules and inorganic species. The change to two ^{1}H signals in the 8w samples can be associated with the re-organization of complexes between water and protein molecules always present in the organic matrix, and/or crystallite growth, which results in diminishing the contribution from surface bound water due to reduction in the specific surface area. This scenario corroborates the recent biomineralization mechanism that manifests clustering and densification into an amorphous precursor phase [2,3,5]. Our results demonstrate that different calcium phosphate minerals are present in the newly formed tissues at various healing stages, and that hydroxyapatite has not been formed in the tissue after two weeks. This observation supports the next stage in the recent biomineralization mechanism, which involves the final crystallization to hydroxyapatite and agrees well with other studies suggesting a duration of primary mineralization of up to 3 months [8,9].

4. Materials and Methods

4.1. Sample Preparation

Six titanium coated polyetheretherketone (PEEK) implants of a special design—to create a defined area between the implant surface and the round drill hole reported in [24,45]—were inserted in the mandibular bone of minipigs. The mandibles were resected after healing periods of two, four, and eight weeks. The details of the animal model are given in Supplementary Materials. The specimens were stored in a phosphate buffer solution containing penicillin, streptomycin, gentamycin, and amphotericin B at 8 °C. Due to tight fixing of the dental screws, contact of the tissues investigated and phosphate solution was avoided. For the solid-state NMR measurements, species of newly formed tissues were isolated from the host bone, placed into MAS zirconia rotors with an o.d. of 4 mm, tightly capped, and immediately measured. Thus, two samples from different animals were examined after each healing period. The bone serving as reference was obtained from mature pristine mandibular bone and ground in a freezer mill. The reference sample of hydroxyapatite was purchased from Merck KGaA (Darmstadt, Germany) and used without purification. Control measurements of both untreated bone blocks and ground bone after extraction and after storage for 4 weeks—a period during which some samples had to be stored before the solid-state NMR following the MRI experiments—were performed to verify that the storage conditions and grinding did not affect the mineral structure. The results of the control experiments are presented in Figures S1 and S2.

4.2. NMR Imaging Measurements

The NMR Imaging was performed on a 7 T Bruker Avance 300 nonclinical NMR spectrometer (Bruker BioSpin, Rheinstetten, Germany) using a linear polarized rf coil of a 15 mm inner diameter and a BrukerBioSpin Micro 2.5 gradient system generating a maximum magnetic field gradient strength of up to 1 T/m on three axes. For the experiment, the whole bone block was immersed into an NMR tube filled with a perfluorinated fluid (Fluorinert FC 77[®], 3M Belgium NV/SA, Zwijndrecht,

Belgium) preventing dehydration and providing a signal-free background. For a proton-density image, a spin-echo multi-slice sequence with a repetition time of 2.5 s, an echo time of 4 ms, a matrix size of 512 × 256 pixel, a slice thickness of 125 μm, and 32 scans was used.

4.3. Solid State NMR

All NMR spectra were obtained on a 7 T Bruker Avance 300 spectrometer operating at resonance frequencies of 300.1 MHz for ^{1}H and 121.5 MHz for ^{31}P. The ^{31}P NMR experiments were acquired at a spinning frequency of 10 kHz employing a BL4 HXY 4 mm MAS probe head. For ^{31}P MAS NMR spectra, a single 90° pulse with a 3.8 μs pulse duration, a recycle delay of 50 s, a proton decoupling of 50 kHz, and up to 512 repetitions were applied. For ^{31}P{^{1}H} CP measurements, a contact time of 0.1 ms, a ^{1}H decoupling (XiX) of 50 kHz, 4096 repetitions, and a recycle delay of 3 s were used. The spectra were fitted using Dmfit [46]. Two-dimensional ^{1}H-^{31}P HETCOR experiments were performed using frequency-switched Lee-Goldburg (FSLG) CP with a contact time of 0.5 ms and 1.5 ms and an LG frequency of 75 kHz. A recycle delay of 3 s and 256 scans per t_1 time increment were used. A total of 128 t_1 slices with a 25.2 μs time increment were acquired. Prior to Fourier transform, exponential multiplication with 20 Hz and 30 Hz line broadenings was used in ^{31}P and ^{1}H dimensions, respectively. The ^{1}H chemical shifts were referenced to tetramethylsilane (TMS) at 0 ppm using poly (vinylidene fluoride) as an external reference at 2.9 ppm; powdered ammonium dihydrogen phosphate was used to reference the ^{31}P spectra at 0.72 ppm relative to 85% phosphoric acid.

5. Conclusions

In the present work, the changes in the mineral phase upon new tissue formation and maturation were investigated on the molecular level by applying solid-state NMR. The tissues grown in the cavity beneath the dental implants inserted into the mandibles of minipigs were investigated after three different healing periods of two, four, and eight weeks. While one-dimensional ^{31}P MAS NMR proved the presence of mineralized tissue, two-dimensional heteronuclear correlation HETCOR spectroscopy enabled the detection and identification of various calcium phosphate minerals and water stages during the bone formation. The increase in mineral content was shown by the ratio between the ^{31}P and ^{1}H signal intensities during the healing period. Thus, after two weeks, we observed pre-stages of biomineralization (rather than a formation of biocrystals, which grow in a straightforward manner) with a broad distribution in structural environment. The chemical composition in the newly formed calcium phosphate species varies with healing time until it achieves a composition close to that of mature bone, at approximately eight weeks after healing starts. Besides ACP and OCP phases, revealed in the reference bone spectrum, brushite-like and β-TCP were observed at early stages of mineralization. At the intermediate stage, hydroxyapatite was formed. Our results provide in vivo confirmation of a recently proposed biomineralization mechanism based on the formation of nanometer-sized calcium phosphate clusters, with subsequent aggregation into an amorphous phase and, finally, crystallization to hydroxyapatite. We proved here that solid-state NMR spectroscopy allows for the identification of inorganic species during the biomineral formation at very early stages, when crystallite particles visible in direct imaging techniques have not yet been formed.

Supplementary Materials: The following materials are available online at www.mdpi.com/2312-7481/3/4/39/s1, Figure S1: ^{1}H-^{31}P heteronuclear correlation (HETCOR) spectra of the bone (a) in block after extraction, (b) cryogenically ground after extraction, (c) in block after storage for 4 weeks in phosphate solution in a fridge and (d) cryogenically ground after storage for 4 weeks in phosphate solution in a fridge. Figure S2: (A) ^{1}H magic angle spinning nuclear magnetic resonance (MAS NMR) spectra, and (B) ^{31}P MAS NMR spectra of the bone in block after extraction, cryogenically ground after extraction, in block after storage for 4 weeks in phosphate solution in a fridge and cryogenically ground after storage for 4 weeks in phosphate solution in a fridge. Figure S3: ^{31}P{^{1}H} CP MAS (10 kHz) spectra of newly formed tissues in the second animal (2w- after two, 4w-four and 8w-eight weeks healing time). The spectra were measured at a contact time of 0.1 ms. Normalization on the sample weight and no line broadening were applied here to demonstrate the sensitivity of the spectra. Figure S4: ^{31}P{^{1}H} CP MAS (2.6 kHz) spectrum of the 2w sample measured at −20 °C.

Acknowledgments: This work has been supported by the German Research Foundation (DFG) in the frame of Transregio TRR67 and the European Social Fund through the Sächsische Aufbaubank. The authors gratefully acknowledge Dirk Zimmerhäckel and the IPF machine shop for manufacturing the dental PEEK implants, Ricardo Bernhardt and Steffen Howitz for sputter coating of dental implants, as well as Bernd Stadlinger and Uwe Eckelt for their support in the preclinical experiments.

Author Contributions: U.S. and R.M. conceived and designed the experiments; M.S. performed the preclinical experiments, C.E. contributed to the sample preparation and performed the MRI experiments, A.V. performed the NMR experiments, analyzed the data, and wrote the paper. All authors contributed to the text.

Conflicts of Interest: The authors declare no conflict of interest.

References

1. Wang, L.; Nancollas, G.H. Pathways to biomineralization and biodemineralization of calcium phosphates: The thermodynamic and kinetic controls. *Dalton Trans.* **2009**, 2665–2672. [CrossRef] [PubMed]
2. Gebauer, D.; Völkel, A.; Cölfen, H. Stable prenucleation calcium carbonate clusters. *Science* **2008**, *322*, 1819–1822. [CrossRef] [PubMed]
3. Pouget, E.M.; Bomans, P.H.H.; Goos, J.A.C.M.; Frederik, P.M.; de With, G.; Sommerdijk, N.A.J.M. The initial stages of template-controlled CaCO$_3$ formation revealed by cryo-TEM. *Science* **2009**, *323*, 1455–1458. [CrossRef] [PubMed]
4. Nudelman, F.; Pieterse, K.; George, A.; Bomans, P.H.H.; Friedrich, H.; Brylka, L.J.; Hilbers, P.A.J.; de With, G.; Sommerdijk, N.A.J.M. The role of collagen in bone apatite formation in the presence of hydroxyapatite nucleation inhibitors. *Nat. Mater.* **2010**, *9*, 1004–1009. [CrossRef] [PubMed]
5. Dey, A.; Bomans, P.H.H.; Müller, F.A.; Will, J.; Frederik, P.M.; de With, G.; Sommerdijk, N.A.J.M. The role of prenucleation clusters in surface-induced calcium phosphate crystallization. *Nat. Mater.* **2010**, *9*, 1010–1014. [CrossRef] [PubMed]
6. Mahamid, J.; Sharir, A.; Addadi, L.; Weiner, S. Amorphous calcium phosphate is a major component of the forming fin bones of zebrafish: Indications for an amorphous precursor phase. *Proc. Natl. Acad. Sci. USA* **2008**, *105*, 12748–12753. [CrossRef] [PubMed]
7. Beniash, E.; Metzler, R.A.; Lam, R.S.; Gilbert, P.U. Transient amorphous calcium phosphate in forming enamel. *J. Struct. Biol.* **2009**, *166*, 133–143. [CrossRef] [PubMed]
8. Boivin, G.; Meunier, P.J. The degree of mineralization of bone tissue measured by computerized quantitative contact microradiography. *Calcif. Tissue Int.* **2002**, *70*, 503–511. [CrossRef] [PubMed]
9. Bala, Y.; Farlay, D.; Delmas, P.D.; Meunier, P.J.; Boivin, G. Time sequence of secondary mineralization and microhardness in cortical and cancellous bone from ewes. *Bone* **2010**, *46*, 1204–1212. [CrossRef] [PubMed]
10. Posner, A.S.; Betts, F. Synthetic amorphous calcium phosphate and its relation to bone mineral structure. *Acc. Chem. Res.* **2002**, *8*, 273–281. [CrossRef]
11. Crane, N.J.; Popescu, V.; Morris, M.D.; Steenhuis, P.; Ignelzi, M.A. Raman spectroscopic evidence for octacalcium phosphate and other transient mineral species deposited during intramembranous mineralization. *Bone* **2006**, *39*, 434–442. [CrossRef] [PubMed]
12. Tseng, Y.-H.; Tsai, Y.-L.; Tsai, T.W.T.; Chao, J.C.H.; Lin, C.-P.; Huang, S.-H.; Mou, C.-Y.; Chan, J.C.C. Characterization of the Phosphate Units in Rat Dentin by Solid-State NMR Spectroscopy. *Chem. Mater.* **2007**, *19*, 6088–6094. [CrossRef]
13. Huang, S.-J.; Tsai, Y.-L.; Lee, Y.-L.; Lin, C.-P.; Chan, J.C.C. Structural Model of Rat Dentin Revisited. *Chem. Mater.* **2009**, *21*, 2583–2585. [CrossRef]
14. Duer, M.J.; Friscić, T.; Murray, R.C.; Reid, D.G.; Wise, E.R. The mineral phase of calcified cartilage: Its molecular structure and interface with the organic matrix. *Biophys. J.* **2009**, *96*, 3372–3378. [CrossRef] [PubMed]
15. Cho, G.; Wu, Y.; Ackerman, J.L. Detection of hydroxyl ions in bone mineral by solid-state NMR spectroscopy. *Science* **2003**, *300*, 1123–1127. [CrossRef] [PubMed]
16. Kolodziejski, W. Solid-state NMR studies of bone. *Top. Curr. Chem.* **2005**, *246*, 235–270. [CrossRef] [PubMed]
17. Wilson, E.E.; Awonusi, A.; Morris, M.D.; Kohn, D.H.; Tecklenburg, M.M.J.; Beck, L.W. Three structural roles for water in bone observed by solid-state NMR. *Biophys. J.* **2006**, *90*, 3722–3731. [CrossRef] [PubMed]

18. Jaeger, C.; Groom, N.S.; Bowe, E.A.; Horner, A.; Davies, M.E.; Murray, R.C.; Duer, M.J. Investigation of the Nature of the Protein−Mineral Interface in Bone by Solid-State NMR. *Chem. Mater.* **2005**, *17*, 3059–3061. [CrossRef]

19. Tseng, Y.-H.; Mou, C.-Y.; Chan, J.C.C. Solid-state NMR study of the transformation of octacalcium phosphate to hydroxyapatite: A mechanistic model for central dark line formation. *J. Am. Chem. Soc.* **2006**, *128*, 6909–6918. [CrossRef] [PubMed]

20. Nassif, N.; Martineau, F.; Syzgantseva, O.; Gobeaux, F.; Willinger, M.; Coradin, T.; Cassaignon, S.; Azaïs, T.; Giraud-Guille, M.M. In Vivo Inspired Conditions to Synthesize Biomimetic Hydroxyapatite. *Chem. Mater.* **2010**, *22*, 3653–3663. [CrossRef]

21. Ndao, M.; Ash, J.T.; Stayton, P.S.; Drobny, G.P. The Role of Basic Amino Acids in the Molecular Recognition of Hydroxyapatite by Statherin using Solid State NMR. *Surf. Sci.* **2010**, *604*, L39–L42. [CrossRef] [PubMed]

22. Vyalikh, A.; Simon, P.; Kollmann, T.; Kniep, R.; Scheler, U. Local Environment in Biomimetic Hydroxyapatite−Gelatin Nanocomposites As Probed by NMR Spectroscopy. *J. Phys. Chem. C* **2011**, *115*, 1513–1519. [CrossRef]

23. Vyalikh, A.; Simon, P.; Rosseeva, E.; Buder, J.; Kniep, R.; Scheler, U. Intergrowth and interfacial structure of biomimetic fluorapatite-gelatin nanocomposite: A solid-state NMR study. *J. Phys. Chem. B* **2014**, *118*, 724–730. [CrossRef] [PubMed]

24. Korn, P.; Elschner, C.; Schulz, M.C.; Range, U.; Mai, R.; Scheler, U. MRI and dental implantology: Two which do not exclude each other. *Biomaterials* **2015**, *53*, 634–645. [CrossRef] [PubMed]

25. Aue, W.P.; Roufosse, A.H.; Glimcher, M.J.; Griffin, R.G. Solid-state phosphorus-31 nuclear magnetic resonance studies of synthetic solid phases of calcium phosphate: Potential models of bone mineral. *Biochemistry* **2002**, *23*, 6110–6114. [CrossRef]

26. Combes, C.; Rey, C. Amorphous calcium phosphates: Synthesis, properties and uses in biomaterials. *Acta Biomater.* **2010**, *6*, 3362–3378. [CrossRef] [PubMed]

27. Yesinowski, J.P.; Eckert, H. Hydrogen environments in calcium phosphates: Proton MAS NMR at high spinning speeds. *J. Am. Chem. Soc.* **1987**, *109*, 6274–6282. [CrossRef]

28. Vyalikh, A.; Simon, P.; Rosseeva, E.; Buder, J.; Scheler, U.; Kniep, R. An NMR Study of Biomimetic Fluorapatite—Gelatine Mesocrystals. *Sci. Rep.* **2015**, *5*, 15797. [CrossRef] [PubMed]

29. Santos, R.A.; Wind, R.A.; Bronnimann, C.E. ^{1}H CRAMPS and ^{1}H-^{31}P HetCor Experiments on Bone, Bone Mineral, and Model Calcium Phosphate Phases. *J. Magn. Reson. Ser. B* **1994**, *105*, 183–187. [CrossRef]

30. Rothwell, W.P.; Waugh, J.S.; Yesinowski, J.P. High-resolution variable-temperature phosphorus-31 NMR of solid calcium phosphates. *J. Am. Chem. Soc.* **1980**, *102*, 2637–2643. [CrossRef]

31. Bak, M.; Thomsen, J.K.; Jakobsen, H.J.; Petersen, S.E.; Petersen, T.E.; Nielsen, N.C. Solid-state ^{13}C and ^{31}P NMR analysis of urinary stones. *J. Urol.* **2000**, *164*, 856–863. [CrossRef]

32. Kaflak-Hachulska, A.; Samoson, A.; Kolodziejski, W. ^{1}H MAS and ^{1}H→^{31}P CP/MAS NMR study of human bone mineral. *Calcif. Tissue Int.* **2003**, *73*, 476–486. [CrossRef] [PubMed]

33. Legrand, A.P.; Sfihi, H.; Lequeux, N.; Lemaître, J. ^{31}P Solid-State NMR study of the chemical setting process of a dual-paste injectable brushite cements. *J. Biomed. Mater. Res. B* **2009**, *91*, 46–54. [CrossRef] [PubMed]

34. Davies, E.; Duer, M.J.; Ashbrook, S.E.; Griffin, J.M. Applications of NMR crystallography to problems in biomineralization: Refinement of the crystal structure and ^{31}P solid-state NMR spectral assignment of octacalcium phosphate. *J. Am. Chem. Soc.* **2012**, *134*, 12508–12515. [CrossRef] [PubMed]

35. Obadia, L.; Deniard, P.; Alonso, B.; Rouillon, T.; Jobic, S.; Guicheux, J.; Julien, M.; Massiot, D.; Bujoli, B.; Bouler, J.-M. Effect of Sodium Doping in β-Tricalcium Phosphate on Its Structure and Properties. *Chem. Mater.* **2006**, *18*, 1425–1433. [CrossRef]

36. Wu, Y.; Ackerman, J.L.; Strawich, E.S.; Rey, C.; Kim, H.-M.; Glimcher, M.J. Phosphate ions in bone: Identification of a calcium-organic phosphate complex by ^{31}P solid-state NMR spectroscopy at early stages of mineralization. *Calcif. Tissue Int.* **2003**, *72*, 610–626. [CrossRef] [PubMed]

37. Fletcher, J.P.; Kirkpatrick, R.J.; Howell, D.; Risbud, S.H. ^{31}P Magic-angle spinning nuclear magnetic resonance spectroscopy of calcium phosphate glasses. *J. Chem. Soc. Faraday Trans.* **1993**, *89*, 3297. [CrossRef]

38. Neuman, W.F.; Bareham, B.J. Evidence for the presence of secondary calcium phosphate in bone and its stabilization by acid production. *Calcif. Tissue Res.* **1975**, *18*, 161–172. [CrossRef] [PubMed]

39. Kodaka, T.; Ohohara, Y.; Debari, K. Scanning electron microscopy and energy-dispersive X-ray microanalysis studies of early dental calculus on resin plates exposed to human oral cavities. *Scanning Microsc.* **1992**, *6*, 475–486. [PubMed]

40. LeGeros, R.Z.; Orly, I.; LeGeros, J.P.; Gomez, C.; Kazimiroff, J.; Tarpley, T.; Kerebel, B. Scanning electron microscopy and electron probe microanalyses of the crystalline components of human and animal dental calculi. *Scanning Microsc.* **1988**, *2*, 345–356. [PubMed]

41. Jin, Y.; Yip, H.-K. Supragingival Calculus: Formation and Control. *Crit. Rev. Oral Biol. Med.* **2002**, *13*, 426–441. [CrossRef] [PubMed]

42. Loong, C.-K.; Rey, C.; Kuhn, L.T.; Combes, C.; Wu, Y.; Chen, S.-H.; Glimcher, M.J. Evidence of hydroxyl-ion deficiency in bone apatites: An inelastic neutron-scattering study. *Bone* **2000**, *26*, 599–602. [CrossRef]

43. Vyalikh, A.; Mai, R.; Scheler, U. OH$^-$ deficiency in dental enamel, crown and root dentine as studied by ^{1}H CRAMPS. *Bio-Med. Mater. Eng.* **2013**, *23*, 507–512. [CrossRef]

44. Vyalikh, A.; Emmler, T.; Grünberg, B.; Xu, Y.; Shenderovich, I.; Findenegg, G.H.; Limbach, H.-H.; Buntkowsky, G. Hydrogen Bonding of Water Confined in Controlled-Pore Glass 10–75 Studied by ^{1}H-Solid State NMR. *Z. Phys. Chem.* **2007**, *221*, 155–168. [CrossRef]

45. Stadlinger, B.; Hintze, V.; Bierbaum, S.; Möller, S.; Schulz, M.C.; Mai, R.; Kuhlisch, E.; Heinemann, S.; Scharnweber, D.; Schnabelrauch, M.; et al. Biological functionalization of dental implants with collagen and glycosaminoglycans-A comparative study. *J. Biomed. Mater. Res. B* **2012**, *100*, 331–341. [CrossRef] [PubMed]

46. Massiot, D.; Fayon, F.; Capron, M.; King, I.; Le Calvé, S.; Alonso, B.; Durand, J.-O.; Bujoli, B.; Gan, Z.; Hoatson, G. Modelling one- and two-dimensional solid-state NMR spectra. *Magn. Reson. Chem.* **2002**, *40*, 70–76. [CrossRef]

magnetochemistry

MDPI

Article

NMR-Assisted Structure Elucidation of an Anticancer Steroid-β-Enaminone Derivative

Donald Poirier [1,2,*] and René Maltais [1]

[1] Laboratory of Medicinal Chemistry, Endocrinology and Nephrology Unit, CHU de Québec—Research Center, Québec, QC G1V 4G2, Canada; rene.maltais@crchul.ulaval.ca

[2] Department of Molecular Medicine, Faculty of Medicine, Université Laval, Québec, QC G1V 0A6, Canada

* Correspondence: donald.poirier@crchul.ulaval.ca; Tel.: +1-418-654-2296

Received: 31 October 2017; Accepted: 17 November 2017; Published: 21 November 2017

Abstract: The fortuitous modification of a quinoline-proline-piperazine side chain linked to a steroid in the presence of lithium (trimethylsilyl) acetylide has generated an unknown product that is more active than its precursor. After having characterized two β-enaminones (two-carbon homologation compounds) that were generated from a simplified model side chain, we have identified the unknown product as being the β-enaminone steroid derivative **1**. NMR analysis, especially two-dimensional (2D) experiments (correlation spectroscopy (COSY), NOE spectroscopy (NOESY), heteronuclear single-quantum correlation (HSQC) and heteronuclear multiple-bond correlation (HMBC)) provided crucial information that was found essential in the characterization of enaminone **1**. We also proposed a mechanism to rationalize the formation of this biologically active compound.

Keywords: enaminone; steroid; NMR; ethynylation; rearrangement; anti-cancer agent

1. Introduction

Cancer is still a major cause of death worldwide [1,2]. It is therefore urgent to develop new molecules to counter the uncontrolled proliferation of cancer cells [3,4]. Steroid derivatives having a (1-quinolin-2-ylcarbonyl)-L-proline-piperazine side chain at position C2 of estra-1,3,5(10)-trien-3,17β-diol [5] or 5α-androstane-3α,17β-diol [6–8] represent a new family of anticancer agents. They were found to reduce the proliferation of various human cancer cell lines [8–10], as well as to block tumor growth when tested in mouse tumor (xenografts) models of human cancers (pancreas, ovary, breast, and leukemia) [7,11]. During our work to increase their metabolic stability, especially for oral administration, one of our strategies was to introduce an ethynyl group at position C17 of the steroid nucleus to generate a tertiary alcohol instead of a secondary alcohol [12]. In the field of steroid drugs, this kind of transformation is known to increase metabolic stability by avoiding the oxidation of 17β-OH into a ketone by the metabolism of phase-I enzymes of the CYP family [13,14].

Trying to introduce the ethynyl group at the last step of the chemical synthesis gave unsatisfactory low yields (<30%) of the expected 17α-ethynyl/17β-OH derivative RM-133 from RM-128 (Figure 1), and the former was contaminated with unidentified side products. This more metabolic stable compound was alternatively obtained by another strategy that introduced the ethynyl group earlier in the reaction sequence, thus providing RM-133 in good yield and purity [7], but the negative results of the first chemical strategy provided an unexpected finding. In fact, the addition of lithium (trimethylsilyl)acetylide to the C17-carbonyl of RM-128 generated a new compound, which after purification was found to be very active as an antiproliferative agent on a variety of human cancer cell lines (OVCAR-3, PANC-1, MCF-7, T-47D), being about two times more potent than RM-133 in a comparative assay (unpublished results). The biological potential of this unknown compound prompted us to elucidate its chemical structure.

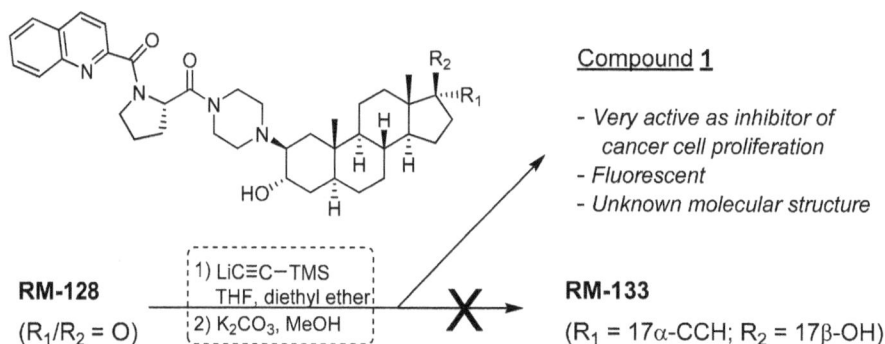

Figure 1. Unexpected formation of an unknown biologically active steroid derivative.

2. Results and Discussion

2.1. Characteristic of Unknown Compound and Hypothesis

The major difficulty for the characterization of unknown compound **1** was the complex nature of the starting aminosteroid derivative RM-128 [8], which is well-known to exist as a mixture of two conformers in different proportions according to the solvent used for NMR analysis [7]. In fact, the presence of two amide bonds linking a piperazine, a proline, and a quinoline explain these two conformers, but the combinaison of this side chain and a steroid backbone with 19 carbons greatly complexifies the identification of such steroid derivatives by NMR. Moreover, as observed with other aminosteroids of similar structure, it is not possible to obtain a crystal for X-ray analysis. However, some conclusions can be drawn by the comparison of ^{1}H NMR and ^{13}C NMR spectra of the starting ketone (RM-128) and the unknown compound **1**. In fact, the steroid backbone was not modified, except at C17, where an ethynyl group seems present, and the quinoline moiety was not altered either, thus suggesting a modification of the side chain close to the proline, although the two amide groups also seem to be present. Mass spectra analysis showed a peak at 679.5 m/z (M+H) instead of the expected peak at 652.4 m/z, suggesting the presence of another group in addition to the expected ethynyl group. The infrared (IR) analysis only confirmed the presence of amide and alcohol, the C≡C acetylenic signal being too small to be significant. Finally, an interesting observation is the strong fluorescence of this unknown compound (excitation λ_{max} = 410 nm; emission λ_{max} = 494 nm), which supports a modification of the side chain that is linked to the steroid nucleus.

2.2. Synthesis of Side Chain **6** as a Model for the Formation of Unknown Compounds

Given the limited information that is generated by the analysis of the spectral data of the unknown compound **1**, as well as the complex structure of this type of aminosteroid derivative, we decided to try the same kind of reaction on the side chain model, as represented by compound **6**. This later was easily prepared in three steps (Figure 2), by 1) a coupling of 2-quinaldic acid (**2**) with proline-*t*-butylester (**3**) using benzotriazol-1-yl-oxytripyrrolidinophosphonium hexafluorophosphate (PyBOP), 2) a hydrolysis of the ester group to the corresponding acid, and 3) a coupling of this acid with *N*-methyl-piperazine using 2-(1*H*-benzotriazol-1-yl)-1,1,3,3-tetramethyluronium hexafluorophosphate (HBTU) as reagent. When reacting compound **6** with lithium (trimethylsilyl)acetylide, we obtained a mixture of products that was purified by flash chromatography, leading to two unknown compounds in 15 and 30% yields, respectively.

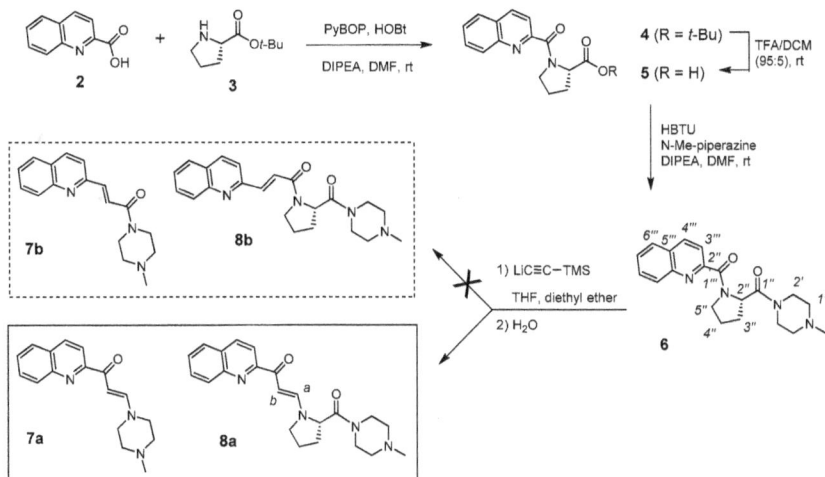

Figure 2. Synthesis of model side chain (compound **6**) and two rearranged side chains (compounds **7a** and **8a**) resulting from the ethynylation reaction.

*2.3. NMR Characterization of Unknown Side Chain (Enaminones **7** and **8**)*

Comparison of the NMR data of compound **6** (Figure 3A) with those of the two unknown side chains generated from **6** provided interesting information. In fact, the ^{1}H NMR analysis of the minor compound (Figure 3B) demonstrated the presence of *N*-methylpiperazine and quinoline groups, but surprisingly, the absence of the proline moiety. These observations were also confirmed by mass analysis ([M+H]$^+$ = 282.1 *m*/*z*). For the major compound, however, the proline was clearly present in NMR spectra (Figure 3C) and mass analysis ([M+H]$^+$ = 379.1 *m*/*z*). In NMR spectra of both compounds, there is no evidence of peak splitting for the N-CH$_3$ of piperazine, the CH of proline and some quinoline signals. This duplication is typical of the presence of the two rotamers observed for a side chain like **6** [7], especially obvious for the CH-2'' (Figure 3A). New signals (*a* and *b*) integrating for 2H were also detected in the aromatic or vinylic region (6.80 and 7.91 ppm for **7** and 6.45/6.70 and 7.90/8.08 ppm for **8**; correlations in correlation spectroscopy (COSY) spectra).

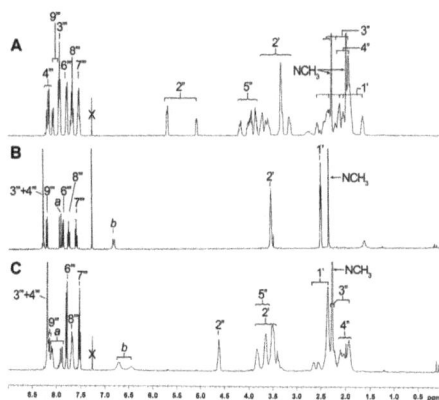

Figure 3. Comparison of ^{1}H NMR spectra in CDCl$_3$ of known side chain **6** (**A**) and unknown side chains **7a** (**B**) and **8a** (**C**).

In ^{13}C NMR, the carbonyl peaks at 166.1 and 166.6 ppm (C-1''') disappeared and a weak signal at 187.6 or 187.4 ppm for **7** or **8**, respectively. Moreover, two new weak peaks (CH) at 91.0/153.0 or 93.7/150.0 ppm appeared for **7** or **8**, respectively (Table 1). Two CH-aromatic signals were also deshielded, while the CH-3''' signal was slightly shielded, but the other signals were roughly the same. Taking into account NMR and mass data, especially an additional 26-mass units that were observed for the major compound, we can therefore assume the presence of a CH=CHCO group between the quinoline and proline moieties, which compounds could be represented by **7a** or **7b**, for the minor compound, and **8a** or **8b**, for the major compound.

Table 1. Chemical shifts (δ in ppm) and assignation of carbons from compounds **6**, **7a**, **8a**, and **1**.

#	6	7a	8a	1 (Side Chain)	1 (Steroid; C #)
CH$_2$-4''	22.3, 25.2	–	23.4, 23.9	23.6, 23.9	12.9 (CH$_3$-18)
CH$_2$-3''	28.8, 31.5	–	30.1, 30.6	30.1, 30.7	17.2 (CH$_3$-19)
CH$_2$-2'	41.5, 42.0	46.6 [1]	42.1, 45.1	41.9, 42.7	21.0 (CH$_2$-11)
	44.7, 45.5		45.5, 45.7	45.8, 46.2	
N-CH$_3$	45.6, 45.9	46.0	46.0	–	23.0 (CH$_2$-15)
CH$_2$-5''	48.1, 49.8	–	48.1, 53.3	48.1, 53.4	28.1 (CH$_2$-6)
CH$_2$-1'	54.2, 54.6, 54.9	54.6 [1]	54.5, 54.8	46.8, 48.0, 48.6	31.1 (CH$_2$-7)
CH-2''	57.4, 59.1	–	57.9, 62.7	57.7, 62.7	32.7 (CH$_2$-12)
CH-3'''	120.9, 121.5	119.0	119.1	119.1	32.8 (CH$_2$-1)
CH-7'''	127.4	127.4	127.4	127.0	34.6 (CH$_2$-4)
CH-6'''	127.7	127.6	127.6	127.4	35.7 (C-10)
C-5'''	128.1, 128.3	129.0	129.0	129.1	36.1 (CH-8)
CH-9'''	129.2, 129.8	130.1	129.3, 130.3	129.9, 130.3	38.4 (CH-5)
CH-8'''	129.6	129.4	129.4	129.4	38.9 (CH$_2$-16)
CH-4'''	136.6	136.6	136.5	136.5	47.0 (C-13)
C-10'''	145.8, 146.3	147.0	147.1	147.1	50.3 (CH-14)
C-2'''	153.4, 154.1	155.9	156.1	156.1	55.6 (CH-9)
C-1'''	166.0, 166.6	187.6	187.4 [2]	187.5	63.7, 63.8 (CH-3)
C-1''	169.0, 170.1	–	168.6, 169.4	168.9, 169.2	64.8, 64.9 (CH-2)
CH-*b*	–	91.0	93.7 [2]	93.6	73.9 (C-20)
CH-*a*	–	153.0	150.0 [2]	149.7	79.8 (C-17)
–	–	–	–	–	87.6 (C-21)

[1] The CH$_2$ signals of piperazine ring are very weak. [2] The δ values estimated by ChemDraw 14.0 are 187.0, 153.8 and 92.7 ppm for C-1''', CH-*a* and CH-*b*, respectively. These signals are also very weak.

We next focused on the major compound **8** because its side chain is similar to the side chain that is present in the bioactive steroid compound **1**. IR analysis does not make it possible to discriminate between the two possible regioisomers **8a** and **8b** (same carbonyl band at 1643 cm^{-1}), but in ^{13}C NMR, the peak at 187.4 ppm corresponds very well to a conjugated ketone rather than a conjugated amide. Moreover, the experimental and calculated (ChemDraw 14.0) chemical shifts for this conjugated ketone are identical (Table 1). The HMBC spectrum of **8** showed J3-coupling between the vinyl CH-*a* and the carbonyl C-1''' at 187.4 ppm, as well as with two proline signals (CH-2'' and CH-5''') (Figure 4A). A J3-coupling between the CH-3''' of quinoline and the carbonyl at 187.4 ppm was also observed. The NOESY spectrum does not show correlations between a CH-2'' of proline and any CH of the quinoline nucleus, but correlations between CH-2'' and the vinylic CH-*a* and CH-*b* were found, thus supporting the β-enaminone structure of **8a** (Figure 4B). The coupling constants (*J* = 13.0 Hz), as well as no NOE correlation between CH-*a* and CH-*b* confirm the *trans*-configuration of the alkene **8a**. After the characterization of **8a**, the minor unknown side chain was found to be the β-enaminone **7a**. In addition to confirm the structures of **7a** and **8a** as the minor and major compounds resulting from the ethynylation of **6**, two-dimensional (2D) NMR analyzes also allowed for the complete assignment of protons and carbons of **6, 7a,** and **8a** (Table 1). These data will be crucial for the elucidation of the unknown steroidal compound **1**.

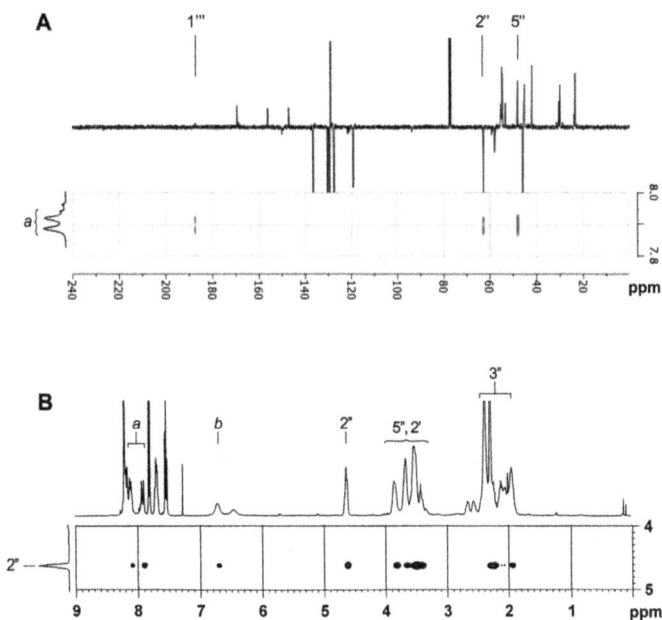

Figure 4. Partial HMBC (**A**) and NOE spectroscopy (NOESY) (**B**) results supporting the structure of β-enaminone **8a**.

2.4. NMR Characterization of Unknown Steroidal Compound 1

After we characterized the enaminone **8a** from an experiment with a simplified model of side chain, it was possible to make a direct comparison of the NMR data of **1** and **8a**. In ^{1}H NMR, the quinoline and enaminone signals were found to be identical, while the CH-2'' and CH$_2$-5'' signals of proline were at the same chemical shifts (4.6 and 3.4–3.9 ppm). However, a comparison of the other signals was not possible because of the presence of numerous protons of steroid backbone, but two methyl groups (CH$_3$-18 and CH$_3$-19) and an acetylenic proton (2.58 ppm) are also present. A comparison of the ^{13}C NMR data clearly demonstrated the similarity of the chemical shifts for the β-enaminone **8a** and the side chain at position 2β of the steroid. In fact, the only disparities are related to the presence of a CH$_3$ instead of the steroid nucleus on a piperazine nitrogen, which affects the chemical shifts of the two CH$_2$-1'. Analysis of the ^{13}C NMR data also makes it possible to completely confirm the 5α-androstan-3β,17β-diol backbone, as well as the presence of an ethynyl group (73.9 and 87.6 ppm) at C-17α (79.8 ppm) [7]. Finally, mass analysis ([M+H]$^+$ = 679.5 *m/z*) is in agreement with the proposed structure for compound **1**.

2.5. Mechanism of β-Enaminone Formation

After an exhaustive research in literature for such transformation, we found that the preparation of an enaminone by a two-carbon homologation of amides with lithium(triphenylsilyl)acetylide was already described by Suzuki et al. in the eighties [15]. A brief explanation of the reaction mechanism was given, suggesting a sequence of events, including an initial formation of a silylalkynone, followed by a Michael-type addition of in situ-formed lithium amide, and a subsequent protiodesilylation [16]. Interestingly, a D$_2$O-quenching experiment showed the double incorporation of deuterium in the generated double bond [15]. Inspired by the mechanistic explanations provided by Suzuki et al., and to better visualize this interesting transformation, we hereby suggest a stepwise mechanism to explain the β-enaminone formation from the addition of lithium (trimethylsilyl)acetylide to the quinoline-proline-piperazine side chain of compound **6** (Figure 5). In a first step, the lithium (trimethylsilyl)acetylide attacks the carbonyl of the amide that is located between the quinoline and proline. The resulting intermediate **9** undergoes a rearrangement following events *1–3*, which is driven by the electrophilic nature of acetylenic carbons. The rearranged intermediate **10** is then protonated by a molecule of water to afford **11**, which thereafter undergoes a protiodesilylation leading to the β-enaminone **8a**. Using the proposed mechanism it is now possible to explain the synthesis of compound **1**, from RM-128 (C17-ketone) or RM-133 (17α-ethynyl). In the first case, the carbonyls C-17 and C-1′′′ of RM-128 will react with lithium(trimethylsilyl)acetylide to introduce the 17α-ethynyl and to produce the β-enaminone side chain. In the second case, however, only the C-1′′′ of RM-133 is involved in the reaction producing **1**.

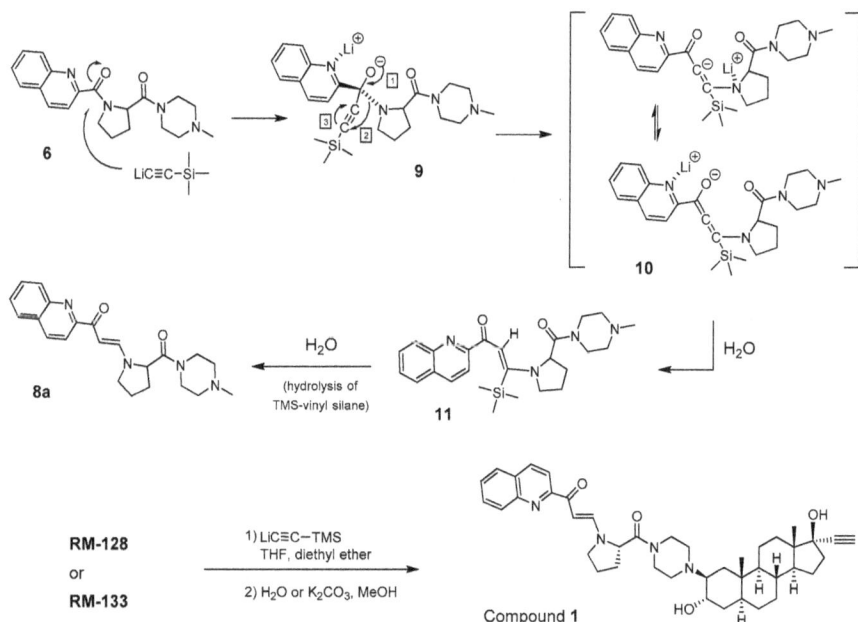

Figure 5. Proposed mechanism of reaction leading to β-enaminones **8a** and **1**.

3. Materials and Methods

3.1. General

Chemical reagents and solvents were purchased from commercial suppliers and were used as received. Phase separator syringes were purchased from Biotage (Isolute phase separator, 6 mL). Thin-layer chromatography (TLC) and flash-column chromatography were performed on 0.20-mm silica gel 60 F254 plates (E. Merck; Darmstadt, Germany) and with 230–400 mesh ASTM silica gel 60 (Silicycle, Québec, QC, Canada), respectively. Infrared (IR) spectra were recorded with a Horizon MB 3000 ABB FTIR spectrometer (Québec, QC, Canada). Nuclear magnetic resonance (NMR) spectra were recorded at room temperature in CDCl$_3$ with a 5-mm NMR tube on a Bruker AVANCE 400 spectrometer (Billerica, MA, USA). ^{1}H and ^{13}C NMR chemical shifts were referenced to the residual peak of CHCl$_3$ (7.26 and 77.0 ppm, respectively). For characterization, we used the following experiments: attached proton test (APT), correlation spectroscopy (COSY), homonuclear two-dimensional NOE spectroscopy (NOESY), heteronuclear single-quantum correlation (HSQC), and heteronuclear multiple-bond correlation (HMBC). These NMR experiments were performed according to the manufacturer's instructions. Low-resolution mass spectra (LRMS) were recorded on a Shimadzu Prominence apparatus (Kyoto, Japan) equipped with a Shimadzu LCMS-2020 mass spectrometer and an APCI probe.

3.2. Synthesis of Side Chain 6

The details of the chemical synthesis of **6** were previously published [6].

3.3. Synthesis of β-Enaminones **7a** and **8a**

To a solution of trimethylsilylacetylene (1.35 mL, 9.55 mmol) in anhydrous tetrahydrofuran (THF) (100 mL) at 0 °C under an atmosphere of argon was added dropwise methyl lithium (4.8 mL, 7.68 mmol; 1.6 M in diethyl ether) over a period of 10 min. This solution was left to return to room temperature and was stirred for 1 h. The solution was then cooled again at 0 °C and compound **6** (870 mg 1.91 mmol) was added in anhydrous THF (20 mL). The solution was stirred at room temperature for 1 h. The resulting solution was then poured into water (750 mL) and was extracted twice with dichloromethane (DCM). The organic layer was washed with water, dried over phase separator syringe, and evaporated under reduced pressure to give 750 mg of a brown amorphous solid. The crude compound was purified by flash chromatography using a gradient of DCM/MeOH (95:5) to DCM/MeOH (9:1) to give β-enaminones **7a** (117 mg, 15%) and **8a** (226 mg, 30%) as green and yellow amorphous solid, respectively.

(E)-3-(4-methylpiperazin-1-yl)-1-(quinolin-2-yl)prop-2-en-1-one **(7a)**: ^{1}H NMR (CDCl$_3$) δ in ppm: 2.35 (s, 3H, NCH$_3$), 2.51 (t, *J* = 5.0 Hz, 4H, CH$_2$-1'), 3.54 (s broad, 4H, CH$_2$-2'), 6.80 (d, *J* = 12.7 Hz, 1H, CH-*b*), 7.58 (t, *J* = 7.1 Hz, 1H, CH-7'''), 7.74 (t, *J* = 8.4 Hz, 1H, CH-8'''), 7.85 (d, *J* = 8.1 Hz, 1H, CH-6'''), 7.91 (d, *J* = 12.9 Hz, 1H, CH-*a*), 8.18 (d, *J* = 8.5 Hz, 1H, CH-9'''), 8.26 (s, 2H, CH-3''' and CH-4'''). ^{13}C NMR (CDCl$_3$) δ in ppm: 46.6 (CH$_2$-2'), 46.0 (NCH$_3$), 54.6 (CH$_2$-1'), 91.0 (CH-*b*), 119.0 (CH-3'''), 127.4 (CH-7'''), 127.6 (CH-6'''), 129.0 (CH-5'''), 129.4 (CH-9'''), 130.1 (CH-8'''), 136.6 (CH-4'''), 147.0 (C-10'''), 153.0 (CH-*a*), 155.9 (C-2'''), 187.6 (C-1'''). LRMS for C$_{17}$H$_{20}$N$_3$O [M + H]$^+$ 282.2 (calc), 282.1 (found). *(S,E)-3-(2-(4-methylpiperazine-1-carbonyl)pyrrolidin-1-yl)-1-(quinolin-2-yl)prop-2-en-1-one*

(8a): ^{1}H NMR (CDCl$_3$) δ in ppm: 1.90–2.70 (m, 8H, CH$_2$-3'', CH$_2$-4'', 2 x CH$_2$-1'), 2.27 (s, 3H, NCH$_3$), 3.41, 3.51, 3.65 and 3.84 (4m, 6H, CH$_2$-2' and CH$_2$-5''), 4.62 (s broad, 1H, CH-2''), 6.45 and 6.70 (2s broad, 1H, CH-*b*), 7.53 (t, *J* = 7.1 Hz, 1H, CH-7'''), 7.69 (t, *J* = 7.0 Hz, 1H, CH-8'''), 7.79 (d, *J* = 8.1 Hz, 1H, CH-6'''), 7.90 and 8.08 (2d, *J* = 13.0 Hz, 1H, CH-*a*), 8.14 (m, 1H, CH-9'''), 8.19 (s, 2H, CH-3''' and CH-4'''). ^{13}C NMR (CDCl$_3$) δ in ppm: 23.4 and 23.9 (CH$_2$-4''), 30.1 and 30.6 (CH$_2$-3''), 42.1, 45.1, 45.5 and 45.7 (CH$_2$-2'), 46.0 (NCH$_3$), 48.1 and 53.3 (CH$_2$-5''), 54.5 and 54.8 (CH$_2$-1''), 57.9 and 62.7 (CH$_2$-2''), 93.7 (CH-*b*), 119.1 (CH-3'''), 127.4 (CH-7'''), 127.6 (CH-6'''), 129.0 (CH-5'''), 129.3 and 129.4 (CH-9'''), 130.3 (CH-8'''), 136.5 (CH-4'''), 147.1 (C-10'''), 150.0 (CH-*a*), 156.1 (C-2'''), 168.6 and 169.4 (C-1''), 187.4 (C-1'''). LRMS for C$_{22}$H$_{27}$N$_4$O$_2$ [M + H]$^+$ 379.2 (calc), 379.1 (found).

3.4. Synthesis of Enaminone **1**

To a solution of trimethylsilylacetylene (216 μL, 1.52 mmol) in anhydrous THF (5 mL) at 0 °C under an atmosphere of argon, was added dropwise methyl lithium (860 μL, 1.38 mmol; 1.6 M in diethyl ether) over a period of 10 min. This solution was left to return to room temperature and was stirred for 1 h. The solution was then cooled again at 0 °C and compound RM-133 (100 mg 0.15 mmol) was added in anhydrous THF (20 mL). The solution was stirred at room temperature overnight. The resulting solution was then poured into water (200 mL) and extracted two times with EtOAc. The organic layer was washed with brine, dried over sodium sulfate, filtered, and evaporated under reduced pressure to give 110 mg of a yellow amorphous solid. The crude compound was purified by flash chromatography using a gradient of DCM/MeOH (97:3) to DCM/MeOH (9:1) to give β-enaminone **1** (31 mg, 30%) as fluorescent pale green amorphous solid.

(S,E)-3-(2-(4-[(3β,5α,17α)-17-hydroxypregn-20-yn-3-yl]piperazine-1-carbonyl)pyrrolidin-1-yl)-1-(quinolin-2 -yl)prop-2-en-1-one (**1**): ^{1}H NMR (CDCl$_3$) δ in ppm: 0.75 (m, 1H, CH-9), 0.83 (s, 3H, CH$_3$-18), 0.85 (s, 3H, CH$_3$-19), 0.88 (m, 1H of CH$_2$-7), 1.10–2.35 (residual CH and CH$_2$), 2.40–2.75 (m, 5H, CH-2α and 2 x CH$_2$-1'), 2.56 (s, 1H, CH-21), 3.35–3.40 (m, 6H, CH$_2$-5'' and 2 x CH$_2$-2'), 3.86 (m, 1H, CH-3β), 4.66 (s broad, 1H, CH-2''), 6.48 and 6.73 (2s broad, 1H, CH-*b*), 7.57 (t, *J* = 7.4 Hz, 1H, CH-7'''), 7.73 (t, *J* = 7.2 Hz, 1H, CH-8'''), 7.84 (d, *J* = 8.1 Hz, 1H, CH-6'''), 7.94 and 8.10 (2d, *J* = 13.0 Hz, 1H, CH-*a*), 8.16 (m, 1H, CH-9'''), 8.13 (s, 2H, CH-3''' and CH-4'''). ^{13}C NMR (CDCl$_3$) δ in ppm: 23.6, 23.9 (CH$_2$-4''), 30.1 and 30.7 (CH$_2$-3''), 41.9 and 42.7, 45.8 and 46.2 (CH$_2$-2'), 48.1 and 53.4 (CH$_2$-5''), 46.8, 48.0 and 48.6 (CH$_2$-1'), 57.7 and 62.7 (CH$_2$-2''), 93.6 (CH-*a*), 119.1 (CH-3'''), 127.0 (CH-7'''), 127.4 (CH-6'''), 129.1 (CH-5'''), 129.9 and 130.3 (CH-9'''), 129.4 (CH-8'''), 136.5 (CH-4'''), 147.1 (C-10'''), 149.7 (CH-*b*), 156.1 (C-2''), 168.9 and 169.2 (C-1''), 187.5 (C-1'''). IR (film) ν in cm-1: 3480 (OH), 1643 (C=O), 1543 (C=C). LRMS for C$_{42}$H$_{54}$N$_4$O$_4$ [M + H]$^+$ 679.4 (calc), 679.5 (found).

4. Conclusions

The combination of a steroid nucleus and a quinoline-proline-piperazine side chain has resulted in the discovery of a new family of anticancer drugs that induce the apoptosis of cancer cells via endoplasmic reticulum stress. The fortuitous modification of this chain in the presence of lithium (trimethylsilyl) acetylide having generated an unknown product that is more active than the precursor, it was important to determine its structure. After having determined the formation of two homologation products from a model side chain, thanks to NMR analysis, we have been able to identify the unknown product as being β-enaminone **1**. We also proposed a mechanism to rationalize the formation of this biologically active compound. The optimization of the experimental conditions remains to be done, but this type of rearrangement could be favorably used in a diversity-oriented synthesis strategy to obtain structurally different substituted enaminones, such as **7a** and **8a**, from the same amide unit.

Acknowledgments: We are grateful to the Canadian Institutes of Health Research for financial support (POP-I program), to Jean-Yves Sancéau for helpful discussions, to Marie-Claude Trottier and Sophie Boutin for NMR analyses and to Micheline Harvey for careful reading of this manuscript.

Author Contributions: D.P. and R.M. conceived and designed the experiments; R.M. performed the experiments; D.P. analyzed the data; D.P. and R.M. wrote the paper.

Conflicts of Interest: The authors declare no conflict of interest. D.P. and R.M. have patent rights on US8653054 and CA2,744,369 (2-(*N*-Substituted piperazinyl) steroid derivatives) as well as on PCT/CA2017/000140 (Aminosteroid derivatives and process for producing same).

References

1. Siegel, R.L.; Miller, K.D.; Jemal, A. Cancer statistics, 2017. *CA Cancer J. Clin.* **2017**, *67*, 7–30. [CrossRef] [PubMed]

2. Gabriele, L.; Buoncervello, M.; Ascione, B.; Bellenghi, M.; Matarrese, P.; Care, A. The gender perspective in cancer research and therapy: Novel insights and on-going hypotheses. *Ann. Ist. Super. Sanit.* **2016**, *52*, 213–222.

3. Gupta, A.; Kumar, B.S.; Negi, A.S. Current status on development of steroids as anticancer agents. *J. Steroid Biochem. Mol. Biol.* **2013**, *137*, 242–270. [CrossRef] [PubMed]

4. Atkins, J.H.; Gershell, L.J. Selective anticancer drugs. *Nat. Rev. Drug Discov.* **2002**, *1*, 491–492. [CrossRef] [PubMed]

5. Perreault, M.; Maltais, R.; Roy, J.; Dutour, R.; Poirier, D. Design of a mestranol 2-*N*-piperazino-substituted derivative showing potent and selective in vitro and in vivo activities in MCF-7 breast cancer models. *ChemMedChem* **2017**, *12*, 177–182. [CrossRef] [PubMed]

6. Perreault, M.; Maltais, R.; Dutour, R.; Poirier, D. Explorative study on the anticancer activity, selectivity and metabolic stability of related analogs of aminosteroid RM-133. *Steroids* **2016**, *115*, 105–113. [CrossRef] [PubMed]

7. Maltais, R.; Hospital, A.; Delhomme, A.; Roy, J.; Poirier, D. Chemical synthesis, NMR analysis and evaluation on a cancer xenograft model (HL-60) of the aminosteroid derivative RM-133. *Steroids* **2014**, *82*, 68–76. [CrossRef] [PubMed]

8. Ayan, D.; Maltais, R.; Hospital, A.; Poirier, D. Chemical synthesis, cytotoxicity, selectivity and bioavailability of 5alpha-androstane-3alpha,17beta-diol derivatives. *Bioorg. Med. Chem.* **2014**, *22*, 5847–5859. [CrossRef] [PubMed]

9. Jegham, H.; Roy, J.; Maltais, R.; Desnoyers, S.; Poirier, D. A novel aminosteroid of the 5α-androstane-3α,17β-diol family induces cell cycle arrest and apoptosis in human promyelocytic leukemia HL-60 cells. *Investig. New Drugs* **2012**, *30*, 176–185. [CrossRef] [PubMed]

10. Jegham, H.; Maltais, R.; Roy, J.; Doillon, C.; Poirier, D. Biological evaluation of a new family of aminosteroids that display a selective toxicity for various malignant cell lines. *Anti-Cancer Drugs* **2012**, *23*, 803–814. [CrossRef] [PubMed]

11. Kenmogne, L.C.; Ayan, D.; Roy, J.; Maltais, R.; Poirier, D. The aminosteroid derivative RM-133 shows in vitro and in vivo antitumor activity in human ovarian and pancreatic cancers. *PLoS ONE* **2015**, *10*, e0144890. [CrossRef] [PubMed]

12. Talbot, A.; Maltais, R.; Poirier, D. New diethylsilylacetylenic linker for parallel solid-phase synthesis of libraries of hydroxy acetylenic steroid derivatives with improved metabolic stability. *ACS Comb. Sci.* **2012**, *14*, 347–351. [CrossRef] [PubMed]

13. Huber, M.M.; Ternes, T.A.; von Gunten, U. Removal of estrogenic activity and formation of oxidation products during ozonation of 17alpha-ethinylestradiol. *Environ. Sci. Technol.* **2004**, *38*, 5177–5186. [CrossRef] [PubMed]

14. Thorpe, K.L.; Cummings, R.I.; Hutchinson, T.H.; Scholze, M.; Brighty, G.; Sumpter, J.P.; Tyler, C.R. Relative potencies and combination effects of steroidal estrogens in fish. *Environ. Sci. Technol.* **2003**, *37*, 1142–1149. [CrossRef] [PubMed]

15. Suzuki, K.; Ohkuma, T.; Tsuchihashi, G. Preparation of enaminones by two-carbon homologation of amides with lithium (triphenylsilyl)acetylide. *J. Org. Chem.* **1987**, *52*, 2929–2930. [CrossRef]

16. Newman, H. Preparation of α,β-unsaturated aldehydes from acid chlorides. *J. Org. Chem.* **1973**, *38*, 2254–2255. [CrossRef]

magnetochemistry

MDPI

Article

Low Field NMR Determination of pKa Values for Hydrophilic Drugs for Students in Medicinal Chemistry

Aleksandra Zivkovic [1], Jan Josef Bandolik [1], Alexander Jan Skerhut [1], Christina Coesfeld [1], Miomir Raos [2], Nenad Zivkovic [2], Vlastimir Nikolic [3] and Holger Stark [1],*

[1] Institute of Pharmaceutical and Medicinal Chemistry, Heinrich Heine University Düsseldorf, Universitätsstr. 1, 40225 Duesseldorf, Germany; aleksandra.zivkovic@hhu.de (A.Z.); Jan.Bandolik@uni-duesseldorf.de (J.J.B.); alexander.skerhut@uni-duesseldorf.de (A.J.S.); christina-coesfeld@gmx.de (C.C.); stark@hhu.de (H.S.)

[2] Faculty of Occupational Safety, University of Nis, Carnojevica 10A, 18000 Nis, Serbia; miomir.raos@znrfak.ni.ac.rs (M.R.); nenad.zivkovic@znrfak.ni.ac.rs (N.Z.)

[3] Faculty of Mechanical Engineering, University of Nis, Aleksandra Medvedeva 14, 18000 Nis, Serbia; vnikolic@masfak.ni.ac.rs

* Correspondence: stark@hhu.de; Tel.: +49-211-811-0478

Received: 6 September 2017; Accepted: 19 September 2017; Published: 20 September 2017

Abstract: For an interdisciplinary approach on different topics of medicinal and analytical chemistry, we applied a known experimental pKa value determination method on the field of the bench top nuclear magnetic resonance (NMR) spectrometry of some known biologically active pyridine-based drugs, i.e., pyridoxine hydrochloride, isoniazid, and nicotine amide. The chemical shifts of the aromatic ring protons in the ^{1}H NMR spectrum change depending on the protonation status. The data were analyzed on dependence of the chemical shifts by different pH (pD) environments and then the pKa values were calculated. The pKa values obtained were in agreement with the literature data for the compounds, searched by the students on web programs available at our university. The importance of the pKa values in protein-ligand interactions and distribution etc. of drugs was brought up to the students' attention. In addition, by the use of a free web application for pKa values prediction, students calculated the predicted modeled pKa value. The experimental and in-silico approaches enhance the tool box for undergraduate students in medicinal chemistry.

Keywords: pKa value; low field NMR; molecular properties; pharmacokinetics; undergraduate pharmacy students

---

1. Introduction

The protonation status of drugs under physiological and pathophysiological conditions is of mandatory importance for the understanding of drug action concerning pharmacodynamics and the pharmacokinetic mechanism of action. NMR experiments deliver a versatile experimental method for the determination of pKa values and thereby increase the knowledge on NMR techniques. Due to the complex theoretical background, it has remained a challenge to teach the basics in NMR spectroscopy. Although this analytical technique has been one of the most important techniques for structure identification for decades, it has been difficult to teach this as practical course containing students' experimental NMR spectroscopy because of the price of the instruments and the handling. The development of bench-top NMR spectrometers has made this kind of teaching possible [1–3]. Interest for complex and high molecular weight protein structures has been an inspiration for the amazing developments in the high-resolution nuclear magnetic resonance. This kind of structural analysis is usually performed in the strong magnetic field generated up to 1.2 GHz by huge magnets.

The technological progress allowed the development of the smaller but more powerful superconducting magnets. For many reasons, the student experiments done in instrumental analytic courses on sophisticated equipment were mostly limited to spectra interpretation. There is still a strong need to teach undergraduate students this attractive technique through experimentation.

The NMR improvement in strength, size, homogeneity, and temperature stability endorsed the production of small, portable NMRs. There are a few low-field spectrometers available on the market today. There is no need to use the cryogens, and there are no additional maintenance costs [4]. There are other experiments developed for the student lab involving hydrogen bond dynamics determination, kinetic determination, oil spill determinations, quantitative determination, etc. In order to introduce our students to ^{1}H NMR spectroscopy, we offer the possibility to solve a physico-chemical problem (determination of pKa values of drugs) employing the newly obtained knowledge on NMR spectroscopy [5–8]. We focused on developing the experiments that stress the importance of the NMR techniques in pharmaceutical analytics and drug design. We previously published two experiments that deepened the understanding of structure determination [2] and quantification [1] using NMR spectroscopy. Generally, NMR spectroscopy in a student lab is regarded as a technique for structural determination. It is rarely considered as reliable method for experimental pKa value determination, although the method is common in science, especially in protein determination [9]. However, with the full set of experiments, we are trying to show our students how NMR can be also employed for determination of physiochemical characteristics of the compounds. The significance of acid/base properties in drug discovery and pharmacokinetics research is of the great importance in medicinal chemistry. The pKa value of the drug influences drug lipophilicity, solubility, protein binding, and permeability. All these affect pharmacokinetic characteristics of the drugs, such as adsorption, distribution, metabolism, and excretion (ADME) [10,11]. Furthermore, the pKa value and acidic/basic characteristics can help us determine if the drug is suitable for oral application or not. The determination of those characteristics early in the drug development process allows pharmaceutical companies to reduce the number of failed studies. It is common knowledge that the pH of different body compartments is very different, varying from roughly pH of 1 to pH of 9. Depending on the pKa value of the drug, an optimal place for drug absorption can be predicted [12].

We designed an experiment for undergraduates with simple drug structures but tried to make our students familiar with the possibility that the same approach can also be applied on more complex structures. To foster research conditions, our goal is that students search for the drugs and the acidity/basicity of the drugs and model the data on their own using web applications. The assistant's role is to encourage the students to research and ask questions about the method and the results and to discuss the conclusions with the students. The main learning goals are (i) to deepen the understanding of the ^{1}H NMR technique, (ii) to understand the connection in-between the chemical characteristics and chemical shifts, (iii) to understand the effects that pKa value has on pharmacokinetics, and (iv) to process the data on their own using MNOVA Software.

We applied Peer-Led Team Learning (PLTL), benefiting the student collaboration. We supplied the group leaders with the publications determining the pKa values of the solvents and let them form groups of three to discuss the approach and understand how it works [9]. Afterward, the students were asked to cross-reference the Pharmacopoea Europaea (Ph. Eur.) (searching for drugs containing a pyridine substructure) and find compounds that could potentially be suitable for determination using this approach. After cross-referencing their list with the drug inventory list we supplied, they decided on their own which compound they wanted to use for the determination. The work of the students was facilitated with pre-leaders and instructors. The small group learning setting is very beneficial [13].

The pH-dependent change in chemical shift can be followed by ^{1}H NMR measurements and used for pKa value determination. For reasons of comparison, we took advantage of the chemical and therefore magnetic changes due to protonation of the pyridine ring on related drugs (Figure 1). We determined the pKa values, corresponding the nitrogen in the heterocyclic pyridine ring in isoniazid, pyridoxine hydrochloride, and nicotinamide (Figure 2). The NMR-based experiments on the physicochemical properties of known drugs enables the students to combine the protonation status to ^{1}H NMR spectra interpretation (*cf.* Supporting Information).

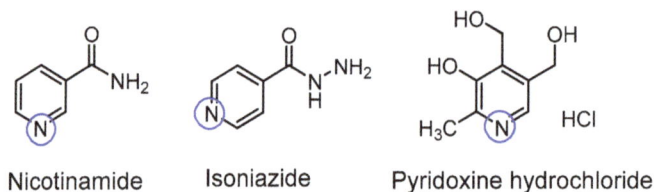

Figure 1. p*Ka* values of the nicotinamide, isoniazid, and pyridoxine hydrochloride were determined.

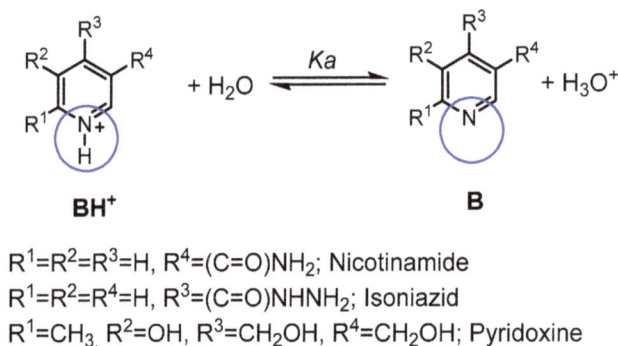

$R^1=R^2=R^3=H$, $R^4=(C=O)NH_2$; Nicotinamide
$R^1=R^2=R^4=H$, $R^3=(C=O)NHNH_2$; Isoniazid
$R^1=CH_3$, $R^2=OH$, $R^3=CH_2OH$, $R^4=CH_2OH$; Pyridoxine

Figure 2. Protonation changes of the drugs in water.

2. Results

2.1. Method Adoption

In a previously published experiment (developed in 1973 and adopted in 2012 [9,14]), pyridine, picoline, and lutidine solvents were used for the determination. However, for our approach, it was important to show students that simple, basic scientific methods can be transferred and applied to the more complex structures they focus on in their studies, as the pharmacokinetics of drugs is one of the most important issues in drug development. In the classic high field NMR machines, a deuterated solvent with internal standard is used, whereas our instrument uses an external lock and can work in non-deuterated solvents and without internal standards. The method itself was only slightly modified. The internal standard was omitted, whereas the pH was adjusted in a related manner as described in ref. [9]. These procedures have been applied to water-soluble pyridine-based drugs (Figure 1, *cf.* Supplementary Materials Figures S1–S7, Tables S1–S3).

We observed the following protonation changes of the drugs in water:

The ionization constant for the equilibrium is given in the classic Equation (1).

$$Ka = \frac{[H_3O^+]\ [B]}{[BH^+]} \qquad (1)$$

By transformation into a logarithmic function, Equation (1) is reformed into the Equation (2) [9].

$$pKa = pH + \log \frac{[BH^+]}{[B]} \qquad (2)$$

The proton chemical shifts of the cationic and charged molecules are very different. The main reasons for this difference are anisotropy and electron density. When the proton dissociation is faster than the NMR time scale, there is only one signal to be observed representing the average position of the signal, defined by the Equation (3) [9].

$$\delta_{obs.} = \delta_{BH^+}P_{BH^+} + \delta_B P_B \tag{3}$$

Equation (3) can be further simplified, as previously reported in detail in [9] (here shown with the Equations (4)–(6)). This is used to connect the NMR experiment with this physicochemical experiment.

$$P_{B.} = \frac{\delta_{pH=1} - \delta}{\delta_{pH=1} - \delta_{pH=13}} \tag{4}$$

$$P_{B.} + P_{BH^+} = 1 \tag{5}$$

$$pKa = pH + \log\frac{P_{BH^+}}{P_B} \tag{6}$$

For our measurements, we used the mixture of the D_2O and H_2O solutions at a startiing concentration of approximately 20 mg/mL. We measured so called pH* by directly reading the D_2O-solution of the water-calibrated pH-meter. The conversion of the pH* into the pD is done by adding a constant of 0.4. The pKa (H) values were calculated with Equation (7) from a determined pKa (D) [15].

$$pKa\,(D) - pKa\,(H) = 0.076pK\,(H) - 0.05 \tag{7}$$

2.2. Student Experimental Results

This experiment was done by approximately 70 students attending an instrumental analytics course for pharmacists, divided into three-membered groups (*cf.* Supporting Information). As an example, we showed the students a data plot of the pyridoxine NMR titration. Each of aromatic protons can be used for the processing of NMR data to determine the pKa value. Enlarging the region with the protons of interest was helpful when analyzing data (Figure 3).

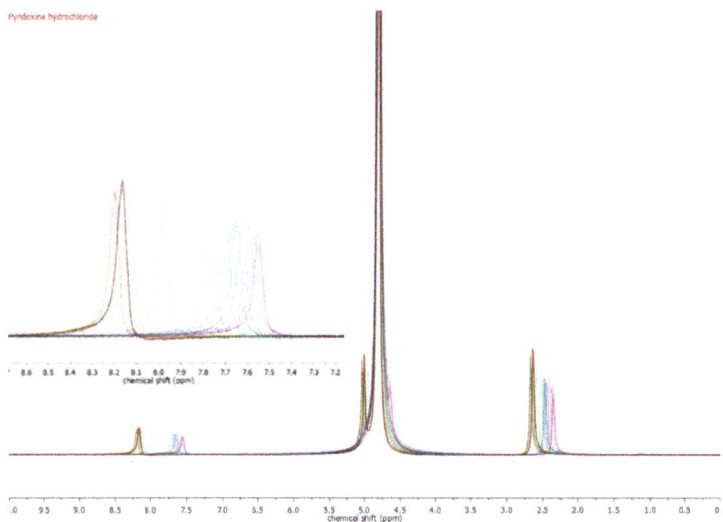

Figure 3. Stacked ^{1}H NMR spectra of pyridoxine at different pH values (starting conc. 20 mg/mL) showing full-scale and enlarged region of interest.

The collected ^{1}H NMR data were processed and plotted against the pD values (i.e., value related to pH in D$_2$O) (Figure 4). The curves obtained can be compared to regular titration curves. If so, the pKa can be determined as the equivalence point by titration, as determined using a concentrated arc method (Tubbs' method, [16]) or as the first derivation of the function (Excel). Therefore, the determined point can be recalculated using Equation (7) (*cf.* Supporting Information Chart S1). Working with undergraduate students applying of graphical method has advantages as they usually do not have a lot of experience in creating charts. The pH* is measured with a glass electrode upon the addition of either the KOH-solution or the HCl-solution. All experimental details and data can be found in the supporting information.

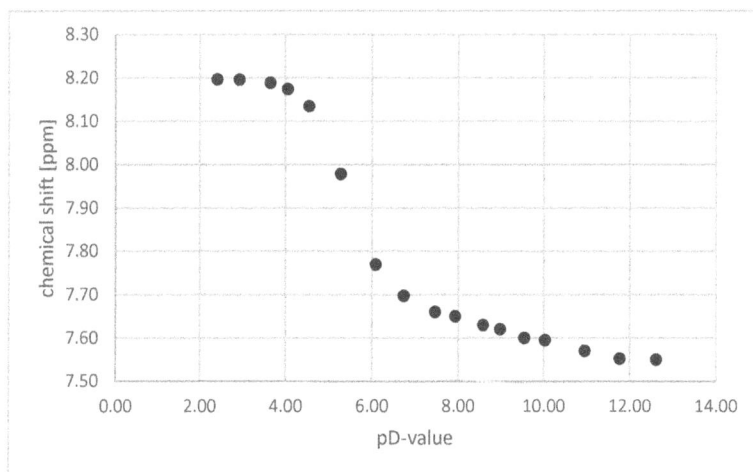

Figure 4. An example plot of the ^{1}H NMR chemical shift of a ring proton of pyridoxine hydrochloride as a function of pD.

The results obtained by students, the corresponding literature values, and the values obtained by the computational approaches are shown in Table 1.

Table 1. The determined pKa values by NMR experiments, literature data, and predicted values based on computational approaches.

Compound	pKa (NMR Method) [b]	pKa [c] (Literature)	pKa (Web-Based Calculation) [d]
Nicotinamide	3.54 ± 0.02	3.35 [17]	3.60
Isoniazid	3.65 ± 0.19	3.50 [18]	3.20
Pyridoxine [a]	5.24 ± 0.15	5.20 [19]	5.60

All pKa values correspond to the pKa values of the heterocyclic ring nitrogen. [a] Used as hydrochloride salt. [b] The student data are the mean value of two determinations and recalculation according to Equation (7). Values for one standard deviation are also included (upper t probability t of 0.10); [c] Ref. given for each value; [d] Calculated using CE and JChem acidity and basicity calculator, https://epoch.uky.edu/ace/public/pKa.jsp.

3. Discussion

With the positive feedback from the students, we evaluated the learning outcome. Working in groups with defined group leader was successful, and the discussions were fruitful. Each group was able to apply the concept and successfully determine the pKa value of interest, which confirms that the methodology is simple enough to be done by students. The applied methodology allowed the students to learn important features of the NMR spectroscopy on their own and employ their learning with better success than when solving imagined problems in a theoretical course. Hands-on learning in small, peer-lead groups was a new experience for our students. Even though the process was guided by assistants due to the cost issue (list of matches is limited), this gave them an important feeling of working in a scientific setting. Searching through the literature on their own and using even a very simple modeling assignment showed them the possibilities and issues in the science. Stressing the importance of the physicochemical characteristics in the process of drug discovery and in adsorption/distribution/metabolism/elimination (ADME) gave our students a better perspective on how the learning of elemental basics (acidity/basicity, p*Ka* values) is of use with regard to those complex topics. In the final discussion, the importance of the p*Ka* values of drugs in pharmacodynamics and pharmacokinetic mechanisms is generally discussed.

In addition, in a written exam, the students obtained in average 15% more points in the NMR assignments and in electrochemistry assignments involving potentiometric titrations. When asked what were the two most positive aspects of this experiment, the student answered that the most attractive elements for them were to getter on their own the information from the different sources and then to apply those in the experiment and to measure the NMR spectra on their own. To conclude, the modern PLTL-method, combined with molecular modeling and a hands-on NMR experience, had a very positive influence on the knowledge and motivation of the students in the lab.

4. Materials and Methods

4.1. Materials and Instrumentation

A variety of low field instruments are on the market and are suitable for use in this experiment. We used Magritek Spinsolve NMR Benchtop (Magritek, Aachen, Germany) (42.5 MHz) with standard 5 mm NMR-tubes. The measurement used was a 1D PROTON Powerscan measurement using a single 90-degree excitation pulse. The standard 1D measurement used 90-degree excitation pulse as this maximizes the signal in the x,y-plane where the signal is detected [20].

4.2. Prelab Exercise

The students worked in groups of three, and three 6 h lab periods were required for the experiment. According to the PLTL, we supped group leaders with reference [5] and allow them one lab period (6 h) to discuss the method with the group. Then, they chose the drugs from Ph. Eur. that could be used for the determination according to protonation and expected hydrophilicity. In the third step, they needed to cross-reference their drug list with our drug inventory (supplied by the supervisor) and decide on the drug they wanted to use for the determination. At this point (1 h to the end of the first lab period), they discussed the method and the drug of their selection with the instructor. For the drug selected, they researched the application and probable absorption body compartment, as well as all the details concerning the expected p*Ka* (for example, ionization levels at different pH of the body compartments). All the important chemical structures involved had to be drawn and discussed. All the drugs student findings by cross-referencing are given in the Figure 1. Nevertheless, it is possible to extend this on more complex compounds with more experienced students.

4.3. Procedure

In the second lab period (6 h), the students performed the actual experimental measurements. In general, the solutions of the drugs were measured by ^{1}H NMR at different pH values of the solvent, and a classical titration curve was obtained by the chemical shift vs. the pH value. All pyridine-containing drugs in our experiments had at least one aromatic proton that could be easily taken as a reference signal. The students are free to take the proton of their choice or, in advanced groups, to compare the results while using different protons. The p*Ka* value can simply be determined by graphical evaluation or by calculation. The desired value could be determined from the titration curves (e.g., Tubbs' concentric arcs method of determination of equivalence point) [16,21]. We decided to apply this graphical determination, as up to this point in the students' education, this was the most common one. Certainly, there are other methods that can be applied, e.g., non-linear regression method, derivative method, etc. [9,16,21]. Another possibility is to calculate the mean values of the p*Ka* value determination where the chemical shifts are rapidly changing. There is no internal standard used, as the measurements are referenced to the H^2HO chemical shift, even though this is temperature-dependent at 20–25 °C, as can be found at 4.78 ppm [22]. All of the prepared solutions are adjusted using HCl and KOH solutions, as briefly described in the literature and Supporting Information [9].

The third lab period was left for the group to do literature recherché (www.reaxys.com and www.scifinder.org) and use web-based calculator to obtain the predicted p*Ka* values (https://epoch.uky.edu/ace/public/pKa.jsp). For the literature search, any web-based literature search programs could be used. Our students used sophisticated, campus-licensed programs from Heinrich Heine University, as mentioned above. They wrote a protocol and discussed their results, the literature value, and the value they obtained from the molecular modeling webpage.

5. Conclusions

The NMR-based experimental determination of p*Ka* values of different drugs in comparison to in-silico calculation with increments resulted in an increased interest and a critical scientific view of the students for data material and for physicochemical properties. The students in medicinal chemistry have achieved a deeper understanding on the importance of protonation status as well as the use of NMR as a general and versatile tool in structural determination.

Supplementary Materials: Supporting information is available online at www.mdpi.com/2312-7481/3/3/29/s1. Figure S1: 1H NMR data of Nicotinamide (extraction), Figure S2: Titration curve of Nicotinamide, Figure S3: 1H NMR data of Pyridoxine hydrochloride, Figure S4: 1H NMR data of Pyridoxine hydrochloride (extraction), Figure S5: Titration curve of Pyridoxine hydrochloride, Figure S6: 1H NMR data of Isoniazid (extraction), Figure S7: Titration curve of Isoniazid, Table S1: Chemical shifts and p*Ka* values from Nicotinamid determination, Table S2: Chemical shifts and p*Ka* values from Pyridoxine hydrochloried determination, Table S3: Chemical shifts and p*Ka* values from Isoniazid determination.

Acknowledgments: Funding for this project was kindly provided by a fund for teaching support (Lehrförderfonds) from Heinrich Heine University in Düsseldorf number L020401/15. We acknowledge the support of the Serbian Ministry of Science and Technological Development, project number III43014.

Author Contributions: A.Z. and H.S. conceived and designed the experiments; A.Z., J.J.B., A.J.S. and C.C. performed the experiments and analyzed the data; M.R., N.Z. and V.N. contributed important discussion points and technical notes; A.Z. and H.S. wrote the paper; All authors have revised the paper and agreed to publication.

Conflicts of Interest: The authors declare no conflict of interest.

References

1. Zivkovic, A.; Bandolik, J.J.; Skerhut, A.J.; Coesfeld, C.; Prascevic, M.; Zivkovic, L.; Stark, H. Quantitative analysis of multicomponent mixtures of over-the-counter pain killer drugs by low-field NMR spectroscopy. *J. Chem. Educ.* **2017**, *94*, 121–125. [CrossRef]

2. Zivkovic, A.; Bandolik, J.J.; Skerhut, A.J.; Coesfeld, C.; Zivkovic, N.; Raos, M.; Stark, H. Introducing students to NMR methods using low-field 1H NMR spectroscopy to determine the structure and the identity of natural amino acids. *J. Chem. Educ.* **2017**, *94*, 115–120. [CrossRef]

3. Bonjour, J.L.; Pitzer, J.M.; Frost, J.A. Introducing high school students to NMR spectroscopy through percent composition determination using low-field spectrometers. *J. Chem. Educ.* **2015**, *92*, 529–533. [CrossRef]

4. Danieli, E.; Perlo, J.; Blümich, B.; Casanova, F. Small magnets for portable NMR spectrometers. *Angew. Chem. Int. Ed.* **2010**, *49*, 4133–4135. [CrossRef] [PubMed]

5. Periyannan, G.R.; Lawrence, B.A.; Egan, A.E. 1H NMR Spectroscopy-Based Configurational Analysis of Mono- and Disaccharides and Detection of β-Glucosidase Activity: An Undergraduate Biochemistry Laboratory. *J. Chem. Educ.* **2015**, *92*, 1244–1249. [CrossRef]

6.   Morton, J.G.; Joe, C.L.; Stolla, M.C.; Koshland, S.R.; Londergan, C.H.; Schofield, M.H. NMR Determination of Hydrogen Bond Thermodynamics in a Simple Diamide: A Physical Chemistry Experiment. *J. Chem. Educ.* **2015**, *92*, 1086–1090. [CrossRef]

7.   Mobley, T.A. NMR kinetics of the SN2 reaction between BuBr and I: An introductory organic chemistry laboratory exercise. *J. Chem. Educ.* **2015**, *92*, 534–537. [CrossRef]

8.   Simpson, A.J.; Mitchell, P.J.; Masoom, H.; Liaghati Mobarhan, Y.; Adamo, A.; Dicks, A.P. An oil spill in a tube: An accessible approach for teaching environmental nmr spectroscopy. *J. Chem. Educ.* **2015**, *92*, 693–697. [CrossRef]

9.   Gift, A.D.; Stewart, S.M.; Bokashanga, P.K. Experimental Determination of p*Ka* Values by Use of NMR Chemical Shifts, Revisited. *J. Chem. Educ.* **2012**, *89*, 1458–1460. [CrossRef]

10. Manallack, D.T. The p*Ka* Distribution of Drugs: Application to Drug Discovery. *Perspect. Med. Chem.* **2007**, *1*, 25–38. [CrossRef]

11. Manallack, D.T.; Prankerd, R.J.; Yuriev, E.; Oprea, T.I.; Chalmers, D.K.; David, K. The significance of acid/base properties in drug discovery. *Chem. Soc. Rev.* **2014**, *42*, 485–496. [CrossRef] [PubMed]

12. Chillistone, S.; Hardman, J.G. Factors affecting drug absorption and distribution. *Anaesth. Intensive Care Med.* **2014**, *15*, 309–313. [CrossRef]

13. Repice, M.D.; Sawyer, R.K.; Hogrebe, M.C.; Brown, P.J.; Luesse, S.B.; Gealy, D.J.; Frey, R.F. Talking Through the Problems: A Study of Discourse in Peer-Led Small Groups. *Chem. Educ. Res. Pract.* **2016**, *17*, 555–568. [CrossRef]

14. Handloser, C.S.; Chakrabarty, M.R.; Mosher, M.W. Experimental Determination of p*Ka* Values by Use of NMR Chemical Shifts. *J. Chem. Ed.* **1973**, *50*, 510–511. [CrossRef]

15. Krężel, A.; Bal, W. A formula for correlating p*Ka* values determined in D_2O and H_2O. *J. Inorg. Biochem.* **2004**, *98*, 161–166. [CrossRef] [PubMed]

16. Ebel, S.; Glaser, E.; Kantelberg, R.; Reyer, B. Auswertung digitaler Titrationskurven nach dem Tubbs-Verfahren. *Fresenius' Z. Anal. Chem.* **1982**, *312*, 604–607. [CrossRef]

17. Perrin, D.D. Dissociation contants of inorganic acids and bases in aqueous solution. *Pure Appl. Chem.* **1969**, *20*, 133–236. [CrossRef]

18. Becker, C.; Dressman, J.B.; Amidon, G.L.; Junginger, H.E.; Kopp, S.; Midha, K.K.; Shah, V.; Stavchansky, S.; Barends, D.M. Biowaiver Monographs for Immediate Release Solid Oral Dosage Forms: Isoniazid. *J. Pharm. Sci.* **2007**, *96*, 522–531. [CrossRef] [PubMed]

19. Santos, T.D.A.D.D.; Da Costa, D.O.; Da Rocha Pita, S.S.; Semaan, F.S. Potentiometric and conductimetric studies of chemical equilibria for pyridoxine hydrochloride in aqueous solutions: Simple experimental determination of p*Ka* values and analytical applications to pharmaceutical analysis. *Eclet. Quim.* **2010**, *35*, 81–86. [CrossRef]

20. Friebolin, H. *Ein- und Zweidimensionale NMR-Spektroskopie*, 5th ed.; Wiley-VCH: Weinheim, Germany, 2013; ISBN 978-3-527-33492-6.

21. Tubbs, C.F. Determination of Potentiometric Titration Inflection Point by Concentric Arcs Method. *Anal. Chem.* **1954**, *26*, 1670. [CrossRef]

22. Fulmer, G.R.; Miller, A.J.M.; Sherden, N.H.; Gottlieb, H.E.; Nudelman, A.; Stoltz, B.M.; Bercaw, J.E.; Goldberg, K.I. NMR chemical shifts of trace impurities: Common laboratory solvents, organics, and gases in deuterated solvents relevant to the organometallic chemist. *Organometallics* **2010**, *29*, 2176–2179. [CrossRef]

magnetochemistry

MDPI

Article

Solid-State NMR Study of New Copolymers as Solid Polymer Electrolytes

Jean-Christophe Daigle [1], Alexandre A. Arnold [2], Ashok Vijh [1] and Karim Zaghib [1,*]

[1] Center of Excellence in Transportation Electrification and Energy Storage (CETEES), Hydro-Québec, 1806, Lionel-Boulet blvd., Varennes, QC J3X 1S1, Canada; daigle.jean-christophe@ireq.ca (J.-C.D.); vijh.ashok@ireq.ca (A.V.)

[2] Department of Chemistry, Université du Québec à Montréal (UQAM), Succ. Centre-Ville—C.P. 8888, Montréal, QC H3C 3P8, Canada; arnold.alexandre@uqam.ca

* Correspondence: zaghib.karim@ireq.ca; Tel.: +1-450-652-8019

Received: 4 December 2017; Accepted: 11 January 2018; Published: 18 January 2018

Abstract: We report the analysis of comb-like polymers by solid-state NMR. The polymers were previously evaluated as solid-polymer-electrolytes (SPE) for lithium-polymer-metal batteries that have suitable ionic conductivity at 60 °C. We propose to develop a correlation between ^{13}C solid-state NMR measurements and phase segregation. ^{13}C solid-state NMR is a perfect tool for differentiating polymer phases with fast or slow motions. ^{7}Li was used to monitor the motion of lithium ions in the polymer, and activation energies were calculated.

Keywords: solid-state NMR; comb-like copolymer; solid polymer electrolyte; lithium-polymer -metal battery

1. Introduction

Global warming is a major challenge of the twenty-first century, and finding solutions is critical for the future of humanity. The widespread implementation of a fully electric transportation network must be a part of the solution. Electrical vehicles (EV) using lithium-ion batteries remain one of the most promising avenues in the short term. At present, Li-ion batteries containing liquid electrolytes are the most prevalent cells used as the electrical propulsion power in EV; however, their performance and safety are not sufficient to fully compete with gasoline-powered motors [1,2]. Considering the safety issue, liquid electrolytes are unavoidably problematic due to their flammability and the toxic fumes released during overcharging/abuse or over heating [3]. Solid electrolytes are currently getting much attention from the scientific community [4]. Based on the discovery of ionic conduction in poly(ethylene oxide) (PEO) by Wright in 1975 [5] and its first application to batteries by Armand and co-workers in 1979 [6], solid polymer electrolytes (SPE) have been investigated by many groups as a safety-improving solution. An acceptable SPE needs excellent lithium conductivity in a wide temperature range [7] and good mechanical strength [8] in order to limit dendrite growth from the lithium metal anode, which is critical for the production of efficient Li-polymer batteries. More importantly, simple and economically viable processing as a thin film is a major advantage. In 2007, Balsara and co-workers introduced a new SPE, poly(ethylene oxide-*b*-polystyrene). Poly(styrene) acts as a reinforcement block in SPE [8] to prevent the formation of dendrites by presenting an impenetrable wall [9,10].

Recently, Bouchet et al. developed single-ion triblock copolymers that demonstrated excellent ionic and mechanical performance at 80 °C [11]. Using elegant polymer architectures with rigid and soft blocks to create phase separation is now a trend in the development of a new SPE [12–15]. The development of SPEs involves complex architectures that influence the interaction between the soft and the hard blocks. The investigation of Li$^+$ ion mobility by different techniques appears to be a fascinating field of research. Several articles were published on the investigation of SPE by solid-state

NMR [16–19], especially scrutinizing the mobility of lithium coordinated with the polymer chains; however, the examination of the mobility of the organic part has not been extensively studied. In this work, we describe the use of a combination of ^{13}C and ^{7}Li high-resolution solid-state NMR results to establish a complete portrait of high-performance comb-like copolymers.

2. Results and Discussion

2.1. Description of Polymers Investigated by Solid-State NMR

Comb-like copolymers with soft and hard blocks, i.e., with low and high T_gs, display phase segregation when prepared as films [12]. This behaviour is key to promote good ionic conductivity and to prevent dendrite growth [20]. Comb-like copolymers are based on a poly(styrene) (PS) backbone obtained by anionic polymerization. PS backbone is used as a reinforcement block for preventing dendrite growth during cell operation, phenyl groups are very rigid structures. The backbone is a hard block with a reported Young Modulus of 3 GPa [8]. Grafting poly(ethylene glycol) methyl ether methacrylate (PEGMA) with a M_n = 500 was done by atom-transfer radical polymerization (ATRP). Poly(ethylene glycol) is known to be a very good polymer for Li$^+$ transportation, moreover short chains have a highest conductivity due by the lack of crystallinity. PEGMA is polymerized by soft radical polymerization while ethylene oxide is polymerized by a less safe process only. This monomer enables the formation of highly branched structures as a polymer graft. Both techniques allow narrow poly(dispersity), which facilitates the structure-properties relationship, and therefore better control of the desired properties. Scheme 1 shows the structure of the polymers investigated in this article.

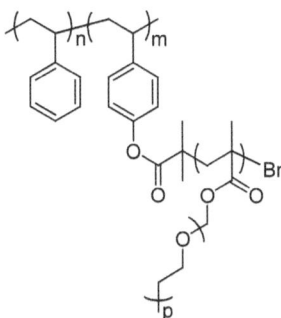

Scheme 1. Structure of the polymers investigated by Solid-state NMR.

In order to make a comprehensive comparison between the structures, different ratios of soft/hard blocks were selected. Table 1 reports the different characteristics of the polymers that are reported in ref. [12]. The PEGMA/PS ratio is important because it has been directly related to the electrochemical performance of the cells [12]. Polymer 3 was not used as SPE because the high molecular weight of the PEGMA block made it impossible to dissolve in a reasonable amount of solvent. All the polymers were doped with bis(trifluoromethane)sulfonimide lithium salt (LiTFISI) to promote ionic conductivity. Consequently, the study of this polymer by solid-state NMR is an interesting tool for elucidating the lithium motion. The rigidity of the structure is assessed by solid-state ^{13}C NMR, and an improved understanding of the polymer micro-structure is obtained, as reported in the case of cross-linked polymers [21]. The purpose of this study is to develop a relationship between the lithium mobility and the electrochemical results.

Table 1. Polymer Characteristics.[a]

ID	Ratio PEGMA/PS [b]	M_n (g mol^{-1}) [c]	T_{g1} (K) [d]	T_{g1Li} (K) [d]	σ at 60 °C (10^{-4}S cm^{-1}) [e]	Capacity (mA hg^{-1}) [f]
1	2.6	140,500	211	224	2.54	146
2	3.9	103,000	216	227	1.44	144
3	30	1,200,000	212	225	nd	nd

[a] Values reported from ref. [12]. [b] Composition determined by ^{1}H NMR and GPC. [c] Determined by GPC at 40 °C in THF. [d] Determined by DSC. [e] Determined by AC Impedance. [f] Values recorded at 80 °C and C/24. nd: not determined.

2.2. Characterization of Copolymers by ^{13}C Solid-State NMR

Solid-state ^{13}C NMR is invaluable for the characterization of polymers [22–25]. Qualitative information is obtained for the dynamics in heterogeneous systems. The signal is obtained using different polarization schemes, which preferentially excite rigid or dynamic molecular segments. In this work, cross-polarization is used to excite the rigid regions, while a simple 90° pulse excites the rigid and mobile segments. Experimental details are reported in Appendix A. Figure 1 shows the results for three different polymers with differing PEGMA to PS ratios.

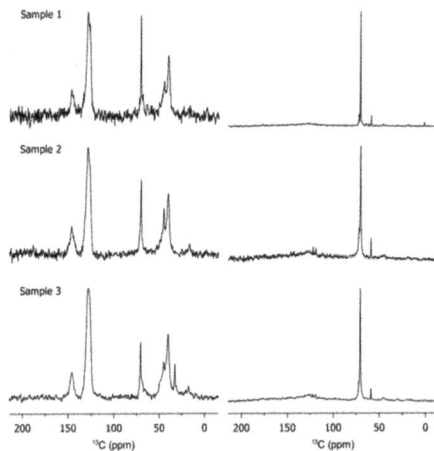

Figure 1. Cross-polarization solid-state ^{13}C NMR (left column) and direct pulse solid-state ^{13}C NMR excitation (right column) of samples 1 to 3 at 298 K. All spectra are normalized by maximum peak intensity; the change in noise level thus reflects the amount of rigid segments in the polymer.

The cross-polarization spectra (left column in Figure 1) are dominated by the PS main chain (peaks at 150 and 125 ppm) while the direct-pulse with low-power decoupling spectra (right column in Figure 1) are dominated by the poly(ethylene glycol) chains at 70 ppm. A fraction of PEGMA between 25–75 ppm is observed on the cross-polarization spectra, which indicates that the PEGMA chains display considerable rigidity. More specifically, the PEGMA backbones (25–50 ppm) appear to be less mobile than the pendant groups (70 ppm), which have lower intensity. However, it should be noted that the pendant groups are still rigid enough to show up in the cross-polarization spectra. This partial rigidity of pendant groups possibly results from their coordination with lithium salts. Furthermore, the ^{13}C NMR results show that the phase segregation reported by Daigle et al. [12] is not complete and that a model consisting of two perfectly separated blocks has to be refined. As the PS fraction is reduced, the efficiency of the cross-polarization decreases. Finally, an observation of the relative intensities between cross-polarization spectra gives an immediate diagnostic of the relative rigidity of a given copolymer series. For example, based on the relative spectral intensities shown in Figure 1, sample 3 appears to have a stronger rigid phase component compared to samples 1 and 2. We believe that some fluid parts of sample 3 have to be considered as rigid because the lithium ions strongly coordinated the PEGMA chains for forming "ionomer" as demonstrated in previous publication [12]; those parts are "frozen" and the ratio reported in Table 1 is based on GPC analysis.

The same methodology is used to determine changes in internal dynamics with temperature. Figure 2 shows the evolution of ^{13}C spectra for sample 3 with temperature.

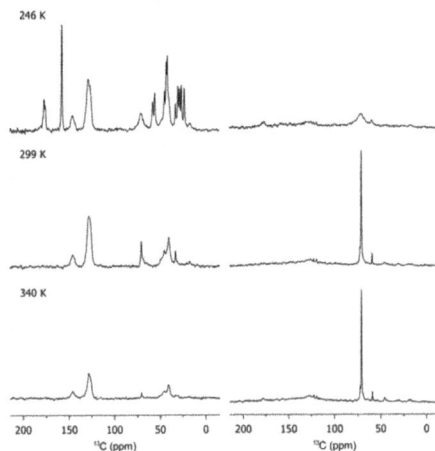

Figure 2. Evolution of cross-polarization (left column) and direct-pulse excitation (right column) of sample 3 with temperature. Note the decrease in intensity of the rigid segments as temperature is increased (left column).

The effect of temperature of the PEGMA phase is unambiguous: as temperature decreased, the mobility of the PEGMA chains is reduced as expected. The very low signal on the pendant chains on the DP spectrum, and the appearance of the carbonyl peak at 175 ppm on the CP spectrum, reflects the strong stiffening of the chains at 246 K. This effect is directly correlated to the poor ionic conductivity at low temperatures, especially near the glass transition point (225 K). The glass transition of poly(styrene) is ca. 373 K, and the signal at 30 ppm is characteristic of the poly(styrene) backbone, indicating that PS is already rigid at the highest temperatures studied in this work. The poly(styrene) phase was not observed in the DP spectra because the temperatures were too far from the glass transition. No major changes in the PEGMA segments are observed between 299 K and 340 K since they are not affected by the melting of chains.

2.3. Lithium Diffusion in the Membranes by ^{7}Li Solid-State NMR

The ^{13}C spectra provide information on the polymer rigidity, and ^{7}Li NMR is useful to monitor the mobility of Li^{+} ions, which are qualitatively correlated to the conductivity of the material. Due to the very high mobility of lithium in our samples, ^{7}Li spectra consist of a single sharp peak with full-widths at half height as low as 12 Hz, which reach a maximum at low temperatures of 120 Hz (Figure 3). This is almost two orders of magnitude lower than the linewidths reported for polyurethane-poly(dimethylsiloxane) copolymers [17]. The highest ionic conductivity reported in Table 1 is 2.54×10^{-4} Scm^{-1} at 60 °C, which is higher (about 3 times) than those reported earlier [17], so we can evaluate qualitatively the ionic conductivity of a polymer by this method. Interestingly, ^{1}H decoupling appears to have no effect on the ^{7}Li linewidth in our samples: thus, the mobility of lithium ions is sufficient to completely eliminate ^{1}H–^{7}Li dipolar couplings.

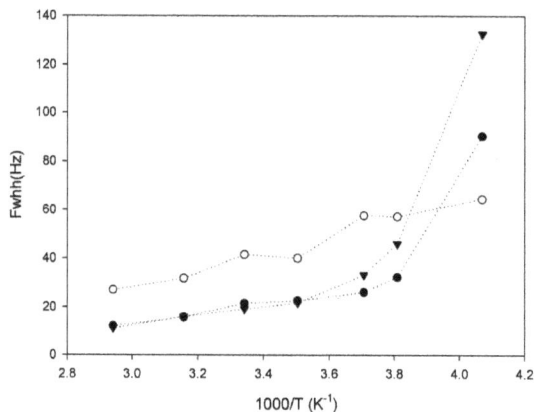

Figure 3. Evolution of ^{7}Li linewidth as a function of inverse temperature for sample 1 (black triangles), sample 2 (open circles) and sample 3 (black circles).

While ^{7}Li linewidths are influenced by motions with correlation times shorter than µs, faster motions with correlation times in the nanoseconds will contribute to longitudinal relaxation (T_1) of the NMR signal. Thus, T_1 characterizes fast motion that facilitates lithium diffusion.

Sample 2 in Figure 4 has the fastest lithium motion (shorter T_1 relaxation times), which is not in agreement with the ionic conductivity reported in Table 1. Also, it appears that temperature had less influence on the lithium mobility in this polymer. Mobility in samples 1 and 3 dropped around 263 K, which is related to the crystallization of poly(ethylene glycol) pendant chains in graft copolymers. Figure 2 shows that low temperatures have a great effect on solidification of the polymer chains (PEGMA). The close correlation between the mobility of lithium ions and the motions of PEGMA pendant chains suggests a strong association between the lithium ions and PEGMA groups.

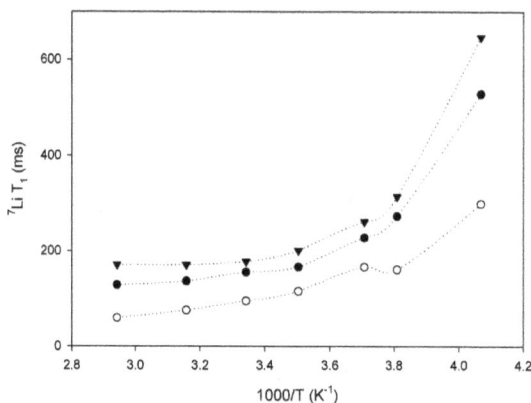

Figure 4. ^{7}Li longitudinal relaxation (T_1) as a function of inverse temperature for sample 1 (black triangles), sample 2 (open circles) and sample 3 (black circles).

Sample 3 shows the lowest lithium mobility (see Figure 4), despite the highest ratio of PEGMA. We explain this behavior as resulting from the strangling of the polymeric chains due to the high molecular weight (1,000,000 g mol^{-1}), which hinders the motion of lithium [12].

Lin et al. [17] calculated the activation energy (E_a) of lithium diffusion from the slope of the curve at low temperatures by the Arrhenius relationship. This information is relevant for sample 3 because it is impossible to obtain using AC impedance measurements because this polymer cannot be prepared as a thin film. Table 2 reports the results using the two methods. A qualitative correlation is observed with an approximate difference of a factor of 2 between the two methods, the same factor was also reported in reference [17]. The relaxation time depends on the size of the mobile segments of the polymer electrolyte; one should note that there would be a range of sizes of the mobile polymeric segments and the relaxation time would be some sort of average value. Thus it would be a rough and approximate measure. The activation energy and conductivity as determined by the impedance measurements would be more significant in comparing the conductivity mechanisms of different polymers. Moreover, AC impedance allowed the measurements of long-scale motion while the solid-state NMR ^{7}Li measurements are related with local motion of Li$^+$, so that can contribute also for the difference of values measured for conductivity and by consequence E_a. Nevertheless, the NMR measurements indicate that the activation energy of polymers 2 and 3 is similar. It also important to note that the result obtained for sample 1 by solid-state NMR is within the same magnitude of the normal thermal fluctuation (4 kT = 9.6 kJ), thus the value is probably none applicable in this case.

Table 2. Energies of activation recorded by AC Impedance and ^{7}Li longitudinal relaxation (T_1).

ID	E_a (kJ mol^{-1}) [a]	E_a (kJ mol^{-1})
1	29	11
2	45	21
3	nd	20

[a] Values reported from ref. [12] and recorded by AC Impedance. nd: not determined.

3. Conclusions

The development of well-define micro-structured polymers offers a new trend in solid polymer electrolytes. Solid-state NMR is a perfect tool for identifying the mobile and rigid parts in polymers. A combined ^{13}C and ^{7}Li NMR approach to correlate polymers and lithium motilities was investigated and successfully applied to comb-like copolymers. Our results strongly suggest that there is an interaction between the PEGMA pendant groups and lithium ions. Solid-state NMR is very useful when materials cannot be prepared in a form amenable to classical techniques such as AC impedance because the activation energy can be calculated in the solid state. This NMR technique is useful to evaluate the viability of polymers as SPEs without preparing films, and thus can be used for preliminary testing.

Acknowledgments: This work was supported by Hydro-Québec.

Author Contributions: Jean-Christophe Daigle and Alexandre A. Arnold conceived and designed the experiments; Alexandre A. Arnold performed the experiments; all the authors analysed the data and wrote the paper.

Conflicts of Interest: The authors declare no conflict of interest.

Appendix A

The spectra were recorded on a Bruker Avance III HD operating at frequencies of 400.03, 100.60 and 155.47 MHz for ^{1}H, ^{13}C and ^{7}Li, respectively, using a triple resonance 1.9 mm MAS probe in double resonance mode. The samples had a mass of 15 mg and the s spinning frequency was 20 kHz. ^{13}C spectra were obtained using 1.5 ms cross-polarization ramped from 70 to 100% of the maximum amplitude or a 90° pulse. In both cases, the ^{13}C radio-frequency field was ca. 80 kHz. The cross-polarization spectra of rigid segments were obtained under high-power ^{1}H TPPM decoupling with a 100 kHz radio-frequency field. The direct-pulse excitation spectra of the mobile segments were obtained under low-power GARP decoupling at a ^{1}H radio-frequency field of 3.5 kHz. In both cases, the acquisition times were 25 ms and the recycle delays were 5 s. ^{7}Li spectra were produced using 2.5 µs long 90° pulses with a 200 ms acquisition time and a 5 s recycle delay. The ^{7}Li longitudinal relaxation times were measured using the inversion recovery pulse sequence incrementing the relaxation delays in 12 steps between 1 ms and 5 s. All spectra were processed and fitted with Topspin3.5.2 and the dynamics center module. Note that sample temperature was calibrated taking into account additional heating due to rotor friction or radio-frequency fields.

References

1. Murata, K.; Izuchi, S.; Yoshihisa, Y. An overview of the research and development of solid polymer electrolyte batteries. *Electrochim. Acta* **2000**, *45*, 1501–1508. [CrossRef]
2. Tarascon, J.M.; Armand, M. Issues and challenges facing rechargeable lithium batteries. *Nature* **2001**, *414*, 359. [CrossRef] [PubMed]
3. Hammami, A.; Raymond, N.; Armand, M. Runaway risk of forming toxic compounds. *Nature* **2003**, *424*, 635. [CrossRef] [PubMed]
4. Manthiram, A.; Yu, X.; Wang, S. Lithium battery chemistries enabled by solid-state electrolytes. *Nat. Rev. Mater.* **2017**, *2*, 16103. [CrossRef]
5. Wright, P.V. Electrical conductivity in ionic complexes of poly(ethylene oxide). *Br. Polym. J.* **1975**, *7*, 319–327. [CrossRef]
6. Armand, M.B.; Chabagno, J.M.; Duclot, M.J. *Polyethers as Solid Electrolytes*; Elsevier: Amsterdam, The Netherlands, 1979; pp. 131–136.
7. Marzantowicz, M.; Dygas, J.R.; Krok, F.; Florjańczyk, Z.; Zygadło-Monikowska, E. Influence of crystalline complexes on electrical properties of peo: Litfsi electrolyte. *Electrochim. Acta* **2007**, *53*, 1518–1526. [CrossRef]
8. Singh, M.; Odusanya, O.; Wilmes, G.M.; Eitouni, H.B.; Gomez, E.D.; Patel, A.J.; Chen, V.L.; Park, M.J.; Fragouli, P.; Iatrou, H.; et al. Effect of molecular weight on the mechanical and electrical properties of block copolymer electrolytes. *Macromolecules* **2007**, *40*, 4578–4585. [CrossRef]
9. Monroe, C.; Newman, J. Dendrite growth in lithium/polymer systems: A propagation model for liquid electrolytes under galvanostatic conditions. *J. Electrochem. Soc.* **2003**, *150*, A1377–A1384. [CrossRef]
10. Monroe, C.; Newman, J. The impact of elastic deformation on deposition kinetics at lithium/polymer interfaces. *J. Electrochem. Soc.* **2005**, *152*, A396–A404. [CrossRef]
11. Bouchet, R.; Maria, S.; Meziane, R.; Aboulaich, A.; Lienafa, L.; Bonnet, J.-P.; Phan, T.N.T.; Bertin, D.; Gigmes, D.; Devaux, D.; et al. Single-ion bab triblock copolymers as highly efficient electrolytes for lithium-metal batteries. *Nat. Mater.* **2013**, *12*, 452–457. [CrossRef] [PubMed]
12. Daigle, J.-C.; Vijh, A.; Hovington, P.; Gagnon, C.; Hamel-Pâquet, J.; Verreault, S.; Turcotte, N.; Clément, D.; Guerfi, A.; Zaghib, K. Lithium battery with solid polymer electrolyte based on comb-like copolymers. *J. Power Sources* **2015**, *279*, 372–383. [CrossRef]
13. Villaluenga, I.; Inceoglu, S.; Jiang, X.; Chen, X.C.; Chintapalli, M.; Wang, D.R.; Devaux, D.; Balsara, N.P. Nanostructured single-ion-conducting hybrid electrolytes based on salty nanoparticles and block copolymers. *Macromolecules* **2017**, *50*, 1998–2005. [CrossRef]
14. Sun, J.; Stone, G.M.; Balsara, N.P.; Zuckermann, R.N. Structure–conductivity relationship for peptoid-based peo–mimetic polymer electrolytes. *Macromolecules* **2012**, *45*, 5151–5156. [CrossRef]
15. Devaux, D.; Glé, D.; Phan, T.N.T.; Gigmes, D.; Giroud, E.; Deschamps, M.; Denoyel, R.; Bouchet, R. Optimization of block copolymer electrolytes for lithium metal batteries. *Chem. Mater.* **2015**, *27*, 4682–4692. [CrossRef]
16. Gorecki, W.; Jeannin, M.; Belorizky, E.; Roux, C.; Armand, M. Physical properties of solid polymer electrolyte PEO(LiTFSI) complexes. *J. Phys. Condens. Matter* **1995**, *7*, 6823. [CrossRef]
17. Lin, C.-L.; Kao, H.-M.; Wu, R.-R.; Kuo, P.-L. Multinuclear solid-state nmr, dsc, and conductivity studies of solid polymer electrolytes based on polyurethane/poly(dimethylsiloxane) segmented copolymers. *Macromolecules* **2002**, *35*, 3083–3096. [CrossRef]
18. Hayamizu, K.; Akiba, E.; Bando, T.; Aihara, Y.; Price, W.S. Nmr studies on poly(ethylene oxide)-based polymer electrolytes with different cross-linking doped with LIN(SO2CF3)2. Restricted diffusion of the polymer and lithium ion and time-dependent diffusion of the anion. *Macromolecules* **2003**, *36*, 2785–2792. [CrossRef]

19. Yang, L.-Y.; Wei, D.-X.; Xu, M.; Yao, Y.-F.; Chen, Q. Transferring lithium ions in nanochannels: A PEO/Li+ solid polymer electrolyte design. *Angew. Chem. Int. Ed.* **2014**, *53*, 3631–3635. [CrossRef] [PubMed]
20. Harry, K.J.; Hallinan, D.T.; Parkinson, D.Y.; MacDowell, A.A.; Balsara, N.P. Detection of subsurface structures underneath dendrites formed on cycled lithium metal electrodes. *Nat. Mater.* **2014**, *13*, 69–73. [CrossRef] [PubMed]
21. Daigle, J.-C.; Asakawa, Y.; Vijh, A.; Hovington, P.; Armand, M.; Zaghib, K. Exceptionally stable polymer electrolyte for a lithium battery based on cross-linking by a residue-free process. *J. Power Sources* **2016**, *332*, 213–221. [CrossRef]
22. Spiess, H.W. Molecular motion, phase separation and internal surfaces in rubber-elastic polymers. *Angew. Makromol. Chem.* **1992**, *202–203*, 331–342. [CrossRef]
23. Spiess, H.W.; Schmidt-Rohr, K. Multidimensional solid-state NMR studies of chain motions in polymers. *Polym. Prepr. (Am. Chem. Soc. Div. Polym. Chem.)* **1992**, *33*, 68–69.
24. Bluemich, B.; Bluemich, P.; Guenther, E.; Jansen, J.; Schauss, G.; Spiess, H.W. *NMR Imaging of Polymers: Methods and Applications*; Wiley-VCH Verlag GmbH & Co. KGaA: Weinheim, Germany, 1992.
25. Bluemich, B.; Bluemler, P.; Guenther, E.; Spiess, H.W. Methods and applications of NMR imaging in polymer research. *Polym. Prepr. (Am. Chem. Soc. Div. Polym. Chem.)* **1992**, *33*, 759–760.

magnetochemistry

MDPI

Article

Local and Average Structural Changes in Zeolite A upon Ion Exchange

Lisa Price [1], Ka Ming Leung [2] and Asel Sartbaeva [1,*]

[1]   Department of Chemistry, University of Bath, Bath BA2 7AY, UK; lisa.a.price@btinternet.com
[2]   Inorganic Chemistry Lab, South Parks Road, University of Oxford, Oxford OX1 3QR, UK;
      ka.leung.thomas@gmail.com
*   Correspondence: a.sartbaeva@bath.ac.uk; Tel.: +44-(0)-1225-38-5410

Received: 21 August 2017; Accepted: 7 December 2017; Published: 12 December 2017

Abstract: The infamous 'structure–property relationship' is a long-standing problem for the design, study and development of novel functional materials. Most conventional characterization methods, including diffraction and crystallography, give us a good description of long-range order within crystalline materials. In recent decades, methods such as Solid State NMR (SS NMR) are more widely used for characterization of crystalline solids, in order to reveal local structure, which could be different from long-range order and sometimes hidden from long-range order probes. In particular for zeolites, this opens a great avenue for characterization through studies of the local environments around Si and Al units within their crystalline frameworks. In this paper, we show that some structural modifications occur after partially exchanging the extraframework Na^+ ions with monovalent, Li^+, K^+, Rb^+ and NH_4^+ and divalent, Ca^{2+} cations. Solid state NMR is deployed to study the local structure of exchanged materials, while average stricture changes can be observed by powder diffraction (PXRD). To corroborate our findings, we also employ Fourier Transform Infrared spectroscopy (FT-IR), and further characterization of some samples was done using Scanning Electron Microscopy (SEM) and Energy-Dispersive X-ray spectroscopy (EDX).

Keywords: zeolite; solid state NMR; ion exchange; synthesis; characterization

---

1. Introduction

Zeolites are aluminosilicate porous minerals. Many zeolites occur in nature as aluminosilicate minerals. To date, we can make more than 200 synthetic zeolites in the laboratory [1,2]. They are classed as porous materials as they possess cages, channels and open void spaces within their highly crystalline frameworks. Each zeolite framework has a unique structure, and because there is such a variety of zeolite structures, there is also a very wide diversity of zeolite applications. Synthetic zeolites are used as green, re-usable catalysts in industrial processes as heterogeneous catalysts for processes that involve hydro-cracking, acrylation, oxidation and reforming. Most zeolite syntheses employ Organic Structure Directing Agents (OSDAs), such as TMA-OH (Tetramethylammonium Hydroxide) or crown-ether, which act as templates to guide the formation of particular types of zeolite pores and channels [3–6]. This reduces the chance of producing competing zeolite phases. However, due to the high manufacturing costs of producing these organic materials, which cannot be recovered after calcination, current research is becoming more concerned with optimising synthesis conditions in order to produce pure zeolites in the absence of OSDAs [3,7]. Here, we performed a low temperature synthesis of small zeolite A (Na-A) crystals without the use of OSDAs, and the corresponding LTA framework (Linde Type A) is shown in Figure 1.

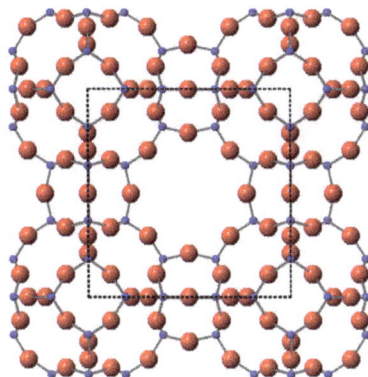

Figure 1. Linde Type A (LTA) framework with the α cage as the unit cell in the middle (black square). Red spheres: oxygens; blue: Si/Al atoms.

Na-A is a commercially important zeolite used in industry for catalysis, adsorption and industrial gas separations [8]. More recently, its sustainability as a drug delivery system has been investigated [9]. One of its greatest applications, however, is ion exchange; in particular, rapid Na^+/Ca^{2+} exchange. Consequently, Na-A is very effective in water softening, and one of its main functions is in washing powders as a detergent builder [10].

Ion exchange can also produce zeolites with different properties. For example, K-A is commonly used in the ethanol drying processes [11] and partially-exchanged K/Na-A is used to separate CO_2 from CO_2/N_2 dry mixtures [12]. Ca-A zeolites are important in industry, where they selectively adsorb linear alkanes from a mixture of branched alkanes [13], and Na^+/NH_4^+ exchange is useful in minimising environmental pollution and eutrophication [14–17]. Li-A was proposed as a possible delivery material for pharmacological studies [18]. Li-exchanged zeolites are also used for the separation of nitrogen from air [19]. In this work, we carried out aqueous ion exchange of monovalent alkali metals: Li^+, K^+, Rb^+ and NH_4^+ and divalent Ca^{2+} into zeolite A crystals produced from a low temperature and organic template-free synthesis and performed analysis using solid state NMR and other methods to determine the structural effects of the exchange. Previously, solid state NMR has been used to study other zeolites [20–28].

2. Results and Discussion

2.1. Na-A, Li-A, K-A and Rb-A

The powder patterns of the hydrated Na-A samples are consistent with those recorded in the literature for pure phase zeolite A crystals [29,30]. Figure 2 shows the indexed pattern for Na-A, synthesised at 40 °C for 24 h. Reflection peaks corresponding to both Na-A and Faujasite (FAU) phases [30] were observed when the zeolites were synthesised at 50 °C for 24 h and at 40 °C for 48 h, as shown in Figure 3. It is evident that, if given enough time or if heated above a certain temperature, there is a tendency for the metastable LTA framework structure to subsequently transform into the FAU framework [7,31]. Careful control of the synthesis conditions is, therefore, necessary to avoid the formation of the unwanted polymorph, which is a ubiquitous problem in synthesising zeolites in the absence of OSDAs. The low temperature of 40 °C limits the presence of these competing phases.

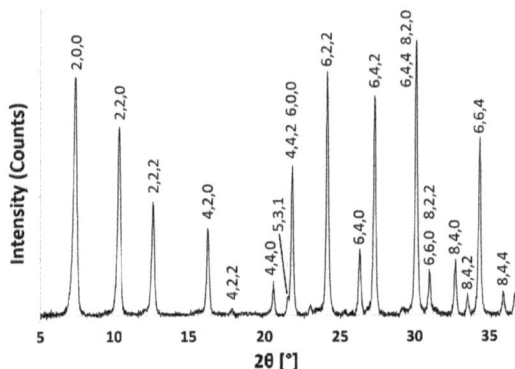

Figure 2. PXRD pattern for Na-A synthesised at 40 °C for 24 h, indexed as the LTA framework.

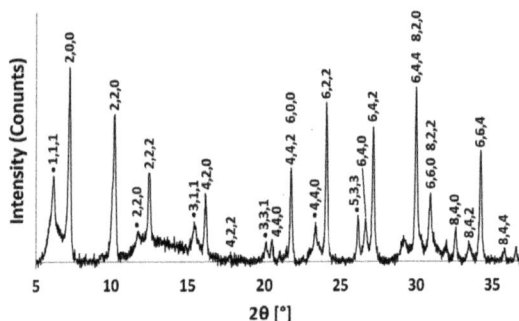

Figure 3. PXRD pattern for Na-A synthesised at 40 °C for 48 h. Characteristic Faujasite (FAU) framework peaks are indexed with black dots.

In this investigation, we decided to study all samples after one ion exchange only. The extent of exchange was quantified from the filtrates using a sodium ion selective electrode (ISE). For each sample, even after one exchange, we see good, although not complete, ion exchange. This is not surprising as sometimes as many as 8–10 steps are required for a complete exchange to occur [32]. Results show that the extent of exchange decreases with increasing cation size, $Li^+ > K^+ > Rb^+$. The steric restrictions of the zeolite pores make full exchange difficult to obtain, particularly for those ions with large ionic radii. Energy Dispersive X-ray (EDX) elemental analysis was also carried out. In all samples, it was evident that partial exchange had taken place, as residual Na^+ ions were detected.

PXRD and FT-IR analyses show that there is no significant alteration to either the long-range crystal order or the local framework structure of Na-A after exchange with Li^+, K^+ and Rb^+ ions. These monovalent alkali cations vary in their ionic radii, and changes in the PXRD peak intensities are expected to occur as a result of these cations occupying slightly different sites in the pores. Figure 4 shows the PXRD patterns for the alkali metal exchanged zeolites. For K-A, the (4,4,0) reflection almost disappears, whereas the (4,2,2) and (8,0,0) peaks increase in intensity. These results are in agreement with those observed by Lührs et al., where complete exchange with K and Ca was studied using diffraction and structure refinement [33]. The PXRD pattern for Rb-A is also slightly different. The characteristic intensities of the first four Na-A reflections are altered; most noticeable is the increase in the intensity of the (2,2,0) reflection.

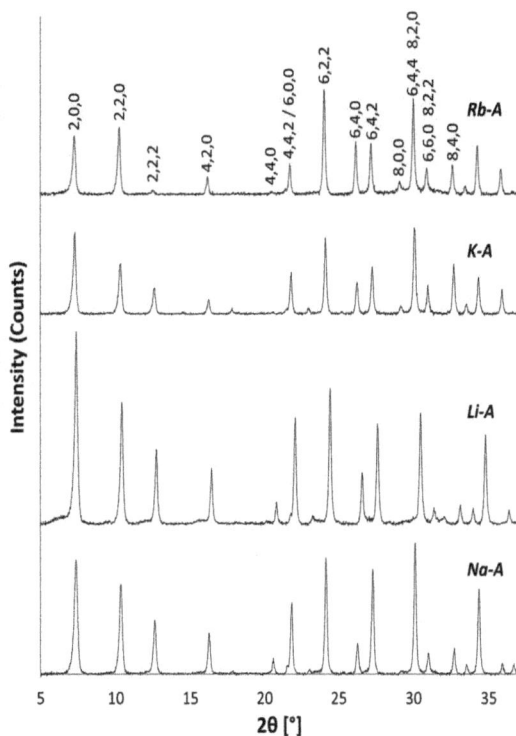

Figure 4. PXRD patterns for Na-A, Li-A, K-A and Rb-A.

The unit cell parameters (a) for the exchanged zeolites were calculated from the PXRD peak positions and Miller indices using the program UnitCell [34] and are shown in Table 1. All samples have cubic symmetry, and it can be seen that the unit cell size decreases by about 1% on exchanging larger Na^+ (1.02 Å) for smaller Li^+ (0.59 Å) ions and increases slightly on exchange with larger K^+ (1.38 Å) and Rb^+ (1.49 Å) ions [35]. Correspondingly, an increase in the lattice parameter is progressive from Li < Na < K < Rb-A, in accordance with the increasing ionic radii of the monovalent cations.

Table 1. Unit cell parameters for the exchanged zeolites.

	Li-A	Na-A	K-A	Rb-A
Unit cell parameter, a (Å)	24.151 (2)	24.435 (2)	24.454 (2)	24.486 (3)

The FT-IR spectra for Li-A, Na-A, K-A and Rb-A all display the fundamental zeolite framework vibration ν_{max}/cm^{-1} at 959, 968, 972 and 969, respectively, corresponding to the asymmetric $\nu_{(O-T-O)}$ stretch, shown in Figure 5. In addition, the framework symmetric $\nu_{(O-T-O)}$ stretch occurs at ν/cm^{-1} 684, 664, 663 and 666 for Li-A, Na-A, K-A and Rb-A, respectively. The broad peaks at 3350–3450 cm^{-1} and the weaker peaks at 1650 cm^{-1} observed in all spectra correspond to the $\nu_{(O-H)}$ stretching and $\delta_{(H-O-H)}$ bending vibrations of water molecules in the hydrated samples. The only noticeable differences in the spectra are that the framework $\nu_{(O-T-O)}$ asymmetric stretch shifts to slightly a higher wavenumber and the $\nu_{(O-T-O)}$ symmetric stretch shifts to slightly a lower wavenumber, with increasing cation size, $Li^+ < Na^+ < K^+$.

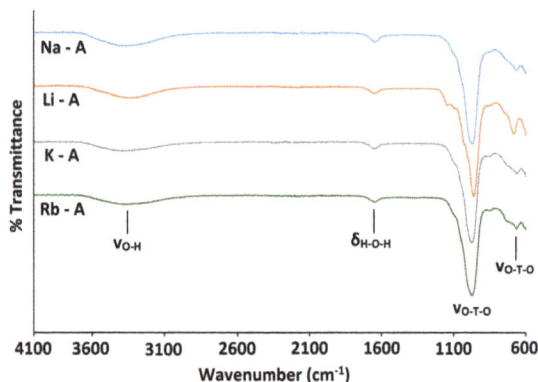

Figure 5. FT-IR spectra for Na-A (blue line) and cation-exchanged Li-A (orange line); K-A (grey line) and Rb-A (green line).

For the parent Na-A zeolites, ^{29}Si and ^{27}Al MASNMR spectra contain one peak at δ −89.54 ppm and δ 58.41 ppm, respectively, confirming that the Si/Al ratio of the framework is one. This is in agreement with data reported by Thomas et al. [24] for a single silicon, Si(OAl)$_4$, and aluminium environment, Al(OSi)$_4$, in Linde Type A zeolites. ^{27}Al MAS NMR spectra for the exchanged zeolites also display one sharp peak between δ 57 ppm and 60 ppm, pertaining to Al(4Si) units. Likewise, ^{29}Si spectra for Li-A, K-A and Rb-A are also dominated by a single sharp peak at δ −87.23 ppm, −89.83 ppm and −89.87 ppm, respectively. These peaks all lie within the chemical shift range for which Si(4Al) units can occur (δ −80.0 ppm−−90.5 ppm from TMS) [28]. Some small, low intensity peaks at δ −84.02 ppm, −87.96 ppm and −85.03 ppm are observed in the Li, K and Rb-A samples, respectively. These are identified as silanol peaks. From deconvolution of the Li, K and Rb-A ^{29}Si NMR spectra as shown in Figure 6, the Si/Al ratios were calculated to be one using Equation (1) [22,28]:

$$\frac{Si}{Al} = \frac{\sum_{n=0}^{n=4} I_{Si(nAl)}}{\sum_{n=0}^{n=4} \left(\frac{n}{4}\right) I_{Si(nAl)}} \tag{1}$$

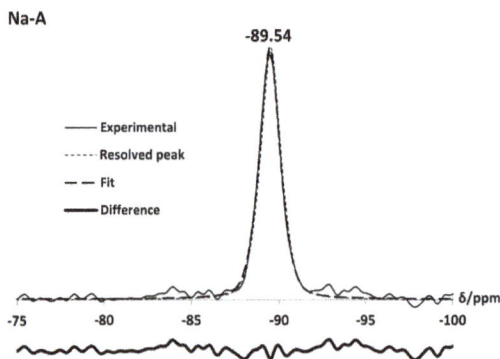

Figure 6. Cont.

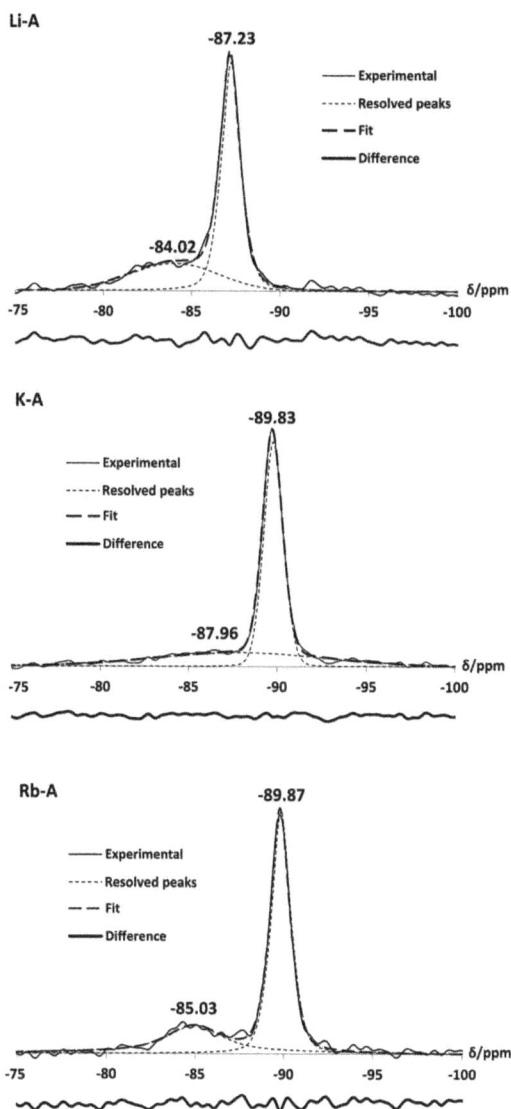

Figure 6. ^{29}Si NMR spectrum of the as-synthesised Na-A and monovalent cation-exchanged Li-A, K-A and Rb-A.

It is also interesting to note that the Si(4Al) peak for Li-A is significantly shifted to a lower field, centred at δ −87.23 ppm, in comparison to that of the parent Na-A peak, at δ −89.54 ppm. The linear relationship between the average Si-O-T framework bond angles, (α) and ^{29}Si chemical shifts can offer an explanation for this difference. Table 2 shows the Si/Al ratios and average T-O-T bond angles that were calculated from the deconvoluted ^{29}Si NMR spectra using Equation (2) [36].

$$\delta = {}^{29}\text{Si}(\text{ppm}) = -5.230 - 0.570\alpha \tag{2}$$

Table 2. ^{29}Si NMR data: Si/Al ratios and average T-O-T bond angles for Li, Na, K and Rb-A. The angular values are estimated based on NMR data using the regression relationship from [36], and we estimate the accuracy on angles up to ±2 degrees.

	Li-A	Na-A	K-A	Rb-A
Si/Al	1	1	1	1
∠T-O-T [α (°)]	143.9	147.9	148.6	148.4

Na-A zeolites, synthesised at 40 °C for 24 h, are shown to have good cation exchange ability with Li$^+$, K$^+$ and Rb$^+$ ions. The only noticeable differences between these exchanged zeolites are the sizes of the unit cells and average framework T-O-T bond angles, which increase accordingly with increasing cation size Li < Na < K < Rb.

2.2. NH$_4$-A

Exchange with NH$_4$$^+$ ions compromises some of the long-range order of the zeolite A crystals, inferred by the broad PXRD peaks and low signal to noise ratio, as seen in Figure 7. NMR and FT-IR spectra also indicate that the local framework environment is affected. Figure 8 shows a weak peak at $v_{(NH)}$/cm^{-1} 1453 in the FT-IR spectrum, which confirms that exchange has taken place. The asymmetric $v_{(O-T-O)}$ stretch is, however, weaker and broader than that of the parent spectrum and is shifted toward a higher wavenumber, occurring at v_{max}/cm^{-1} 987 [37,38]. As Si-O bonds (1.64 Å) are shorter than Al-O bonds (1.73 Å) [35], the force constant is higher for the former. Therefore, the shift to higher frequencies indicates that a loss of some aluminium from the framework has occurred.

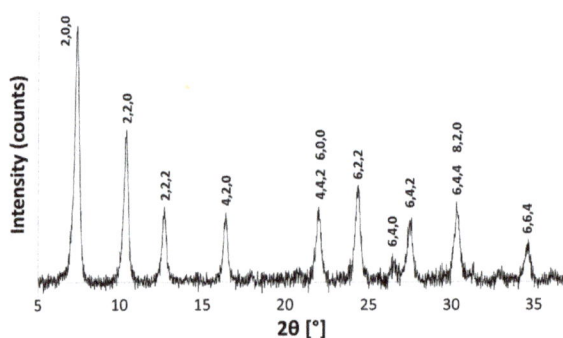

Figure 7. PXRD pattern for NH$_4$-A.

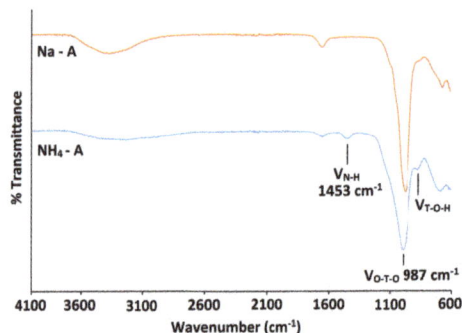

Figure 8. FT-IR spectra for the parent Na-A zeolite (orange line) and NH$_4$-A (blue line).

Furthermore, deconvolution of the ^{29}Si MAS NMR, as shown in Figure 9, confirms the presence of both Si(3Al) and Si(4Al) environments in the framework, with peaks occurring at characteristic chemical shifts of δ −91.41 ppm and −89.08 ppm, respectively. Another peak at δ −85.04 ppm is present in the spectrum due to silanol species. This is in line with the appearance of the $\nu_{(T-O-H)}$ stretch at 868 cm^{-1} in the FT-IR spectrum. Using Equation (1), the Si/Al ratio of NH$_4$-A was calculated to be 1.04. The loss of some aluminium from the framework is further confirmed in the ^{27}Al MAS NMR spectrum, which displays a resonance signal at δ 1.29 ppm, attributed to extraframework octahedrally-coordinated aluminium [23,28]. This peak is broad and overlaps with the main signal at δ 58.89 ppm (for a tetrahedral Al coordination). The broadening and overlapping of these peaks can be attributed to severely distorted six coordinated and four coordinated aluminium environments. The presence of six coordinated extraframework Al species in NH$_4$-exchanged zeolite A has been previously reported by Klinowski et al. [22,24], Sartbaeva et al. [15] and M. Dyballa et al. [39]. It is evident that some dealumination of the zeolite A framework occurs upon exchange with NH$_4^+$ ions.

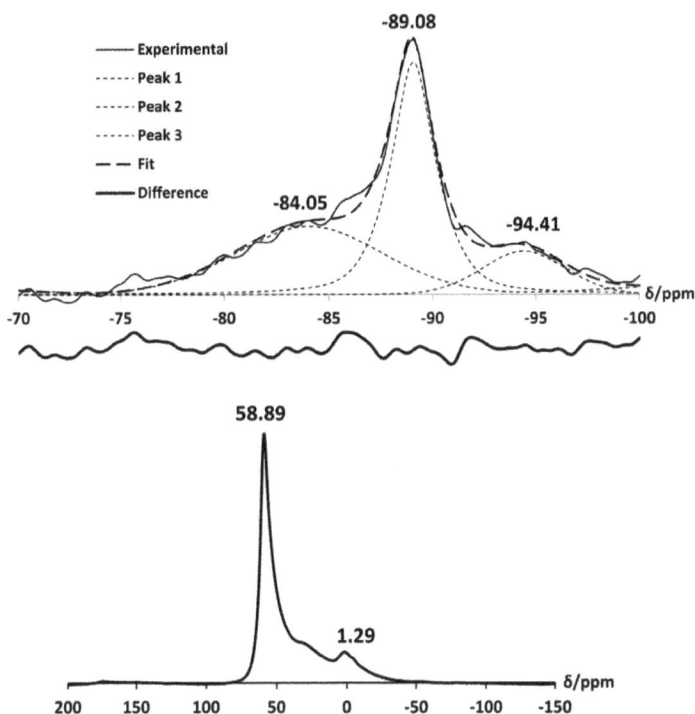

Figure 9. ^{29}Si (top) and ^{27}Al (bottom) NMR spectrum of the monovalent cation-exchanged NH$_4$-A.

2.3. Ca-A

There is no considerable alteration to the long-range crystal order of zeolite A after exchange with Ca^{2+}. The increase in intensity of the (4,0,0) reflection and the decrease of the (4,4,0) reflection is consistent with the results obtained by Lührs et al. [33]. The peak in the PXRD pattern at 2θ 29.57°, as shown in Figure 10, is characteristic of calcite (CaCO$_3$), which must have formed from the Ca(OH)$_2$ exchange solution. Further characterisation with ^{13}C Cross-Polarisation (CP) NMR (Figure 11) shows a distinct peak centred at δ 168 ppm, which confirms the presence of CO$_3^{2-}$ [26]. Furthermore, the split bands in the FT-IR spectrum at 1413 cm^{-1} and 1459 cm^{-1} identify the carbonate as a monodentate species, as shown in Figure 12 [40].

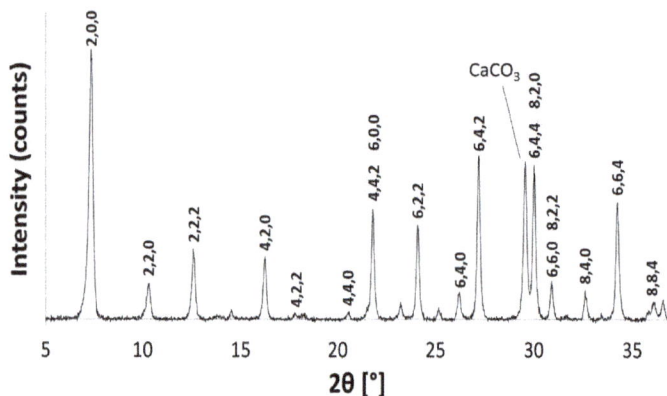

Figure 10. PXRD pattern for Ca-A. The characteristic calcite peak is identified.

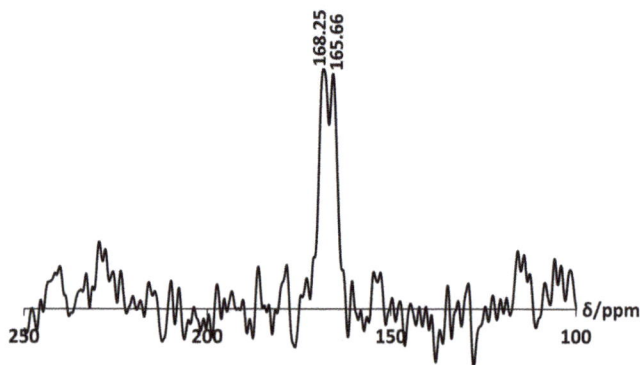

Figure 11. ^{13}C CPNMR of Ca-A.

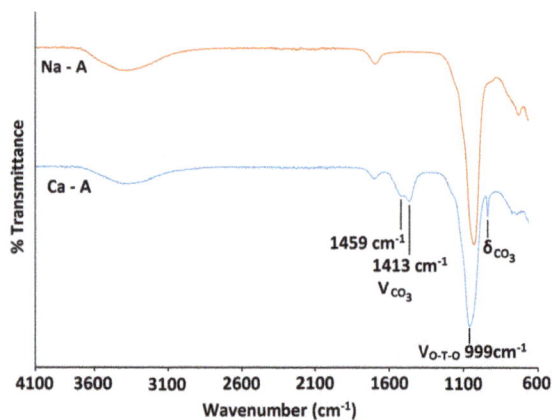

Figure 12. FT-IR spectra for the parent Na-A zeolite (orange) and Ca-A (blue).

Exchange with Ca^{2+} has a small effect on the local framework environment of zeolite A. The asymmetric $v_{(O-T-O)}$ stretch in the FT-IR spectrum is less intense and is shifted toward a higher wavenumber, occurring at v_{max}/cm^{-1} 999 [37,38]. The ^{29}Si MAS NMR spectrum, as shown in Figure 13, is dominated by an intense peak at δ −91.41 ppm, due to the expected Si(4Al) units of the framework. However, deconvolution of the spectrum also confirms the presence of Si(3Al), with a very small peak occurring at δ −96.35 ppm. Another peak at δ −85.55 ppm is also present due to some silanol species. Using Equation (1), the Si/Al ratio of Ca-A was calculated to be 1.01. Similar to the exchange with NH_4^+ ions, it is evident that some dealumination of the zeolite A framework has occurred upon exchange with Ca^{2+} ions. This loss of aluminium is confirmed in the ^{27}Al NMR spectrum, which displays a broad peak centred at δ 50.11 ppm, characteristic of Al(4Si) units and also a slight peak centred at δ 9.10 ppm, which can be attributed to a distorted octahedral aluminium environment. The broadness of the peaks in the spectra can be attributed to quadrupolar interactions of ^{27}Al nuclei in these distorted environments.

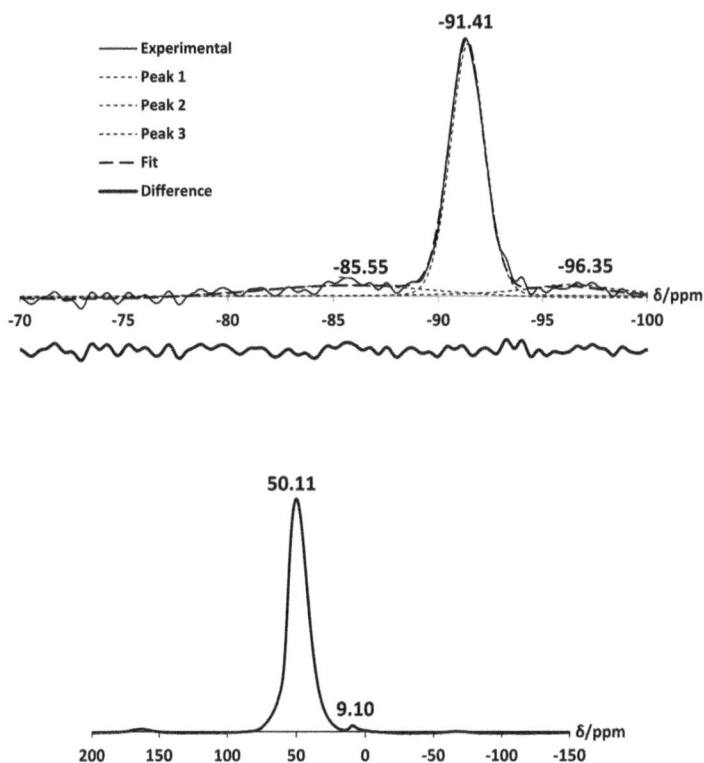

Figure 13. ^{29}Si (top) and ^{27}Al (bottom) NMR spectrum of the divalent cation-exchanged Ca-A.

2.4. SEM

The surface morphologies of the zeolite crystals were observed using Field Emission Scanning Electron Microscopy (FESEM). Particle sizes were calculated from the scale bars on the FESEM micrographs, using the ImageJ processing and analysis program. The crystallite sizes were also calculated from the broadening of the most intense PXRD peaks, in this case the (6,2,2) and (6,4,4) reflections, using the Scherrer Equation (3), where L is the crystallite size, $\beta_{(hkl)}$ is the full width at half maximum for the major peak of the PXRD pattern subtracting the instrumental contribution to the broadening, K is the Scherrer constant, which is 0.9, λ is the wavelength of the X-rays in nm and θ is the Bragg angle of the incident X-rays [41].

$$L = \frac{K\lambda}{\beta_{(hkl)}\cos\theta} \tag{3}$$

Table 3 shows the average zeolite particle sizes. For Na-A and Ca-A zeolites, there is good consistency between the sizes calculated from the Scherrer equation and FESEM data.

Figure 14 shows the FESEM micrograph for the parent Na-A zeolite. A variety of crystallite shapes and sizes can be observed, somewhere in between cubic and spherical and ranging from 122–354 nm. The visibly larger crystals with well-defined edges display the typical cubic morphology of LTA zeolites [42,43]. However, the low temperature conditions employed in this synthesis appear to slow down crystal growth, instead favouring nucleation in the initial stages [44]. Particles that are more rounded in shape and significantly smaller in diameter are evident in the FESEM images, indicating that some of the crystals have not had enough time to form completely. These results are in agreement with those reported by Dimitrov et al. [45] and Smaihi et al. [46]. The introduction of Ca^{2+} into the zeolites is accompanied by noticeable changes in the surface morphology, as shown in Figure 15.

Table 3. Crystallite sizes (nm) calculated from FESEM data using Equation (3) compared to crystal size observed by FESEM. Statistical analysis shows a large spread of values from the Scherrer equation up to ±28 nm; for FESEM data, the standard deviation is up to ±14 nm.

	Na-A	Ca-A
Scherrer Equation (nm)	242	270
FESEM (nm)	241	290

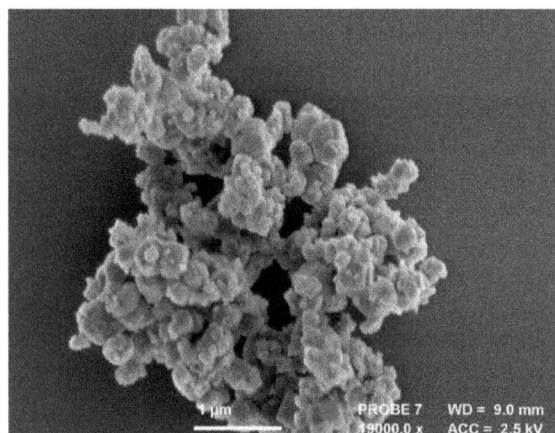

Figure 14. FESEM Na-A.

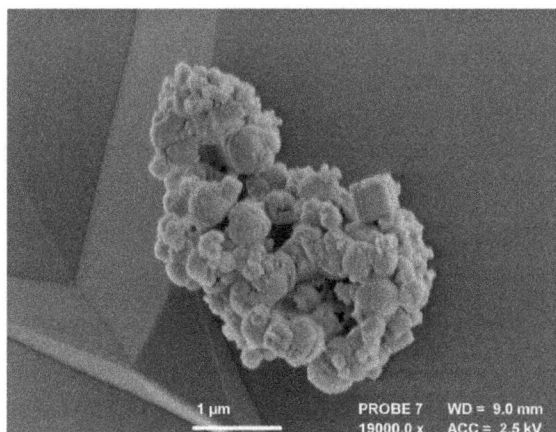

Figure 15. FESEM Ca-A.

3. Materials and Methods

3.1. Synthesis of Na-A: Low Temperature and Organic Template-Free

Zeolite A, with chemical composition $[Na_{12}[(AlO_2)_{12}(SiO_2)_{12}]\cdot 27H_2O$, was prepared using a method described by Leung et al. [4]. The crystallisation process was carried out at 40 °C in the absence of an organic template. Six-point-seven-five grams of NaOH pellets (100% NaOH, Fischer, Ried im Innkreis, Austria) were dissolved in 40 cm^3 of deionised water, and the solution was divided into two equal volumes. Zero-point-nine-eight-eight grams of $NaAlO_2$ (100% Al, Aldrich, St. Louis, MO, USA) were added into one bottle, and 2 cm^3 of colloidal silica (Ludox HS-30, 30 wt % SiO_2, Aldrich, St. Louis, MO, USA) were added to the other. Both solutions were left to stir at room temperature for 90 min until clear. The silica solution was then poured slowly into the aluminium solution with gentle stirring. A thick gel with batch composition of $2SiO_2:Al_2O_3:15Na_2O:400H_2O$ formed, and this was shaken or stirred vigorously for 15 min, either by hand or with a magnetic stirrer, and was put in the oven at 40 °C for 24 h. The zeolite crystals were filtered three times with deionized water, until the pH of the filtrate was 7. The product was then dried in an oven at 100 °C for a further 24 h. In order to see the effect of temperature and time on the synthesis, the method above was varied slightly. For some samples the crystallisation temperature was increased to 50 °C for 24 h, and for other samples, the crystallisation time was increased to 48 h at 40 °C.

3.2. Ion Exchange

The cation sources used for ion exchange were LiOH, KCl, RbOH, Ca(OH)$_2$ and NH$_4$Br. The mass of each cation source was calculated so that the exchange solutions contained a surplus of the exchanging cation that was twice the theoretical amount of Na$^+$ in the pores of 1 g of Na-A. The appropriate cation source was dissolved in 50 cm^3 of deionized water. One gram of the parent Na-A zeolite was added, and the solution was stirred for 6 h at 65 °C. This was then filtered and washed with deionized water and the exchanged zeolite left in an oven to dry overnight at 100 °C. The filtrates were kept in order to quantify the amounts of exchange that had taken place using the sodium ion selective electrode.

3.3. Product Characterisation

3.3.1. Powder X-ray Diffraction

PXRD patterns were recorded at room temperature using a BRUKER AXS D8 Advance diffractometer equipped with a Vantec-1 detector using Cu-Kα radiation (λ = 1.54) in flat plate mode. The scan range was from $3° < 2\theta < 60°$ over 20 min. The unit cell parameters for the samples were calculated using the program UnitCell [34].

3.3.2. Solid State Nuclear Magnetic Resonance

^{29}Si, ^{27}Al, ^{23}Na and ^{13}C NMR spectra were measured by a VARIAN VNMRS 400 spectrometer using the Direct Excitation (DE) method, with neat tetramethylsilane (TMS), 1 M aqueous Al(NO$_3$)$_3$ and 1 M aqueous NaCl as references. ^{13}C Cross-Polarisation (CP) MAS-NMR was used for Ca-A to enhance the signal. The spinning rates of ^{29}Si and ^{13}C NMR were 6.0 kHz and for ^{27}Al and ^{23}Na NMR were 12 kHz. The frequency of ^{29}Si NMR was 79.438 MHz and for ^{27}Al 104.199 MHz. The spectral width of ^{29}Si NMR was 40,322.6 Hz and for ^{27}Al 416.7 Hz. Solid-state NMR spectra were obtained at the EPSRC U.K. National Solid-state NMR Service at Durham.

3.3.3. Scanning Electron Microscopy and Energy Dispersive X-ray Spectroscopy

Low resolution micrographs were taken using a JEOL SEM6480LV scanning electron microscope. EDX data were acquired using an Oxford INCA X-ray analyser attached to the microscope with an acceleration voltage of 20 kV.

High resolution micrographs were taken using a JEOL FESEM6301F field emission scanning electron microscope (FESEM). The powder samples for FESEM were coated with 5 nm chromium to prevent surface charging after EDX data were collected.

3.3.4. Fourier Transform Infrared Spectroscopy

FT-IR spectra for the powder samples were recorded using a Perkin Elmer 100 FT-IR Spectrometer in the range of 4000–600 cm^{-1}.

3.3.5. Sodium Ion Selective Electrode

The concentration of Na$^+$ in the exchange filtrates was calculated using a Cole-Parmer double junction, combination sodium Ion Selective Electrode (ISE) filled with a reference solution of 0.1 M NH$_4$Cl and connected to a pH/mV meter.

4. Conclusions

Solid state NMR revealed changes to the local structure of the LTA framework upon ion exchange with NH$_4$$^+$ and Ca^{2+}. Exchange with Li$^+$, K$^+$ and Rb$^+$ ions does not significantly affect the long-range crystal order. Exchange with NH$_4$$^+$ ions compromises some of the long-range order of the zeolite A crystals due to the loss of some framework aluminium as can be seen from X-ray data. Exchange with divalent Ca^{2+} ions introduces some monodentate carbonate species into the framework, but no alteration to the long-range crystal order is observed. This study confirms that using a local probe such as SS NMR alongside PXRD and other long-range methods to study zeolites can reveal an extra level of information about the structure of those useful minerals, which will further their use as potential catalysts and ion exchange materials.

Acknowledgments: A.S. would like to acknowledge the Royal Society for funding of URF. We thank John Mitchel and Ursula Potter for help with collecting SEM data and Fraser Markwell for collecting SS NMR data at the EPSRC-funded SS NMR facility at Durham University.

Author Contributions: A.S. conceived the study. K.M.L. designed experiments. L.P. and K.M.L. have collected data and performed data analysis. All authors contributed to writing and editing the manuscript.

Conflicts of Interest: The authors declare no conflict of interest.

References

1. Baerlocher, C.; McCusker, L.B.; Olson, D.H. *Atlas of Zeolite Framework Types*; Elsevier: Amsterdam, The Netherlands, 2017.

2. Sartbaeva, A.; Wells, S.A.; Treacy, M.M.J.; Thorpe, M.F. The flexibility window in zeolites. *Nat. Mat.* **2006**, *5*, 962–965, doi:10.1038/nmat1784.

3. Conato, M.T.; Oleksiak, M.D.; McGrail, B.P.; Motkuri, R.K.; Rimer, J.D. Framework Stabilization of Si-Rich LTA Zeolite Prepared in Organic-Free Media. *Chem. Commun.* **2014**, *51*, 269–272, doi:10.1039/C4CC07396G.

4. Leung, K.M.; Edwards, P.P.; Jones, E.; Sartbaeva, A. Microwave synthesis of LTN framework zeolite with no organic structure directing agents. *RSC Adv.* **2015**, *5*, 35580–35585, doi:10.1039/C4RA16583G.

5. Nearchou, A.; Raithby, P.; Sartbaeva, A. Systematic approaches towards template-free synthesis of EMT-type zeolites. *Microporous Mesoporous Mater.* **2018**, *255*, 261–270, doi:10.1016/j.micromeso.2017.08.036.

6. Nearchou, A.; Sartbaeva, A. Influence of alkali metal cations on the formation of zeolites under hydrothermal conditions with no organic structure directing agents. *CrystEngComm* **2008**, *17*, 2496–2503, doi:10.1039/C4CE02119C.

7. Maldonado, M.; Oleksiak, M.D.; Chinta, S.; Rimer, J.D. Controlling Crystal Polymorphism in Organic-Free Synthesis of Na-Zeolites. *J. Am. Chem. Soc.* **2012**, *135*, 2641–2652, doi:10.1021/ja3105939.

8. Nicholas, C.P. *Zeolites in Industrial Separation and Catalysis*; Wiley VCH: Weinheim, Germany, 2010; pp. 355–402.

9. Amorim, R.; Vilaça, N.; Martinho, O.; Reis, R.M.; Sardo, M.; Rocha, J.; Fonseca, A.M.; Baltazar, F.; Neves, I.C. Zeolite Structures Loading with an Anticancer Compound As Drug Delivery Systems. *J. Phys. Chem. C* **2012**, *116*, 25642–25650, doi:10.1021/jp3093868.

10. Chaves, T.F.; Soares, F.; Cardoso, D.; Carneiro, R.L. Monitoring of the crystallization of Zeolite LTA using Raman and Chemometric tools. *Analyst* **2015**, *140*, 854–859, doi:10.1039/C4AN00913D.

11. Lalik, E.; Mirek, R.; Rakoczy, J.; Groszek, A. Microcalorimetric study of sorption of water and ethanol in zeolites 3A and 5A. *Catal. Today* **2006**, *114*, 242–247.

12. Liu, Q.; Mace, A.; Bacsik, Z.; Sun, J.; Laaksonen, A.; Hedin, N. NaKA sorbents with high CO_2-over-N_2 selectivity and high capacity to adsorb CO_2. *Chem. Comm.* **2010**, *46*, 4502–4504, doi:10.1039/C000900H.

13. Sun, H.; Wu, D.; Guo, X.; Shen, B.; Liu, J.; Navrotsky, A. Energetics of Confinement of n-Hexane in Ca-Na Ion Exchanged Zeolite A. *J. Phys. Chem. C* **2014**, *118*, 25590–25596, doi:10.1021/jp508514e.

14. Kwakye-Awuah, B.; Labik, L.K.; Nkrumah, I.; Williams, C. Removal of ammonium ions by laboratory synthesized zeolite linde type A adsorption from water samples affected by mining activities in Ghana. *J. Water Health* **2014**, *12*, 151–160, doi:10.2166/wh.2013.093.

15. Sartbaeva, A.; Rees, N.H.; Edwards, P.P.; Ramirez-Cuesta, A.J.; Barney, E. Local probes show that framework modification in zeolites occurs on ammonium exchange without calcination. *J. Mater. Chem. A* **2013**, *1*, 7415–7421, doi:10.1039/C3TA10243B.

16. Seel, A.G.; Sartbaeva, A.; Edwards, P.P.; Rammirez-Cuesta, A.J. Inelastic neutron scattering of Na-zeolite A with in situ ammoniation: An examination of initial coordination. *Phys. Chem. Chem. Phys.* **2010**, *12*, 9661–9666.

17. Watanabe, Y.; Yamada, H.; Tanaka, J.; Komatsu, Y.; Moriyoshi, Y. Ammonium Ion Exchange of Synthetic Zeolites: The Effect of Their Open Window Sizes, Pore Structures, and Cation Exchange Capacities. *Sep. Sci. Technol.* **2005**, *39*, 2091–2104, doi:10.1081/SS-120039306.

18. Navarrete-Casas, R.; Navarrete-Guijosa, A.; Valenzuela-Calahorro, C.; López-González, J.D.; García-Rodríguez, A. Study of lithium ion exchange by two synthetic zeolites: Kinetics and equilibrium. *J. Colloid Interface Sci.* **2007**, *306*, 345–353, doi:10.1016/j.jcis.2006.10.002.

19. Kim, H.S.; Choi, S.Y.; Lim W.T. Complete Li^+ exchange into zeolite X (FAU, Si\Al = 1.09) from undried methanol solution. *J. Porous Mater.* **2013**, *20*, 1449–1456, doi:10.1007/s10934-013-9731-1.

20. Bignami, G.P.; Dawson, D.M.; Seymour, V.R.; Wheatley, P.S.; Morris, R.E.; Ashbrook, S.E. Synthesis, Isotopic Enrichment, and Solid-State NMR Characterization of Zeolites Derived from the Assembly, Disassembly, Organization, Reassembly Process. *J. Am. Chem. Soc.* **2017**, *139*, 5140–5148.

21. Brouwer, D.H.; Darton, R.J.; Morris, R.E.; Levitt, M.H. A solid-state NMR method for solution of zeolite crystal structures. *J. Am. Chem. Soc.* **2005**, *127*, 10365–10370, doi:10.1021/ja052306h.

22. Fyfe, C.A.; Feng, Y.; Grondey, H.; Kokotailo, G.T.; Gies, H. One- and two-dimensional high-resolution solid-state NMR studies of zeolite lattice structures. *Chem. Rev.* **1991**, *91*, 1525–1543, doi:10.1021/cr00007a013.

23. Fyfe, C.A.; Gobbi, G.C.; Hartman, J.S.; Klinowski, J.; Thomas, J.M. Solid-state magic-angle spinning. Aluminum-27 nuclear magnetic resonance studies of zeolites using a 400-MHz high-resolution spectrometer. *J. Phys. Chem.* **1982**, *86*, 1247–1250, doi:10.1021/j100397a006.

24. Fyfe, C.A.; Thomas, J.M.; Klinowski, J.; Gobbi, G.C. Magic-Angle-Spinning NMR (MAS-NMR) Spectroscopy and the Structure of Zeolites. *Angew. Chem. Int. Ed. Engl.* **1983**, *22*, 259–275, doi:10.1002/anie.198302593.

25. Lippmaa, A.; Maedi, M.; Samoson, A.; Tarmak, M.; Engelhardt, G. Investigation of the structure of zeolites by solid-state high-resolution silicon-29 NMR spectroscopy. *J. Am. Chem. Soc.* **1981**, *103*, 4992–4996, doi:10.1021/ja00407a002.

26. Nebel, H.; Neumann, M.; Mayer, C.; Epple, M. On the Structure of Amorphous Calcium Carbonate–A Detailed Study by Solid-State NMR Spectroscopy. *Inorg. Chem.* **2008**, *47*, 7874–7879, doi:10.1021/ic8007409.

27. Park, M.B.; Vicente, A.; Fernandez, C.; Hong, S.H. Solid-state NMR study of various mono- and divalent cation forms of the natural zeolite natrolite. *Phys. Chem. Chem. Phys.* **2013**, *15*, 7604–7612; doi:10.1039/C3CP44421J.

28. Thomas, J.M.; Klinowsky, J.; Ramadas, S.; Anderson, M.W.; Fyfe, C.A.; Gobbi, G.C. New Approaches to the Structural Characterization of Zeolites: Magic-Angle Spinning NMR (MASNMR). In *Intrazeolite Chemistry*; Galen, D.S., Francis, G.D., Eds.; ACS Publications: Washington, DC, USA, 1983; pp. 159–180, ISBN 9780841207745.

29. Gramlich, V.; Meier, W.M. The crystal structure of hydrated NaA: A detailed refinement of a pseudosymmetric zeolite structure. *Z. Krist.* **1971**, *10*, 134–149, doi:10.1524/zkri.1971.133.133.134.

30. Treacy, M.M.J.; Higgins, J.B. *Collection of Simulated XRD Powder Patterns for Zeolites*; Elsevier: Amsterdam, The Netherlands, 2007; pp. 252–253.

31. Alfaro, S.; Rodriguez, C.; Valenzuela, M.A.; Bosch, P. Aging time effect on the synthesis of small crystal LTA zeolites in the absence of organic template. *Mater. Lett.* **2007**, *61*, 4655–4658, doi:10.1016/j.matlet.2007.03.009.

32. Carey, T.; Tang, C.C.; Hriljac, J.A.; Anderson, P.A. Chemical Control of Thermal Expansion in Cation-Exchanged Zeolite A. *Chem. Mater.* **2014**, *26*, 1561–1566, doi:10.1021/cm403312q.

33. Lührs, H.; Derr, J.; Fischer, R.X. K and Ca exchange behavior of zeolite A. *Microporous Mesoporous Mater.* **2012**, *151*, 457–465, doi:10.1016/j.micromeso.2011.09.025.

34. Holland, T.J.B.; Redfern, S.A.T. Unit cell refinement from powder diffraction data–the use of regression diagnostics. *Miner. Mag.* **1997**, *61*, 65–77.

35. O'Keeffe, M.; Navrotsky, A.; Some Aspects of the Ionic Model of Crystals. In *Structure and Bonding in Crystals*; O'Keeffe, M., Ed.; Academic Press: Cambridge, MA, USA, 1981; pp. 299–322.

36. Ramdas, S.; Klinowski, J. A simple correlation between isotropic 29Si-NMR chemical shifts and T-O-T angles in zeolite frameworks. *Nature* **1984**, *308*, 521–523, doi:10.1038/308521a0.

37. Edith, M.F.; Hassan, K.; Herman, A.S. Infrared Structural Studies of Zeolite Frameworks. In *Molecular Sieve Zeolites-I*; Edith M.F., Leonard B.S., Eds.; ACS Publications: Washington, DC, USA, 1974; pp. 201–229, ISBN 9780841201149.

38. Milkey, R.G. Infrared Spectra of some Tectosilicates. *Am. Mineral.* **1960**, *45*, 990–1007.

39. Dyballa, M.; Obenaus, U.; Lang, S.; Gehring, B.; Traa, Y.; Koller, H.; Hunger, M. Brønsted sites and structural stabilization effect of acidic low-silica zeolite A prepared by partial ammonium exchange. *Microporous Mesoporous Mater.* **2015**, *212*, 110–116, doi:10.1016/j.micromeso.2015.03.030.

40. Ucun, F. An Infrared Study of the CaA Zeolite Reacted with CO_2. *Z. Naturforsch.* **2002**, *57*, 283.

41. Holzwarth, U.; Gibson, N. The Scherrer equation versus the 'Debye-Scherrer equation'. *Nat. Nano* **2011**, *6*, 534.

42. Cubillas, P.; Gebbie, J.T.; Stevens, S.M.; Blake, N.; Umemura, A.; Terasaki, O.; Anderson, M.W. Atomic Force Microscopy and High Resolution Scanning Electron Microscopy Investigation of Zeolite A Crystal Growth. Part 2: In Presence of Organic Additives. *J. Phys. Chem. C* **2014**, *118*, 23092–23099, doi:10.1021/jp506222y.

43. Kliewer, C.E. Electron Microscopy and Imaging. In *Zeolite Characterization and Catalysis*; Chester, A.W., Derouane, E.G., Eds.; Springer: Dordrecht, The Netherlands, 2009; pp. 169–196.

44. Mintova, S.; Olson, N.H.; Valtchev, V.; Bein, T. Mechanism of Zeolite A Nanocrystal Growth from Colloids at Room Temperature. *Science* **1999**, *283*, 958–960, doi:10.1126/science.283.5404.958.

45. Dimitrov, L.; Valtchev, V.; Nihtianova, D.; Kalvachev, Y. Submicrometer Zeolite A Crystals Formation: Low-Temperature Crystallization Versus Vapor Phase Gel Transformation. *Cryst. Growth Des.* **2011**, *11*, 4958–4962, doi:10.1021/cg2008667.

46. Smaihi, M.; Barida, O.; Valtchev, V. Investigation of the Crystallization Stages of LTA-Type Zeolite by Complementary Characterization Techniques. *Eur. J. Inorg. Chem.* **2003**, *2003*, 4370–4377, doi:10.1002/ejic.200300154.

magnetochemistry

|MDPI|

Article

Separation of the *α*- and *β*-Anomers of Carbohydrates by Diffusion-Ordered NMR Spectroscopy

Takashi Yamanoi [1,*], Yoshiki Oda [2] and Kaname Katsuraya [3]

1   Faculty of Pharmacy and Pharmaceutical Sciences, Josai University, 1-1 Keyakidai, Sakado,
    Saitama 350-0295, Japan
2   Technology Joint Management Office, Tokai University, 4-1-1 Kitakaname, Hiratsuka,
    Kanagawa 259-1292, Japan; oy287363@tsc.u-tokai.ac.jp
3   Department of Human Ecology, Wayo Women's University, Chiba 272-8533, Japan; katsuraya@wayo.ac.jp
*   Correspondence: yamanoi1119@gmail.com; Tel.: +81-49-271-7958

Received: 31 October 2017; Accepted: 20 November 2017; Published: 22 November 2017

Abstract: This article describes the successful application of the DOSY method for the separation and analysis of the α- and β-anomers of carbohydrates with different diffusion coefficients. In addition, the DOSY method was found to effectively separate two kinds of glucopyranosides with similar aglycon structures from a mixture.

Keywords: DOSY; diffusion coefficient; anomer separation; carbohydrate isomer; glycoside

---

1. Introduction

High-resolution nuclear magnetic resonance (NMR) spectroscopy has become an excellent established tool for determining the molecular structures and conformations of compounds while preserving the sample integrity. In addition, pulsed gradient spin echo (PGSE) NMR is recognized as a powerful technique for the determination of diffusion coefficients and the separation of different species in a mixture on the basis of their diffusion coefficients [1]. Diffusion-ordered NMR spectroscopy (DOSY), which displays the PGSE NMR data in a two-dimensional spectrum, is a practical experiment for separating the ^{1}H NMR spectra of different species [2]. In addition to the DOSY separation of a mixture, the DOSY method has been widely used for the characterization of high-molecular-weight polymeric compounds and the identification of supramolecular structures [3–8]. However, the DOSY method has generally failed to identify the isomeric species of similar size and structure because of their similar diffusion coefficients. Therefore, recent studies on the DOSY technique have focused on developing strategies for the separation of isomeric species [9–14]. The DOSY method has been also applied to carbohydrate chemistry as a tool for the separation and analysis of mono-, di-, oligo-, and polysaccharides, as well as for the structural analysis of metal-complexed carbohydrates [15–20]. However, reports on the application of the DOSY method for the separation of carbohydrate anomeric isomers are still scarce. With an aim of increasing the utility and applicability of the DOSY method in carbohydrate chemistry, we tackled its evaluation in the separation and analysis of the α- and β-anomers of carbohydrates.

2. Results and Discussion

The DOSY analysis for the isomer separation in a mixture of α- and β-anomers was investigated using several kinds of carbohydrate derivatives of glucopyranosides and glucopyranoses, as shown in Figure 1.

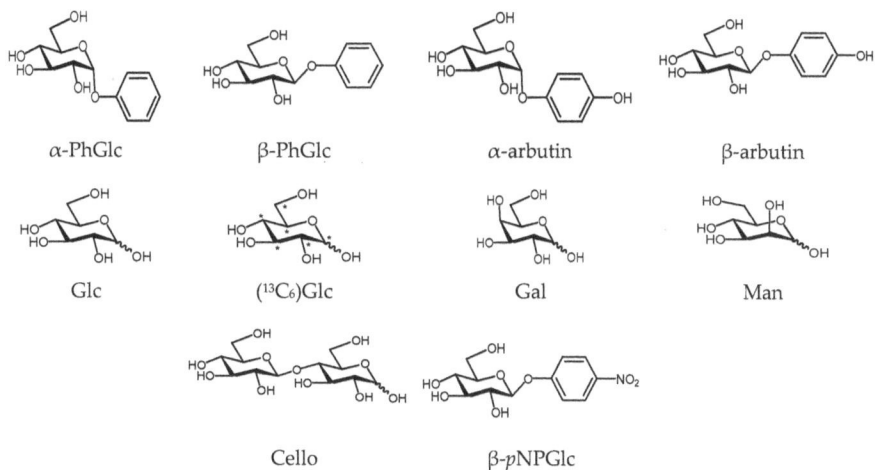

Figure 1. Carbohydrate derivatives used for DOSY analysis on anomer isomers.

2.1. DOSY Separation of the α- and β-Anomeric Isomers of Glycopyranosides

The DOSY separation of the anomeric isomers in a mixture of 10 mM phenyl β-glucopyranoside (β-PhGlc) and 10 mM phenyl α-glucopyranoside (α-PhGlc) was firstly investigated. Figure 2 shows the DOSY spectrum of the mixture of α-PhGlc and β-PhGlc in D_2O at 30 °C, together with the individual ^1H NMR spectra of β-PhGlc and α-PhGlc in D_2O. In the DOSY spectrum, two different species with diffusion coefficients (D) of 5.9×10^{-10} $m^2 \cdot s^{-1}$ and 5.6×10^{-10} $m^2 \cdot s^{-1}$ could be identified, whose resonances corresponded to the ^1H NMR spectrum of β-PhGlc and α-PhGlc, respectively. It was thereby found that the apparent difference between the diffusion coefficients of β-PhGlc and α-PhGlc allow the DOSY separation of these glucopyranoside anomers.

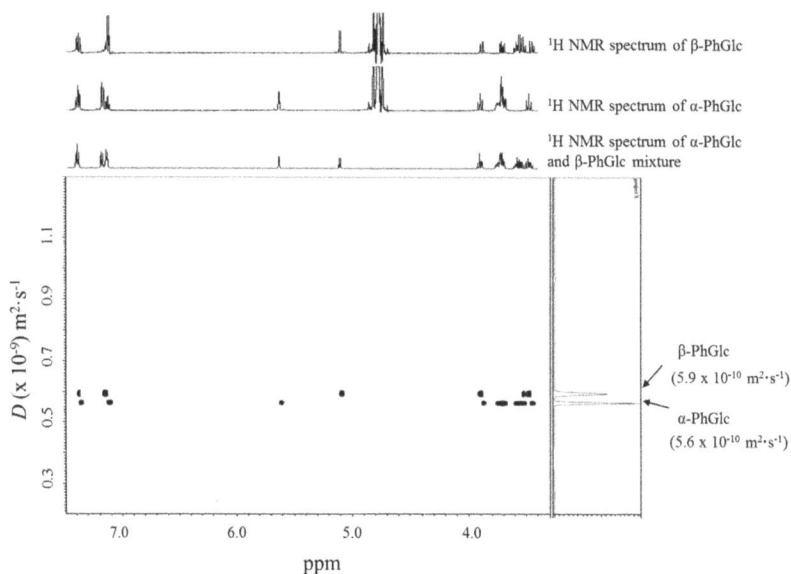

Figure 2. DOSY spectrum of a 10 mM β-PhGlc and 10 mM α-PhGlc mixture.

Next, the DOSY separation of the anomeric isomers in a mixture of 10 mM α-arbutin (*p*-hydroxyphenyl α-glucopyranoside) and 10 mM β-arbutin in D$_2$O at 30 °C was similarly investigated. The two glucopyranoside anomers—which exhibited diffusion coefficients (*D*) of 5.9×10^{-10} m^2·s^{-1} (α-arbutin) and 5.8×10^{-10} m^2·s^{-1} (β-arbutin), respectively—were also successfully separated by the DOSY technique, as shown in Figure S1.

2.2. DOSY Separation of the α- and β-Anomeric Isomers of Glycopyranoses

Glycopyranoses are known to undergo mutarotation; they interconvert their α- and β-anomers in water and an equilibrium mixture of the two forms is achieved. Figure 3 displays the mutarotation of D-glucopyranose (Glc). The ^{1}H NMR spectrum of Glc at a concentration of 20 mM in D$_2$O indicated that the anomer ratio of Glc was ca. 1:1. We investigated whether the DOSY method could separate the individual ^{1}H NMR spectra of the anomeric isomers in an equilibrium mixture of α-Glc and β-Glc. The DOSY spectrum of a 20 mM solution of Glc in D$_2$O at 30 °C revealed that two species with different diffusion coefficients (*D*) of 7.6×10^{-10} m^2·s^{-1} and 5.8×10^{-10} m^2·s^{-1} were present, the former corresponding to the ^{1}H NMR spectrum of α-Glc, and the latter to the ^{1}H NMR spectrum of β-Glc, as can be seen in Figure 4. We found that the difference between the diffusion coefficients of α-Glc and β-Glc was sufficient to separate the two anomeric isomers of Glc by using the DOSY technique.

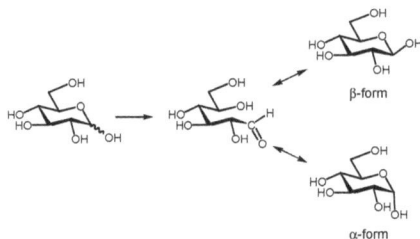

Figure 3. Mutarotation of Glc to interconvert between α-anomer and β-anomer in water.

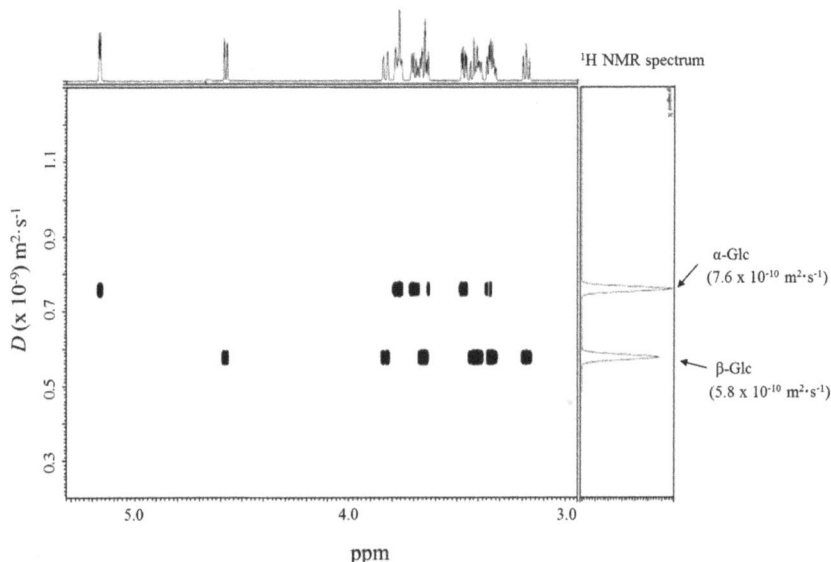

Figure 4. DOSY spectrum of 20 mM Glc.

In order to confirm the applicability of the DOSY method for the determination of the diffusion coefficients of both anomers of glycopyranose showing mutarotation, the DOSY spectra of several kinds of glycopyranoses were measured. The DOSY measurements were performed using 20 mM solutions of ($^{13}C_6$)-D-glucopyranose (($^{13}C_6$)Glc), D-galactopyranose (Gal), D-mannopyranose (Man), and cellobiose (Glcβ(1→4)Glc, Cello) in D_2O at 30 °C. We had previously confirmed the presence of the α- and β-anomers of these glycopyranoses in D_2O by ^1H NMR measurements. The individual ^1H NMR spectra of the α- and β-anomers were successfully separated in all the DOSY spectra, and their corresponding diffusion coefficients (*D*) were thereby obtained. The DOSY spectra are shown in Figures S2–S5, and the α- and β-anomers of these glycopyranoses are summarized in Table 1, together with the average diffusion coefficients of some of the α- and β-glycopyranose mixtures previously reported. Since the DOSY technique was able to separate the α- and β-anomers of glycopyranoses, it seems to be a reliable method for the estimation of their individual diffusion coefficients.

Table 1. Diffusion coefficients of carbohydrate anomers measured in this study and their reported values.

Entry	Carbohydrate (Conditions, 30 °C, D_2O)	Diffusion Coefficient D ($\times 10^{-10}$) $m^2 \cdot s^{-1}$	Reported Diffusion Coefficient D ($\times 10^{-10}$) $m^2 \cdot s^{-1}$ (Conditions) [Lit.]
1	α-PhGlc (10 mM)	5.6	-
	β-PhGlc (10 mM)	5.9	-
2	α-arbutin (10 mM)	5.9	-
	β-arbutin (10 mM)	5.8	-
3	Glc (20 mM)	-	6.3 (100 mM, H_2O) [21]
	α-anomer	7.6	-
	β-anomer	5.8	-
4	($^{13}C_6$)Glc (20 mM)	-	
	α-anomer	9	
	β-anomer	5.4	
5	Gal (20 mM)	-	8.7 (100 mM, H_2O) [21]
	α-anomer	10.9	-
	β-anomer	5.8	-
6	Man (20 mM)	-	7.0 (25 °C, H_2O) [22]
	α-anomer	7.4	-
	β-anomer	6.85	-
7	Cello (20 mM)	-	5.2 (25 °C, H_2O) [23]
	α-anomer	5. 5	-
	β-anomer	5.95	-
8	α-PhGlc (10 mM)	5.77	-
	α-arbutin (10 mM)	5.49	-
9	β-arbutin (10 mM)	7.2	-
	β-pNPGlc (10 mM)	5.6	-

2.3. DOSY Separation of a Mixture of Two Kinds of Glycopyranosides Having Similar Aglycon Structures

We also investigated the DOSY separation of a mixture of two kinds of glycopyranosides having a similar aglycon structures. The DOSY spectrum of a mixture of 10 mM α-PhGlc and 10 mM α-arbutin in D_2O at 30 °C clearly separated the two kinds of glycopyranoside species, which exhibited different diffusion coefficients (*D*) of 5.77×10^{-10} $m^2 \cdot s^{-1}$ (α-PhGlc) and 5.49×10^{-10} $m^2 \cdot s^{-1}$ (α-arbutin), as shown in Figure S6. The DOSY spectrum in Figure S7 also evinces the successful separation of a mixture of 10 mM β-arbutin and 10 mM β-pNPGlc, whose diffusion coefficients (*D*) were 7.2×10^{-10} $m^2 \cdot s^{-1}$ and 5.6×10^{-10} $m^2 \cdot s^{-1}$, respectively.

3. Materials and Methods

D-glucose, D-galactose, D-mannose, D-cellobiose, β-arbutin, phenyl α-glucopyranoside, phenyl β-glucopyranoside, and *p*NP β-glucopyranoside were purchased from Tokyo Chemical Industry Co., Ltd. (Chuo-ku, Tokyo, Japan) and ($^{13}C_6$)-D-glucose was purchased from Cambridge Isotope Laboratories, Inc. (Andover, MA, USA). α-Arbutin was purchased from Wako Pure Chemical Industries, Ltd. (Chuo-Ku, Osaka, Japan). D_2O was purchased from Kanto Chemical Co., INC. (99.8% minimum in D, Chuo-ku, Tokyo, Japan).

The NMR spectra were obtained using a JEOL ECA-600 spectrometer (JEOL Ltd., Akishima, Tokyo, Japan), using 20 mM concentrations for the individual samples and 10 mM concentrations for the mixtures. The ^1H NMR spectra were recorded at 600 MHz. The chemical shifts were referenced to the solvent values (δ 4.70 ppm for HOD). The spectra were analyzed after 16 scans and 4 dummy scans. The 2D DOSY experiments were performed at 30 °C using the bipolar pulse pair and longitudinal eddy current delay sequence. Gradient amplitudes were 20–247 mT/m. The spectral width was 6000 Hz. Bipolar rectangular gradients were used with total durations of 1 to 2 ms. Gradient recovery delays were 0.1 ms. Diffusion times were between 50 and 200 ms. The relaxation delay was 7.0 s. The spectra were analyzed after 256 scans and 16 dummy scans. The spectral analyses were processed by the Delta NMR processing software version 4.3.6. (JEOL USA, Inc., Peabody, MA, USA).

4. Conclusions

This article describes the evaluation of the DOSY method for the separation and analysis of the α- and β-anomers of carbohydrates. We found that the α- and β-anomers of carbohydrates having different diffusion coefficients can be separated by using the DOSY technique, and their individual diffusion coefficients can be determined. In addition, the DOSY method was also applicable to the separation of two kinds of glucopyranosides having similar aglycon structures from a mixture.

Supplementary Materials: The following are available online at www.mdpi.com/2312-7481/3/4/38/s1, Figure S1. DOSY spectrum of a 10 mM α-arbutin and 10 mM β-arbutin mixture in D_2O at 30 °C, Figure S2. DOSY spectrum of 20 mM ($^{13}C_6$)Glc in D_2O at 30 °C, Figure S3. DOSY spectrum of 20 mM Gal in D_2O at 30 °C, Figure S4. DOSY spectrum of 20 mM Man in D_2O at 30 °C, Figure S5. DOSY spectrum of 20 mM Cello in D_2O at 30 °C, Figure S6. DOSY spectrum of a 10 mM α-PhGlc and 10 mM α-arbutin mixture in D_2O at 30 °C, Figure S7. DOSY spectrum of a 10 mM β-arbutin and 10 mM β-*p*NPGlc mixture in D_2O at 30 °C.

Author Contributions: T.Y. conceived idea of the article and wrote the paper. Y.O. performed the experiments. K.K. conceived and designed the experiments. All authors approved the manuscript.

Conflicts of Interest: The authors declare no conflict of interest.

References

1. Johnson, C.S. Diffusion ordered nuclear magnetic resonance spectroscopy: Principles and applications. *Prog. Nucl. Magn. Reson. Spectrosc.* **1999**, *34*, 203–256. [CrossRef]
2. Antalek, B. Using Pulsed Gradient Spin Echo NMR for Chemical Mixture Analysis: How to Obtain Optimum Results. *Concepts Magn. Reson.* **2002**, *14*, 225–258. [CrossRef]
3. Kavakka, J.S.; Kilpeläinen, I.; Heikkinen, S. General Chromatographic NMR Method in Liquid State for Synthetic Chemistry: Polyvinylpyrrolidone Assisted DOSY Experiments. *Org. Lett.* **2009**, *6*, 1349–1352. [CrossRef] [PubMed]
4. Kavakka, J.S.; Parviainen, V.; Wahala, K.; Kilpeläinen, I.; Heikkinen, S. Enhanced chromatographic NMR with polyethyleneglycol. A novel resolving agent for diffusion ordered spectroscopy. *Magn. Reson. Chem.* **2010**, *48*, 777–781. [CrossRef] [PubMed]
5. Huang, S.; Wu, R.; Bai, Z.; Yang, Y.; Li, S.; Dou, X. Evaluation of the separation performance of polyvinylpyrrolidone as a virtual stationary phase for chromatographic NMR. *Magn. Reson. Chem.* **2014**, *52*, 486–490. [CrossRef] [PubMed]
6. Xu, J.; Tan, T.; Kenne, L.; Sandström, C. The use of diffusion-ordered spectroscopy and complexation agents to analyze mixtures of catechins. *New J. Chem.* **2009**, *33*, 1057–1063. [CrossRef]

7.   Oda, Y.; Kobayashi, N.; Yamanoi, T.; Katsuraya, K.; Takahashi, K.; Hattori, K. β-Cyclodextrin Conjugates with Glucose Moieties Designed as Drug Carriers: Their Syntheses, Evaluations Using Concanavalin A and Doxorubicin, and Structural Analyses by NMR Spectroscopy. *Med. Chem.* **2008**, *4*, 244–255. [CrossRef] [PubMed]

8.   Oda, Y.; Matsuda, S.; Yamanoi, T.; Murota, A.; Katsuraya, K. Identification of the inclusion complexation between phenyl β-D-($^{13}C_6$)-glucopyranoside and α-cyclodextrin using 2D ^1H or ^{13}C DOSY spectrum. *Supramol. Chem.* **2009**, *21*, 638–642. [CrossRef]

9.   Haiber, S.; Nilsson, M.; Morris, G.A. Matrix-assisted diffusion-ordered spectroscopy: Application of surfactant solutions to the resolution of isomer spectra. *Magn. Reson. Chem.* **2012**, *50*, 458–465.

10. Reile, I.; Aspers, R.L.E.G.; Feiters, M.C.; Rutjes, F.P.J.T.; Tessari, M. Resolving DOSY spectra of isomers by methanol-d4 solvent effects. *Magn. Reson. Chem.* **2017**, *55*, 759–762. [CrossRef] [PubMed]

11. Gramosa, N.V.; Ricardo, N.M.S.P.; Adams, R.W.; Morris, G.A.; Nilsson, M. Matrix-assisted diffusion-ordered spectroscopy: Choosing a matrix. *Magn. Reson. Chem.* **2016**, *54*, 815–820. [CrossRef] [PubMed]

12. Adams, R.W.; Aguilar, J.A.; Cassani, J.; Morris, G.A.; Nilsson, M. Resolving natural product epimer spectra by matrix-assisted DOSY. *Org. Biomol. Chem.* **2011**, *20*, 7062–7064. [CrossRef] [PubMed]

13. Codling, D.J.; Zheng, G.; Stait-Gardner, T.; Yang, S.; Nilsson, M.; Price, W.S. Diffusion Studies of Dihydroxybenzene Isomers in Water−Alcohol Systems. *J. Phys. Chem. B* **2013**, *117*, 2734–2741. [CrossRef] [PubMed]

14. Chaudhari, S.R.; Suryaprakash, N. Diffusion ordered spectroscopy for resolution of double bonded cis, trans-isomers. *J. Mol. Struct.* **2012**, *1017*, 106–108. [CrossRef]

15. Viel, S.; Capitani, D.; Mannina, L.; Segre, A. Diffusion-Ordered NMR Spectroscopy: A Versatile Tool for the Molecular Weight Determination of Uncharged Polysaccharides. *Biomacromolecules* **2003**, *4*, 1843–1847. [CrossRef] [PubMed]

16. Groves, P.; Rasmussen, M.O.; Molero, M.D.; Samain, E.; Cañada, F.J.; Driguez, H.; Jiménez-Barbero, J. Diffusion ordered spectroscopy as a complement to size exclusion chromatography in oligosaccharide analysis. *Glycobiology* **2004**, *14*, 451–456. [CrossRef] [PubMed]

17. Cao, R.; Komura, F.; Nonaka, A.; Kato, T.; Fukumashi, J.; Matsui, T. Quantitative analysis of D-(+)-glucose in fruit juices using diffusion ordered-^1H nuclear magnetic resonance spectroscopy. *Anal. Sci.* **2014**, *30*, 383–388. [CrossRef] [PubMed]

18. Politi, M.; Groves, P.; Chávez, M.I.; Cañada, F.J.; Jiménez-Barberoa, J. Useful applications of DOSY experiments for the study of mushroom polysaccharides. *Carbohydr. Res.* **2006**, *341*, 84–89. [CrossRef] [PubMed]

19. Kählig, H.; Dietrich, K.; Dorner, S. Analysis of Carbohydrate Mixtures by Diffusion Difference NMR Spectroscopy. *Monatsh. Chem.* **2002**, *133*, 589–598. [CrossRef]

20. Díaz, M.D.; Berger, S. Studies of the complexation of sugars by diffusion-ordered NMR spectroscopy. *Carbohydr. Res.* **2000**, *329*, 1–5. [CrossRef]

21. Nagy, L.; Gyetvai, G.; Nagy, G. Determination of the Diffusion Coefficient of Monosaccharides with Scanning Electrochemical Microscopy (SECM). *Electroanalysis* **2009**, *21*, 542–549. [CrossRef]

22. Mori, N.; Sugai, E.; Fuse, Y.; Funazukuri, T. Infinite Dilution Binary Diffusion Coefficients for Six Sugars at 0.1 MPa and Temperatures from (273.2 to 353.2). *J. Chem. Eng. Data* **2007**, *52*, 40–43.

23. Ihnat, M.; Goring, D.A.I. Shape of the cellodextrins in aqueous solution at 25 °C. *Can. J. Chem.* **1967**, *45*, 2353–2361. [CrossRef]

magnetochemistry

MDPI

Review

Multi-Quanta Spin-Locking Nuclear Magnetic Resonance Relaxation Measurements: An Analysis of the Long-Time Dynamical Properties of Ions and Water Molecules Confined within Dense Clay Sediments

Patrice Porion * and Alfred Delville *

Interfaces, Confinement, Matériaux et Nanostructures (ICMN), UMR 7374, CNRS and Université d'Orléans, 45071 Orléans CEDEX 2, France
* Correspondence: porion@cnrs-orleans.fr (P.P.); delville@cnrs-orleans.fr (A.D.)

Received: 5 September 2017; Accepted: 31 October 2017 ; Published: 14 November 2017

Abstract: Solid/liquid interfaces are exploited in various industrial applications because confinement strongly modifies the physico-chemical properties of bulk fluids. In that context, investigating the dynamical properties of confined fluids is crucial to identify and better understand the key factors responsible for their behavior and to optimize their structural and dynamical properties. For that purpose, we have developed multi-quanta spin-locking nuclear magnetic resonance relaxometry of quadrupolar nuclei in order to fill the gap between the time-scales accessible by classical procedures (like dielectric relaxation, inelastic and quasi-elastic neutron scattering) and obtain otherwise unattainable dynamical information. This work focuses on the use of quadrupolar nuclei (like ^{2}H, ^{7}Li and ^{133}Cs), because quadrupolar isotopes are the most abundant NMR probes in the periodic table. Clay sediments are the confining media selected for this study because they are ubiquitous materials implied in numerous industrial applications (ionic exchange, pollutant absorption, drilling, waste storing, cracking and heterogeneous catalysis).

Keywords: diffusion in porous media; NMR relaxation; multi-quanta relaxometry; quadrupolar nuclei; clay sediments

---

1. Introduction

In the last few decades, numerous experimental [1] and theoretical [2] studies have been devoted to solid/liquid interfaces in order to understand and predict the influence of confinement on the structural, thermodynamical and dynamical properties of fluids. In that context, clay-water solid/liquid interfaces [3–9] were frequently investigated for two reasons. First, from a theoretical point of view, clay platelets are flat and atomically smooth with a well-characterized structure and atomic composition, leading to ideal models of solid/liquid interfacial systems. Second, natural and synthetic clays are used in a large variety of industrial applications (drilling, heterogeneous catalysis [8], waste storing [9], food, paint and cosmetic industries), exploiting their various physico-chemical properties (gelling, thixotropy, surface acidity, high specific surface and ionic exchange capacity, water and polar solvent adsorption, swelling). Optimizing applications such as heterogeneous catalysis and waste storing requires quantifying the mobility of solvent molecules and neutralizing counterions inside the porous network of clay minerals. For that purpose, numerous experimental studies have been performed to determine the mobility of confined fluids over a broad range of diffusing time. At short time-scales (between pico-seconds and 100 nano-seconds), the mobility of confined water molecules was successfully investigated by classical Inelastic (INS) [10–12] and Quasi-Elastic Neutron

Scattering (QENS) [13,14] experiments. By contrast, the long-time mobility of bulk fluids is generally investigated by pulsed gradient spin echo NMR spectroscopy [15] to probe time-scales larger than the millisecond. Unfortunately, the mobility of neutralizing counterions is difficult to measure by neutron scattering experiments, and the presence of paramagnetic impurities within the solid network significantly enhances the NMR relaxation rates of confined fluids, strongly limiting the use of pulsed gradient spin echo NMR spectroscopy. For that purpose, NMR relaxation measurements were frequently performed [16–21] to extract dynamical information on the mobility of the diffusing NMR probes.

A general trend of the NMR relaxation property of confined fluids is the large difference between the longitudinal (R_1) and transversal (R_2) relaxation rates. While the bulk fluids have generally the same NMR relaxation rates ($R_1 \sim R_2$), confinement drastically enhances their transverse relaxation rate leading to $R_2 \gg R_1$. Two different phenomena may be responsible for the above-mentioned difference between the longitudinal and transverse relaxation rates: either chemical exchange of the NMR probes between various environments under the so-called "moderately rapid exchange" condition [22] or a slow modulation of the NMR relaxation mechanisms [21,23,24] induced by the molecular motions of the confined fluids. One can differentiate between these two interpretations without modifying the sample's environment (temperature, concentration, composition) by measuring both relaxation rates as a function of the static magnetic field B_0: an increase of their difference ($R_2 - R_1$) as a function of the field's strength is the fingerprint of an intermediate exchange [22,25], while the opposite trend results from the slow modulation of the NMR relaxation mechanisms. That condition is generally fulfilled by reducing the fluid temperature or after complexation of the NMR probe by a macromolecule [26–28].

In that framework, confinement was recently shown to induce, at room temperature, the slow modulation of the NMR relaxation mechanism of fluid [29–31]. As a consequence, numerous theoretical and experimental studies were devoted to that problem in order to quantify the influence of the geometrical and thermodynamical properties [31–39] of the porous media on the NMR relaxation mechanisms of their confined fluids. In addition to NMR relaxation measurements performed at a limited number of available magnetic fields [16,33], spin-locking relaxation measurements [16,33,40–43] were initially performed to extend the investigation of the dispersion curves to lower magnetic fields. These two complementary procedures lead however to a large gap within the dispersion curves that was successfully filled recently by field cycling NMR relaxometry [29–32,44–48]. Unfortunately, in the case of confined quadrupolar nuclei, the enhancement of the transverse relaxation rate of the confined quadrupolar probes prohibits the use of field cycling NMR relaxometry because of the time required to switch the magnetic field. In that context, we have developed multi-quanta spin-locking NMR relaxometry to probe the dynamical properties of confined quadrupolar nuclei that pertain to a large class of observable NMR isotopes within the periodic table [49]. To test the potentiality of that new approach, we selected the clay/water interface because natural and synthetic clays are well characterized and exploited in numerous industrial applications. In that context, we used multi-quanta NMR relaxometry to quantify the mobility of water molecules (heavy water D_2O) [40,50–53] and neutralizing counterions (^7Li [54], ^{133}Cs [55]) diffusing within the porous network of clay sediments.

The swelling clays used in this study (montmorillonite, hectorite, beidellite and laponite) pertain to the class of smectites. Their elementary platelets result from the sandwiching of one layer of octahedral metallic oxides (Al^{III} or Mg^{II}) between two layers of tetrahedral silica. Atomic substitutions of some metals in these octahedral or tetrahedral layers by less charged metals lead to a net negative charge of the clay network neutralized by cations. These exchangeable cations are localized within the interlamellar space between individual clay platelets and are responsible for the water affinity of the clay network. Furthermore, the mechanical behavior (swelling versus setting) of the clay/water interface is monitored by the nature and valance of the neutralizing counterions, the number of substitution sites and their localization within the clay network.

As displayed in Figure 1, clay sediment exhibits a multiscale structure. At short distances, the sediment is composed of highly anisotropic platelets (thickness $\sim$7 Å, diameter $\sim$300–3000 Å).

At intermediate distances, microscopic domains result from the stacking of numerous (10–100) parallel clay platelets. Depending on the hygrometry, the interlamellar space between these platelets is partially or totally filled by adsorbed water molecules [56–58] in addition to the neutralizing counterions. Finally, at the largest scale, clay sediment results from the juxtaposition of micro-domains with different orientations. We have used multi-quanta spin-locking NMR relaxometry measurements to determine the average residence time of the water molecules and some neutralizing counterions within the interlamellar space between the clay platelets inside each micro-domain. Furthermore, two-time stimulated echo NMR spectroscopy [59] was used to quantify the time-scale required by the water molecules to probe micro-domains with different orientations [52,53,60].

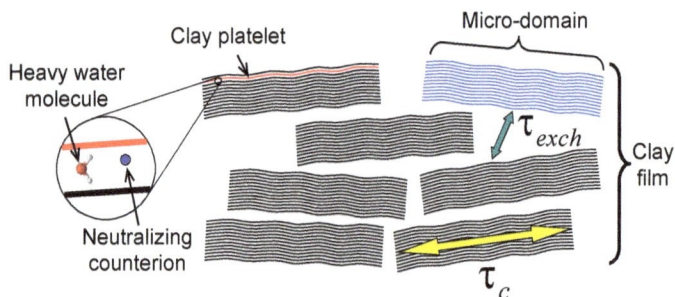

Figure 1. Schematic view of multiscale organization of the clay sediment resulting from the coexistence of clay aggregates with various orientations of the platelet directors. Reprinted with permission from [53]. Copyright (2014) American Chemical Society.

In addition to these experimental investigations, multi-scale numerical simulations were performed to determine the structure of the confined fluids and their mobility. Grand Canonical Monte Carlo (GCMC) simulations [56–58] were first performed to determine the number of confined water molecules as a function of the water partial pressure and the interlamellar distance. These numerical simulations also illustrate the organization of the water molecules and neutralizing counterions confined between the clay platelets (see Figure 2a,b). This organization of the confined water molecules significantly contributes to the X-ray and neutron scattering spectra [56–58] of oriented clay sediments. Nevertheless, the same confined water molecules exhibit a large mobility in the direction parallel to the clay surface, as detected by QENS [13,14]. This local mobility of the confined probes was determined by numerical simulations of molecular dynamics and directly compared to the QENS spectra [14]. Numerical simulations of Brownian dynamics [51] are then required to propagate at a larger time-scale the water mobility predicted by MD simulations in order to interpret the residence time determined by multi-quanta NMR relaxometry. Finally, a set of macroscopic differential equations [60] were solved to describe the exchange of the water molecules between differently-oriented micro-domains in order to interpret the echo attenuation detected by two-time stimulated echo NMR spectroscopy.

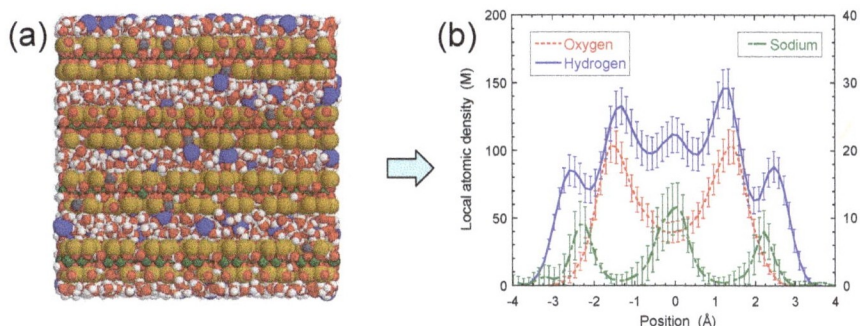

Figure 2. (**a**) Snapshot illustrating one Grand Canonical Monte Carlo (GCMC) equilibrium configuration of confined water molecules and neutralizing sodium counterions; (**b**) concentration profiles of sodium counterions and oxygen and hydrogen atoms pertaining to water molecules confined between two beidellite clay lamellae. Reprinted with permission from [53]. Copyright (2014) American Chemical Society.

2. Sample Preparation and Experimental Setup

All natural clay samples used in this study were purified according to the classical procedure [61], and the neutralizing cations were exchanged leading predominantly to mono-ionic clay samples. The clay platelets were further selected according to their size by centrifugation [61]. Transmission Electronic Microscopy (TEM) was used to determine their size distribution [61]. Self-supporting clay films were obtained by ultrafiltration under nitrogen pressure of dilute clay dispersions. The clay films were further dried under nitrogen flux before equilibration with a reservoir of heavy water at fixed chemical potential by using saturated salt solutions. The partial pressure of D_2O is selected to obtain hydrated clay samples with mainly one or two hydration layers in accordance with the water adsorption isotherm [56,57]. A lamella (30×5.5 mm^2) is cut into the clay film and inserted into a sealed glass cylinder, which fit the gap into a home-made solenoid coil [51] used for the NMR measurements (Figure 3). The sample holder can rotate into the coil in order to perform NMR experiments with different orientations of the clay film (denoted θ^{LF}) by reference to the static magnetic field B_0. An important point is that a home-made detection coil is not required to perform multi-quanta spin-locking relaxometry measurements. The only requirement is the use of a solenoidal coil. Note however that spectra and relaxation measurements must be recorded at various orientations of the clay sample by reference to the static magnetic field. As a consequence, the detection coil must be modified to measure the sample orientation with good accuracy.

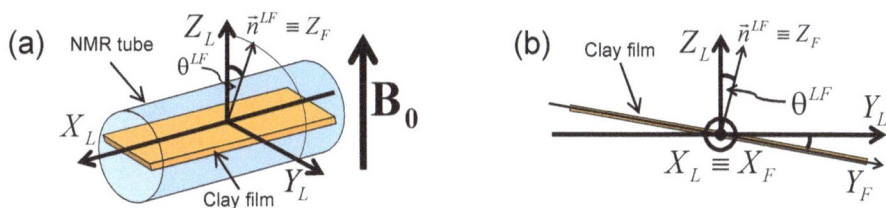

Figure 3. Schematic view of: (**a**) the film orientation within the NMR tube used to insert the clay film sample into the detection coil (see the text); and (**b**) the different Euler angles characterizing the orientation of the clay film by reference to the static magnetic field B_0. Reprinted with permission from [53]. Copyright (2014) American Chemical Society.

NMR measurements were performed on a DSX360 Bruker spectrometer operating at a field of 8.465 Tesla equipped with home-made detection coils. For this experimental setup, the typical duration for the inversion of the longitudinal magnetization varies between 15 and 30 microseconds depending on the nature of the NMR probes (^{2}H, ^{7}Li and ^{133}Cs). Because of the efficiency of the NMR relaxation of these NMR quadrupolar probes, a fast detection mode was selected with a sampling time varying between 0.25 and one microsecond. Pulse sequences where selected to optimize the magnetization transfers required by multi-quanta NMR relaxation measurements. Details on the theory of NMR relaxation and multi-quanta spin-locking measurements are given in Appendices A–D.

3. Results and Discussion

3.1. NMR Spectra

Figures 4 and 5 exhibit the variation of the NMR spectra as a function of the orientation of the clay film into the static magnetic film B_0. The doublet detected by ^{2}H NMR spectroscopy of the water molecules confined within hectorite [53] (Figure 4) results from two phenomena: a good alignment of the various clay platelets with respect to the lamella director and a specific orientation of the water molecules confined within the interlamellar space of the clay platelets. The principal component of the tensor quantifying the Electric Field Gradient (EFG) monitoring the quadrupolar Hamiltonian (see Appendix A, Equations (A1)–(A2b)) responsible for the NMR relaxation of the deuterium atoms of the water molecule is directed along the $\overrightarrow{OD}$ director [62]. If the water molecule reorients freely, this $\overrightarrow{OD}$ director samples uniformly all orientations with respect to the static magnetic field, canceling the average quadrupolar coupling felt by the deuterium atoms (Equations (A5)–(A6)). By contrast, confined water molecules are strongly structured with specific orientations (see Figure 2a,b) leading to a non-vanishing average of the order parameter quantifying the orientation of the $\overrightarrow{OD}$ director with respect to the clay director (see Equation (A6)). Furthermore, if the clay directors are randomly oriented into the static magnetic field, a powder spectrum should be detected [63], partially masking the doublets reported in Figure 4a. Since the order of magnitude of the quadrupolar coupling felt by the ^{2}H atom within the water molecule is known (180–200 kHz) [62], the maximum splitting measured for an orientation of the clay director parallel to the static magnetic field may be used to evaluate the degree of alignment of the confined water molecules. Finally, the variation of the ^{2}H doublet as a function of the orientation of the clay film in the static magnetic field (see Figure 4b) perfectly matches the expected relationship (Equation (A6)) with annulation of the doublet at the so-called magic-angle ($\theta^{LF} = 54.74°$).

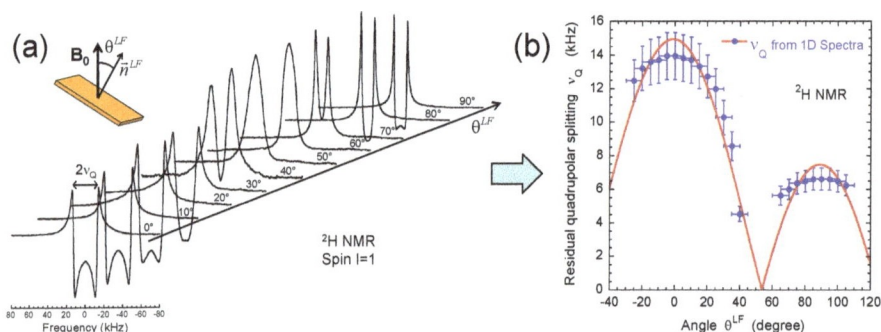

Figure 4. (a) ^{2}H NMR spectra as a function of the film orientation θ^{LF} into the static magnetic field B_0; (b) variation of the residual quadrupolar coupling ν_Q extracted from the ^{2}H NMR spectra as a function of the film orientation θ^{LF} into the static magnetic field B_0. Reprinted with permission from [53]. Copyright (2014) American Chemical Society.

^{7}Li NMR spectra (Figure 5a) recorded for lithium counterions confined within Laponite sediment [64] exhibit the same behavior except that the previous doublet is replaced by a triplet, as expected for 3/2 spin nuclei [63]. As determined by numerical simulations [65], the principal axis of the tensor describing the EFG felt by the neutralizing counterions is oriented parallel to the clay director. As a consequence, the residual quadrupolar coupling is also monitored by the film orientation into the static magnetic field.

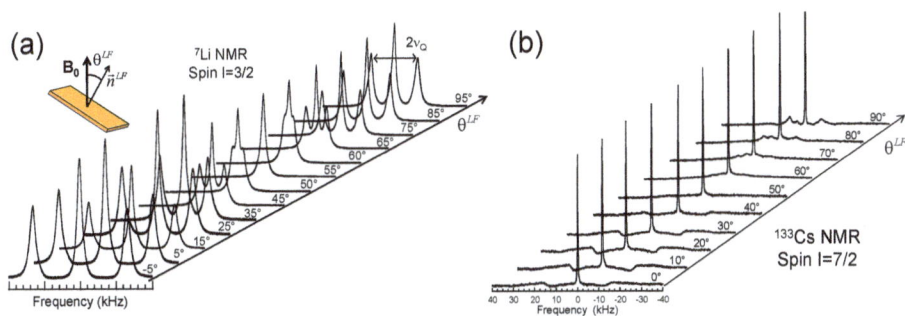

Figure 5. NMR spectra recorded as a function of the film orientation θ^{LF} of the clay sediment into the static magnetic field B_0: (**a**) ^{7}Li NMR spectra measured within laponite concentrated clay dispersions; (**b**) ^{133}Cs NMR spectra measured within hectorite clay sediment. Reprinted with permission from [55]. Copyright (2015) American Chemical Society.

In the case of ^{133}Cs, a sextuplet is expected to occur since this isotope is a 7/2 spin nucleus [63]. Unfortunately, only the first satellites [55] are detected (see Figure 5b) because the fast relaxation of this confined nuclei partially masks the theoretical quadrupolar structure.

3.2. Multi-Quanta NMR Relaxation Rates

Because of the presence of paramagnetic impurities within the clay network [18], two mechanisms are expected to monitor the NMR relaxation of confined quadrupolar probes, i.e., quadrupolar and heteronuclear dipolar couplings (see Appendix A). Theoretical details on the contributions of these two relaxation mechanisms are given in Appendix B. In that framework, a complete basis set [66–68] is required to fully understand the time evolution of the magnetization under the influence of the relaxation mechanisms, the pulse sequences and the static residual quadrupolar and heteronuclear dipolar couplings. For that purpose, we used the irreducible tensor operators [66–68] (see Appendix C) whose number increases as a function of the spin I of the nucleus. Nuclei with I = 1/2 spin are fully described by the identity (labeled T_{00}) and a row of first-order operators describing the three components of the magnetization (labeled T_{1-1}, T_{10} and T_{11} see Figure A1). The well-known Pauli matrices are another irreducible representation of these four operators, also called coherences. Quadrupolar I = 1 spin nuclei (like ^{2}H) require another set of five second-order coherences [68] (labeled T_{2-2}, T_{2-1}, T_{20}, T_{21}, T_{22}; see Figure A2) in order to describe the quadrupolar coupling. In the same manner, I = 3/2 spin nuclei (like ^{7}Li) require another set of third-order coherences [68] (labeled T_{3-3}, T_{3-2}, T_{3-1}, T_{30}, T_{31}, T_{32}, T_{33}; see Figure A3) describing the octopolar coupling. In that framework, I = 7/2 spin nuclei (like ^{133}Cs) require a basis set extending up to seventh-order coherences [67] (see Figure A4). Thanks to the completeness of these different basis sets, it becomes possible to describe the time evolution of the various coherences during each step of the pulse sequence by taking implicitly into account the influence of various relaxation mechanisms and residual static couplings (see Appendix D).

In the case of heavy water molecules confined within the clay sediments, we have measured by ^{2}H NMR the time evolution of two independent coherences [52] (namely T_{20} and $T_{22}(a,s)$) in addition to the classical longitudinal and transverse magnetizations corresponding respectively to the T_{10} and $T_{11}(a,s)$ coherences (see Figure 6a). Let us call R_{ij} the corresponding relaxation rates. The pulse sequence used to measure these different relaxation rates was detailed in previous publications [52]. A first general feature of these relaxation measurements of confined fluid is the large difference between the transverse and longitudinal relaxation rates (i.e., $R_{11} \gg R_{10}$; see Figure 6a).

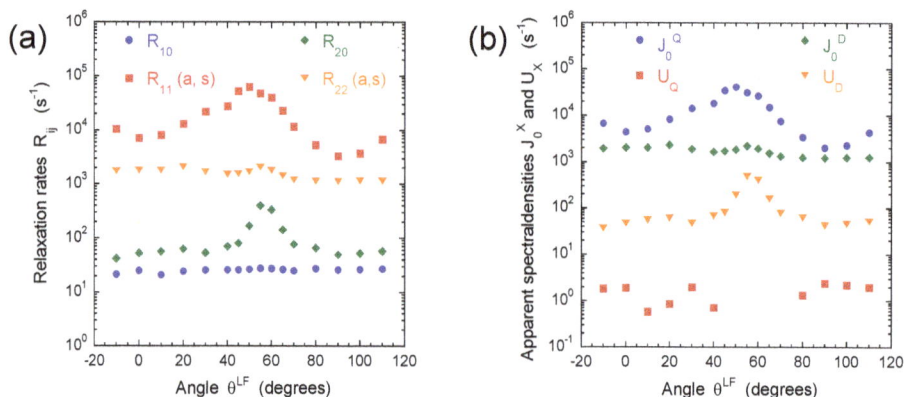

Figure 6. Variations as a function of the film orientation θ^{LF} into the static magnetic field B_0 of: (**a**) the apparent multi-quantum relaxation rates of the T_{10}, $T_{11}(a,s)$, T_{20} and $T_{22}(a,s)$ coherences, denoted R_{10}, $R_{11}(a,s)$, R_{20} and $R_{22}(a,s)$, respectively; and (**b**) the apparent spectral densities $J_0^Q(0)$, $J_0^D(0)$, U_Q and U_D extracted from these R_{ij} values (see Equation (A24)). Reprinted with permission from [52]. Copyright (2013) American Chemical Society.

As evidenced by numerical simulations [18], that difference results from the long time-scale necessary to obtain a complete decorrelation of the couplings monitoring the NMR relaxation. This phenomenon is characteristic of the slow modulation regime occurring when the corresponding time-scale becomes larger than the inverse of the resonance angular velocity (ω_0) [23,63]. Under such conditions, the integral of the decorrelation function (i.e., $J(0)$) becomes much larger than its Fourier transforms evaluated at the resonance angular velocity (i.e., $J(\omega_0)$ and $J(2\omega_0)$):

$$J(0) \gg J(\omega_0) \approx J(2\omega_0) \tag{1}$$

As detailed in Appendix D, the detected difference ($R_{11} \gg R_{10}$) becomes obvious since the longitudinal relaxation rate R_{10} is a linear combination of $J(\omega_0)$ and $J(2\omega_0)$ (see Equations (A21) and (A22e)), while $J(0)$ also contributes to the transverse relaxation rate R_{11} (see Equations (A21) and (A22b)). The contributions of the quadrupolar and heteronuclear dipolar couplings to the various relaxation rates are detailed in Appendix D. By focusing our analysis on the dominant components (see Figure 6b), it becomes possible to distinguish the relative contributions of both the quadrupolar and heteronuclear dipolar relaxation mechanisms by simply performing four independent measurements of the R_{10}, R_{11}, R_{20} and R_{22} relaxation rates (see Equations (A21)–(A24)). As displayed in Figure 6b, the quadrupolar and heteronuclear dipolar couplings contribute significantly to the relaxation of confined water molecules. Figure 6b also exhibits a significant variation of the dominant contribution to the quadrupolar relaxation mechanism (denoted $J_0^Q(0)$) as a function of the orientation of the clay lamella in the static magnetic field, with a large enhancement near the magic angle. As detailed by numerical simulations [64], this behavior results from the organization of the clay platelets within the self-supporting lamella. As explained in Appendix B, one can use the Wigner rotation matrices (Equation (A16a–c)) to extract the intrinsic contributions to both quadrupolar and heteronuclear dipolar relaxation mechanisms. These intrinsic contributions are evaluated in the frame of the clay lamella where molecular diffusion occurs. By this analysis, we can extract, for each relaxation mechanism, labeled X for $X \in \{Q, D\}$, three intrinsic components, called spectral densities and denoted $J_m^X(0)$ for $m \in \{0, 1, 2\}$, respectively. The purpose of our multi-quanta spin locking relaxation measurements is to probe the low frequency variation of this set of six independent spectral densities to obtain dynamical information on the long-time mobility of the confined NMR probes. By contrast with the ^{2}H [52] and ^{7}Li [64] relaxation measurements, the heteronuclear dipolar coupling becomes negligible for confined ^{133}Cs nuclei [55] because of the enhancement of its quadrupolar coupling (see Table A1).

3.3. Multi-Quanta Spin-Locking NMR Relaxometry

The purpose of spin-locking relaxation measurements is to extract the dispersion curve of the spectral densities $J_m^X(\omega)$ in order to quantify the time-scale describing the decorrelation of the quadrupolar and heteronuclear dipolar couplings felt by the confined diffusing probes. This study focuses on the long-time motions responsible for the complete decorrelation of these nuclear couplings. As illustrated by numerical modeling of water diffusion [40], such complete decorrelation occurs only after desorption of the confined probes in order to lose the memory of their residual coupling that is not averaged to zero by the local motions. As a consequence, the dispersion curves are expected to exhibit a transition between a plateau [40], at low angular velocities, and a continuous decrease, at high angular velocities. The inverse of that characteristic angular velocity (ω_c) is a measure of the average residence time ($\tau_c = 1/\omega_c$) of the nuclear probes confined within the interlamellar spaces of the clay sediments. Furthermore, 2D diffusion within the interlamellar space of the clay platelets [40] is the dynamical process responsible for such long-time decorrelation of the nuclear couplings felt by the confined NMR probes. As a consequence, the dispersion curve is expected to exhibit a logarithmic decrease [33,36] at angular velocities larger than ω_c.

Figure 7a–d exhibits the typical time evolution of the $T_{11}(s)$, $T_{21}(a)$, $T_{21}(s)$ and $T_{22}(a)$ coherences (see Appendix C) measured by ^{2}H NMR under spin-locking conditions for heavy water confined within beidellite clay sediment [53]. The irradiation power used for these measurements is quantified by the angular velocity ($\omega_1 = 1.12 \times 10^5$ rad/s) describing free nutation of ^{2}H nuclei under such irradiation. As illustrated by a Fourier transform of the time evolutions (Figure 7e–h), we detect three non-zero characteristic angular velocities ($\lambda_1 = 2.5 \times 10^5$ rad/s, $\lambda_2 = 1.6 \times 10^5$ rad/s and $\lambda_3 = 0.9 \times 10^5$ rad/s) for a single irradiation power, extending significantly the dynamical range probed by this quadrupolar nucleus (see Table 1). As explained in Appendix D (see Equation A20), these three characteristic angular velocities vary not only as a function of the irradiation power, but also the residual quadrupolar coupling felt by the quadrupolar probes. As a consequence, by varying the film orientation into the static magnetic field (i.e., θ^{LF}), it becomes possible to probe a large dynamical range by using a limited number of irradiation powers (see Table 1). As displayed in Figure 8a, the resulting dispersion curve covers two decades, exhibiting a clear transition at the characteristic angular velocity ($\omega_c = (6 \pm 1) \times 10^4$ rad/s), corresponding to an average residence time ($\tau_c = (17 \pm 3)$ μs) of the water molecules confined within the interlamellar space of beidellite. As illustrated in Figure 8b, this result is compatible with the average residence time obtained by simulations of Brownian dynamics exploiting the size of the clay platelets 500 ± 100 nm and the water mobility ($D = 7 \times 10^{-10}$ m^2/s) measured by QENS on equivalent samples [13,14].

Table 1. Set of characteristic angular velocities (λ_1, λ_2, λ_3) detected by multi-quanta spin-locking relaxometry for ^{2}H NMR experiments, varying the irradiation power ω_1 and the angle θ^{LF} (see the text).

θ^{LF}	0°	30°	90°	0°	30°	90°	0°	30°	90°
ω_1 (10^5 rad/s)	λ_1 (10^5 rad/s)			λ_2 (10^5 rad/s)			λ_3 (10^5 rad/s)		
1.122	2.46	2.28	2.22	1.60	1.36	1.23	0.86	0.86	0.86
0.561	1.48	1.29	1.36	1.11	0.86	0.86	0.37	0.37	0.43
0.280	1.05	0.80	0.80	0.92	0.68	0.62	0.18	0.18	0.09
0.140	0.86	0.55	0.55	0.80	0.62	0.49	0.06	0.06	0.09
0.070	0.80	0.55	0.43	0.80	0.55	0.43	0.06	0.06	0.03

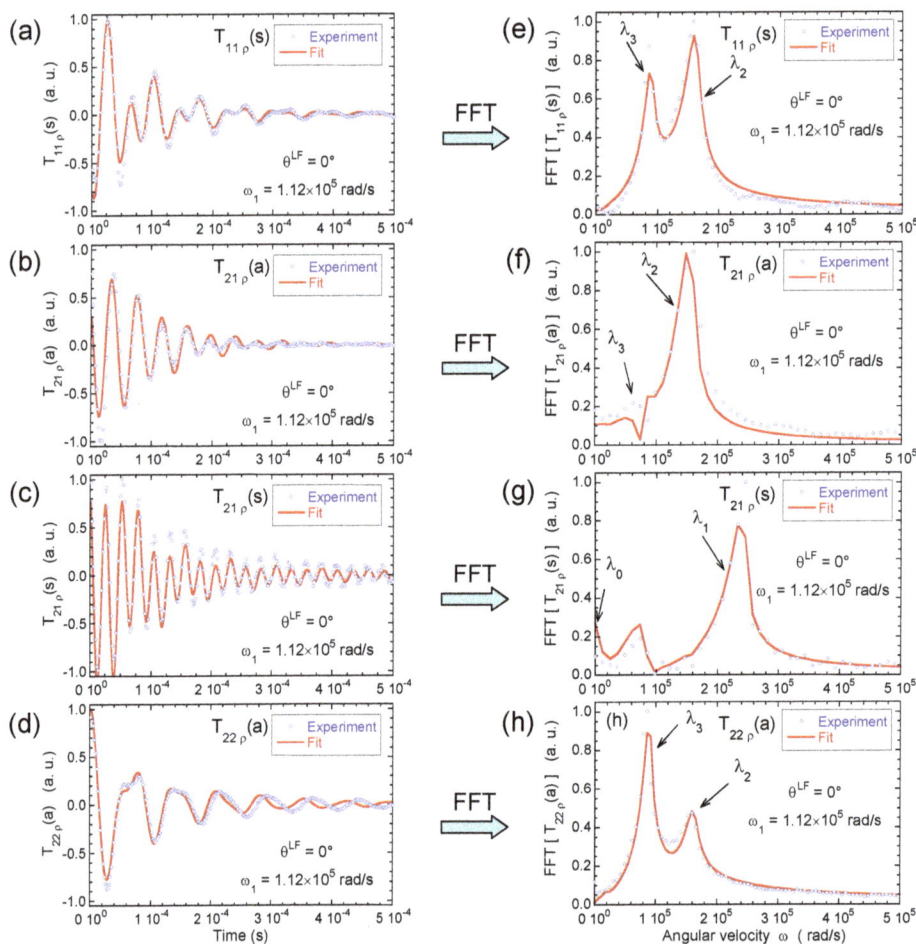

Figure 7. Time evolution of (**a**) $T_{11}(s)$; (**b**) $T_{21}(a)$; (**c**) $T_{21}(s)$ and (**d**) $T_{22}(a)$ coherences measured under spin-lock conditions, denoted $T_{11\rho}(s)$, $T_{21\rho}(a)$, $T_{21\rho}(s)$ and $T_{22\rho}(a)$, respectively, and their Fourier transforms for (**e**) $T_{11}(s)$; (**f**) $T_{21}(a)$; (**g**) $T_{21}(s)$ and (**h**) $T_{22}(a)$ coherences (^{2}H NMR). Reprinted with permission from [53]. Copyright (2014) American Chemical Society.

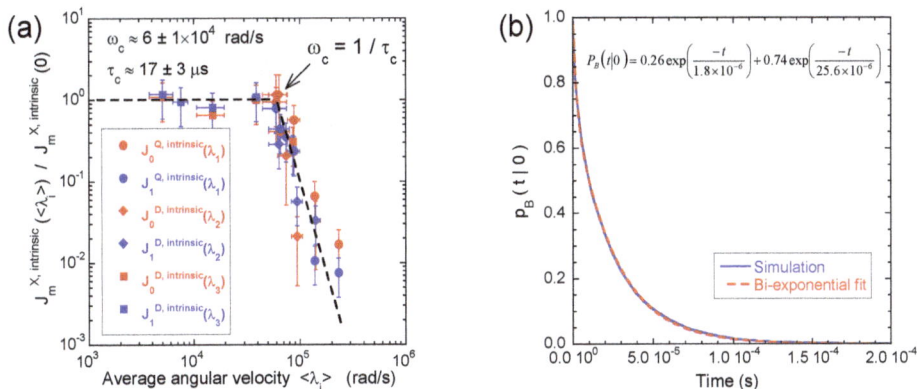

Figure 8. (a) Variation of the intrinsic spectral densities $J_m^{Q,intr}(\lambda_1)$ with $m \in \{0,1\}$, describing quadrupolar coupling, and $J_m^{D,intr}(\lambda_i)$ with $m \in \{0,1\}$ and $i \in \{2,3\}$, describing the heteronuclear dipolar coupling, as a function of the corresponding averaged angular velocities $< \lambda_i >$ (^{2}H NMR); (b) average residence time of the confined water molecules obtained by numerical simulations of Brownian dynamics. Reprinted with permission from [51,53]. Copyright (2014 and 2012) American Chemical Society.

In the case of $3/2$ spin nuclei, six non-zero angular velocities (denoted λ_i) are expected to occur in the time evolution of the various coherences under the spin-locking condition (see Equation (A27)). Spin-locking measurements of the $T_{11}(s)$, $T_{11}(a)$, $T_{33}(s)$ and $T_{33}(a)$ coherences have been performed for ^{7}Li counterions neutralizing laponite synthetic clay [54]. Figure 9a–d exhibits complex time evolutions of the various coherences under spin-locking because of the multiplicity of contributing modes. By contrast, their Fourier transform (see Figure 9e–h) clearly identifies the six expected modes, with a perfect matching between the experimental data and the theoretical analysis [54]. As displayed in Figure 10, a broad range of angular velocities ω is then probed by using only four irradiation powers sampling simply one decade.

In the case of ^{133}Cs neutralizing a synthetic fluoro-hectorite [55], a complete numerical treatment of the time evolution of the coherences is required due to the large size of the basis-set required to describe all the quantum states of this $7/2$ spin nucleus. Our analysis leads to a good agreement between the experimental and calculated data (Figure 11) by setting all sampled spectral densities equal to their high frequency [55] value by assuming:

$$J_0^Q(\lambda_p \neq 0) = J_0^Q(\omega_0) \tag{2}$$

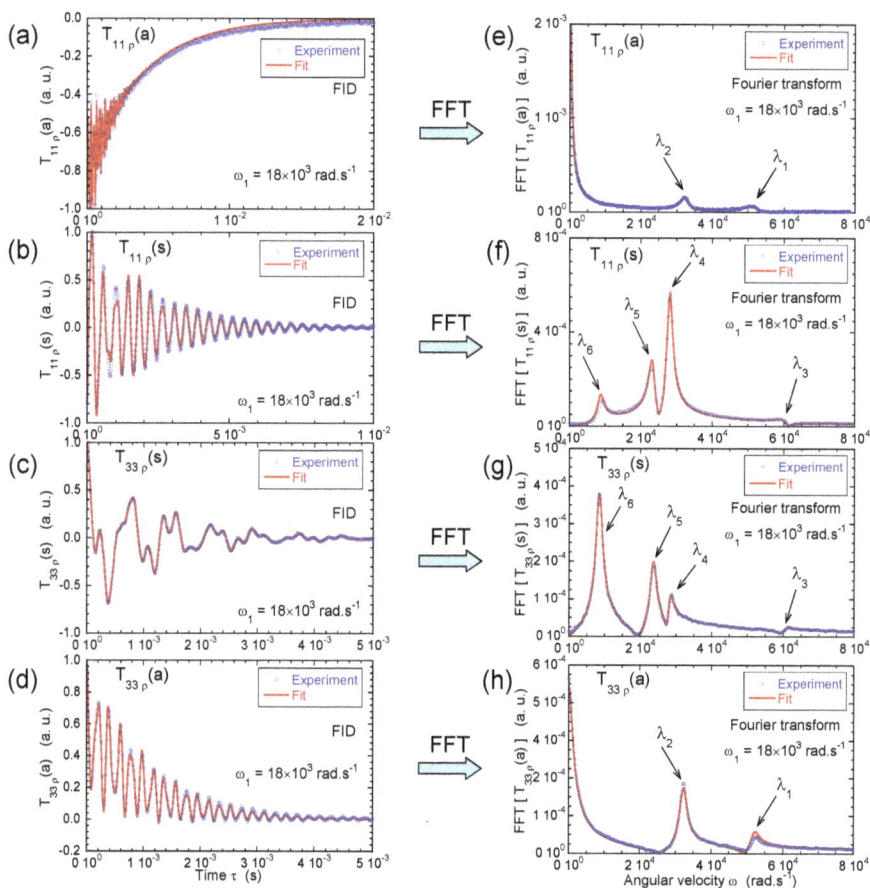

Figure 9. For Li$^+$ ions in concentrated laponite dispersion, the comparison between the experimental and fitted time evolution of the coherences: (**a**) $T_{11}(a)$; (**b**) $T_{11}(s)$; (**c**) $T_{33}(s)$ and (**d**) $T_{33}(a)$ coherences measured under spin-lock conditions, denoted $T_{11\rho}(a)$, $T_{11\rho}(s)$, $T_{33\rho}(s)$ and $T_{33\rho}(a)$, respectively, and their Fourier transforms for (**e**) $T_{11}(a)$; (**f**) $T_{11}(s)$; (**g**) $T_{33}(s)$ and (**h**) $T_{33}(a)$ coherences (^{7}Li NMR). Reprinted with permission from [54]. Copyright (2009) American Chemical Society.

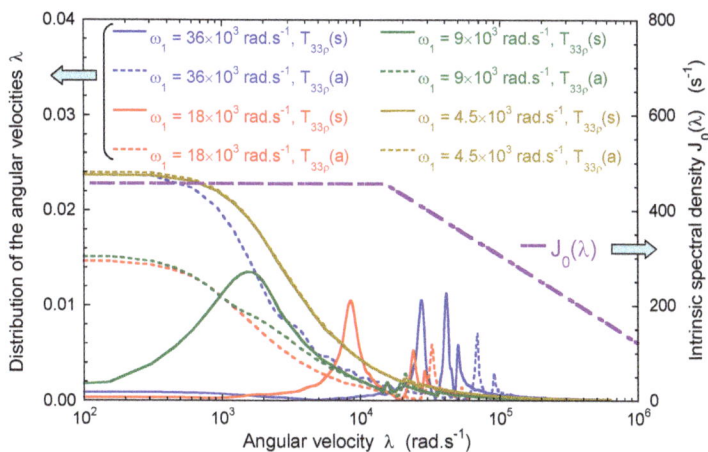

Figure 10. Distribution of the complete set of resonance angular velocities probed by the triple-quantum filtered relaxation measurements under the spin-locking condition performed at four irradiating fields ω_1, evaluated from the intrinsic spectral density $J_0(\lambda)$ (^{7}Li NMR). Reprinted with permission from [54]. Copyright (2009) American Chemical Society.

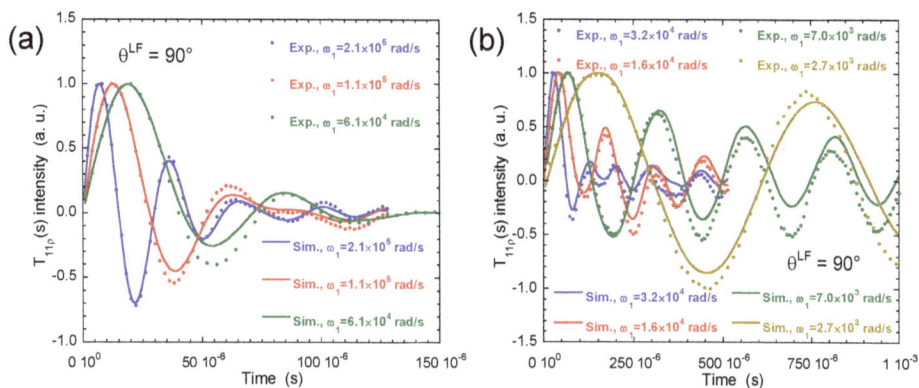

Figure 11. Time evolutions of the spin-locking relaxation measurements (^{133}Cs NMR) for seven irradiation powers corresponding to various angular velocities ω_1: (**a**) decreasing from 2.1×10^5 to 6.1×10^4 rad·s^{-1}; and (**b**) decreasing from 3.2×10^4 to 2.7×10^3 rad·s^{-1} (see the text). Reprinted with permission from [55]. Copyright (2015) American Chemical Society.

Figure 12 illustrates the range of angular velocities that can be sampled by ^{133}Cs spin-locking relaxation measurements induced by the quadrupolar relaxation mechanism that was shown to monitor the relaxation of ^{133}Cs.

As a consequence, the transition between the low frequency plateau and the continuous decrease of the spectral density must occur at angular velocities (ω_c) much smaller than the lowest eigenvalue λ_p^Q probed by these spin-locking measurements, i.e., 10^3 rad/s (see Figure 12). The corresponding average residence time of the confined cesium counterions must be larger than 1 ms [55]. By taking into account the average size of the hectorite platelets determined by TEM ($L \sim 0.4$ μm), we obtain a self-diffusion coefficient ($D \approx L^2/2\tau_c$) smaller than 8×10^{-11} m^2/s. That upper limit is fully compatible with the cesium mobility within clay sediments obtained by numerical simulations of molecular dynamics [55,69–74].

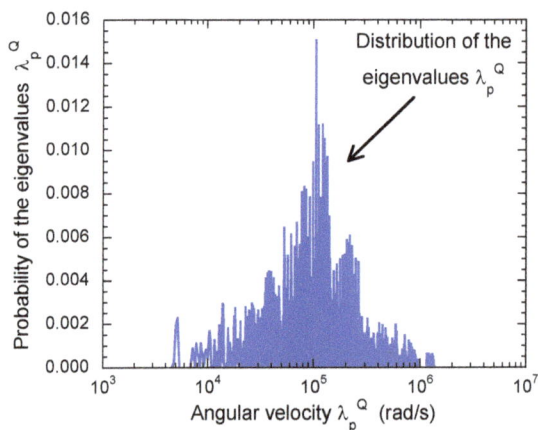

Figure 12. Histograms of the distribution of the angular velocities λ_p^Q corresponding to the quadrupolar relaxation mechanism that may be probed by the spin-locking experiments of $I = 7/2$ spin for irradiation powers ω_1 varying between 2.7×10^3 and 2.1×10^5 rad·s^{-1} (^{133}Cs NMR). Reprinted with permission from [55]. Copyright (2015) American Chemical Society.

3.4. Two-Time Stimulated Echo Attenuation

Two-time stimulated echo NMR spectroscopy [59,60] exploits the heterogeneity of the micro-domain orientations within the clay sediment (see Figure 1). For ^{2}H nuclei, these heterogeneities of the clay platelets' orientation induce heterogeneities of the residual quadrupolar coupling felt by the confined water molecules (Equation (A6)). The pulse sequence displayed in Figure 13 illustrates the experimental procedure [60]: During the first evolution procedure, the transverse magnetization (corresponding to the $T_{1\pm1}$ coherence) of all water molecules pertaining to the micro-domain labeled i oscillates at a specific angular velocity (denoted ω_{Qi}) corresponding to the orientation of their micro-domain. The total transverse magnetization is next transferred into the T_{20} coherence and freely evolves during the mixing time τ_M. The duration of the fourth pulse (ψ) is selected to optimize the double-quanta filtering, by optimizing the transfer of the T_{20} coherence into the T_{22} coherence and minimizing the transfer from the other zero-order coherence, i.e., the T_{10} coherence, into the same T_{22} coherence. During the second evolution period, the transverse magnetization of the confined water molecules again oscillates at the angular velocity (denoted now ω_{Qj}) corresponding to the specific orientation of their actual micro-domain, labeled j. As a consequence, the net magnetization satisfies [59,60]:

$$I(t_e, \tau_M) = \left\langle \cos\left(\omega_{Qi}(0)t_e\right) \times \cos\left(\omega_{Qj}(\tau_M)t_e\right) \right\rangle \times e^{-(R_{20}\tau_M + 2R_{11}t_e)} \tag{3}$$

If the mixing time (τ_M) is smaller than the time (denoted τ_{exch}) required by the confined water molecules to exchange between two micro-domains with different orientation, the statistical average of the product of the cosinus functions within the bracket in Equation (3) reaches its maximum value. By contrast, for mixing times larger than the same exchange time, the water molecules will now probe two micro-domains with different orientations, thus reducing the previous statistical average. Finally, the exponential law of Equation (3) describes the intrinsic attenuation of the magnetization during the evolution time t_e and mixing time τ_M.

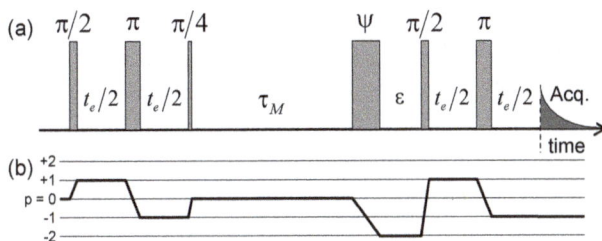

Figure 13. (a) Pulse sequence and (b) coherence pathway used to measure the attenuation of the two-time ^{2}H NMR stimulated echo $I(t_e, \tau_M)$ as a function of the evolution period t_e and the mixing time τ_M. Reprinted with permission from [53]. Copyright (2014) American Chemical Society.

Figure 14a illustrates the resulting attenuation of the two-time stimulated echo as a function of the mixing time [60]. A better illustration is given by simply noting the relative intensity of the first maximum (see Figure 14b), leading to an exchange time of 33 ms, i.e., three orders of magnitude larger than the water residence time is the interlamellar space. That interpretation is fully validated by a simple numerical model [60] describing the exchange of water molecules between neighboring cubic boxes labeled by a set of three indices (i, j, k):

$$\frac{d\sigma_{i,j,k}}{dt} = \left(R_{i,j,k} - 6k_{exch}t \right) \sigma_{i,j,k} + k_{exch}t \left[\sigma_{i+1,j,k} + \sigma_{i-1,j,k} + \sigma_{i,j+1,k} + \sigma_{i,j-1,k} + \sigma_{i,j,k+1} + \sigma_{i,j,k-1} \right] \quad (4)$$

where $R_{i,j,k}$ contains the contributions from the pulses, the local residual quadrupolar couplings and the relaxation mechanisms (see Appendixes A and B), leading to a set of generalized Bloch equations [22,75].

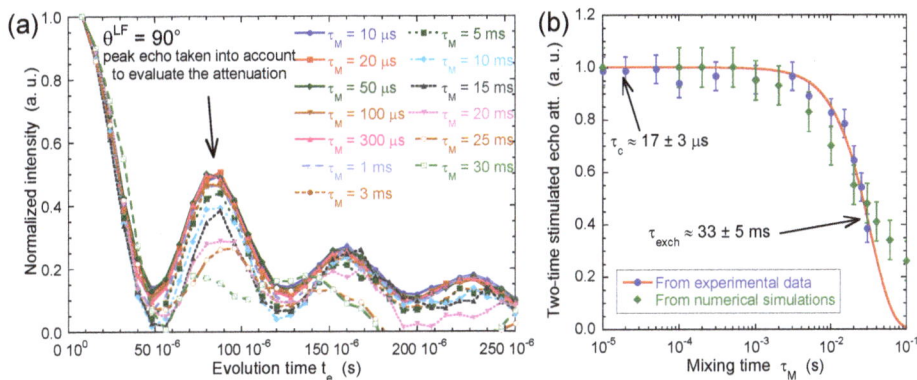

Figure 14. (a) Variation of the two-time stimulated echo attenuation $I(t_e, \tau_M)$ as a function of mixing time τ_M (^{2}H NMR). The data are normalized to take into account the relaxation of the T_{20} coherence during the mixing time τ_M (see Equation (3)). (b) Two-time correlation function extracted from the normalized stimulated echo attenuation as a function of the mixing time τ_M. The red line corresponds to the best fit of a stretched exponential function, $f(t) = A \exp(-t/\tau_{exch})^\alpha$, to determine the exchange time τ_{exch} ($\tau_{exch} = 33 \pm 5$ ms with an exponent α set equal to 1.5), and the green line dots are obtained by numerical modeling (see the text). Reprinted with permission from [53]. Copyright (2014) American Chemical Society.

4. Conclusions

Multi-quanta spin-locking NMR relaxometry of quadrupolar nuclei was shown to be a powerful tool to quantify the average residence time of molecular (D_2O) and ionic (^7Li, ^{133}Cs) probes confined within the interlamellar space of clay lamellae inside dense sediments. Furthermore, two-time stimulated echo NMR attenuation leads to dynamical information on the long-time mobility of the water molecules exchanging between differently-oriented micro-domains constituting dense clay sediments. Multi-scale numerical simulations were performed to better understand the structural and dynamical properties of confined ions and water molecules, improving our analysis of the NMR experiments. These multi-quanta spin-locking NMR relaxometry measurements are expected to be easily extended to study other interfacial systems, including porous silicate, zeolites, cements, etc. A large number of diffusing probes may be used for such investigations since quadrupolar isotopes pertain to a large fraction of detectable NMR isotopes within the periodic table.

Acknowledgments: These works were partially supported by the CNRS Interdisciplinary Project "Nucléaire, Energie, Environnement, Déchets, Société (NEEDS)" through its "Milieux Poreux (MIPOR)" Program, which provided financial support (Projects MULTIDYN and TRANSREAC).

Conflicts of Interest: The authors declare no conflict of interest.

Abbreviations

The following abbreviations are used in this manuscript:

EFG   Electric Field Gradient
GCMC   Grand Canonical Monte Carlo
INS   Inelastic Neutron Scattering
NMR   Nuclear Magnetic Resonance
QENS   Quasi-Elastic Neutron Scattering
TEM   Transmission Electron Microscopy

Appendix A. Quadrupolar and Heteronuclear Dipolar Hamiltonian

The quadrupolar Hamiltonian [23,24,63] is defined by:

$$H_Q = C_Q \sum_{p=-2}^{2} (-1)^p F_{2,-p}^{Q,L} T_{2,p}^{IR} \quad \text{with} \quad C_Q = \sqrt{\frac{3}{2}} \frac{e\,Q\,(1+\gamma_\infty)}{I(2I-1)\,\hbar} \tag{A1}$$

where e is the electron charge, Q is the quadrupolar moment of the nuclei [76] and $(1+\gamma_\infty)$ is the Steinhermer antishielding factor [76]. These three last parameters are detailed in Table A1 for different quadrupolar nuclei.

In the above equation,

$$F_{2,0}^{Q,L} = \frac{1}{2} V_{zz}^L, \quad F_{2,\pm1}^{Q,L} = \mp \frac{1}{\sqrt{6}} \left(V_{xz}^L \pm i V_{yz}^L \right), \quad F_{2,\pm2}^{Q,L} = \frac{1}{2\sqrt{6}} \left(V_{xx}^L - V_{yy}^L \pm 2i V_{xy}^L \right) \tag{A2a}$$

and

$$T_{2,0}^{Q,IR} = \frac{1}{\sqrt{6}} \left(3I_z^2 - I(I+1) \right), \quad T_{2,\pm1}^{Q,IR} = \mp \frac{1}{2} \left(I_z I_\pm + I_\pm I_z \right), \quad T_{2,\pm2}^{Q,IR} = \frac{1}{2} I_\pm^2 \tag{A2b}$$

where $V_{\alpha\beta}^L$ are the components of the EFG evaluated in the laboratory frame (denoted L); $T_{2,\pm p}^{Q,IR}$ (for $p = -2$ to 2) are the second-order irreducible tensor operators; I_x, I_y and I_z are the spin operators and $I_\pm = I_x \pm iI_y$.

Table A1. Parameters monitoring the order of magnitude of the quadrupolar Hamiltonian of some alkali cations (see [76]).

Isotope	Spin I	Q (10^{-24} cm^2)	$1 + \gamma_\infty$	R_{10} in Water (s^{-1})
^{7}Li	3/2	0.042	0.74	0.03
^{23}Na	3/2	0.11	5.1	16.2
^{39}K	3/2	0.09	18.3	24
^{85}Rb	5/2	0.31	48.2	420
^{133}Cs	7/2	0.004	111	0.08

In the presence of a static quadrupolar coupling, the equidistant Zeeman energy levels are modified by the residual quadrupolar coupling, leading to a quadrupolar splitting of the resonance lines according to:

$$\omega_{m-1,m} = \sqrt{\frac{3}{8}} \, C_Q \left\langle V_{zz}^L \right\rangle (1 - 2m) \tag{A3}$$

for m varying between I and $-I + 1$.

During a change of frame, the components of the EFG transform like the second-order spherical harmonics [24]:

$$F_{2,q}^{Q,L} = \sum_{p=-2}^{2} F_{2,p}^{Q,P} \, D_{p,q}^{LP}(\theta, \varphi, \psi) \tag{A4}$$

with $D_{p,q}^{LP}(\theta, \varphi, \psi)$, the components of the Wigner rotation matrices [24] where the set of (θ, φ, ψ) Euler angles defines the orientation, into the static magnetic field, of the principal axis of the tensor describing the EFG felt by the quadrupolar nucleus. Three sets of frames are useful to describe the orientation of the principal component of the EFG: the laboratory frame (denoted L), a frame attached to the dense clay sediment (denoted F) and a frame attached to the individual quadrupolar nucleus (denoted P). The $\mathbf{e}_z$ directors of these different frames are respectively the direction of the static magnetic field B_0 (laboratory frame L), the normal to the clay sediment $\mathbf{n}$ (sediment frame F) and the director of the principal component of the EFG, denoted V_{zz}^P (particle frame P).

The measured quadrupolar splitting is derived from Equations (A3) and (A4):

$$\omega_{m-1,m}^{app} = A_m \, V_{zz}^P \sum_{p=-2}^{2} D_{p,0}^{LF}(\theta^{LF}, \varphi^{LF}, \psi^{LF}) \left\langle D_{0,p}^{FP}(\theta^{FP}, \varphi^{FP}, \psi^{FP}) \right\rangle \tag{A5}$$

$$\text{with} \quad A_m = \frac{3eQ(1 + \gamma_\infty)(1 - 2m)}{4I(2I - 1)\hbar}$$

The angular average is evaluated in Equation (A5) over all the orientations of EFG principal component within the sediment. The first set of the Wigner rotation matrix describes the orientation of the macroscopic clay sample with respect to the magnetic field, and the Wigner rotation matrix in the bracket characterizes the average orientation of EFG principal component within the sediment. For clay sediments with cylindrical symmetry, only the component $p = 0$ contributes to Equation (A5), which reduces to:

$$\omega_{m-1,m}^{app} = A_m \, V_{zz}^P \frac{3\cos^2\theta^{LF} - 1}{2} \left\langle \frac{3\cos^2\theta^{FP} - 1}{2} \right\rangle \tag{A6}$$

In addition to the quadrupolar coupling, the heteronuclear dipolar coupling may also be responsible for the NMR relaxation of the confined probes because of the presence of paramagnetic impurities. The corresponding heteronuclear dipolar Hamiltonian [23,24,63] becomes:

$$H_D(t) = C_D \sum_{m=-2}^{2} (-1)^m \, T_{2,m}^{D,IR} \, F_{2,-m}^{D,L}(t) \quad \text{with} \quad C_D = -\frac{\mu_0}{4\pi} \gamma_I \, \gamma_S \, \hbar \tag{A7}$$

where the C_D is the dipolar coupling constant, and the spin operators become [23,24,63]:

$$T_{2,0}^{D,IR} = \frac{1}{\sqrt{6}} \left(2I_zS_z - \frac{1}{2}(I_+S_- + I_-S_+) \right), \ T_{2,\pm 1}^{D,IR} = \mp \frac{1}{2}(I_zS_\pm + I_\pm S_z) \text{ and } T_{2,\pm 2}^{D,IR} = \frac{1}{2} I_\pm S_\mp \quad \text{(A8)}$$

The functions $F_{2,m}^{D,L}(t)$ in Equation (A7) are related to the second-order spherical harmonics describing the reorientation of the vector joining the two coupled spin (denoted $\vec{r}_{IS}(t)$) by reference to the static magnetic field [23,24,63]:

$$F_{2,-m}^{D,L}(t) = \sqrt{\frac{24\pi}{5}} \frac{Y_{2,-m}(\theta, \varphi)}{r_{IS}^3} \quad \text{(A9)}$$

Appendix B. NMR Relaxation Theory

In the framework of the Redfield theory [77], the time evolution of the spin quantum states, also called coherences, is described by the master equation [23,24,63]:

$$\frac{d\sigma^*}{dt} = -i \ [H_S^*, \sigma^*] + f(\sigma^*) \quad \text{(A10)}$$

As denoted by the asterisk (*), all terms are evaluated in the Larmor frequency rotating frame. The commutator describes the contribution from the static Hamiltonians H_S^*, including the excitation pulses and the residual quadrupolar Hamiltonian. The second term describes the contribution from the fluctuating parts of the quadrupolar and dipolar Hamiltonians:

$$H_{XF}^*(t) = C_X \sum_{m=-2}^{2} (-1)^m \ T_{2,m}^{X,IR} \ e^{im\omega_0 t} \left(F_{2,-m}^{X,L}(t) - \left\langle F_{2,-m}^{X,L}(t) \right\rangle \right) \quad \text{(A11)}$$

where the index X stands for the various relaxation mechanisms (i.e., Q or D). This last contribution to the master equation is given by [23,24,63,77]:

$$f(\sigma^*) = \int_0^{t_{sup}} \left\langle \left[H_{XF}^*(t), \left[e^{-iH_S^*\tau} \ H_{XF}^{*+}(t-\tau) \ e^{iH_S^*\tau}, \sigma^*(t) \right] \right] \right\rangle d\tau \quad \text{(A12)}$$

If the time-scales characterizing the decorrelation of the various Hamiltonians are much smaller than the time-scale sampled by the evolution of the coherences, the upper limit of the integral t_{sup} in Equation (A12) may be set equal to infinity. This hypothesis restricts the validity of the Redfield theory applied to NMR relaxation [23,24,63].

Let us introduce the autocorrelation functions of the fluctuating components of the Hamiltonian:

$$G_m^{X,L}(\tau) = \left\langle \left(F_{2,m}^{X,L}(0) - < F_{2,m}^{X,L} > \right) \times \left(F_{2,m}^{X,L}(\tau) - < F_{2,m}^{X,L} > \right) \right\rangle$$
$$+ \left\langle \left(F_{2,-m}^{X,L}(0) - < F_{2,-m}^{X,L} > \right) \times \left(F_{2,-m}^{X,L}(\tau) - < F_{2,-m}^{X,L} > \right) \right\rangle \text{ with } m \in \{0,1,2\} \quad \text{(A13)}$$

By neglecting the time evolution of the coherences during the irradiation pulses, Equation (A12) becomes then [23,24,63]:

$$f(\sigma^*) = - \sum_{m=0}^{2} \left[T_{2,m}^{X,IR}, \left[T_{2,-m}^{X,IR}, \sigma^* \right] \right] J_m^{X,L}(m\omega_0) \quad \text{(A14)}$$

where the so-called spectral densities $J_m^{X,L}(m\omega_0)$ satisfy the relationship [23,24,63]:

$$J_m^{X,L}(m\omega_0) = -C_X^2 \int_0^{\infty} G_m^{X,L}(t) \ e^{-im\omega_0 t} \ dt \quad \text{(A15)}$$

The above-mentioned approximation is generally valid for classical relaxation measurements because the duration of the detection pulses (typically a few µs) is much shorter than the time evolution of the coherences. Finally, a complete basis set of coherences is required to translate Equation (A14) into a matrix form [18,64,78–80].

In the case of spin-locking relaxation measurements, one cannot neglect the time evolution of the coherences during the irradiation power since it is applied during the entire evolution period of the coherences. As a consequence, the time evolution of the dipolar and quadrupolar couplings under the influence of the static Hamiltonian H_S^* must be taken into account in Equation (A12) as described implicitly by the term $e^{-iH_S^*\tau}\,H_{XF}^{*+}(t-\tau)\,e^{iH_S^*\tau}$ in the double commutator (see Equation (A12)). For that purpose, the static Hamiltonian is also formulated in a matrix form by using the complete basis set of coherences [40,54,55]. After evaluating its eigenvalues (denoted $\pm i\lambda_p$) and corresponding eigenvectors (denoted $\vec{v}_p$), one obtains a new complete basis set. The problem is then easily solved by projecting, into this eigenvectors basis set, the initial basis set of the coherences used to describe the $T_{2,m}^{X,IR}$ spin operators.

By using the Wigner rotation matrices [81] (cf. Equation (A4)), it is possible to relate the derivation of the apparent correlation functions $G_m^{X,L}(\tau)$, evaluated in the laboratory frame (denoted L), with their intrinsic value evaluated in the frame attached to the clay sediment (denoted F) [82]:

$$G_0^{X,L}(\tau) = \frac{(1-3\cos^2\theta^{LF})^2}{4}\,G_0^{X,F}(\tau) + 3\cos^2\theta^{LF}\sin^2\theta^{LF}\,G_1^{X,F}(\tau) + \frac{3(1-3\cos^2\theta^{LF})^2}{4}\,G_2^{X,F}(\tau) \tag{A16a}$$

$$G_1^{X,L}(\tau) = \frac{3\cos^2\theta^{LF}\sin^2\theta^{LF}}{2}\,G_0^{X,F}(\tau) + \frac{1-3\cos^2\theta^{LF}+4\cos^4\theta^{LF}}{2}\,G_1^{X,F}(\tau) + \frac{1-\cos^4\theta^{LF}}{2}\,G_2^{X,F}(\tau) \tag{A16b}$$

$$G_2^{X,L}(\tau) = \frac{3(1-\cos^2\theta^{LF})^2}{8}\,G_0^{X,F}(\tau) + \frac{1-\cos^4\theta^{LF}}{2}\,G_1^{X,F}(\tau) + \frac{1+6\cos^2\theta^{LF}+\cos^4\theta^{LF}}{8}\,G_2^{X,F}(\tau) \tag{A16c}$$

Obviously, the same relationship may be deduced for the corresponding spectral densities thanks to the linearity of the Fourier transform (see Equation (A15)).

Appendix C. Matrix Representation of the Irreducible Tensor Operators

Depending on the spin state, complete orthonormal basis sets may be constructed by using the irreducible tensor operators, also called coherences [66–68]. Symmetric and antisymmetric combinations of the coherences [78–80] are also introduced:

$$T_{lp}(s) = \frac{1}{\sqrt{2}}(T_{l-p} + T_{lp}) \quad\text{and}\quad T_{lp}(a) = \frac{1}{\sqrt{2}}(T_{l-p} - T_{lp}) \tag{A17}$$

By using these new coherences, the three spin operators, I_x, I_y and I_z, become proportional to $T_{11}(a)$, $T_{11}(s)$ and T_{10}, respectively, simplifying the formulation of the Hamiltonians describing the irradiation pulse and the heteronuclear dipolar coupling (see Equation (A8)).

As displayed in Figures A1–A4, the size of the basis set increases significantly as a function of the spin of the nucleus.

(a)

$$T_{00}$$

$$T_{1-1} \quad T_{10} \quad T_{11}$$

(b)

$$\begin{pmatrix} \frac{1}{\sqrt{2}} & 0 \\ 0 & \frac{1}{\sqrt{2}} \end{pmatrix}$$

$$\begin{pmatrix} 0 & 0 \\ 1 & 0 \end{pmatrix} \quad \begin{pmatrix} \frac{1}{\sqrt{2}} & 0 \\ 0 & -\frac{1}{\sqrt{2}} \end{pmatrix} \quad \begin{pmatrix} 0 & -1 \\ 0 & 0 \end{pmatrix}$$

Figure A1. Complete orthogonal basis set (four elements) describing the evolution of the spin $I = 1/2$ (Pauli matrices): (a) symbolic representation T_{ij}; (b) explicit matrix representation of irreducible tensor operators.

(a)

T_{00}

T_{1-1} T_{10} T_{11}

T_{2-2} T_{2-1} T_{20} T_{21} T_{22}

(b)

$$\begin{pmatrix} \frac{1}{\sqrt{3}} & 0 & 0 \\ 0 & \frac{1}{\sqrt{3}} & 0 \\ 0 & 0 & \frac{1}{\sqrt{3}} \end{pmatrix}$$

$$\begin{pmatrix} 0 & 0 & 0 \\ \frac{1}{\sqrt{2}} & 0 & 0 \\ 0 & \frac{1}{\sqrt{2}} & 0 \end{pmatrix} \quad \begin{pmatrix} \frac{1}{\sqrt{2}} & 0 & 0 \\ 0 & 0 & 0 \\ 0 & 0 & \frac{-1}{\sqrt{2}} \end{pmatrix} \quad \begin{pmatrix} 0 & \frac{-1}{\sqrt{2}} & 0 \\ 0 & 0 & \frac{-1}{\sqrt{2}} \\ 0 & 0 & 0 \end{pmatrix}$$

$$\begin{pmatrix} 0 & 0 & 0 \\ 0 & 0 & 0 \\ 1 & 0 & 0 \end{pmatrix} \quad \begin{pmatrix} 0 & 0 & 0 \\ \frac{1}{\sqrt{2}} & 0 & 0 \\ 0 & \frac{-1}{\sqrt{2}} & 0 \end{pmatrix} \quad \begin{pmatrix} \frac{1}{\sqrt{6}} & 0 & 0 \\ 0 & \frac{-2}{\sqrt{6}} & 0 \\ 0 & 0 & \frac{1}{\sqrt{6}} \end{pmatrix} \quad \begin{pmatrix} 0 & \frac{-1}{\sqrt{2}} & 0 \\ 0 & 0 & \frac{1}{\sqrt{2}} \\ 0 & 0 & 0 \end{pmatrix} \quad \begin{pmatrix} 0 & 0 & 1 \\ 0 & 0 & 0 \\ 0 & 0 & 0 \end{pmatrix}$$

Figure A2. Complete orthogonal basis set (nine elements) describing the evolution of the spin I = 1: (a) symbolic representation T_{ij}; (b) explicit matrix representation of irreducible tensor operators.

(a)

T_{00}

T_{1-1} T_{10} T_{11}

T_{2-2} T_{2-1} T_{20} T_{21} T_{22}

T_{3-3} T_{3-2} T_{3-1} T_{30} T_{31} T_{32} T_{33}

(b)

T_{10} $T_{11}(s)$

$T_{21}(a)$ $T_{22}(a)$

T_{30} $T_{31}(s)$ $T_{32}(s)$ $T_{33}(s)$

(c)

$T_{11}(a)$

T_{20} $T_{21}(s)$ $T_{22}(s)$

$T_{31}(a)$ $T_{32}(a)$ $T_{33}(a)$

Figure A3. Complete orthogonal basis set (16 elements) describing the evolution of the spin I = 3/2: (a) symbolic representation T_{ij}; (b) first subset of eight independent coherences $T_{ij}(a,s)$ including the T_{10} coherence; and (c) second subset of seven independent coherences $T_{ij}(a,s)$ including the T_{20} coherence.

(a)

T_{10} $T_{11}(s)$
$T_{21}(a)$ $T_{22}(a)$
T_{30} $T_{31}(s)$ $T_{32}(s)$ $T_{33}(s)$
$T_{41}(a)$ $T_{42}(a)$ $T_{43}(a)$ $T_{44}(a)$
T_{50} $T_{51}(s)$ $T_{52}(s)$ $T_{53}(s)$ $T_{54}(s)$ $T_{55}(s)$
$T_{61}(a)$ $T_{62}(a)$ $T_{63}(a)$ $T_{64}(a)$ $T_{65}(a)$ $T_{66}(a)$
T_{70} $T_{71}(s)$ $T_{72}(s)$ $T_{73}(s)$ $T_{74}(s)$ $T_{75}(s)$ $T_{76}(s)$ $T_{77}(s)$

(b)

$T_{11}(a)$
T_{20} $T_{21}(s)$ $T_{22}(s)$
$T_{31}(a)$ $T_{32}(a)$ $T_{33}(a)$
T_{40} $T_{41}(s)$ $T_{42}(s)$ $T_{43}(s)$ $T_{44}(s)$
$T_{51}(a)$ $T_{52}(a)$ $T_{53}(a)$ $T_{54}(a)$ $T_{55}(a)$
T_{60} $T_{61}(s)$ $T_{62}(s)$ $T_{63}(s)$ $T_{64}(s)$ $T_{65}(s)$ $T_{66}(s)$
$T_{71}(a)$ $T_{72}(a)$ $T_{73}(a)$ $T_{74}(a)$ $T_{75}(a)$ $T_{76}(a)$ $T_{77}(a)$

Figure A4. Orthogonal basis set (63 elements) describing the evolution of the spin I = 7/2: (a) first subset of 32 independent coherences $T_{ij}(a,s)$ including the T_{10} coherence; and (b) second subset of 31 independent coherences $T_{ij}(a,s)$ including the T_{20} coherence.

Appendix D. Application to the Relaxation of Quadrupolar Nuclei

The differential equation describing the time evolution of the coherences (Equations (A10)–(A15)) may be written in a matrix form. To simplify the derivations of these matrices, we selected symmetric and antisymmetric combinations of the coherences [78–80]; see Equation (A17). By using these coherences, the three spin operators, I_x, I_y and I_z, become proportional to $T_{11}(a)$, $T_{11}(s)$ and T_{10}, respectively, simplifying the formulation of the Hamiltonians describing the irradiation pulse and the heteronuclear dipolar coupling (see Equations (A7)–(A9)).

For spin $I = 1$ nuclei, the time evolution of the coherences under the influence of the static Hamiltonian (denoted H_S^* in Equation (A10)) including the residual quadrupolar coupling (ω_Q) and the irradiation pulse (ω_1) becomes [18,78]:

$$
\frac{d}{dt}
\begin{pmatrix}
T_{20} \\
T_{11}(a) \\
T_{21}(s) \\
T_{22}(s) \\
T_{10} \\
T_{11}(s) \\
T_{21}(a) \\
T_{22}(a)
\end{pmatrix}
= i
\begin{pmatrix}
0 & 0 & -\sqrt{3}\,\omega_1 & 0 & 0 & 0 & 0 & 0 \\
0 & 0 & \omega_Q & 0 & 0 & 0 & 0 & 0 \\
-\sqrt{3}\,\omega_1 & \omega_Q & 0 & -\omega_1 & 0 & 0 & 0 & 0 \\
0 & 0 & -\omega_1 & 0 & 0 & 0 & 0 & 0 \\
0 & 0 & 0 & 0 & 0 & -\omega_1 & 0 & 0 \\
0 & 0 & 0 & 0 & -\omega_1 & 0 & \omega_Q & 0 \\
0 & 0 & 0 & 0 & 0 & \omega_Q & 0 & -\omega_1 \\
0 & 0 & 0 & 0 & 0 & 0 & -\omega_1 & 0
\end{pmatrix}
\times
\begin{pmatrix}
T_{20} \\
T_{11}(a) \\
T_{21}(s) \\
T_{22}(s) \\
T_{10} \\
T_{11}(s) \\
T_{21}(a) \\
T_{22}(a)
\end{pmatrix}
\tag{A18}
$$

leading to two independent sub-sets of coherences. Analytical derivation of the eigenvalues $\pm i\lambda_p$ with $p \in \{0, \cdots, 3\}$ and their corresponding eigenvectors $\vec{v}_p$ are used to derive the general solutions of Equation (A18) [40,78]. Among others, two coherences are of practical interest for the interpretation of spin-locking relaxation measurements:

$$
\begin{aligned}
e^{iH_S^*\tau} T_{20} e^{-iH_S^*\tau} = {}& \sqrt{3}\,\omega_1\omega_Q \frac{1 - \cos(\lambda_1\tau)}{\lambda_1^2} T_{11}(a) + \frac{\omega_Q^2 + \omega_1^2(1 - 3\cos(\lambda_1\tau))}{\lambda_1^2} T_{20} \\
& - i\sqrt{3}\,\omega_1 \frac{\sin(\lambda_1\tau)}{\lambda_1} T_{21}(s) + \sqrt{3}\,\omega_1^2 \frac{\cos(\lambda_1\tau) - 1}{\lambda_1^2} T_{22}(s)
\end{aligned}
\tag{A19a}
$$

and

$$
\begin{aligned}
e^{iH_S^*\tau} T_{10} e^{-iH_S^*\tau} = {}& \frac{\lambda_2 \cos(\lambda_3\tau) + \lambda_3 \cos(\lambda_2\tau)}{\lambda_1} T_{10} - i\omega_1 \frac{\sin(\lambda_3\tau) + \sin(\lambda_2\tau)}{\lambda_1} T_{11}(s) \\
& + \omega_1 \frac{\cos(\lambda_3\tau) - \cos(\lambda_2\tau)}{\lambda_1} T_{21}(a) - i \frac{\lambda_2 \sin(\lambda_3\tau) - \lambda_3 \sin(\lambda_2\tau)}{\lambda_1} T_{22}(a)
\end{aligned}
\tag{A19b}
$$

where the characteristic angular velocities $\lambda_0, \lambda_1, \lambda_2$ and λ_3 are defined by:

$$
\lambda_0 = 0, \quad \lambda_1 = \sqrt{\omega_Q^2 + 4\omega_1^2}, \quad \lambda_2 = \frac{\lambda_1 + \omega_Q}{2}, \quad \text{and} \quad \lambda_3 = \frac{\lambda_1 - \omega_Q}{2} \quad \text{respectively.}
\tag{A20}
$$

As a consequence, under the simultaneous influences of the residual quadrupolar coupling (ω_Q) and the irradiation pulse (ω_1), the T_{20} coherence oscillates according to the angular velocity λ_1, while two angular velocities (λ_2 and λ_3) drive the oscillations of the T_{10} coherence.

Straightforward calculations of the set of Equations (A14) and (A15) lead to the contributions of the quadrupolar [78] and heteronuclear dipolar [18] relaxation mechanisms to the time evolution of the coherences.

$$
\frac{d}{dt}
\begin{pmatrix}
T_{20} \\
T_{11}(a) \\
T_{21}(s) \\
T_{22}(s) \\
T_{10} \\
T_{11}(s) \\
T_{21}(a) \\
T_{22}(a)
\end{pmatrix}
= -\operatorname{diag}\,(A, B, C, D, E, B, C, D) \times
\begin{pmatrix}
T_{20} \\
T_{11}(a) \\
T_{21}(s) \\
T_{22}(s) \\
T_{10} \\
T_{11}(s) \\
T_{21}(a) \\
T_{22}(a)
\end{pmatrix}
\tag{A21}
$$

with:

$$A^Q = 3J_1^Q(\omega_1) \text{ and}$$

$$A^D = \frac{J_0^D(\omega_0 - \omega_S)}{3} + J_1^D(\omega_0) + 2J_2^D(\omega_0 + \omega_S)$$

(A22a)

$$B^Q = \frac{3J_0^Q(0)}{2} + \frac{5J_1^Q(\omega_0)}{2} + J_2^Q(2\omega_0) \text{ and}$$

$$B^D = \frac{2J_0^D(0)}{9} + \frac{J_0^D(\omega_0 - \omega_S)}{18} + \frac{J_1^D(\omega_0)}{6} + \frac{J_1^D(\omega_S)}{3} + \frac{J_2^D(\omega_0 + \omega_S)}{3}$$

(A22b)

$$C^Q = \frac{3J_0^Q(0)}{2} + \frac{J_1^Q(\omega_0)}{2} + J_2^Q(2\omega_0) \text{ and}$$

$$C^D = \frac{2J_0^D(0)}{9} + \frac{5J_0^D(\omega_0 - \omega_S)}{18} + \frac{5J_1^D(\omega_0)}{6} + \frac{J_1^D(\omega_S)}{3} + \frac{5J_2^D(\omega_0 + \omega_S)}{3}$$

(A22c)

$$D^Q = J_1^Q(\omega_0) + 2J_2^Q(2\omega_0) \text{ and}$$

$$D^D = \frac{8J_0^D(0)}{9} + \frac{J_0^D(\omega_0 - \omega_S)}{9} + \frac{J_1^D(\omega_0)}{3} + \frac{4J_1^D(\omega_S)}{3} + \frac{2J_2^D(\omega_0 + \omega_S)}{3}$$

(A22d)

$$E^Q = J_1^Q(\omega_0) + 4J_2^Q(2\omega_0) \text{ and}$$

$$E^D = \frac{J_0^D(\omega_0 - \omega_S)}{9} + \frac{J_1^D(\omega_0)}{3} + \frac{2J_2^D(\omega_0 + \omega_S)}{3}$$

(A22e)

Under the slow modulation of the quadrupolar and heteronuclear dipolar couplings, i.e., when:

$$J_0^X(0) >> J_m^X(\omega_0) >> J_m^X(\omega_S) \quad \text{with} \quad m \in \{1,2\} \quad \text{and} \quad X \in \{Q, D\}$$

(A23)

the set of equations in Equation (A22a–e) reduces to [51,52]:

$$A = 3U_Q + U_D$$

$$B = C = \frac{3J_0^Q(0)}{2} + \frac{5U_Q}{2} + \frac{2J_0^D(0)}{9} + \frac{U_D}{2}$$

$$D = 3U_Q + \frac{8J_0^D(0)}{9} + U_D$$

$$E = 5U_Q + \frac{U_D}{3}$$

(A24)

where $U_Q = J_1^Q(\omega_0) \approx J_2^Q(2\omega_0)$ and: $U_D = \frac{J_0^D(\omega_S - \omega_0)}{3} + J_1^D(\omega_0) + 2J_1^D(\omega_S + \omega_0)$

Four independent measurements of the relaxation of the T_{20}, T_{11}, T_{22} and T_{10} coherences lead then to the four dominant contributions (U_Q, $J_0^Q(0)$, U_D and $J_0^D(0)$) quantifying the quadrupolar and heteronuclear dipolar relaxation mechanisms [51,52].

The derivation of the time evolution of the coherences under the spin-locking condition requires taking into account the evolution of the fluctuating part of the quadrupolar [78] and dipolar [40] Hamiltonians under the influence of the static Hamiltonians as described by the term $e^{-iH_S^*\tau} H_{XF}^{*+}(t - \tau) e^{iH_S^*\tau}$ in Equation (A12). In the next approximation, we focus only on the $m = 0$ component of the fluctuating Hamiltonians because their $m = 1$ and $m = 2$ components oscillate at angular velocities (ω_0 and $2\omega_0$) (cf. Equation (A14)) much larger than the characteristic angular velocities (λ_i) (cf. Equations (A19a)–(A20)). By using Equation (A19a,b), that approximation leads to:

$$\frac{d}{dt}\begin{pmatrix} T_{20} \\ T_{11}(a) \\ T_{21}(s) \\ T_{22}(s) \\ T_{10} \\ T_{11}(s) \\ T_{21}(a) \\ T_{22}(a) \end{pmatrix} = -\begin{pmatrix} A & -\sqrt{3}K & 0 & 0 & 0 & 0 & 0 & 0 \\ 0 & B & 0 & 2K^D & 0 & 0 & 0 & 0 \\ 0 & 0 & C & 0 & 0 & 0 & 0 & 0 \\ 0 & -K^Q + K^D & 0 & D & 0 & 0 & 0 & 0 \\ 0 & 0 & 0 & 0 & E & 0 & -K & 0 \\ 0 & 0 & 0 & 0 & 0 & L & 0 & 2K^D \\ 0 & 0 & 0 & 0 & 0 & 0 & M & 0 \\ 0 & 0 & 0 & 0 & 0 & -K^Q + K^D & 0 & D \end{pmatrix} \times \begin{pmatrix} T_{20} \\ T_{11}(a) \\ T_{21}(s) \\ T_{22}(s) \\ T_{10} \\ T_{11}(s) \\ T_{21}(a) \\ T_{22}(a) \end{pmatrix} \quad \text{(A25)}$$

with:

$$A^Q = 3J_1^Q(\omega_0) \quad \text{and}$$
$$A^D = \frac{J_0^D(\omega_0 - \omega_S)}{3} + J_1^D(\omega_0) + 2J_2^D(\omega_0 + \omega_S) \tag{A26a}$$

$$B^Q = \frac{3\omega_Q^2 J_0^Q(0) + 4\omega_1^2 J_0^Q(\lambda_1)}{2\lambda_1^2} + \frac{5J_1^Q(\omega_0)}{2} + J_2^Q(2\omega_0) \quad \text{and}$$
$$B^D = \frac{2\left(\lambda_2 J_0^D(\lambda_3) + \lambda_3 J_0^D(\lambda_2)\right)}{9\lambda_1} + \frac{J_0^D(\omega_0 - \omega_S)}{18} + \frac{J_1^D(\omega_0)}{6} + \frac{J_1^D(\omega_S)}{3} + \frac{J_2^D(\omega_0 + \omega_S)}{3} \tag{A26b}$$

$$C^Q = \frac{3\omega_0^2 J_0^Q(0) + 4\omega_1^2 J_0^Q(\lambda_1)}{2\lambda_1^2} + \frac{J_1^Q(\omega_0)}{2} + J_2^Q(2\omega_0) \quad \text{and}$$
$$C^D = \frac{2\left(\lambda_2 J_0^D(\lambda_3) + \lambda_3 J_0^D(\lambda_2)\right)}{9\lambda_1} + \frac{5J_0^D(\omega_0 - \omega_S)}{18} + \frac{5J_1^D(\omega_0)}{6} + \frac{J_1^D(\omega_S)}{3} + \frac{5J_2^D(\omega_0 + \omega_S)}{3} \tag{A26c}$$

$$D^Q = J_1^Q(\omega_0) + 2J_2^Q(2\omega_0) \quad \text{and}$$
$$D^D = \frac{8J_0^D(0)}{9} + \frac{J_0^D(\omega_0 - \omega_S)}{9} + \frac{J_1^D(\omega_0)}{3} + \frac{4J_1^D(\omega_S)}{3} + \frac{2J_2^D(\omega_0 + \omega_S)}{3} \tag{A26d}$$

$$E^Q = J_1^Q(\omega_0) + 4J_2^Q(2\omega_0) \quad \text{and}$$
$$E^D = \frac{J_0^D(\omega_0 - \omega_S)}{9} + \frac{J_1^D(\omega_0)}{3} + \frac{2J_2^D(\omega_0 + \omega_S)}{3} \tag{A26e}$$

$$L^Q = \frac{3\omega_Q^2 J_0^Q(0) + 6\omega_1^2\left(J_0^Q(0) + J_0^Q(\lambda_1)\right)}{2\lambda_1^2} + \frac{5J_1^Q(\omega_0)}{2} + 2J_2^Q(2\omega_0) \quad \text{and}$$
$$L^D = \frac{2\left(\lambda_2 J_0^D(\lambda_3) + \lambda_3 J_0^D(\lambda_2)\right)}{9\lambda_1} + \frac{J_0^D(\omega_0 - \omega_S)}{18} + \frac{J_1^D(\omega_0)}{6} + \frac{J_1^D(\omega_S)}{3} + \frac{J_2^D(\omega_0 + \omega_S)}{3} \tag{A26f}$$

$$M^Q = \frac{3\omega_Q^2 J_0^Q(0) + 2\omega_1^2\left(J_0^Q(0) + J_0^Q(\lambda_1)\right)}{2\lambda_1^2} + \frac{J_1^Q(\omega_0)}{2} + J_2^Q(2\omega_0) \quad \text{and}$$
$$M^D = \frac{2\left(\lambda_2 J_0^D(\lambda_3) + \lambda_3 J_0^D(\lambda_2)\right)}{9\lambda_1} + \frac{5J_0^D(\omega_0 - \omega_S)}{18} + \frac{5J_1^D(\omega_0)}{6} + \frac{J_1^D(\omega_S)}{3} + \frac{5J_2^D(\omega_0 + \omega_S)}{3} \tag{A26g}$$

$$K^Q = \frac{3\omega_1\omega_Q(J_0^Q(0) - J_0^Q(\lambda_1))}{2\lambda_1^2} \quad \text{and}$$
$$K^D = \frac{2\omega_1\left(J_0^D(\lambda_3) - J_0^D(\lambda_2)\right)}{9\lambda_1} \tag{A26h}$$

As mentioned above, the quadrupolar relaxation mechanism (implying the T_{20} coherence) samples the spectral densities at the angular velocity λ_1, while the heteronuclear dipolar relaxation mechanism (implying the T_{10} coherence) samples the two other angular velocities (λ_2 and λ_3), extending notably the dynamical range probed by spin-locking relaxation measurements.

For spin $I = 3/2$ nuclei, the whole set of coherences splits again into two independent sub-sets [79,80], namely $\{T_{11}(a), T_{20}, T_{21}(s), T_{22}(s), T_{31}(a), T_{32}(a), T_{33}(a)\}$ and $\{T_{10}, T_{11}(s), T_{21}(a)), T_{22}(a), T_{30}, T_{31}(s), T_{32}(s), T_{33}(s)\}$; see Figure A3. The eigenvalues $\pm i\lambda_p$ with $p \in \{0, \cdots, 6\}$ describing their time evolution under the influence of the static quadrupolar Hamiltonian and irradiation pulse [64,79,80] are given by:

$$\lambda_0 = 0 \qquad \lambda_1 = \sqrt{\omega_Q^2 + 2\omega_1\omega_Q + 4\omega_1^2} \qquad \lambda_2 = \sqrt{\omega_Q^2 - 2\omega_1\omega_Q + 4\omega_1^2}$$

$$\lambda_3 = \omega_1 + \sqrt{\frac{\omega_Q^2 + 4\omega_1^2 + \lambda_1\lambda_2}{2}} \qquad \lambda_4 = \omega_1 + \sqrt{\frac{\omega_Q^2 + 4\omega_1^2 - \lambda_1\lambda_2}{2}} \qquad (A27)$$

$$\lambda_5 = \omega_1 - \sqrt{\frac{\omega_Q^2 + 4\omega_1^2 + \lambda_1\lambda_2}{2}} \qquad \lambda_6 = \omega_1 - \sqrt{\frac{\omega_Q^2 + 4\omega_1^2 - \lambda_1\lambda_2}{2}}$$

By contrast, the corresponding eigenvectors must be calculated numerically [83].

For spin $I = 7/2$ nuclei, no analytical solutions are available, and the eigenvalues and eigenvectors describing the time evolution of the 63 coherences must be solved numerically [55].

References

1. Israelachvili, J.N. *Intermolecular and Surface Forces*; Academic Press: New York, NY, USA, 1985.
2. Henderson, D. *Fundamentals of Inhomogeneous Fluids*; M. Dekker: New York, NY, USA, 1992.
3. Jobbagy, M.; Iyi, N. Interplay of charge density and relative humidity on the structure of nitrate layered double hydroxides. *J. Phys. Chem. C* **2010**, *114*, 18153–18158.
4. Lee, S.S.; Fenter, P.; Park, C.; Sturchio, N.C.; Nagy, K.L. Hydrated cation speciation at the Muscovite (001)-water interface. *Langmuir* **2010**, *26*, 16647–16651.
5. Boily, J.F. Water structure and hydrogen bonding at Goethite/water interfaces: Implications for proton affinities. *J. Phys. Chem. C* **2012**, *116*, 4714–4724.
6. Ho, T.A.; Argyris, D.; Cole, D.R.; Striolo, A. Aqueous NaCl and CsCl solutions confined in crystalline slit-shaped silica nanopores of varying degree of protonation. *Langmuir* **2012**, *28*, 1256–1266.
7. Malani, A.; Ayappa, K.G. Relaxation and jump dynamics of water at the mica interface. *J. Chem. Phys.* **2012**, *136*, 194701.
8. Buyukdagli, S.; Blossey, R. Dipolar correlations in structured solvents under nanoconfinement. *J. Chem. Phys.* **2014**, *140*, 234903.
9. Sato, K.; Fujimoto, K.; Dai, W.; Hunger, M. Molecular mechanism of heavily adhesive Cs: Why radioactive Cs is not decontaminated from soil. *J. Phys. Chem. C* **2013**, *117*, 14075–14080.
10. Jiménez-Ruiz, M.; Ferrage, E.; Delville, A.; Michot, L.J. Anisotropy on the collective dynamics of water confined in swelling clay minerals. *J. Phys. Chem. A* **2012**, *116*, 2379–2387.
11. Michot, L.J.; Ferrage, E.; Delville, A.; Jiménez-Ruiz, M. Influence of layer charge, hydration state and cation nature on the collective dynamics of interlayer water in synthetic swelling clay minerals. *Appl. Clay Sci.* **2016**, *119*, 375–384.
12. Jiménez-Ruiz, M.; Ferrage, E.; Blanchard, M.; Fernandez-Castanon, J.; Delville, A.; Johnson, M.R.; Michot, L.J. Combination of inelastic neutron scattering experiments and ab Initio quantum calculations for the study of the hydration properties of oriented Saponites. *J. Phys. Chem. C* **2017**, *121*, 5029–5040.
13. Michot, L.J.; Delville, A.; Humbert, B.; Plazanet, M.; Levitz, P. Diffusion of water in a synthetic clay with tetrahedral charges by combined neutron time-of-flight measurements and molecular dynamics simulations. *J. Phys. Chem. C* **2007**, *111*, 9818–9831.
14. Michot, L.J.; Ferrage, E.; Jiménez-Ruiz, M.; Boehm, M.; Delville, A. Anisotropic features of water and ion dynamics in synthetic Na- and Ca-smectites with tetrahedral layer charge. A combined Quasi-Elastic Neutron-Scattering and Molecular Dynamics simulations study. *J. Phys. Chem. C* **2012**, *116*, 16619–16633.
15. Callaghan, P.T. *Principles of Nuclear Magnetic Resonance Microscopy*; Clarendon Press: Oxford, UK, 1991.
16. Delville, A.; Porion, P.; Faugère, A.M. Ion diffusion within charged porous network as probed by nuclear quadrupolar relaxation. *J. Phys. Chem. B* **2000**, *104*, 1546–1551.

17. Bryar, T.R.; Daughney, C.J.; Knight, R.J. Paramagnetic effects of iron(III) species on nuclear magnetic relaxation of fluid protons in porous media. *J. Magn. Reson.* **2000**, *142*, 74–85.

18. Porion, P.; Michot, L.J.; Faugère, A.M.; Delville, A. Structural and dynamical properties of the water molecules confined in dense clay sediments: A study combining ^{2}H NMR spectroscopy and multiscale numerical modeling. *J. Phys. Chem. C* **2007**, *111*, 5441–5453.

19. Gao, Y.; Zhang, R.; Lv, W.; Liu, Q.; Wang, X.; Sun, P.; Winter, H.H.; Xue, G. Critical effect of segmental dynamics in polybutadiene/clay nanocomposites characterized by solid state ^{1}H NMR spectroscopy. *J. Phys. Chem. C* **2014**, *118*, 5606–5614.

20. Bortolotti, V.; Brizi, L.; Brown, R.J.S.; Fantazzini, P.; Mariani, M. Nano and sub-nano multiscale porosity formation and other features revealed by ^{1}H NMR relaxometry during cement hydration. *Langmuir* **2014**, *30*, 10871–10877.

21. Halle, B. Spin dynamics of exchanging quadrupolar nuclei in locally anisotropic systems. *Prog. Nucl. Magn. Reson. Spectrosc.* **1996**, *28*, 137–159.

22. Woessner, D.E. Nuclear transfer effects in Nuclear Magnetic Resonance pulse experiments. *J. Chem. Phys.* **1961**, *35*, 41–48.

23. Abragam, A. *The Principles of Nuclear Magnetism*; Clarendon Press: Oxford, UK, 1961.

24. Kimmich, R. *NMR: Tomography, Diffusometry, Relaxometry*; Springer: Berlin, Germany, 1997.

25. Delville, A.; Stöver, H.D.H.; Detellier, C. Crown-ether cation decomplexation mechanics. ^{23}Na NMR studies of the sodium cation complexes with dibenzo-24-crown-8 and dibenzo-18-crown-6 in nitromethane and acetonitrile. *J. Am. Chem. Soc.* **1987**, *109*, 7293–7301.

26. Delville, A.; Detellier, C.; Laszlo, P. Determination of the correlation time for a slowly reorienting spin-3/2 nucleus: Binding of Na$^+$ with the 5'-GMP supramolecular assembly. *J. Magn. Reson.* **1979**, *34*, 301–315.

27. Delville, A.; Grandjean, J.; Laszlo, P.; Gerday, C.; Grabarek, Z.; Drabikowski, W. Sodium-23 nuclear magnetic resonance as an indicator of sodium binding to Troponin C and tryptic fragments, in relation to calcium content and attendant conformational changes. *Eur. J. Biochem.* **1980**, *105*, 289–295.

28. Delville, A.; Laszlo, P.; Schyns, R. Displacement of sodium ions by surfactant ions from DNA: A ^{23}Na NMR investigation. *Biophys. Chem.* **1986**, *24*, 121–133.

29. Liu, G.; Li, Y.; Jonas, J. Confined geometry effects on reorientational dynamics of molecular liquids in porous silica glasses. *J. Chem. Phys.* **1991**, *95*, 6892–6901.

30. Korb, J.P.; Whaley-Hodges, M.; Bryant, R.G. Translational diffusion of liquids at surfaces of microporous materials: Theoretical analysis of field-cycling magnetic relaxation measurements. *Phys. Rev. E* **1997**, *56*, 1934–1945.

31. Levitz, P.E. Confined dynamics, forms and transitions in colloidal systems: From clay to DNA. *Magn. Reson. Imaging* **2005**, *23*, 147–152.

32. Korb, J.P. Nuclear magnetic relaxation of liquids in porous media. *New J. Phys.* **2011**, *13*, 035016.

33. Delville, A.; Letellier, M. Structure and dynamics of simple liquids in heterogeneous condition: An NMR study of the clay water interface. *Langmuir* **1995**, *11*, 1361–1367.

34. Avogadro, A.; Villa, M. Nuclear magnetic resonance in a two-dimensional system. *J. Chem. Phys.* **1977**, *66*, 2359–2367.

35. Korb, J.P.; Ahadi, M.; McConnell, H.M. Paramagnetically enhanced nuclear relaxation in lamellar phases. *J. Phys. Chem.* **1987**, *91*, 1255–1259.

36. Korb, J.P.; Delville, A.; Xu, S.; Demeulenaere, G.; Costa, P.; Jonas, J. Relative Role of Surface Interactions and Topological Effects in Nuclear-Magnetic-Resonance of Confined Liquids. *J. Chem. Phys.* **1994**, *101*, 7074–7081.

37. Pasquier, V.; Levitz, P.; Delville, A. ^{129}Xe NMR as a probe of gas diffusion and relaxation in disordered porous media: An application to Vycor. *J. Phys. Chem.* **1996**, *100*, 10249–10256.

38. Korb, J.P.; Malier, L.; Cros, F.; Xu, S.; Jonas, J. Surface dynamics of liquids in nanopores. *Phys. Rev. Lett.* **1996**, *77*, 2312–2315.

39. Faux, D.A.; McDonald, P.J.; Howlett, N.C. Nuclear-magnetic-resonance relaxation due to the translational diffusion of fluid confined to quasi-two-dimensional pores. *Phys. Rev. E* **2017**, *95*, 033116.

40. Porion, P.; Michot, L.J.; Faugère, A.M.; Delville, A. Influence of confinement on the long-range mobility of water molecules within clay aggregates: A ^{2}H NMR analysis using spin-locking relaxation rates. *J. Phys. Chem. C* **2007**, *111*, 13117–13128.

41. Trausch, G.; Canet, D.; Cadène, A.; Turq, P. Separation of components of a ^{1}H NMR composite signal by nutation experiments under low amplitude radio-frequency fields. Application to the water signal in clays. *Chem. Phys. Lett.* **2006**, *433*, 228–233.

42. Fleury, M.; Canet, D. Water orientation in smectites using NMR nutation experiments. *J. Phys. Chem. C* **2014**, *118*, 4733–4740.

43. Volgmann, K.; Epp, V.; Langer, J.; Stanje, B.; Heine, J.; Nakhal, S.; Lerch, M.; Wilkening, M.; Heitjans, P. Solid-state NMR to study translational Li ion dynamics in solids with low-dimensional diffusion pathways. *Z. Phys. Chem.* **2017**, *231*, 1215–1241.

44. Godefroy, S.; Fleury, M.; Deflandre, F.; Korb, J.P. Temperature effect on NMR surface relaxation in rocks for well logging applications. *J. Phys. Chem. B* **2002**, *106*, 11183–11190.

45. Leon, N.; Korb, J.P.; Bonalde, I.; Levitz, P. Universal nuclear spin relaxation and long-range order in nematics strongly confined in mass fractal silica gels. *Phys. Rev. Lett.* **2004**, *92*, 195504.

46. Kruk, D.; Meier, R.; Rössler, E.A. Nuclear magnetic resonance relaxometry as a method of measuring translational diffusion coefficients in liquids. *Phys. Rev. E* **2012**, *85*, 020201(R).

47. Muncaci, S.; Mattea, C.; Stapf, S.; Ardelean, I. Frequency-dependent NMR relaxation of liquids confined inside porous media containing an increased amount of magnetic impurities. *Magn. Reson. Chem.* **2013**, *51*, 123–128.

48. Kimmich, R.; Fatkullin, N. Self-diffusion studies by intra- and inter-molecular spin-lattice relaxometry using field-cycling: Liquids, plastic crystals, porous media, and polymer segments. *Prog. Nucl. Magn. Reson. Spectrosc.* **2017**, *101*, 18–50.

49. Harris, R.K.; Mann, B.E. *NMR and the Periodic Table*; Academic Press: London, UK, 1978.

50. Porion, P.; Delville, A. Multinuclear NMR study of the structure and micro-dynamics of counterions and water molecules within clay colloids. *Curr. Opin. Colloid Interface Sci.* **2009**, *14*, 216–222.

51. Porion, P.; Michot, L.J.; Warmont, F.; Faugère, A.M.; Delville, A. Long-time dynamics of confined water molecules probed by ^{2}H NMR multiquanta relaxometry: An application to dense clay sediments. *J. Phys. Chem. C* **2012**, *116*, 17682–17697.

52. Porion, P.; Faugère, A.M.; Delville, A. Multiscale water dynamics within dense clay sediments probed by ^{2}H multiquanta NMR relaxometry and two-time stimulated echo NMR spectroscopy. *J. Phys. Chem. C* **2013**, *117*, 26119–26134.

53. Porion, P.; Faugère, A.M.; Delville, A. Structural and dynamical properties of water molecules confined within clay sediments probed by deuterium NMR spectroscopy, multiquanta relaxometry, and two-time stimulated echo attenuation. *J. Phys. Chem. C* **2014**, *118*, 20429–20444.

54. Porion, P.; Faugère, A.M.; Delville, A. Long-time scale ionic dynamics in dense clay sediments measured by the frequency variation of the ^{7}Li multiple-quantum NMR relaxation rates in relation with a multiscale modeling. *J. Phys. Chem. C* **2009**, *113*, 10580–10597.

55. Porion, P.; Warmont, F.; Faugère, A.M.; Rollet, A.L.; Dubois, E.; Marry, V.; Michot, L.J.; Delville, A. ^{133}Cs Nuclear Magnetic Resonance relaxometry as a probe of the mobility of cesium cations confined within dense clay sediments. *J. Phys. Chem. C* **2015**, *119*, 15360–15372.

56. Rinnert, E.; Carteret, C.; Humbert, B.; Fragneto-Cusani, G.; Ramsay, J.D.F.; Delville, A.; Robert, J.L.; Bihannic, I.; Pelletier, M.; Michot, L.J. Hydration of a synthetic clay with tetrahedral charges: A multidisciplinary experimental and numerical study. *J. Phys. Chem. B* **2005**, *109*, 23745–23759.

57. Ferrage, E.; Sakharov, B.A.; Michot, L.J.; Delville, A.; Bauer, A.; Lanson, B.; Grangeon, S.; Frapper, G.; Jiménez-Ruiz, M.; Cuello, G.J. Hydration properties and interlayer organization of water and ions in synthetic Na-smectite with tetrahedral layer charge. Part 2. Toward a precise coupling between molecular simulations and diffraction data. *J. Phys. Chem. C* **2011**, *115*, 1867–1881.

58. Dazas, B.; Lanson, B.; Delville, A.; Robert, J.L.; Komarneni, S.; Michot, L.J.; Ferrage, E. Influence of tetrahedral layer charge on the organization of interlayer water and ions in synthetic Na-saturated smectites. *J. Phys. Chem. C* **2015**, *119*, 4158–4172.

59. Böhmer, R.; Qi, F. Spin relaxation and ultra-slow Li motion in an aluminosilicate glass ceramic. *Solid State Nucl. Magn. Reson.* **2007**, *31*, 28–34.

60. Porion, P.; Faugère, A.M.; Delville, A. Long-distance water exchange within dense clay sediments probed by two-time ^{2}H stimulated echo NMR spectroscopy. *J. Phys. Chem. C* **2013**, *117*, 9920–9931.

61. Michot, L.J.; Bihannic, I.; Porsch, K.; Maddi, S.; Baravian, C.; Mougel, J.; Levitz, P. Phase diagrams of Wyoming Na-Montmorillonite clay. Influence of particle anisotropy. *Langmuir* **2004**, *20*, 10829–10837.
62. Edmonds, D.T.; Hunt, M.J.; Mackay, A.L. Deuteron interactions in pure quadrupole-resonance. *J. Magn. Reson.* **1975**, *20*, 505–514.
63. Mehring, M. *Principles of High Resolution NMR in Solids*, 2nd ed.; Springer: Berlin, Germany, 1983.
64. Porion, P.; Faugère, A.M.; Delville, A. ^{7}Li NMR spectroscopy and multiquantum relaxation as a probe of the microstructure and dynamics of confined Li$^+$ cations: An application to dense clay sediments. *J. Phys. Chem. C* **2008**, *112*, 9808–9821.
65. Porion, P.; Faugère, A.M.; Lécolier, E.; Gherardi, B.; Delville, A. ^{23}Na Nuclear quadripolar relaxation as a probe of the microstructure and dynamics of aqueous clay dispersions: An application to Laponite gels. *J. Phys. Chem. B* **1998**, *102*, 3477–3485.
66. Bowden, G.J.; Hutchison, W.D. Tensor operator formalism for multiple-quantum NMR. 1. Spin-1 nuclei. *J. Magn. Reson.* **1986**, *67*, 403–414.
67. Bowden, G.J.; Hutchison, W.D.; Khachan, J. Tensor operator formalism for multiple-quantum NMR. 2. Spins-3/2, spin-2, and spin-5/2 and general I. *J. Magn. Reson.* **1986**, *67*, 415–437.
68. Müller, N.; Bodenhausen, G.; Ernst, R.R. Relaxation-induced violations of coherence transfer selection rules in Nuclear Magnetic Resonance. *J. Magn. Reson.* **1987**, *75*, 297–334.
69. Ngouana, W.B.F.; Kalinichev, A.G. Structural arrangements of isomorphic substitutions in smectites: Molecular simulation of the swelling properties, interlayer structure, and dynamics of hydrated Cs-Montmorillonite revisited with new clay models. *J. Phys. Chem. C* **2014**, *118*, 12758–12773.
70. Marry, V.; Turq, P.; Cartailler, T.; Levesque, D. Microscopic simulation of structure and dynamics of water and counterions in a monohydrated Montmorillonite. *J. Chem. Phys.* **2002**, *117*, 3454–3463.
71. Kosakowski, G.; Churakov, S.V.; Thoenen, T. Diffusion of Na and Cs in Montmorillonite. *Clays Clay Miner.* **2008**, *56*, 190–206.
72. Liu, X.; Lu, X.; Wang, R.; Zhou, H. Effects of layer-charge distribution on the thermodynamic and microscopic properties of Cs-smectite. *Geochim. Cosmochim. Acta* **2008**, *72*, 1837–1847.
73. Bourg, I.C.; Sposito, G. Connecting the molecular scale to the continuum scale for diffusion processes in smectite-rich porous media. *Environ. Sci. Technol.* **2010**, *44*, 2085–2091.
74. Zheng, Y.; Zaoui, A. How water and counterions diffuse into the hydrated Montmorillonite. *Solid State Ion.* **2011**, *203*, 80–85.
75. Cuperlovic, M.; Meresi, G.H.; Palke, W.E.; Gerig, J.T. Spin relaxation and chemical exchange in NMR simulations. *J. Magn. Reson.* **2000**, *142*, 11–23.
76. Hertz, H.G. Magnetic relaxation by quadrupole interaction of ionic nuclei in electrolyte solutions. Part I: Limiting values for infinite dilution. *Ber. Bunsenges. Phys. Chem.* **1973**, *77*, 531–540.
77. Redfield, A.G. On the theory of relaxation processes. *IBM J. Res. Dev.* **1957**, *1*, 19–31.
78. Van der Maarel, J.R.C. The relaxation dynamics of spin I=1 nuclei with a static quadrupolar coupling and a radio-frequency field. *J. Chem. Phys.* **1993**, *99*, 5646–5653.
79. Van der Maarel, J.R.C. Thermal relaxation and coherence dynamics of spin 3/2. I. Static and fluctuating quadrupolar interactions in the multipole basis. *Concepts Magn. Reson. Part A* **2003**, *19A*, 97–116.
80. Van der Maarel, J.R.C. Thermal relaxation and coherence dynamics of spin 3/2. II. Strong radio-frequency field. *Concepts Magn. Reson. Part A* **2003**, *19A*, 117–133.
81. Kimmich, R.; Anoardo, E. Field-cycling NMR relaxometry. *Prog. Nucl. Magn. Reson. Spectrosc.* **2004**, *44*, 257–320.
82. Barbara, T.M.; Vold, R.R.; Vold, R.L. A determination of individual spectral densities in a smectic liquid-crystal from angle dependent nuclear spin relaxation measurements. *J. Chem. Phys.* **1983**, *79*, 6338–6340.
83. Porion, P.; Al-Mukhtar, M.; Meyer, S.; Faugère, A.M.; van der Maarel, J.R.C.; Delville, A. Nematic ordering of suspensions of charged anisotropic colloids detected by ^{23}Na Nuclear Quadrupolar Spectroscopy. *J. Phys. Chem. B* **2001**, *105*, 10505–10514.

magnetochemistry

MDPI

Review

Orientational Glasses: NMR and Electric Susceptibility Studies

Neil Sullivan [1],*, Jaha Hamida [1], Khandker Muttalib [1], Subrahmanyam Pilla [2] and Edgar Genio [3]

[1]   Department of Physics, University of Florida, Gainesville, FL 32611, USA; jaha1953@yahoo.com (J.H.); muttalib@phys.ufl.edu (K.M.)
[2]   Qorvo Inc., Richardson, TX 75080, USA; manyamp@gmail.com
[3]   KETD East Qianjin Road Enterprise Science and Technology Park, Kunshan 215334, China; edgar_genio@hotmail.com
*   Correspondence: sullivan@phys.ufl.edu; Tel.: +1-352-846-3137

Received: 3 October 2017; Accepted: 17 October 2017; Published: 1 November 2017

Abstract: We review the results of a wide range of nuclear magnetic resonance (NMR) measurements of the local order parameters and the molecular dynamics of solid ortho-para hydrogen mixtures and solid nitrogen-argon mixtures that form novel molecular orientational glass states at low temperatures. From the NMR measurements, the distribution of the order parameters can be deduced and, in terms of simple models, used to analyze the thermodynamic measurements of the heat capacities of these systems. In addition, studies of the dielectric susceptibilities of the nitrogen-argon mixtures are reviewed in terms of replica symmetry breaking analogous to that observed for spin glass states. It is shown that this wide set of experimental results is consistent with orientation or quadrupolar glass ordering of the orientational degrees of freedom.

Keywords: orientational glass; magnetic resonance; fluctuations

---

1. Introduction

Understanding the underlying physics and dynamics of glass systems has remained a challenge for contemporary physics despite the large number of experimental and theoretical studies that have shown that a wide variety of glass systems exhibit a number of universal features [1–6]. Considerable progress was made with the discovery of spin glasses [7–12] and the introduction of the concepts of frustration [13–17] and replica symmetry breaking [13,18–20]. Beyond the simple spin glasses, which displayed a random orientation of dipole moments, it was recognized very early in the history of spin glasses that molecular systems, and in particular diatomic molecules that have short-range anisotropic electrostatic interactions, were highly frustrated and that they formed a special class of glass systems with geometrical frustration. These new glass systems are characterized by (i) a broad distribution of order parameters provided that sufficient disorder was introduced and (ii) a strong temperature dependence for the molecular dynamics, often resembling a Fulcher–Vogel dependence [21,22].

The interest in the molecular glasses is that the interactions are accurately known, the frustration can be described in clear terms and the local order parameters can be measured directly by nuclear magnetic resonance (NMR) methods. Only recently has there been success in relating the low temperature properties of the molecular glasses to replica symmetry breaking [23,24], and in this paper, we review the experimental NMR methods that have been used to measure the order parameters of these systems and relate the observations to electric susceptibility studies and the models of replica symmetry.

2. Orientational and Quadrupolar Glasses: Basics

While spin glasses arise from the combination of frustration (of dipolar magnetic interactions) and disorder, the molecular orientational and quadrupolar glasses arise from the frustration of short-range interactions that are dominated by intermolecular electric quadrupole-quadrupole interactions resulting from the non-spherical electrostatic charge distribution of the molecules. The classic examples are ortho-para hydrogen mixtures [21,25–27] and solid N_2-Ar alloys [28,29]. The order parameters are given by two sets of parameters: (i) the local axes (x,y,z) for the mean orientation of the molecular axes and (ii) the degree of alignment or quadrupolarization about those axes. The latter in the classical case is given by the expectation values of the spherical harmonics, $\sigma_{class} = \langle Y_{20}(\theta) \rangle$ and $\eta_{class} = \Re \langle Y_{22}(\theta, \phi) \rangle$. For solid hydrogen, the orbital angular momentum J is a good quantum number. There are two molecular species: ortho-H_2 (with orbital angular momentum $J = 1$ and total nuclear spin $I = 1$) and para-H_2 (with $J = 0$ and $I = 0$). Only the ortho-H_2 molecules have an electrostatic quadrupole moment and are subject to orientational or quadrupole ordering. The para-H_2 molecules provide the disorder. Although the para-H_2 molecules are the true ground state, the ortho-para conversion is a very slow process in pure solid hydrogen, and one can conduct experiments for a wide range of ortho-para mixtures (for a review of the properties of sold hydrogen, see Silvera [30]). For the ortho-H_2 molecules, the order parameters are are given by $\sigma_{qu} = \langle 3J_z^2 - J^2 \rangle$ and $\eta_{qu} = \langle J_x^2 - J_y^2 \rangle$, which are the expectation values of the tensor operator equivalents of the spherical harmonics to within a simple numerical factor. Above a critical concentration of 55% ortho-H_2, the ortho molecules order in a four-sublattice anti-ferro-orientational Pa3 structure [30]. Below 55%, the ortho molecules form a quadrupolar glass with no sudden phase transition [31]. The onset of glass formation was signaled by the observation of a broad distribution of local order parameters at low temperatures. For a simple orientational glass, only the axes (x,y,z) are randomly distributed, while for the quadrupolar glass, both the axes and the quadrupolarizations vary randomly. The distinction between the two types of molecular glasses is illustrated in Figure 1 [31].

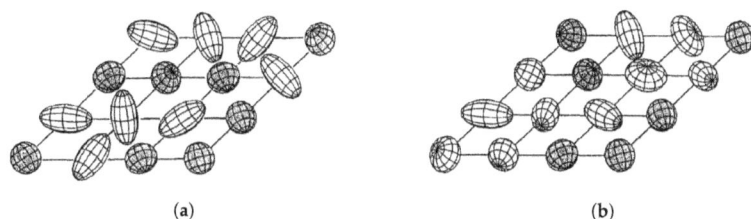

(a)   (b)

Figure 1. Comparison of orientational glass with quadrupolar glasses. Reproduced with permission from [32]. Copyright Springer Nature, 1996. (**a**) Orientational glass; (**b**) quadrupolar glass.

In addition to the static observations of the order parameters, experiments also showed strong temperature dependencies for the dynamics of the molecules (principally their thermal fluctuations about the mean molecular alignments) as inferred from various studies of the nuclear spin relaxation times. Two independent measurements of the relaxation times in solid ortho-para H_2 mixtures are shown in Figure 2. The change in the static order parameter, σ (as defined above and represented by the variable $S \approx (1 - |\sigma|)$ in Figure 2b), is considerably less dramatic than the changes observed in the dynamical parameters that determine the nuclear spin-lattice relaxation times. The main difference in the two measurements shown in Figure 2 is the frequency of the measurements; 22 MHz for Figure 2a (Ishimoto et al. [25]) and 33 MHz for Figure 2b (Cochran et al. [26]). This difference was the first evidence of the characteristic frequency dependence of the response functions observed below the glass formation and will be discussed further in the next section.

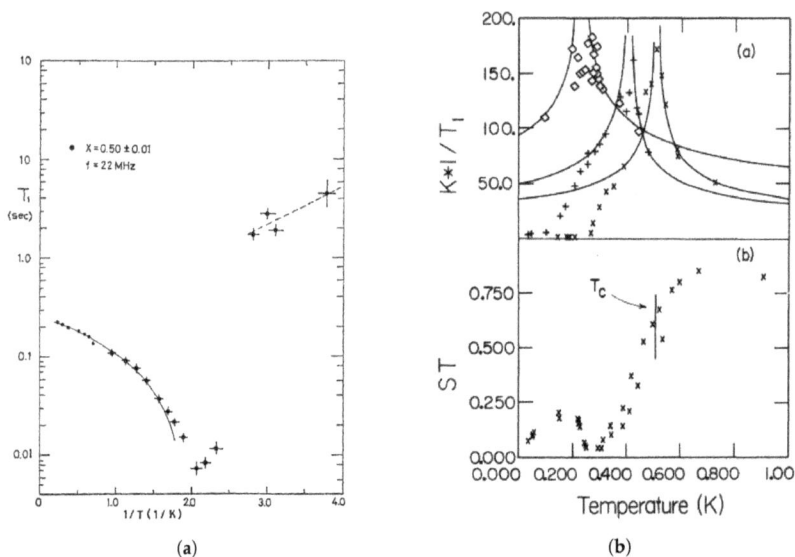

Figure 2. Observed variation of the nuclear spin-lattice relaxation times of solid ortho-para hydrogen mixtures showing the sharp change in the temperature dependence attributed to the onset of quadrupolar glass formation. Figure 3b also shows the variation of the echo amplitudes S, indicating changes in the static order parameters below 0.55 K. In contrast to the variation of T_1, S evolves smoothly with temperature. (**a**) Temperature dependence of relaxation times for 50% ortho H_2 in solid ortho-para H_2 mixtures. Reproduced with permission from Figure 6 of Ishimoto et al. [25]. (**b**) Temperature dependence of relaxation times for 33% (diamonds), 40% (pluses) and 50% (crosses) ortho H_2. Reproduced with permission from [26]. Copyright American Physical Society, 1980.

Studies of solid nitrogen-argon mixtures [33–35] showed similar features to those referenced above for solid ortho-para H_2 mixtures. The randomness for the N_2-Ar solid mixtures is introduced by the replacement of nitrogen molecules with argon atoms, which are spherically symmetrical and carry no electrostatic quadrupole moment. At high nitrogen concentrations (more than 77%), long-range Pa3 ordering is observed similar to pure ortho-H_2. Below that critical concentration, glass behavior is observed at low temperatures. As indicated in Figure 3a,b, a lattice change is observed when the long-rage ordering occurs with the hcp lattice transitioning to an fcc lattice as the orientational ordering occurs. The critical concentration depends on the concentration dependence of the anisotropic interactions. A detailed restricted trace calculation [36] showed that the orientational fluctuations for ortho-para hydrogen mixtures varied as $(2x - 1)$ where x is the ortho concentration leading to a predicted critical concentration of 50%. No detailed calculation has been made for solid N_2-Ar mixtures. For the latter, it is noteworthy that for N_2 concentrations less than 56%, a cubic lattice is maintained at low temperatures. This is important for any analysis of the glass formation because in the hcp structure, the symmetry of the electric quadrupole-quadrupole interaction is incompatible with the non-cubic lattice symmetry. As a consequence, replacing a quadrupole by a spherical diluent is equivalent to adding a local conjugate quadrupolar field so that for the hcp lattice, the glass formation is analogous to that of a spin glass in a static magnetic field, and any phase transition would be rounded out. The solid N_2-Ar mixtures at low concentrations (Phase II of Figure 3b) therefore provide a better testbed for understanding quadrupolar or orientational glass formation.

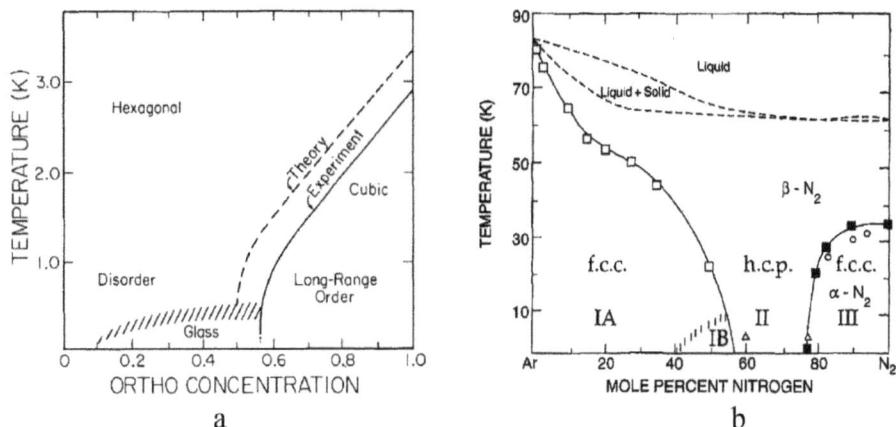

Figure 3. Phase diagrams for orientational and quadrupolar ordering in solid ortho-para H_2 mixtures [21,25,26] and solid N_2-Ar mixtures [28,37]. (**a**) Phase diagram of solid ortho-para H_2 mixtures. Reproduced with permission from Figure 1 of Sullivan [31]. Copyright Canadian Science Publishing, 1998; (**b**) Phase diagram of solid N_2-Ar mixtures. Reproduced with permission from Figure 1 of Hamida et al. [38]. Copyright Springer Nature, 1998.

3. NMR Methods

NMR methods are sensitive to long-range ordering or orientational or quadrupolar glass ordering because the nuclear spin-spin interactions depend on the local order parameters. Both the nuclear dipole-dipole interactions and the anisotropic chemical shift interaction vary with the values of the local order parameter and thus alter the NMR spectra.

The intra-molecular dipole-dipole interaction between the two nuclei of one molecule can be written as [27]:

$$H_{DD} = \hbar D \sum_m T_{2m} N_{2m}^\dagger \tag{1}$$

where we have introduced the irreducible orthonormal tensor operators T_{2m} and N_{2m} to describe the orientational and nuclear spin degrees of freedom, respectively. Under rotations, these operators transform equivalently to the spherical harmonics $Y_{2m}(\theta\phi)$. In the manifold $J = 1$, the orientational operators are [27,39]:

$$T_{20} = \frac{1}{\sqrt{6}}(3J_z^2 - J^2),$$

$$T_{21} = \frac{1}{2}(J_z J_+ + J_+ J_z),$$

$$T_{22} = \frac{1}{2}J_+^2$$

with:

$$T_{lm} = (-)^m T_{l,-m}^\dagger \tag{2}$$

and

$$Tr(T_{lm}(T_{l'm'})^\dagger) = \delta_{ll'}\delta_{mm'}. \tag{3}$$

The strength of the intra-molecular dipole-dipole interaction for H_2 is given by $D/2\pi = 173.06$ kHz. For a high magnetic field, the dipole interaction is only a weak perturbation of the nuclear Zeeman interaction, $H_z = \gamma\hbar I_z B_z$, where γ is the nuclear gyromagnetic ratio and B_z is the magnetic field aligned along the z-axis. For the Zeeman term alone, the energy levels for one ortho-H_2 molecule with total nuclear spin I= 1 consist of three equally-spaced levels corresponding to $I_z = 1, 0, -1$ with energy separations $\Delta E_{01} = \Delta E_{-10} = \hbar \omega_L$ where $\omega_L = \gamma B_z$ is the nuclear Larmor frequency. In the absence of any orientational ordering, the dipolar interaction averages to zero, and this energy separation is unperturbed. However, for a non-trivial orientational ordering, the expectation value of the secular component of H_D (the part that commutes with the Zeeman interaction), $\langle H_{D0} \rangle$, is non-zero, and the energy levels are perturbed as shown in Figure 4.

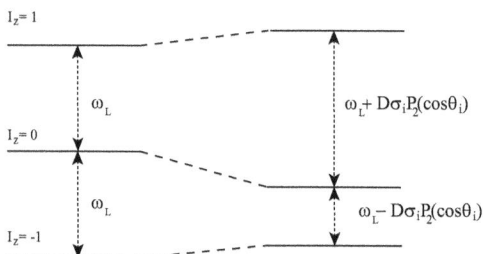

Figure 4. Nuclear Zeeman energy levels for one ortho-H_2 molecule calculated for total molecular nuclear spin I = 1 and angular momentum $J = 1$. Each molecule contributes a doublet with absorption frequency $\omega = \omega_L \pm D\sigma_i P_2(\cos\theta_i)$.

If the orientation of the magnetic field with respect to local axes of molecule i is defined by the polar angles (θ_i, ϕ_i), we have for axial symmetry $\langle (J_x^2 - J_y^2) \rangle = 0$ (which is usually assumed in the literature) [21,27],

$$\langle H_{DD}(i) \rangle = D\sigma_i P_2(\theta_i)\langle (I_z^2 - \frac{2}{3}) \rangle \tag{4}$$

where the local order parameter

$$\sigma_i = \langle (1 - \frac{3}{2}J_Z^2(i)) \rangle \quad \text{and} \quad P_2(\theta_i) = (3\cos^2\theta_i - 1)/2. \tag{5}$$

From Equation (4) and Figure 4, we see that each molecule with a non-zero order parameter σ_i contributes a doublet to the NMR absorption line shape at:

$$\omega = \omega_L \pm D\sigma_i P_2(\theta_i) \tag{6}$$

The experiments are almost always carried out for powdered samples, and we need to sum the contributions to the spectrum over a powder distribution $\mathcal{P}(\Omega)$ of the crystalline orientations $\Omega(\theta, \phi)$. If $\mathcal{I}(\Delta\omega, \sigma)$ is the line intensity at frequency $\Delta\omega$ for a fixed σ, we have $\mathcal{I}(\Delta\omega, \sigma)d(\Delta\omega) = \mathcal{P}(\Omega) = -\frac{1}{2}d(\cos\theta)$. The line shapes therefore consist of a doublet given by:

$$\mathcal{I}_\pm(\omega, \sigma) = \frac{1}{2\sqrt{3}\sigma}\sqrt{\frac{(\pm)2\omega}{D\sigma} + 1}. \tag{7}$$

These are the familiar Pake doublets and are shown in Figure 5 for $\sigma = 1$.

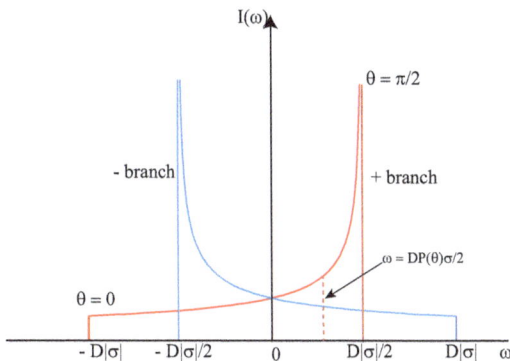

Figure 5. Calculated Pake doublet line shape for the long-range ordered Pa3 structure of solid H_2 for a fixed value of the order parameter σ and a distribution of angles α for the the axes of a powdered sample.

In order to calculate the line shape expected for a quadrupolar glass, we need to consider an approximate form for the distribution of local order parameters. If $\mathcal{P}(\sigma)$ is the probability of finding the value σ in the glass state, then we find $\mathcal{P}(\sigma)$ from the distribution of the values of local entropies. For each molecule with local order parameter σ, the contribution to the entropy is:

$$S(\sigma) = \frac{(2 - 2\sigma)}{3} ln[\frac{(2 - 2\sigma)}{3}] + \frac{(1 + 2\sigma)}{3} ln[\frac{1 + (2\sigma)}{3}]. \tag{8}$$

For small σ, $S(\sigma)$ is linear in σ, and assuming the distribution of entropies, $\mathcal{P}(s)$ is a constant, we have as a first approximation:

$$\mathcal{P}(\sigma) \propto \sigma. \tag{9}$$

Using this approximation and the expressions for the line shapes for the doublets given by Equation (7), the glass line shape is expected to be given by:

$$\mathcal{I}_{glass}(w) = \int \mathcal{P}(\sigma)[\mathcal{I}_+(w,\sigma) + \mathcal{I}_-(w,\sigma)]d\sigma. \tag{10}$$

The calculated glass line shape using Equation (10) is shown in Figure 6. This shape is to be compared with the shape observed at low temperatures. An example is shown in Figure 7 for an ortho-H_2 concentration of $x = 0.23$ and a temperature of 85 mK. The general form is in good agreement with the predictions.

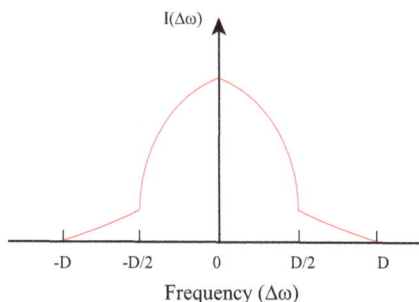

Figure 6. Calculated NMR line shape for quadrupolar glass state of solid H_2 for a broad distribution of order parameters σ. Reproduced with permission from [27]. Copyright Springer Nature, 1998.

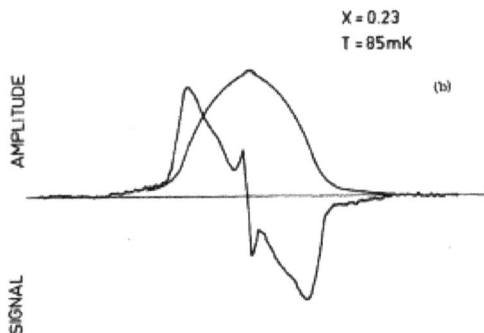

Figure 7. Observed NMR line shape for the quadrupolar glass state of solid H_2 for ortho-H_2 concentration $x = 0.23$ at a temperature of 85 mK. The experiments measured the derivative line shape using traditional lock-in methods, and the symmetrical dome shaped line is obtained by integration. Interestingly, a very small narrow peak is observed at the center that is the contribution from molecules with zero order parameter. Reproduced with permission from [21]. Copyright American Physical Society, 1978.

At lower temperatures and for high magnetic fields, the line shape shows a distinct asymmetry as shown in Figure 8. This asymmetry is due to the finite nuclear spin polarization (16% in the case of Figure 8). For non-trivial nuclear polarizations, the ratio of the amplitudes of the two branches of the Pake doublet is not unity. It is given by:

$$\frac{\mathcal{I}_-}{\mathcal{I}_+} = \exp(-\hbar\omega_L)/k_B T. \tag{11}$$

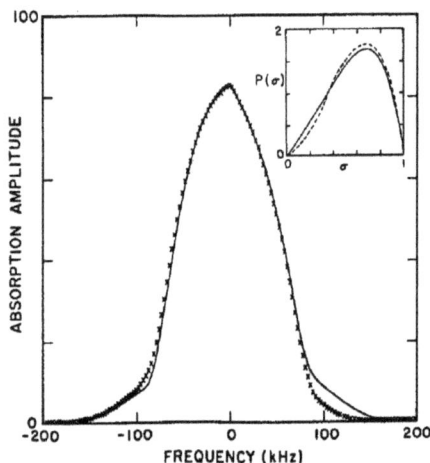

Figure 8. Observed NMR line shape for quadrupolar glass state of solid H_2 for ortho-H_2 concentration $x = 0.23$ at a temperature of 45 mK in a magnetic field of 5 T, showing asymmetry due to a finite nuclear spin polarization. Reproduced with permission from [27]. Copyright Springer Nature, 1998.

where ω_L is the nuclear Larmor frequency and T is the absolute temperature of the sample, obtained by a polynomial fit to the order parameter distribution function using $\mathcal{P}(\sigma) = \sum_n a_N \sigma^n$ to find:

$$\mathcal{I}_{fit} = \sum_n a_n G_n(\omega) \tag{12}$$

using

$$G_n(\omega) = \int d\sigma \sigma^n [\mathcal{I}_+(\sigma,\omega) + \mathcal{I}_-(\sigma,\omega)] \tag{13}$$

The order parameter distribution estimated from the best fits to the NMR line shapes is shown in Figure 9 for ortho-para H_2 samples with 23% ortho hydrogen [27]. There is a smooth variation with temperature with no discontinuity or other evidence of a sudden transition. The relevant parameter for a spin glass transition, however, is the Edwards–Anderson [40] order parameter given by:

$$q_{EA}(T) = [\langle s_i \rangle_T^2]_c \tag{14}$$

where s_i is the component of the spin at site i, $\langle\ \rangle$ is a thermodynamic average and $[\ ]_c$ is an average over configurations c. While the average $[\langle s_i \rangle_T]_c$ vanishes for paramagnetism and for a spin glass, $q_{EA}(T)$ is non-zero for the spin glass.

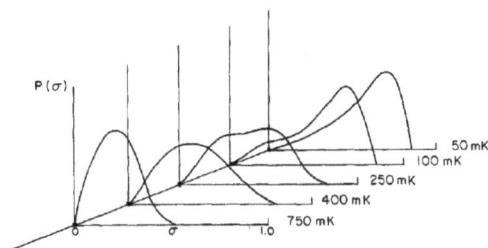

Figure 9. Temperature dependence of the probability distribution $\mathcal{P}(\sigma)$ for order parameters σ as deduced from observed NMR line shapes for solid hydrogen with ortho-H_2 concentrations of 23%. Reproduced with permission from [27]. Copyright Springer Nature, 1998.

For the quadrupolar glass, we therefore consider the glass order parameter:

$$q_{EA}^{Quad} = \int \sigma^2 \mathcal{P}(\sigma) d\sigma \tag{15}$$

The temperature dependence of the quadrupolar glass order parameter q_{EA}^{Quad} is calculated from the order parameter distribution functions $\mathcal{P}(\sigma)$ determined from the NMR line shapes shown in Figure 10 [41]. Although there is no evidence for an abrupt transition in the temperature variation of $q_{EA}^{Quad}(T)$, the observed temperature dependence is much stronger than that predicted for a non-collective simple random field model [42–44] shown by the broken dashed line of Figure 10. The observed variation is in good qualitative agreement with the model for a quadrupolar glass proposed by Lutchinskaia et al. [45] consistent with a collective quadrupolar glass ordering at low temperatures.

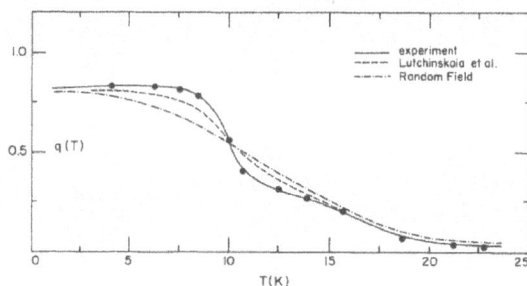

Figure 10. Temperature dependence of the quadrupolar glass order parameter $q_{EA}(T)$. Reproduced with permission from [41]. Copyright Taylor & Francis, 1986.

Classical orientational glass ordering can be observed in solid N_2-Ar mixtures using NMR techniques at low temperatures. Instead of the common isotope ^{14}N, experimenters use ^{15}N, which has nuclear spin 1/2, and the large ($\sim$3 MHz) nuclear quadrupole interaction for ^{14}N vanishes. For the diatomic molecule $^{15}N_2$, one can treat the total molecular nuclear spin as $I = 1$ and use a similar analysis of the NMR spectra as was used for diatomic H_2. Because of the larger mass and large moment of inertia compared to H_2 molecules, the angular momentum is not a good quantum number, and the local order parameters are purely classical and given by:

$$\sigma^{class} = \langle(3cos^2\theta_i - 1)/2\rangle \tag{16}$$

where θ_i is the angle between the instantaneous molecular orientation and the mean orientation. The NMR line shapes for local orientational ordering in solid N_2-Ar mixtures have one major difference compared to those observed for solid hydrogen because there is an extra term in the nuclear spin Hamiltonian that depends on the orientational order parameter. That term is the anisotropic chemical shift resulting from the interaction between the nuclear spin and the magnetic fields of the molecular electron distribution given by:

$$H_i^{acs} = KI_{iz}P_2(\theta_i). \tag{17}$$

$K = (4.0 \ 10^{-4})\omega_L$. The $^{15}N_2$ NMR line shapes in the presence of a finite orientational order parameter form doublets for a powered sample, but unlike those for solid H_2, they are antisymmetric with Equation (6) replaced by:

$$\omega = \omega_L \pm (D \pm K)\sigma_i P_2(\theta_i) \tag{18}$$

as illustrated in Figure 11. A typical line shape is shown in Figure 12 and compared with a calculated line shape for a fixed order parameter $\sigma = 0.86$ that includes a small broadening, as shown by the inset to Figure 12. The agreement is very satisfactory.

Narrow lines consistent with long-range ordering in a Pa3 structure are only observed for N_2 concentrations above 76%, where the lattice structure remains fcc (Part III of Figure 3b). In the region where the lattice remains hcp for $56\% < X_{N_2} < 77\%$ (Part II of Figure 3b), only broad NMR line shapes characteristic of quadrupolar glass-like ordering are observed. The remarkable observation is that in Region IB of the phase diagram (see Figure 3b), one once again observes a relatively narrow line shape (see Figure 13) for $x = 50\%$ [38]. Because of the presence of high substitutional disorder, this narrow line shape is believed to be a true orientational glass (analogous to that depicted in Figure 1a) with a broad distribution of alignment angles, but a relatively narrow distribution of order parameters about those angles and with a mean $\langle\sigma\rangle = 0.74$. On further dilution of X_{N_2} to 40%, only very broad NMR line shapes are observed in Region IA of Figure 3b.

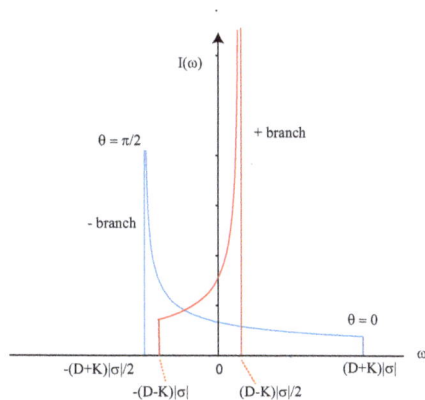

Figure 11. Calculated NMR absorption line shape for long-range Pa3 ordering for pure solid $^{15}N_2$. The line consists of two asymmetric Pake doublets.

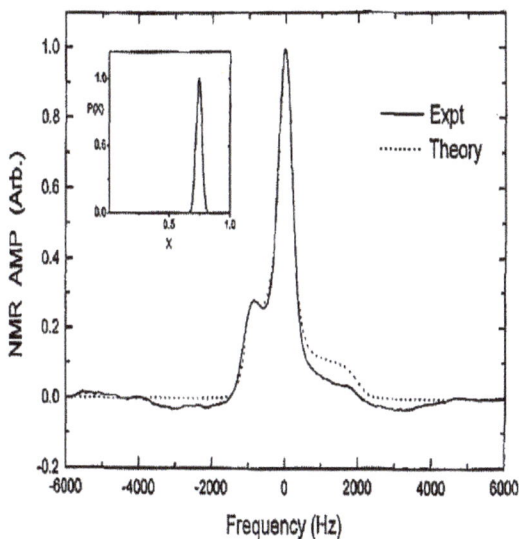

Figure 12. Comparison of observed NMR absorption line shape at 4.2 K with the calculated NMR absorption line shape for long-range Pa3 ordering for pure solid $^{15}N_2$. The inset shows the order parameter distribution $\mathcal{P}(\sigma)$ that was used. Reproduced with permission from [38]. Copyright Springer Nature, 1998.

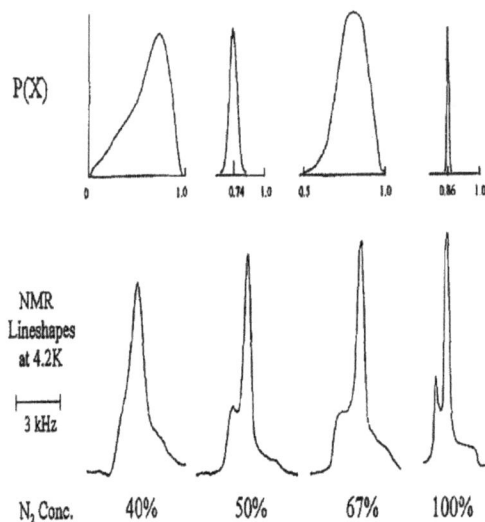

Figure 13. Variation of observed low temperature NMR line shapes with nitrogen concentration in solid N_2-Ar mixtures, showing inferred distributions of order parameters $\mathcal{P}(\sigma)$. Reproduced with permission from [38]. Copyright Springer Nature, 1998.

While the NMR line shapes of solid ortho-para hydrogen mixtures and N_2-Ar mixtures are consistent with quadrupolar and orientational glass states at low temperature and the growth of the quadrupolar order parameter as the temperature is lowered suggests that the local ordering is collective in nature and not driven by fixed local fields, it is important that the dynamics be investigated to determine whether the characteristic rapid freezing of the orientational fluctuations occurs at the onset of glassiness. This is discussed in the next section.

4. Dynamics

NMR methods can be used to investigate the dynamics of molecular fluctuations and in particular the orientational or quadrupolar fluctuations over a wide frequency range [25,26,37]. The sensitivity of NMR methods to motions arises because fluctuations of the molecular alignments or local order parameters render the intra-molecular nuclear dipole-dipole interactions time dependent [46]. Specifically, the nuclear spin-lattice relaxation time, T_{1D}^{-1}, due to the modulation of the dipole-dipole interactions, is given by:

$$T_{1D}^{-1} = \frac{1}{2}D^2 \sum_{|m|=1,2} m^2 J_{2m}(m\omega_0). \tag{19}$$

D measures the strength of the intra-molecular dipole-dipole interaction, and ω_0 is the nuclear Larmor frequency. $J(\omega)$ is the spectral weight of the fluctuations at frequency ω.

$$J_{2m}(\omega) = \int_{-\infty}^{\infty} \langle T_{2m}(t) T_{2m}^{\dagger}(0) \rangle_T \exp(-i\omega t)dt. \tag{20}$$

The T_{lm} are the tensorial operators defined in Equation (2).

The expectation value $\langle T_{2m}(t) T_{2m}(0) \rangle_T$ must be calculated with a quantization axis parallel to the magnetic field. The order parameters, however, are calculated with respect to the local molecular symmetry axis. In order to carryout this evaluation, we therefore consider the rotation:

$$T_{2m} = \sum_{\mu} d^2_{m\mu}(\alpha, \beta) T_{2\mu}. \tag{21}$$

The $d_{m\mu}$ are the rotation matrix elements, and the angles (α, β) are polar angles defining the orientation of the magnetic field in a local reference frame aligned along the molecular equilibrium orientation. One then assumes [47] that the time dependence can be separated from the order parameter variations as:

$$\langle T_{2m}(t) T^{\dagger}_{2m}(0) \rangle = \langle T_{2m}(0) T^{\dagger}_{2m}(0) \rangle_T g_{2m}(t). \tag{22}$$

The original data shown in Figure 14 [48] were interpreted in terms a single correlation time τ_Q. The result, a strong almost exponential temperature dependence, resembles, but is not identical to a Fulcher–Vogel temperature dependence. The data in Figure 14 along with the results observed for the very slow dynamics determined by NMR stimulated echo studies [49] are among the most convincing experimental evidence for traditional glassy dynamics for the quadrupolar glass state of solid ortho-para hydrogen mixtures.

Further studies, however, showed that the nuclear spin relaxation was not a simple exponential, but rather a sum of different exponentials associated with different parts of the NMR line shape, as illustrated in Figure 15.

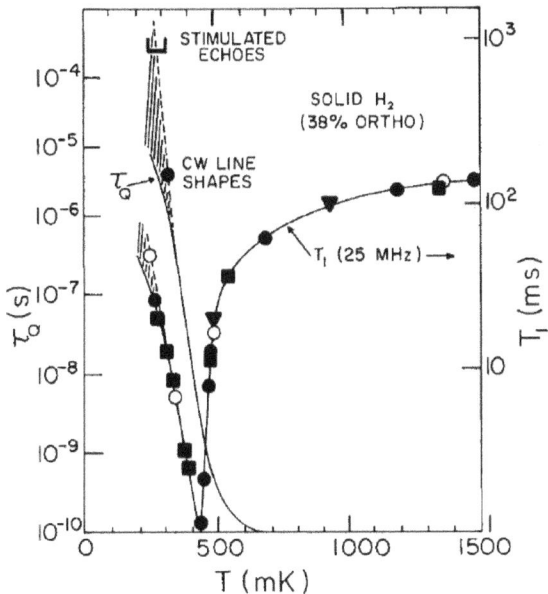

Figure 14. Temperature dependence of the fluctuation rate in quadrupolar glass. Reproduced with permission from Fig. 1 of Ref. [48]. Copyright Elsevier, 1981.

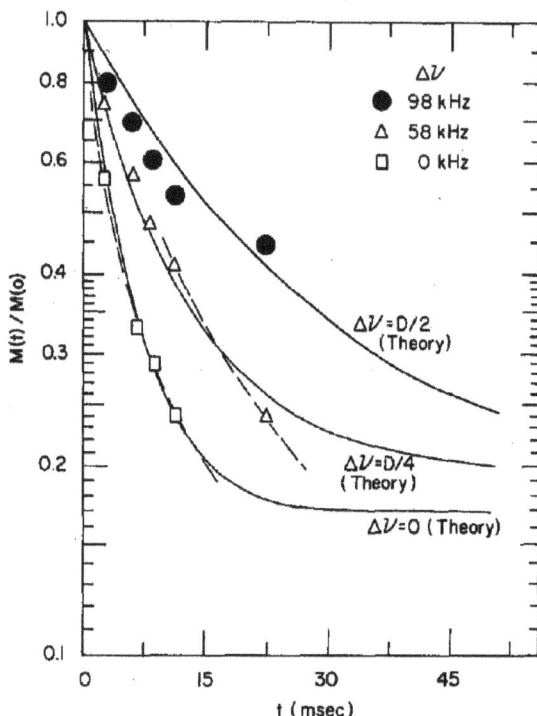

Figure 15. Observed nuclear spin-lattice relaxation in the form of multi-exponentials in the glass state of solid ortho-para hydrogen mixtures. Reproduced with permission from [47]. Copyright Springer Nature, 1986.

This spectral inhomogeneity of the nuclear spin-lattice relaxation can be expected from Equation (22). If one assumes that the relaxation is dominated by the fluctuations of the axial order parameter, σ, then Equation (22) leads to:

$$T_{1D}^{-1} = \frac{1}{12}(2 - \sigma - \sigma^2)D^2 \sum_{m=1,2} m^2 |d_{m0}(\alpha)|^2 g_{20}(m\omega_0) \tag{23}$$

for the contribution to the relaxation by the dipole terms. The factor $(2 - \sigma - \sigma^2)$ is the mean square variation of the operator T_{20}. The angular terms are associated by different isochromats of the NMR line shape, and if one assumes Lorentzian forms with correlation time τ_Q for the spectral densities $g(m\omega_0)$, we have [47]:

$$T_{1D}^{-1} = \frac{1}{12}(2 - \sigma - \sigma^2)D^2 \langle 1 - (\frac{2\Delta\nu}{D\sigma})^2 \rangle / (\omega_0^2 \tau_Q) \tag{24}$$

$\Delta\nu$ is the frequency of an isochromat with respect to the center of the NMR absorption line. The spin-rotational (SR) interaction also contributes to the relaxation rate:

$$T_{1SR}^{-1} = \frac{1}{18}C^2(2 - |\sigma|)(1 \pm \frac{2\Delta\nu}{D|\sigma|})/(\omega_0^2 \tau_Q) \tag{25}$$

where $C = 100$ kHz for a hydrogen molecule. These two contributions to the relaxation rate from the intra-molecular dipole-dipole and spin-rotation interactions are additive, and the overall relaxation given by $T_{1(calc)}^{-1} = T_{1D}^{-1} + T_{1SR}^{-1}$ is shown in Figure 16. The agreement with experiment is satisfactory.

Yu et al. [50] also noted that cross-relaxation, T_{12}^{-1}, between different isochromats will tend to smooth out the inhomogeneities in the relaxation rates, and the observed relaxation will be determined by $T_1^{-1} = T_{1(calc)}^{-1} + T_{12}^{-1}$. The resulting variation of the relaxation as a function of $\Delta \nu$ is shown in Figure 16.

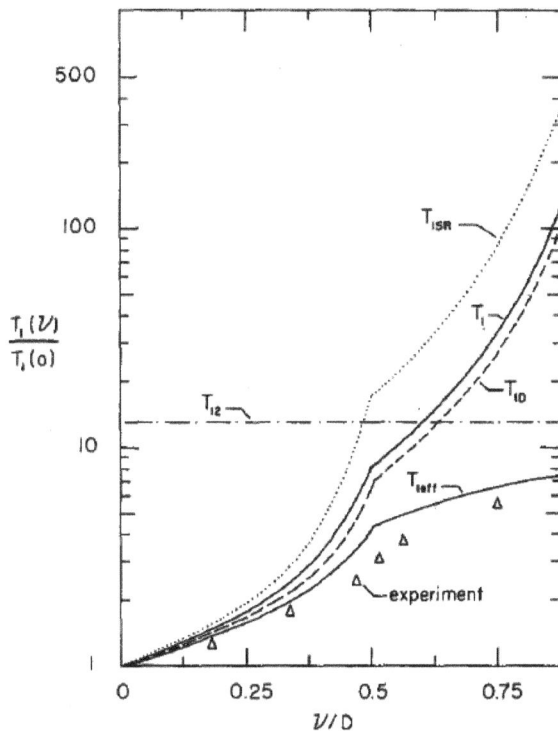

Figure 16. Variation of the nuclear spin-lattice relaxation with frequency for a glass ordering in solid ortho-para hydrogen mixtures. Reproduced with permission from [47]. Copyright Springer Nature, 1986.

5. Thermodynamic Measurements

Measurements of the heat capacities of solid ortho-para hydrogen [51,52] and N_2-Ar mixtures [53,54] showed rather featureless temperature variations and were at first sight disappointing, but studies of the very slow dynamics of solid hydrogen with dilute ortho-H_2 concentrations were able to provide a uniform picture for the very slow motions and the thermodynamic behavior. The low frequency dynamics was investigated using NMR stimulated echo techniques employing three RF pulses [49]. After a first pulse, the nuclear magnetization is rotated to a transverse axis (x-axis), where it evolves under the intra-molecular dipole interactions, which depend on the local order parameter. After a time τ, a second pulse stores the transverse magnetization along the original z-axis. The value stored depends on the original value of the order parameter. After a waiting time t_w, a third pulse returns the stored magnetization back to the x-axis, and after a further time τ, a stimulated echo is formed, provided that the order parameter varies only very slowly in the waiting time t_w. A variety of different pulse sequences can be considered but the simplest $(\pi/2 - \tau - \pi/2 - t_w - \pi/2)$ about the same axes produces an echo signal:

$$S_{2\tau + t_w} = -\frac{\hbar \omega_0}{k_B T} \langle \langle \cos[D\sigma_i(0)\tau] \cos[D\sigma_i(t_w)\tau] \rangle \rangle. \tag{26}$$

$\langle\langle.....\rangle\rangle$ designates a configurational average over sites i. The dependence on t_w allows one to explore slow temporal fluctuations of the molecular order parameters over time τ. The results (Figure 17) were surprising, as they showed a unique logarithmic decay.

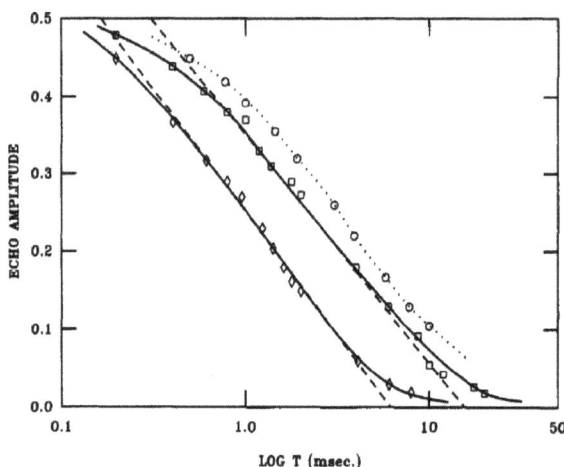

Figure 17. Decay of NMR stimulated echos for the glass state of ortho-para hydrogen mixtures. Reproduced with permission from [39]. Copyright Elsevier, 1969.

The logarithmic decay can be understood in terms of the droplet model proposed by Fisher and Huse [55,56]. Fisher and Huse considered the density of low energy states for connected clusters and argued that because the clusters are rare, they could be treated as "non-interacting two level systems". For a free energy barrier E_B, the tunneling rate at temperature T is:

$$\Gamma(E_B) = \Gamma_0 \exp(-E_b/k_b T) \tag{27}$$

where Γ_0 is an attempt frequency. For long times, Γ_0 is well defined because it is associated with a characteristic cluster size. The relaxation is hierarchical as low-energy barriers must be overcome before larger barriers can be crossed. The dynamics of of the faster degrees of freedom therefore constrain the slow motions. As a result, in time t, the barriers overcome will be those with energy less than $E_{max}(t)$ where:

$$E_{max}(t) = k_B T ln(\Gamma_0 t). \tag{28}$$

For all energy barriers traversed in time t, the order parameters change significantly, and the correlation functions $\langle\langle cos[D\sigma_i(0)\tau]cos[D\sigma_i(t)\tau]\rangle\rangle$ will vanish after the barriers are crossed. If the density of low energy excitations is $P(E_B)$, the stimulated echo amplitude will be given by:

$$S(t) = (\frac{\hbar\omega_0}{2k_B T})[1 - \int_0^{E_{max}} P(E_B)dE_B]. \tag{29}$$

At low temperatures, we only need consider the density of states near $E \approx 0$, and the echo amplitude becomes:

$$S(t) = (\frac{\hbar\omega_0}{2k_B T})[1 - k_B T P(0)ln(t/t_0)]. \tag{30}$$

t_0 is determined by the attempt frequency Γ_0. Since the order parameter changes by the order of unity in time Γ_0^{-1}:

$$t_0 \sim \pi/(D\tau\Gamma_0). \tag{31}$$

The decay of the stimulated echos is therefore directly related to the density of states at low energy, $P(0)$, which is expected to be constant and can be related to the thermodynamic measurements. From Figure 17, we find $P(0) = 0.59 \pm 0.05 \ K^{-1}$.

Knowledge of the density of energy levels $P(E)$ allows one to calculate the heat capacity and test the interpretation of the dynamics against the experimental results for heat capacity studies [51–54]. Using the energy scheme of Figure 18, the heat capacity is calculated from the sum over the distribution of energy states $S(\Delta)$ as:

$$\frac{C_V}{NXR} = \frac{2}{3}k_B T \int_0^\infty \frac{z^2 dz}{4e^{-z} + 4 + e^z} S(\Delta) \tag{32}$$

with $z = 3\Delta/k_B T$. X is the ortho-H_2 concentration. The result of the calculation shown in Figure 19 gives a good description of the characteristic linear temperature dependence at low temperatures, and an excellent fit is obtained using small deviations from a constant density of energy states shown by the solid line of Figure 19.

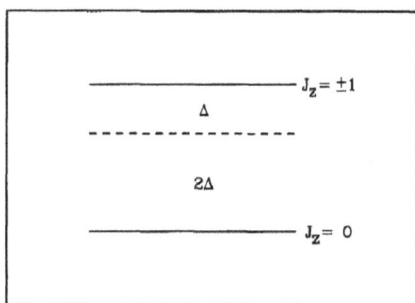

Figure 18. Sketch of the energy levels of an individual ortho-H_2 molecule with angular momentum $J = 1$ with the degenerate states $J_z = \pm 1$ separated from the state $J_Z = 0$ by an energy gap 3Δ. Reproduced with permission from [39]. Copyright Elsevier, 1969.

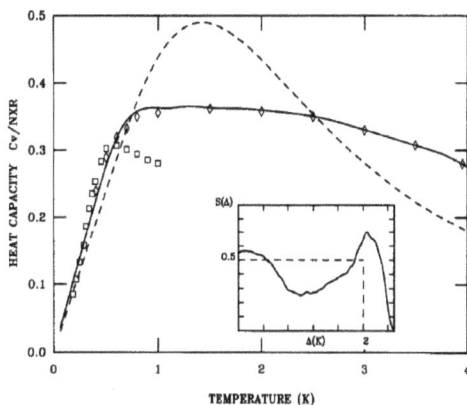

Figure 19. Heat capacity of the glass state of solid ortho-para hydrogen mixtures calculated from density of energy levels $S(\Delta)$. The diamonds represent the data of Ricketson [51] and the squares that of Haase et al. [54]. Reproduced with permission from [39]. Copyright Elsevier, 1969.

6. Dielectric Measurements

While the analysis of the NMR line shapes of solid ortho-para H_2 and solid N_2-Ar mixtures with high disorder (that is, low concentrations of the active quadrupole bearing molecule (ortho-H_2 or N_2)) show clear evidence of the formation of glassy states at low temperatures and the temperature dependence of the dynamics cannot be understood except in terms of local collective ordering, the characteristic hysteresis of the susceptibility (the parameter conjugate to the local order parameters) remained elusive until the studies of Pilla et al. [57,58] using electric susceptibility measurements. These measurements were carried out on dilute solid N_2-Ar mixtures for concentrations where the lattice remained cubic. In this case, the concern that there exist local electric field gradients due to the introduction of substitutional disorder is greatly lessened, and the system is much closer to the class of the familiar spin glasses.

The use of a high sensitivity capacitance bridge [59] to measure dielectric susceptibilities ϵ showed distinct hystereses in the observed values of the dielectric susceptibility on thermal cycling [57]. The break in the observed values of ϵ at 9.8 ± 0.1 K (Figure 20) clearly marks a well-defined dynamical transition for solid N_2-Ar mixtures with 51% N_2 that was absent in the NMR measurements.

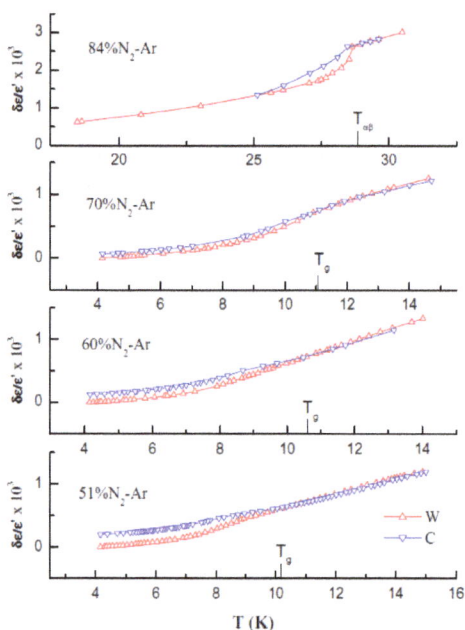

Figure 20. Thermal cycling of dielectric susceptibility measurements of solid N_2-Ar mixtures by Pilla et al. [57]. Reproduced with permission from [57]. Copyright Institute of Physics, 2001.

The really significant consequence of the dielectric susceptibility measurements for the molecular glasses is that they can be used to test the applicability of replica symmetry breaking theories to this class of glass states [14,60]. For simple replica symmetry breaking, Cugliandolo and Kurchan [18] developed a generalized fluctuation dissipation theorem for the generalized susceptibility:

$$C(t, t_w) = \langle q_i(t_w) q_i(t + t_w) \rangle \tag{33}$$

where q_i is a generalized coordinate. The essential result is that below the characteristic glass freezing temperature of the dynamics, T_{dyn}, the change in the response function depends on the waiting time t_w

because of the changes in the landscape of free energy barriers that occur in the time t_w. The response function is defined by:

$$R_i(t, t_w) = \frac{1}{k_B T} \int \chi(C) dC \tag{34}$$

where $\chi(C)$ is the relevant susceptibility. Pilla et al. [23] showed that as a consequence of this generalized fluctuation dissipation theorem, $\chi \propto T/T_{dyn}$ for large excursions compared to $\chi \sim 1$ (the classical result) for small δT. The results are shown in Figure 21.

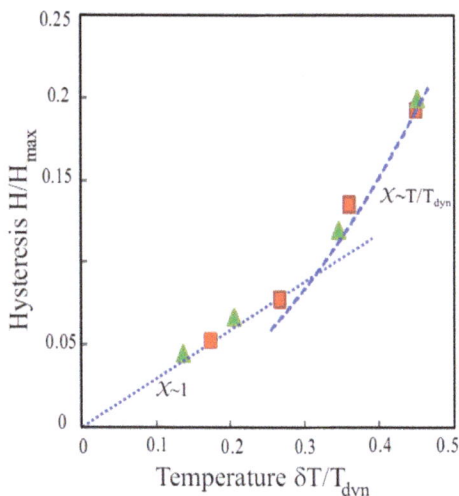

Figure 21. Comparison of the hysteresis observed for the frustrated sold N_2-Ar mixtures and solid O_2 as a function of the temperature swing of the hysteresis. H_{max} is the maximum variation of the hysteresis. Reproduced with permission from [23]. Copyright Institute of Physics, 2009.

7. Discussion

The experimental results for the NMR studies of the line shapes of solid ortho-para H_2 and solid N_2-Ar mixtures using continuous wave techniques show that the quadrupole-bearing molecules in these mixtures have a broad distribution $P(\sigma)$ of local order parameters σ at low temperatures. These parameters consist of two kinds: (i) quadrupolar, referring to the deviation from spherical symmetry of the average molecular alignments; and (ii) orientational, referring only to the orientation of the average alignment of the molecular axes. Ortho-para H_2 mixtures for ortho concentration less than 55% and N_2-Ar mixtures for $56\% < X(N_2) < 77\%$ form quadrupolar glasses with both the quadrupolar and orientational parameters broadly distributed. Solid N_2-Ar mixtures for $40\% < X(N_2) < 56\%$ form orientational glasses for which the local quadrupolar parameters are constant (but not necessarily zero), and only the orientational axes vary. The equivalent of the Edwards–Anderson order parameter, $q(T) = \langle\langle \sigma_i^2 \rangle\rangle$, shows a rapid variation with temperature (stronger than non-collective para-orientational models would predict), but with no sharp transition for studies down to the lowest temperatures (35 mK for solid ortho-para H_2 mixtures and 1.5 K for solid N_2-Ar mixtures).

The dynamics associated with the freezing of the molecular orientational and quadrupolar fluctuations in the quadrupolar glass states has been determined using pulsed NMR techniques, either using standard $\frac{\pi}{2} - \tau - \frac{\pi}{2}$ pulse sequences or $\frac{\pi}{2} - \tau - \frac{\pi}{2} - t_{wait} - \frac{\pi}{2}$ stimulated echo sequences. The latter is particularly useful for providing information about the very slow dynamics in the glass states. If the relaxation is interpreted in terms of a characteristic fluctuation time τ_Q (which does give a good description of the data), τ_Q increases rapidly with decreasing temperature below a temperature T_{dyn} characteristic of a given ortho-H_2 or N_2 concentration. In addition, the unique logarithmic decay of the stimulated echoes provides a measure of the density of energy states $P(E)$ near $E = 0$. This density of states can be used to calculate the heat capacity C_V as a function of temperature. Good agreement is found especially for the magnitude of the characteristic linear dependence on temperature for $T < T_{dyn}$.

Finally, studies of the dielectric susceptibility that depends on the orientational parameters for the molecules have been made at very low audio frequencies, and the experiments show unmistakable hystereses on thermal cycling. Pilla and colleagues [57] have been able to show that this is consistent with the replica symmetry breaking models of Parisi [13], and these results put these disordered quadrupolar molecular solids in the same class as other frustrated systems that exhibit spin-glass-like behavior at low temperatures. Further studies of the dynamics for very low quadrupolar concentrations in solid N_2-Ar mixtures as one varies the concentration through the lattice percolation concentration would provide a valuable test of the underlying physics since the belief is that it is the frustration of the short-range interactions that lie at the origin of the formation of these glass states.

8. Conclusions

NMR studies of the line shapes of solid ortho-para-H_2 and solid N_2-Ar mixtures showed that at low temperatures the quadrupole bearing molecules (ortho-H_2 and N_2) were characterized by a broad range of local order parameters. This characterization lead to the interpretation of the low temperature behavior in terms of a quadrupolar glass analogous to a spin glass. The transition to the glass state as a function of temeprature was, however, smooth with no sudden phase transition. The temperature evolution of the Edwards-Anderson order parameter was nevertheless much stronger than that predicted for non-collective effects. Measurements of the nuclear spin relaxation times provided more insight into the local dynamics and implied a rapid freezing of the dynamical degrees of freedom. A deeper understanding of the underlying physics was provided by analysis of the decay of stimulated NMR echoes designed to explore ultra-slow motions that characterize the thermal behavior of spin glass materials as the system is governed by the crossing of energy barriers in a complex configuration landscape. This analysis provided an unambiguous measure of the density of low energy excitations and lead to a quantitative fit for the results of heat capacity measurements. Finally, the temperature of the quadrupolar susceptibility deduced from measurements of the dielectric susceptibility revealed hysteresis effects that are consistent with the replica symmetry breaking models of generalized spin glass theories.

Acknowledgments: This research has been supported by the National Science Foundation's Division of Materials Science through DMR-1303599 and DMR-1157490).

Author Contributions: N.S., J.H., S.P. and E.G. conceived of and designed the experiments. J.H., S.P. and E.G. performed the experiments.; N.S, J.H., K.M., S.P. and E.G. analyzed the data. N.S. and J. H. wrote the paper.

References

1. Pohl, R.O.; Salinger, G.L. The anomalous thermal properties of glasses at low temperatures. *Ann. N. Y. Acad. Sci.* **1976**, *279*, 150–172.
2. Anderson, P.W.; Halperin, B.I.; Varma, C.M. Anomalous low-temperature thermal properties of glasses and spin glasses. *Philos. Mag.* **1972**, *25*, 1–9.
3. Phillips, J.C. Stretched exponential relaxation in molecular and electronic glasses. *Rep. Prog. Phys.* **1996**, *59*, 1133.
4. Kokshenev, B.B.; Borges, P.D.; Sullivan, N.S. Moderately and strongly supercooled liquids: A temperature-derivative study of the primary relaxaton time scale. *J. Chem. Phys.* **2005**, *122*, 114510.

5. Mézard, M.; Parisi, G.; Virasoro, M.A. SK Model: The Replica Solution without Replicas. *EPL (Europhys. Lett.)* **1986**, *1*, 77.
6. Binder, K.; Kolb, W. *Glassy Materials and Disordered Solids; An Introduction to Their Statistical Mechanics*; World Scientific Publishing: Singapore, 2005.
7. Cannella, V.; Mydosh, J.A. Magnetic Ordering in Gold-Iron Alloys. *Phys. Rev. B* **1972**, *6*, 4220–4237.
8. Rivier, N.; Adkins, K. Resistivity of spin glasses. *J. Phys. F Met. Phys.* **1975**, *5*, 1745.
9. Edwards, S.F.; Anderson, P.W. Theory of spin glasses. II. *J. Phys. F Met. Phys.* **1976**, *6*, 1927.
10. Sherrington, D. A transparent theory of the spin glass. *J. Phys. C Solid State Phys.* **1975**, *8*, L208.
11. Murani, A.P. High-Temperature Spin Dynamics of Cu-Mn Spin-Glasses. *Phys. Rev. Lett.* **1978**, *41*, 1406–1409.
12. Binder, K.; Young, A.P. Spin glasses: Experimental facts, theoretical concepts, and open questions. *Rev. Mod. Phys.* **1986**, *58*, 801–976.
13. Parisi, G. Testing replica predictions in experiments. *Il Nuovo Cimento D* **1998**, *20*, 1221–1228.
14. Mézard, M.; Parisi, G.; Sourlas, N.; Toulouse, G.; Virasoro, M. Nature of the Spin-Glass Phase. *Phys. Rev. Lett.* **1984**, *52*, 1156–1159.
15. Holdsworth, P.C.W.; Gingras, M.J.P.; Bergersen, B.; Chan, E.P. Random bonds and random fields in two-dimensional orientational glasses. *J. Phys. Condens. Matter* **1991**, *3*, 6679.
16. Dzyaloshinsk, I.; Volovik, G. On the concept of local invariance in the theory of spin glasses. *J. Phys.* **1978**, *39*, 693–700.
17. Chowdhury, D. *Spin Glasses and Other Frustrated Systems*; Princeton University Press: Princeton, NJ, USA, 1987.
18. Cugliandolo, L.F.; Kurchan, J. Analytical solution of the off-equilibrium dynamics of a long-range spin-glass model. *Phys. Rev. Lett.* **1993**, *71*, 173–176.
19. Cugliandolo, L.F.; Kurchan, J. Mean-field theory of temperature cycling experiments in spin glasses. *Phys. Rev. B* **1999**, *60*, 922–930.
20. Parisi, G. Spin glasses and fragile glasses: Statics, dynamics and complexity. *Proc. Nat. Acad. Sci. USA* **2006**, *103*, 7948–7955.
21. Sullivan, N.S.; Devoret, M.; Cowan, B.P.; Urbina, C. Evidence for quadrupolar glass phases in solid hydrogen at reduced ortho concentrations. *Phys. Rev. B* **1978**, *17*, 5016–5024.
22. Treusch, J. (Ed.) Theory of spin glasses: A brief review. In *Festkörperprobleme 17*; Plenary Lectures of the Divisions "Semiconductor Physics" "Metal Physics" "Low Temperature Physics" "Thermodynamics and Statistical Physics" "Crystallography" "Magnetism" "Surface Physics" of the German Physical Society Münster, 7–12 March 1977; Springer: Berlin/Heidelberg, Germany, 1977; pp. 55–84.
23. Pilla, S.; Hamida, J.A.; Muttalib, K.A.; Sullivan, N.S. Generalized fluctuation-dissipation and thermal hysteresis of dielectric susceptibility in frustrated molecular orientational glasses. *J. Phys. Conf. Ser.* **2009**, *50*, 042192.
24. Sullivan, N.S.; Hamida, J.A.; Pilla, S.; Muttalib, K.A.; Genio, E. Molecular glasses: NMR and dielectric susceptibility measurements. *J. Struct. Chem.* **2016**, *57*, 301–307.
25. Ishimoto, H.; Nagamine, K.; Kimura, Y.; Kumagai, H. New Anomalous Phenomena in NMR Absorption and Spin Lattice Relaxation of Solid Hydrogen. *J. Phys. Soc. Jpn.* **1976**, *40*, 312–318, doi:10.1143/JPSJ.40.312.
26. Cochran, W.T.; Gaines, J.R.; McCall, R.P.; Sokol, P.E.; Patton, B.R. Study of the Quadrupolar-Glass Region in Solid D_2 via Proton Magnetic Resonance. *Phys. Rev. Lett.* **1980**, *45*, 1576–1580.
27. Edwards, C.M.; Zhou, D.; Lin, Y.; Sullivan, N.S. Local ordering in dilute ortho-para-hydrogen mixtures at low temperatures. *J. Low Temp. Phys.* **1988**, *72*, 1–24.
28. Hamida, J.A.; Sullivan, N.S.; Evans, M.D. Orientational Ordering of Frustrated Molecular Quadrupoles: NMR Studies of N_2-Ar Solid Mixtures. *Phys. Rev. Lett.* **1994**, *73*, 2720–2723.
29. Ward, L.G.; Saleh, A.M.; Haase, D.G. Specific heat of solid nitrogen-argon mixtures: 50 to 100 mol N_2. *Phys. Rev. B* **1983**, *27*, 1832–1838.
30. Silvera, I.F. The solid molecular hydrogens in the condensed phase: Fundamentals and static properties. *Rev. Mod. Phys.* **1980**, *52*, 393–452.
31. Sullivan, N.S. Orientational order-disorder transitions in solid hydrogen. *Can. J. Chem.* **1988**, *66*, 908–914.
32. Hamida, J.A.; Genio, E.B.; Sullivan, N.S. NMR studies of the orientational ordering in dilute solid N_2-Ar mixtures. *J. Low Temp. Phys.* **1996**, *103*, 49–70.
33. Scott, T.A. Solid and Liquid Nitrogen. *Phys. Rep.* **1976**, *27*, 89–157.

34. Ishol, L.; Scott, T. Anisotropy of the chemical shift tensor for solid nitrogen. *J. Magn. Reson. (1969)* **1977**, *27*, 23–28.

35. Barrett, C.S.; Meyer, L. Argon—Nitrogen Phase Diagram. *J. Chem. Phys.* **1965**, *42*, 107–112, doi:10.1063/1.1695654.

36. Sullivan, N.S. Orientational ordering in solid hydrogen. *J. Phys. Fr.* **1976**, *37*, 981–989.

37. Estève, D.; Sullivan, N.S.; Devoret, M. Orientational ordering in a dilute system of classical interacting quadrupoles : (N_2-Ar) solid mixtures. *J. Phys. Lett.* **1982**, *43*, 793–799.

38. Hamida, J.A.; Pilla, S.; Sullivan, N.S. The Orientational Ordering in Dilute Solid N_2–Ar Mixtures. *J. Low Temp. Phys.* **1998**, *111*, 365–370.

39. Lin, Y.; Sullivan, N. Long-time relaxation behavior in orientational glasses. *J. Mag. Reson. (1969)* **1990**, *86*, 319–337.

40. Edwards, S.F.; Anderson, P.W. Theory of spin glasses. *J. Phys. F Met. Phys.* **1975**, *5*, 965.

41. Sullivan, N.S.; Edwards, C.M.; Brookeman, J.R. Orientational Ordering in Solid N_2-Argon Mixtures: Collective Versus Non-Collective Behavior. *Mol. Cryst. Liq. Cryst.* **1986**, *139*, 365–375,

42. Kokshenev, V.B. On the quadrupole galss phase in solid hydrogen. *Solid State Commun.* **1982**, *44*, 1593–1595.

43. Kokshenev, V. On the theory of quadrupole glass. Model Hamiltonian of the classical analogue of ortho-para mixtures of hydrogen. *Solid State Commun.* **1985**, *55*, 143–146.

44. Kokshenev, V.B. Microscopic approach to the quadrupolar-glass problem. *Phys. Rev. B* **1996**, *53*, 2191–2194.

45. Lutchinskaia, E.A.; Ryzhov, V.N.; Tareyeva, E.E. Solvable model of a quadrupolar glass. *J. Phys. C Solid State Phys.* **1984**, *17*, L665.

46. Sullivan, N.S.; Devoret, M.; Estève, D. Correlation functions in the quadrupolar glass phase of solid hydrogen. *Phys. Rev. B* **1984**, *30*, 4935–4945.

47. Lin, Y.; Sullivan, N.S. Spectral inhomogeneity of nuclear spin-lattice relaxation in orientationally ordered solid hydrogen. *J. Low Temp. Phys.* **1986**, *65*, 1–11.

48. Sullivan, N.S.; Esteve, D. Critical slowing-down in spin glasses: Quadrupoalr glass phase of solid H_2. *Physica* **1981**, *107B*, 189–190.

49. Sullivan, N.; Esteve, D.; Devoret, M. NMR Pulse Studies of Molecular Solids, $^{15}N_2$ and H_2; II. Stimulated Echoes and Slow Rotational Motions. *J. Phys. C (Solid State)* **1982**, *15*, 4895–4911.

50. Yu, I.; Washburn, S.; Calkins, M.; Meyer, H. Pulsed NMR studies in solid H_2. II. Stimulated echoes and cross relaxation times. *J. Low Temp. Phys.* **1983**, *51*, 401–422.

51. Ricketson, B.W.A. Specific Heats and Entropies at Low Temperatures. unpublished Thesis, Oxford University, Clarendon, Oxford, UK, 1956,

52. Jarvis, J.F.; Meyer, H.; Ramm, D. Measurement of $\left(\frac{\partial P}{\partial T}\right)_V$ and Related Properties in Solidified Gases. II. Solid H_2. *Phys. Rev.* **1969**, *178*, 1461–1471.

53. Haase, D.G.; Saleh, A. Measurement of (dP/dT)v in solid hydrogen. *Physica B+C* **1981**, *107*, 191–192.

54. Haase, D.G.; Perrell, L.R.; Saleh, A.M. Specific heat, pressure, and the Grüneisen relation in solid hydrogen. *J. Low Temp. Phys.* **1984**, *55*, 283–296.

55. Fisher, D.S.; Huse, D.A. Ordered Phase of Short-Range Ising Spin-Glasses. *Phys. Rev. Lett.* **1986**, *56*, 1601–1604.

56. Fisher, D.S.; Huse, D.A. Nonequilibrium dynamics of spin glasses. *Phys. Rev. B* **1988**, *38*, 373–385.

57. Pilla, S.; Hamida, J.A.; Mutalib, K.A.; Sullivan, N.S. Molecular solid glasses: New insights into frustrated systems. *New J. Phys.* **2001**, *3*, 17–30.

58. Pilla, S.; Hamida, J.; Muttalib, K.; Sullivan, N. Frustration-induced glass behavior in solid N_2? Audio frequency dielectric measurements. *Physica B* **2000**, *284–288*, 1125–1126.

59. Pilla, S.; Hamida, J.A.; Sullivan, N.S. Very high sensitivity capacitance bridge for dielectric studies at low temperatures. *Rev. Sci. Instrum.* **1999**, *70*, 4055–4058.

60. Mydosh, J.A. Disordered magnetism and spin glasses. *J. Magn. Magn. Mater.* **1996**, *157–158*, 606–610.

magnetochemistry

MDPI

Article

Spatially Resolved Measurements of Crosslinking in UV-Curable Coatings Using Single-Sided NMR

Madeline Brass [1], Frances Morin [2] and Tyler Meldrum [1,*]

[1] Department of Chemistry, The College of William & Mary, Williamsburg, VA 23185, USA; mrbrass@email.wm.edu
[2] Department of Chemistry, University of Pittsburgh, Pittsburgh, PA 15260, USA; fjm22@pitt.edu
* Correspondence: tkmeldrum@wm.edu; Tel.: +1-757-221-2561

Received: 21 November 2017; Accepted: 22 December 2017; Published: 9 January 2018

Abstract: The UV-driven photocuring of coatings results in a crosslinked polymeric network. The degree of crosslinking in these coatings is typically assessed via optical spectroscopy; unfortunately, optical methods are typically limited in their maximum depth access. Alternatively, single-sided nuclear magnetic resonance (NMR) can be used to probe the crosslinking of UV-curable coatings in a spatially sensitive manner. Relaxation measurements, which correlate with crosslinking, can be done with a spatial resolution on the order of microns throughout the depth dimension of the coating, regardless of optical transparency of the material. These results can be visualized via a relaxation cross-section that shows the depth at which a particular relaxation value is observed. These measurements are used to probe the effect of a scavenger molecule that is added to the coating mixture, allowing for efficient crosslinking despite the presence of atmospheric oxygen. This method may find purchase in evaluating systems whose crosslinking properties are intentionally varied throughout its thickness; using NMR, these systems, up to approximately one hundred microns thick, can be measured without repositioning or rastering.

Keywords: NMR; coatings; single-sided; crosslinking; photocuring

---

1. Introduction

Coatings that are cured using ultraviolet light ("UV-curables") find use in a wide range of applications as they can be rapidly cured without the evolution of unpleasant or noxious gases. UV-curables contain a certain amount of photoinitiator that strongly absorbs in the UV range. When this photoinitiator absorbs UV light, radicals are formed that drive the formation of crosslinks between monomers to form a polymeric coating [1]. Greater UV absorbance, either from more efficient photoinitiators or by more intense UV light, results in more radicals that, in turn, promote curing of the coating [2,3]. One aspect of these reactions must frequently be addressed: atmospheric oxygen can interfere with radical-driven crosslinking reactions at the coating surface. In our experiments, oxygen reacts rapidly with radicals produced during photoinitiation to produce peroxyl radicals that are slow to react with double bonds in monomers, thus retarding the rate of polymerization [4]. Sometimes curing can be performed in an oxygen-free atmosphere, minimizing this interference. Alternatively, when atmospheric regulation is impractical, antioxidant scavengers can be added to the coating mixture to mitigate the effects of (atmospheric) oxygen radicals, allowing the main crosslinking reaction to proceed [5]. One efficient scavenger is triphenylphosphine (PPh$_3$) which, in our experiments, reacts with a peroxyl radical to form triphenylphosphine oxide and a reactive radical monomer than can participate in chain propagation [4]. Details of these curing processes are described elsewhere [6].

The extent of crosslinking in UV-curable coatings can be quantified in several ways. Fourier-transform infrared spectroscopy (FTIR) can show the conversion of chemical bonds from monomer to polymer, indicating the extent of crosslinking [1]. Although widely used, surface FTIR (attenuated total reflection, or ATR-FTIR) is typically limited to a very shallow depth of the coating—the exact depth is quantitatively dependent on indices of refraction and on the incident angle of the IR light, but typical experimental values range from 0.6 to 2 μm [7]. (Certain research applications may use infrared spectroscopy in a transmission mode, but that approach is inappropriate for coatings applied to IR-opaque substrates.) Depending on the sample, the penetration depth of ATR-FTIR may be appropriate to measure curing at the surface of the coating, but may not represent the curing throughout the entire thickness of the coating. Other spectroscopic techniques, including Raman spectroscopy, ultimately face the same question of penetration depth. Penetration depth is, of course, also affected by the opacity of the sample at the particular wavelength of light used in the experiment.

For many applications, nuclear magnetic resonance (NMR) offers an attractive alternative to traditional optical spectroscopy for quantifying crosslinking in polymeric films, including UV-curable coatings. While traditional, high-field NMR typically requires samples to be in a 5-mm diameter NMR tube that is inserted into the bore of the superconducting magnet, single-sided NMR devices can directly probe materials at surfaces without sampling or sample preparation [8,9]. These devices, while unable to record chemical-shift–resolved spectra, can measure relaxation of nuclear spin states, which is directly affected by the degree of crosslinking [10]. In this regard, single-sided magnets perform similarly to optical spectroscopy, albeit with much lower sensitivity. However, the design of these magnets imparts a fixed and strong magnetic field gradient that, in analogy to clinical MRI, can be used in an imaging modality via Fourier transformation. In this way, the extent of curing can be probed in a spatially selective manner throughout the entire thickness of the sample, even for samples over 100 μm thick, with a spatial resolution on the order of a few microns [11].

In this report, we describe the use of single-sided magnets to investigate the extent of crosslinking in a UV-curable polymer system comprising poly(ethylene glycol) diacrylate (PEGDA), in a spatially sensitive manner. The effects of the scavenger molecule PPh_3, the UV intensity, and the thickness of the coating on the extent of curing are reported. In addition, we report the use of sequential Fourier and inverse Laplace transformations to visualize the relaxation cross-section of a sample that describes the extent of crosslinking throughout the sample thickness. Spatially selective NMR measurements of coatings have only been lightly reported in the literature [12], and those reports have not leveraged Fourier transformations of the NMR data to optimize spatial resolution. These steps improve the utility of single-sided NMR in investigating (polymeric) films via their spatially variant relaxation times.

2. Results

NMR data collected using single-sided magnets cannot capture chemical shift information. When using these devices, typical experiments are made with a single excitation pulse followed by a train of refocusing pulses; interleaved between refocusing pulses are acquisition windows that each capture the refocused magnetization "echo": a record of the signal intensity that no longer contains chemical shift information. This Carr–Purcell–Meiboom–Gill (CPMG) pulse sequence is crucial to the success of single-sided NMR, as its refocusing sustains measurable signals long enough to be successfully recorded in a highly inhomogeneous magnetic field [13,14]. Consequently, the CPMG pulse sequence can be used to evaluate time-dependent signal attenuation that is ultimately related to physical properties of materials. In a CPMG measurement, each echo, like traditional free-induction decay (FID) signals, is made of some number of complex data points each measured over some dwell time. In our data processing, each echo is individually subjected to Fourier transformation, converting the time domain echo signal to a frequency domain signal that, because of the permanent magnetic field gradient, is proportional to a spatial domain signal. Once every echo has been transformed into the spatial domain, the signal decay (over the many successive echoes) at each position is subjected to an inverse Laplace transformation (ILT), the result of which is a relaxation "spectrum" at each position. An ILT produces a spectrum showing the intensity of various relaxation rates that, when added, best represent the overall signal attenuation in the signal: smaller T_2 relaxation times (to the left of the ILT spectrum) indicate greater relaxation which, in turn, suggests more crosslinking [15]. The final product of this data processing approach is a relaxation cross-section (position, z, vs. T_2) that shows differences in local crosslinking at different depths in the sample. The details of ILTs are described in detail elsewhere [16], and the entire data processing procedure is outlined in Figure 1.

The relaxation cross-section (or z-T_2 map) highlights the T_2 relaxation values at different positions within the sample. As an example, Figure 2 shows a z-T_2 map of a PEGDA sample that was not spun, and remains relatively thick (~80 μm). The UV light penetrates only the top ~20 μm of this thick sample (the top of the sample is in the negative z direction), which becomes crosslinked (smaller T_2, around 10^{-3} s), while the lower ~60 μm of the coating (away from the UV light, towards the positive z direction) remain less crosslinked (larger T_2, around 10^{-2} s). This method of data presentation highlights spatial variations in relaxation times and is used throughout this report.

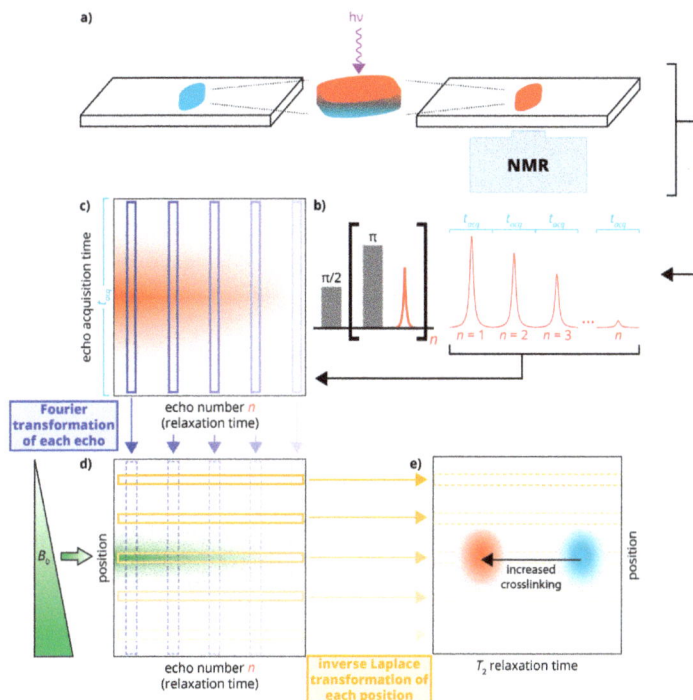

Figure 1. Schematic of the experimental setup and data processing: (**a**) An uncured sample of poly(ethylene glycol) diacrylate (PEGDA)/photoinitiator is applied to a glass slide and cured using UV irradiation. This sample is placed atop the single-sided NMR for measurements; (**b**) The NMR relaxation data is recorded using a Carr-Purcell-Meiboom-Gill (CPMG) pulse sequence [13,14] as a series of echoes, each recorded over an acquisition window of length t_{acq}; (**c**) The echoes are arranged in a 2D data set that is subject to Fourier transformation (**d**); producing a spatially dependent signal that decays in time; (**e**) This spatially dependent signal is subject to inverse Laplace transformation, changing the time domain (signal decay) to a relaxation domain. This produces a "relaxation cross-section" or z-T_2 map, in which crosslinking can be monitored (via T_2) in a spatially sensitive manner.

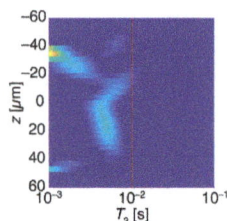

Figure 2. A z-T_2 map of a thick (~100 μm) coating sample containing 1% triphenylphosphine (PPh$_3$). The top surface, exposed to the most intense UV irradiation, is more completely cured and shows greater crosslinking (smaller T_2 values). UV does not penetrate well into the lower regions of the sample (towards the positive z direction), as indicated by the larger T_2 relaxation times.

Figure 3 highlights the correspondence between ATR-FTIR and T_2 measurements of crosslinking in the PEGDA network. The IR spectra are normalized to the peak at 1726 cm^{-1}, corresponding to the acrylic carbonyl group; these carbonyl groups are largely unaffected during curing and their IR intensities should remain fixed. In contrast, the terminal CH=CH$_2$ bonds on the PEGDA monomers produce a peak at 1410 cm^{-1} [17–20]. A decrease in the intensity of the 1410 cm^{-1} peak when the sample is exposed to UV light indicates curing. Simultaneously, the measured T_2 value decreases after the sample has been exposed to UV irradiation. Crosslinking is associated with reduced T_2 values, as reported elsewhere [15].

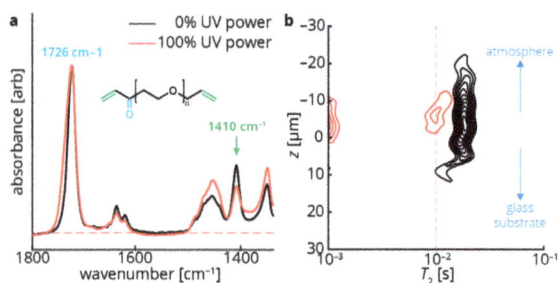

Figure 3. ATR-FTIR (attenuated total reflection–Fourier transform infrared spectroscopy) and single-sided NMR measurements of coating samples both uncured (black) and cured (red). (**a**) ATR-FTIR measurements indicate a peak at 1726 cm^{-1}, matching the acrylate carbonyl (cyan in the poly(ethylene glycol) diacrylate (PEDGA) structure) stretch. This peak is expected to remain constant throughout curing as the carbonyl is mostly unaffected by polymerization. The signals are normalized to this peak. In contrast, the peak at 1410 cm^{-1}, matching the terminal CH=CH$_2$ groups (green in the PEGDA structure), diminish with increased crosslinking; (**b**) z-T_2 map showing that increased crosslinking also correlates with a decrease in the measured T_2 values (the peak range shifts to the left). In addition, an additional peak appears in the 100% UV power film (red) at T_2 ~10^{-3} s. This may indicate a second T_2 component in the material that exhibits rapid relaxation, often due to extensive crosslinking. This peak could not be accurately probed with our instrumentation, but other single-sided NMR devices may be able to access this information [21]. In addition to differences in T_2, the two films in (**b**) are of different thicknesses—the uncured sample is ~30 μm thick, while the cured sample is ~10 μm thick.

When radical-driven curing of PEGDA polymers is performed in the presence of atmospheric oxygen, the oxygen can form radicals that interfere with crosslinking in the polymer network [4]. Several methods can be employed to circumvent this, including curing in an anaerobic environment (requiring engineering controls) or adding a scavenger molecule (PPh$_3$ in our samples) that mitigates the deleterious effects of oxygen [4,5]. Figure 4 shows z-T_2 maps of several samples that are cured in (an oxygen-containing) lab atmosphere. The leftmost column of Figure 4 (Figure 4a–f) lacks PPh$_3$. Their z-T_2 maps highlight that there is no change in T_2 (i.e., no crosslinking) in the samples regardless of the UV power deposited into the sample. Furthermore, there is little evidence of a curing gradient throughout the thickness of the sample (a spatially dependent T_2), suggesting that the oxygen-driven inhibition effect takes place, as expected, at the sample/air interface. Possible exceptions to this are in Figure 4e,f, in which "tails" with shorter T_2s appear near the bottom (positive z direction, near the glass interface) of the sample. In these samples, exposed to the highest UV power, it is possible that some curing occurs ~15 µm below the sample surface. This suggests that, although atmospheric oxygen inhibits curing at the coating surface, some crosslinking may still occur in regions of the sample inaccessible to oxygen provided there is sufficient UV power.

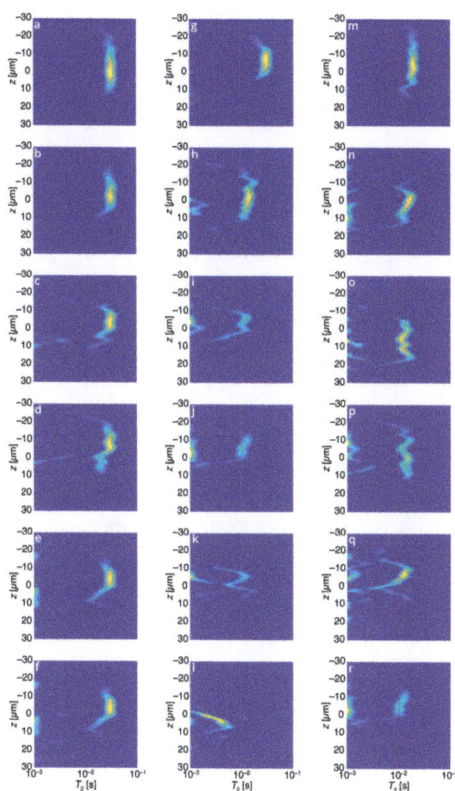

Figure 4. Relaxation cross-linking (or z–T_2) maps for various coating samples, all with identical horizontal and vertical axes: (**a–f**) 0% triphenylphosphine (PPh$_3$), UV powers of (top to bottom) 0%, 20%, 40%, 60% 80%, and 100% for 30 s curing. The T_2 values are unaffected by UV intensity, suggesting that without the scavenger PPh$_3$, no measurable curing of the sample occurs due to interferences from atmospheric oxygen; (**g–l**) Varying concentrations of scavenger PPh$_3$ (top to bottom: 0%, 0.2%, 0.5%, 1%, 2%, 5% w/w) with constant (100%) UV intensity. The measured T_2 values decrease with increasing PPh$_3$ content, showing the scavenger's ability to facilitate curing despite atmospheric interference from oxygen; (**m–r**) One percent PPh$_3$, UV powers of (top to bottom) 0%, 20%, 40%, 60% 80%, and 100% for 30 s curing. In contrast to the leftmost column, when 1% PPh$_3$ is added to the sample as a radical scavenger, increased UV intensity effects greater crosslinking in the sample, as manifest by a decrease in the measured T_2 values.

The second column of Figure 4 (Figure 4g–l) shows the effect of increasing amounts of the scavenger PPh$_3$, from 0–5% w/w, on crosslinking in the PEGDA sample. At low concentrations of PPh$_3$, the sample retains a large T_2 value (less crosslinking), and as PPh$_3$ increases, T_2 decreases, indicating more crosslinking. Using these data, we elected to use 1% PPh$_3$ (Figure 4j) as a standard sample to evaluate the extent of crosslinking as a function of UV power. This power-dependent series (0–100% UV power, 1% PPh$_3$) are shown in the rightmost column of Figure 4 (Figure 4m–r). As expected, measured T_2 values decrease with increasing UV power; again, this is consistent with increased crosslinking in the sample.

3. Discussion

Single-sided NMR is highly conducive to in situ measurements of coatings on surfaces. Along with optical methods (e.g., ATR-FTIR and Raman), NMR can probe the extent of crosslinking within a sample; in magnetic resonance contexts, this is most easily measured as a change in the spin-spin relaxation time T_2. However, NMR offers a much larger spatial range (in the depth dimension) for measurements than do non-confocal optical techniques. In this regard, we observe a crosslinking "gradient" throughout the thickness of a particularly thick UV-cured coating sample (see Figure 2). NMR is successful in evaluating crosslinking differences that are due to both different UV intensities and to different chemical compositions (different amounts of scavenger PPh$_3$ in our case). In other words, single-sided NMR is insensitive to the cause of differences in crosslinking, which may prove useful where in situ analyses are intended only to measure the extent of crosslinking.

Despite its ability to probe thick samples, single-sided NMR measurements show poor spatial resolution relative to many optical techniques, both laterally and in the depth dimension. Single-sided NMR devices use surface coils that are typically several square centimeters in area, meaning that relaxation measurements reflect a lateral spatial average over that area. This is less problematic for homogeneous materials (such as coatings), but needs to be considered during experimental design. The spatial resolution in the depth dimension is fundamentally related to the magnetic field gradient and the acquisition parameters: strong field gradients and large acquisition times (both possible with single-sided devices) result in high spatial resolution. Practically, the coplanarity of the sample with the magnet's "sweet spot" restricts the resolution in the depth dimension: less-than-perfect coplanarity between the sample and the sweet spot will blur the resolution of the measurement. Generally, the depth resolution of single-sided NMR measurements is on the order of 10 μm, though careful calibration and experimental setup can improve this several-fold [11].

Simple future experiments with single-sided NMR extends this work to coatings of other compositions, including different effective molecular weights of PEGDA. A more sophisticated approach may be to exploit dipolar coupling in crosslinked materials to more quantitatively probe crosslinking in a material. This method has been demonstrated elsewhere on elastomers [22,23] and may circumvent some of the limitations of T_2 determinations from CPMG data, in particular a strong dependence on both equipment-specific and user-selected experimental parameters. Furthermore, single-sided NMR may be particularly well suited to evaluating the extent of crosslinking within interpenetrating polymer networks (IPNs) [24]; these coatings consist of two or more phases that in tandem produce specific material properties.

4. Materials and Methods

4.1. Samples

Samples comprised different ratios of poly(ethylene glycol) diacrylate (Mn = 700 g mol^{-1}), photoinitiator 1-hydroxycyclohexyl phenyl ketone (HPCK, 2% w/w), and triphenylphosphine (PPh$_3$). All chemicals were used as received without further purification (Sigma-Aldrich, St. Louis, MO, USA). HCPK and PPh$_3$ were dissolved into acetone prior to being added to the sample. The amount of PPh$_3$ varied per experiment from 0% to 5% by weight with the total amount of acetone fixed at

7% *w/v*. Slides were prepared with a piece of labeling tape in the middle with a ~1 cm diameter circle cut out in the tape; this tape was designed to contain the sample in a region that could be uniformly cured by the UV gun and that would fit in the sensitive region of the NMR magnet. Ten microliters of the sample were placed on a glass slide in the center of the circle. Most slides were spun at a rate of ~3000 rpm for 30 s using a 3D-printed spin coater driven by a computer hard drive motor. (One sample slide was intentionally left thick and was not spun.) While spinning, the sample, at a distance of ~1.4 cm from the UV source, was cured using a ThorLabs CS2010 LED UV curing system (ThorLabs, Inc., Ann Arbor, MI, USA). At 100%, the UV system outputs 185 mW cm^{-2} at 365 nm (instrument-internal calibration). The power of the UV system was adjusted to 20%, 40%, 60%, 80%, and 100% of the maximum power—the output varied linearly with nominal power. After spinning and curing, samples had a thickness of ~20 μm, determined using a Profilm3D surface profiler (Filmetrics, Inc.; San Diego, CA, USA).

4.2. NMR Measurements

All NMR measurements were made with a PM5 NMR-MOUSE ($B_0 = 0.4$ T) operated by a Kea2 spectrometer (Magritek, Ltd., Wellington, New Zealand). An initial CPMG [13,14] sequence was run on each slide to center the magnet to find the most signal. The B_1 frequency was adjusted by a maximum of 50 kHz (~50 μm) to center the maximum signal in the acquisition window. (The rf coil has a bandwidth of ~350 kHz, so these adjustments do not substantially affect the coil sensitivity). This short CPMG had 512 scans, a 1 μs dwell time and 128 complex points. Subsequently, a final CPMG was run for data collection with 4096 scans, 10 μs dwell time, and 32 complex points with all other parameters staying constant between slides. NMR measurements were taken directly after curing to minimize outside factors, such as extra exposure to UV light.

4.3. FTIR Measurements

FTIR data were collected using a Digital Labs FTS 7000 ATR-FTIR (Varian, Inc., Palo Alto, CA, USA). The samples were made of PEGDA with 2% HCPK and 2% PPh$_3$ (*w/w*) and were prepared in the same manner as those for NMR data collection. A blank microscope slide was used for the background scan; all measurements were made from 600 cm^{-1} to 2000 cm^{-1} with 1 cm^{-1} resolution for 64 scans.

4.4. Data Processing

All data processing was performed in Matlab (Mathworks, Inc., Natick, MA, USA) using custom processing scripts. These scripts for importing and Fourier transforming the NMR data were made in house and are available at the institutional repository of the corresponding author. Inverse Laplace transformation was also performed in Matlab using scripts provided by Petrik Galvosas (University of Wellington, New Zealand) [25].

Acknowledgments: Acknowledgment is made to the Donors of the American Chemical Society Petroleum Research Fund for support of this research.

Author Contributions: All authors conceived and designed the experiments; M.B. and F.M. conducted experiments and analyzed the data; and M.B. and T.M. wrote the paper.

Conflicts of Interest: The authors declare no conflict of interest.

References

1. Young, R.J.; Lovell, P.A. *Introduction to Polymers*, 3rd ed.; CRC Press: Boca Raton, FL, USA, 2011; ISBN 9780849339295.
2. Decker, C. Kinetic Study and New Applications of UV Radiation Curing. *Macromol. Rapid Commun.* **2002**, *23*, 1067–1093. [CrossRef]
3. Scherzer, T. Depth Profiling of the Degree of Cure during the Photopolymerization of Acrylates Studied by Real-Time FT-IR Attenuated Total Reflection Spectroscopy. *Appl. Spectrosc.* **2002**, *56*, 1403–1412. [CrossRef]
4. Ligon, S.C.; Husár, B.; Wutzel, H.; Holman, R.; Liska, R. Strategies to reduce oxygen inhibition in photoinduced polymerization. *Chem. Rev.* **2014**, *114*, 557–589. [CrossRef] [PubMed]
5. Husár, B.; Ligon, S.C.; Wutzel, H.; Hoffmann, H.; Liska, R. The formulator's guide to anti-oxygen inhibition additives. *Prog. Org. Coat.* **2014**, *77*, 1789–1798. [CrossRef]
6. Hoyle, C.E. Photocurable Coatings. In *Radiation Curing of Polymeric Materials*; ACS Symposium Series; American Chemical Society: Washington, DC, USA, 1990; Volume 417, pp. 1–16, ISBN 9780841217300.
7. Urbaniak-Domagala, W. The use of the spectrometric technique FTIR-ATR to examine the polymers surface. In *Advanced Aspects of Spectroscopy*; InTech: Rijeka, Croatia, 2012.
8. Blümich, B.; Perlo, J.; Casanova, F. Mobile single-sided NMR. *Prog. Nucl. Magn. Reson. Spectrosc.* **2008**, *52*, 197–269. [CrossRef]
9. *Single-Sided NMR*; Casanova, F.; Perlo, J.; Blümich, B. (Eds.) Springer: Berlin/Heidelberg, Germany, 2011; p. 244, ISBN 9783642163067.
10. Litvinov, V.M. Characterisation of Chemical and Physical Networks in Rubbery Materials Using Proton NMR Magnetisation Relaxation. In *Spectroscopy of Rubber and Rubbery Materials*; Litvinov, V.M., De, P.P., Eds.; Rapra Technology Limited: Shropshire, UK, 2002; pp. 353–400, ISBN 9781859572801.
11. Perlo, J.; Casanova, F.; Blümich, B. Profiles with microscopic resolution by single-sided NMR. *J. Magn. Reson.* **2005**, *176*, 64–70. [CrossRef] [PubMed]
12. Küchel, J. *NMR Probing of the Polymer Structure in UV Curable Acrylate Coatings*; Eindhoven University of Technology: Eindhoven, The Netherlands, 2006.
13. Carr, H.Y.; Purcell, E.M. Effects of Diffusion on Free Precession in Nuclear Magnetic Resonance Experiments. *Phys. Rev.* **1954**, *94*, 630–638. [CrossRef]
14. Meiboom, S.; Gill, D. Modified Spin-Echo Method for Measuring Nuclear Relaxation Times. *Rev. Sci. Instrum.* **1958**, *29*, 688. [CrossRef]
15. Herrmann, V.; Unseld, K.; Fuchs, H.B.; Blümich, B. Molecular dynamics of elastomers investigated by DMTA and the NMR-MOUSE®. *Colloid Polym. Sci.* **2002**, *280*, 758–764. [CrossRef]
16. Venkataramanan, L.; Song, Y.-Q.; Hürlimann, M.D. Solving Fredholm integrals of the first kind with tensor product structure in 2 and 2.5 dimensions. *IEEE Trans. Signal Process.* **2002**, *50*, 1017–1026. [CrossRef]
17. Witte, R.P.; Blake, A.J.; Palmer, C.; Kao, W.J. Analysis of poly(ethylene glycol)-diacrylate macromer polymerization within a multicomponent semi-interpenetrating polymer network system. *J. Biomed. Mater. Res. A* **2004**, *71*, 508–518. [CrossRef] [PubMed]
18. Lin, H.; Kai, T.; Freeman, B.D.; Kalakkunnath, S.; Kalika, D.S. The Effect of Cross-Linking on Gas Permeability in Cross-Linked Poly(Ethylene Glycol Diacrylate). *Macromolecules* **2005**, *38*, 8381–8393. [CrossRef]
19. Lin, H.; Wagner, E.V.; Swinnea, J.S.; Freeman, B.D.; Pas, S.J.; Hill, A.J.; Kalakkunnath, S.; Kalika, D.S. Transport and structural characteristics of crosslinked poly(ethylene oxide) rubbers. *J. Membr. Sci.* **2006**, *276*, 145–161. [CrossRef]
20. Sagle, A.C.; Ju, H.; Freeman, B.D.; Sharma, M.M. PEG-based hydrogel membrane coatings. *Polymer* **2009**, *50*, 756–766. [CrossRef]
21. Oligschläger, D.; Glöggler, S.; Watzlaw, J.; Brendel, K.; Jaschtschuk, D.; Colell, J.; Zia, W.; Vossel, M.; Schnakenberg, U.; Blümich, B. A Miniaturized NMR-MOUSE with a High Magnetic Field Gradient (Mini-MOUSE). *Appl. Magn. Reson.* **2015**, *46*, 181–202. [CrossRef]
22. Wiesmath, A.; Filip, C.; Demco, D.E.; Blümich, B. Double-Quantum-Filtered NMR Signals in Inhomogeneous Magnetic Fields. *J. Magn. Reson.* **2001**, *149*, 258–263. [CrossRef] [PubMed]

23. Wiesmath, A.; Filip, C.; Demco, D.E.; Blümich, B. NMR of Multipolar Spin States Excited in Strongly Inhomogeneous Magnetic Fields. *J. Magn. Reson.* **2002**, *154*, 60–72. [CrossRef] [PubMed]
24. Fouassier, J.P.; Lalevée, J. Photochemical Production of Interpenetrating Polymer Networks; Simultaneous Initiation of Radical and Cationic Polymerization Reactions. *Polymers* **2014**, *6*, 2588–2610. [CrossRef]
25. Callaghan, P.T.; Arns, C.H.; Galvosas, P.; Hunter, M.W.; Qiao, Y.; Washburn, K.E. Recent Fourier and Laplace perspectives for multidimensional NMR in porous media. *Magn. Reson. Imaging* **2007**, *25*, 441–444. [CrossRef] [PubMed]

magnetochemistry

MDPI

Article

Quantification of Squalene in Olive Oil Using [13]C Nuclear Magnetic Resonance Spectroscopy

Anne-Marie Nam, Ange Bighelli, Félix Tomi, Joseph Casanova and Mathieu Paoli *

Équipe Chimie et Biomasse, UMR 6134 SPE, Université de Corse-CNRS, Route des Sanguinaires, 20000 Ajaccio, France; annemarie.nam@gmail.fr (A.-M.N.); ange.bighelli@univ-corse.fr (A.B.); felix.tomi@univ-corse.fr (F.T.); joseph.casanova@wanadoo.fr (J.C.)
* Correspondence: mathieu.paoli@univ-corse.fr; Tel.: +33-4-2020-2169

Received: 11 October 2017; Accepted: 31 October 2017; Published: 6 November 2017

Abstract: In the course of our ongoing work on the chemical characterization of Corsican olive oil, we have developed and validated a method for direct quantification of squalene using [13]C Nuclear Magnetic Resonance (NMR) spectroscopy without saponification, extraction, or fractionation of the investigated samples. Good accuracy, linearity, and precision of the measurements have been observed. The experimental procedure was applied to the quantification of squalene in 24 olive oil samples from Corsica. Squalene accounted for 0.35–0.83% of the whole composition.

Keywords: squalene; olive oil; ^{13}C NMR analysis; quantification; Corsica

1. Introduction

Squalene—(*E*)-2,6,10,15,19,23-hexamethyl-2,6,10,14,18,22-tetracosahexaene—is a natural acyclic symmetrical triterpene. It is a key intermediate in the biosynthesis of sterols [1]. In the human body, squalene is synthesized and then converted into cholesterol. In medicine, squalene plays a major role in the reduction of cancer risks, particularly with regard to cancer of the pancreas and colon in rodents [2–4]. Squalene increases the stability of various emulsions (vaccines, pharmaceutical formulations) [5,6]. It is also useful at the surface of the skin, playing the role of protective barrier against Ultra-Violet (UV) radiations [7]. Hydrogenated squalene (i.e., squalane) is appreciated in cosmetics as emollient agent in creams and capillary serums [8].

The largest source of squalene for industrial purposes is from animal origin, provided by various species of shark [9]. According to the species, squalene represents up to 80% of the shark liver oil [10]. Various species of shark are now endangered as a result of their overexploitation.

Squalene is also widespread in the vegetable kingdom. Indeed, it is present in oil seeds and in green vegetables [11]. In olive oil, squalene represents 0.3% to 0.7% of the whole mass, accounting for 60–75% of the unsaponifiable fraction [12]. The presence of squalene confers to olive oil a great stability against auto-oxidation and photo-oxidation [13].

The Association of Official Analytical Chemists [14] recommended a method for extraction of squalene from natural matrices. Analytical techniques used in quantification of squalene in edible oils, in the presence of acylglycerols, fatty acids, phytosterols, and tocopherols have been recently reviewed [15]. Methods using a preliminary fractionation of samples, procedure that simplifies the analysis have been developed. Analysis of squalene in edible oils is predominantly achieved by chromatographic techniques (Gas Chromatography (GC) or Reversed-Phase High-Performance Liquid Chromatography (RP-HPLC)), after saponification of triglycerides, solvent extraction of the unsaponifiable fraction, and eventually isolation of the hydrocarbon fraction by Column Chromatography (CC) or Thin Layer Chromatography (TLC) [16–21]. The direct injection of olive oil in the injector port has been applied [22], as well as HPLC coupled with GC [23] or HPLC coupled to electrospray tandem mass spectrometry [24].

In parallel, ^{1}H and ^{13}C NMR have been widely used for identification and quantitative evaluation of triglycerides in olive oil (saturated fatty acid chains, mono-unsaturated, poly-unsaturated, stereochemistry of the double bonds, etc.) and for quality assessment and authentication [25–27]. Using the fingerprint technique, characteristic resonances of individual components of the unsaponifiable fraction (sterols, alcohols, tocopherol) have been identified using ^{1}H NMR, and the results allowed the determination of geographical origin of olive oil [28]. Similarly, the ratio of squalene vs. the other minor components of olive oil has been evaluated, and statistical analysis of the results gave useful information on the quality, authenticity, and origin of the investigated olive oil samples [25,29]. The content of squalene in human sebum (containing low proportion of triglycerides) has been measured using a 600 MHz spectrometer equipped with a cryoprobe [30]. Otherwise, quantitative analyses of two structurally close triterpenoid acids, as well as that of positional and geometric isomers of octadecadienoic acid with conjugated double bonds, have been performed using 2D NMR [31,32].

In previous works carried out in our laboratory, we demonstrated ^{13}C NMR spectroscopy was a powerful tool for the identification and quantitative determination of terpenes in natural matrices, mono and sesquiterpenes in essential oils [33] and fixed oil [34], diterpenes in cedar resins [35], triterpenes in solvent extracts from cork [36], or leaves from olive tree [37]. Taking into account that chromatographic techniques used to quantify squalene in olive oil needed laborious and time-consuming fractionation steps, the aim of the present study was to develop a method, based on ^{13}C NMR, and using a routine spectrometer (9.4 Tesla), that allowed the quantitative determination of squalene in olive oil, avoiding the fractionation steps.

2. Results and Discussion

2.1. ^{13}C NMR Data of Squalene and Olive Oil

The ^{13}C NMR spectrum of squalene displayed 15 resonances belonging to quaternary carbons (135.11; 134.90 and 131.26 ppm), ethylenic methines (124.42; 124.32 and 124.28 ppm), allylic methylenes (39.77; 39.74, 28.29; 26.78; 26.67 ppm) and methyl groups (25.71; 17.69; 16.05 and 16.01 ppm). The chemical shift values of our recorded spectrum (Table 1) fitted perfectly with previous data reported by Pogliani et al. [38]. However, it could be noted a difference of 1 ppm for carbon C3, probably due to a misprint in the paper [38].

Table 1. Structure, ^{13}C NMR chemical shifts, and longitudinal relaxation times (T_1) of carbons of squalene.

C	δ (SQ)	T_1	C	δ (SQ)	T_1
1	25.71	1.9	9	39.77 *	0.7
2	131.26	10.0	10	135.11	5.2
3	124.42	2.5	11	124.32 #	1.7
4	26.78	0.9	12	28.29	0.4
5	39.74 *	0.7	13	17.69	4.5
6	134.90	4.1	14	16.05 †	2.8
7	124.28 #	1.2	15	16.01 †	3.3
8	26.67	0.6			

δ (SQ): Chemical shift of carbons of squalene (ppm vs. tetramethylsilane (TMS)). Assignment has been done according to Pogliani et al. [38]. *, # and †: chemical shifts may be inversed. T_1: longitudinal relaxation times in s.

The ^{13}C NMR spectrum of a commercially available olive oil is more complex (Figure 1). Four parts may be distinguished: 172–174 ppm, esters; 124–134 ppm, ethylenic carbons; 60–72 ppm, carbons of glycerol; and 13–35 ppm, aliphatic carbons. In that spectrum, all the resonances with high intensity belong to the triglycerides. Twelve out of 15 resonances of squalene were observed. They were perfectly resolved, and therefore they could be used for quantitative determination of squalene in olive oil.

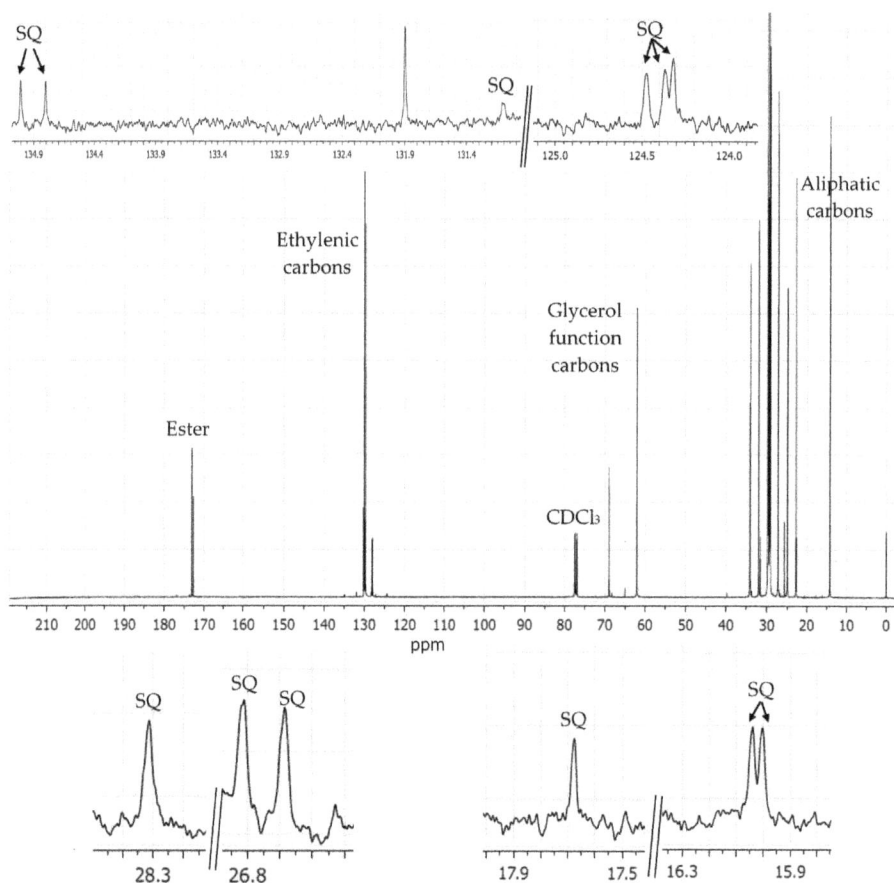

Figure 1. ^{13}C NMR spectrum of a commercially available virgin olive oil. SQ: squalene.

2.2. Validation of the Experimental Procedure for Quantitative Determination of Squalene Using ^{13}C NMR

In order to approach the physico-chemical properties of olive oil (viscosity for instance) the experiments for validation of the experimental procedure have been carried out using know quantities of squalene in trioleine (glyceryl tris octadec-9-enoate) that is the major triglyceride of olive oil accounting for 48–62% of the whole composition [39].

Several techniques have been developed for quantification of individual components of a natural mixture based on ^{13}C NMR spectroscopy. The standard sequence combines a 90° pulse angle, gated decoupling technique and requires waiting a period of $5T_1$ of the longest T_1 value, before applying another pulse. This sequence provides accurate result but is really time consuming. Otherwise, use of a paramagnetic relaxation reagent allows decrease of experimental time but induces a line width broadening. Quantitative determination can be led using a rapid train of short pulses because a small flip angle provides less difference in the steady-state magnetization than a larger one in the presence of carbons having different T_1 values.

Owing to our experience in the analysis of complex natural mixtures containing nuclei with different T_1 values, a good approach is a compromise between the aforementioned procedures. For instance, quantification of various compounds has been performed in our laboratories, using this approach: carbohydrates in ethanol extract of *Pinus* species [40], triterpenes in cork extract [36] and olive leaf extract [37], and taxanes in leaf extract of *Taxus baccata* [41]. Quantitative determination of a component in a natural mixture is achieved by internal standardization by comparison of the areas of the resonances of that compound with those of an internal standard. In these conditions, it is obvious that quantitative estimation will be led from not fully relaxed spectra and that validation of the method should be performed before applying it to the analysis of mixtures [42]. The best conditions for the pulse sequence are those that reduce as far as possible the difference in the steady-state magnetization of nuclei with different T_1 values and that simultaneously allow a good S/N ratio in a short period of time. They could be selected using Becker's equation that allows the calculation of the S/N ratio as a function of the pulse angle and the ratio of longitudinal relaxation time to total recycling time [43].

Then, the theoretical parameters (precision, accuracy, linearity of measures) should be validated using pure squalene in trioleine before application of the method to the quantification of squalene in genuine olive oils. To carry out the validation of the method:

- $CDCl_3$ has been conserved as solvent and trioleine has been used as a model for olive oil;
- Longitudinal relaxation times have been measured for carbons of squalene by the inversion-recovery method. They ranged from 0.4 to 10.0 s, the highest values (4.1–10.0 s) being measured, as expected, for quaternary carbons (Table 1). T_1 values of vinylic methines and allylic methylenes ranged from 1.2 s to 2.5 s and from 0.4 s to 0.9 s, respectively. Finally, T_1s of the four methyl groups ranged from 1.9 s to 4.5 s. Quantitative analysis has been conducted with resonances of carbons not overlapped, perfectly resolved and with T_1 values comprised from 0.7 s to 4.5 s;
- Di-2-methoxyethyl oxide (diglyme) has been chosen as internal standard (T_1 value of its methylenes = 3.8 s) since its resonances do not overlap with those of triglycerides contained in olive oil.

The parameters of the pulse sequence have been determined using formula (1) for various T_1 values (0.7–4.5 s), and for a repetition delay of 3.7 s (acquisition time = 2.7 s; relaxation delay = 1.0 s) required for a 128 K data table. According to Becker et al. [43], we determined and plotted the percentage of recovered signal, expressed as S/N (%), as a function of the pulse angle α, using formula (1). Using a pulse angle of 30°, this procedure provided a small difference (3.6%) in the steady-state magnetization between carbons exhibiting different T_1 values and a reasonable time of analysis in spite of the utilization of a medium field spectrometer (3000 scans in less than 3 h) (Figure 2).

$$\frac{S}{N} = \frac{M_0 \times [1 - e^{(-D/T_1)}] \times \sin \alpha}{\sqrt{D} \times [1 - e^{(-D/T_1)} \cos \alpha]} \tag{1}$$

S/N: signal-to-noise ratio, M_0: initial magnetization, D: time between two pulses (in seconds), T_1: longitudinal relaxation time (in seconds), and α: pulse angle (in degrees).

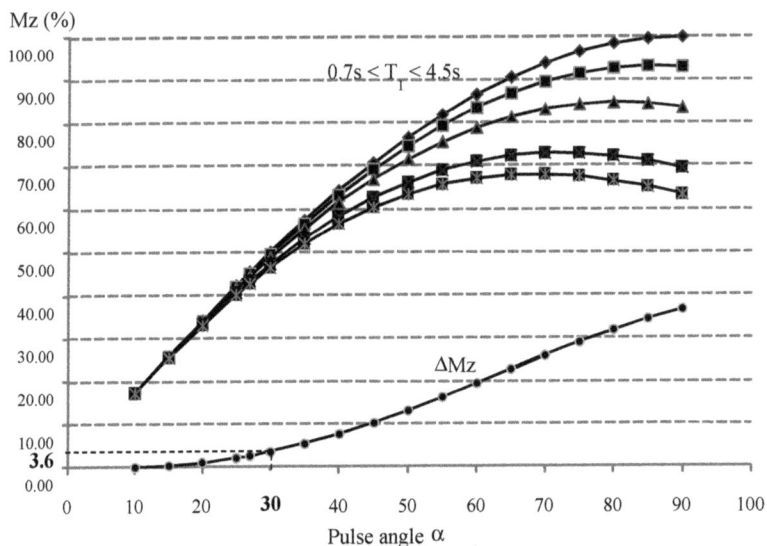

Figure 2. S/N (%) vs. flip angle α, plotted from formula (1) according to Becker et al. [43], for selected values of T_1 (0.7 s< T_1 < 4.5 s) and a minimum total recycling time τ of 3.7 s using a 128 K data table (acquisition time = 2.7 s and relaxation delay = 1.0 s).

Accuracy, precision and response linearity of this method have been validated by various experiments carried out on pure squalene by comparing the weighted quantities (0.37–1.66 mg) with those measured by NMR. From the ^{13}C NMR spectrum, the mass of squalene m_{SQ} (mg) was calculated using Formula (2). Relative errors between weighted and calculated masses are comprised between 0.0% and 10.3%, and therefore they demonstrated good accuracy of measurements (Table 2).

$$mSQ = 2 \times \frac{A_{SQ} \times M_{SQ} \times m_D}{A_D \times M_D} \times p_{SQ} \times p_D \tag{2}$$

The area A_{SQ} taken into account was the mean value of the areas of selected protonated carbons. A_D is the mean value of the areas of the two methylenes of diglyme. M_{SQ} is the molecular weight of squalene. M_D is the molecular weight of diglyme and m_D is the amount of diglyme. p_{SQ} and p_D: purity of squalene and of diglyme, respectively.

Table 2. Quantitative determination of squalene by ^{13}C NMR spectroscopy using diglyme as internal reference.

A_D	0.9901	0.9955	0.9959	1.0074	1.0011	0.9845	1.0168	1.0139
A_{SQ}	0.0734	0.1220	0.1668	0.1970	0.2253	0.2689	0.3129	0.3431
m_w (mg)	0.37	0.55	0.74	0.92	1.10	1.29	1.47	1.66
m_c (mg)	0.33	0.56	0.74	1.00	1.03	1.22	1.44	1.58
ER (%)	10.3	−0.7	0.0	−9.1	6.8	5.5	2.2	4.8

A_D and A_{SQ}: Mean areas of selected carbons of diglyme and squalene, respectively; Mass of diglyme (m_D): 1.49 mg; m_w: weighted mass of squalene (mg); m_c: calculated mass of squalene (mg) using formula (2); ER: relative error (%) between m_c and m_w; Molecular weight of squalene: 410.7 g·mol^{-1}.

Then, we drew the calibration line for the quantification of squalene. The straight line was plotted by expressing the ratio of the mean value of areas of the resonances of squalene selected carbons (A_{SQ})

with those of diglyme (A_D) as a function of the weighed mass of squalene (m_w). We observed a good linearity of the measurements because the linear determination factor (R^2) is 0.996 (Figure 3).

Finally, the spectrum of the sample containing 0.55 mg of squalene has been recorded five times. The repeatability, calculated with a confidence interval of 99% (Student's *t*-test) was equal to 0.56 mg $\pm$ 0.04 mg, i.e., 0.56 mg $\pm$ 6.8% which indicates a good precision of measurements.

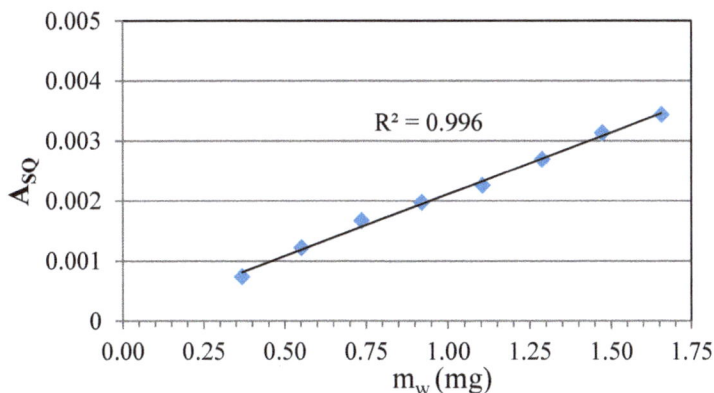

Figure 3. Calibration line of squalene. m_w = weighted mass of squalene.

The experimental procedure developed to quantify squalene in triolein exhibited good accuracy, precision and linearity of measurements. Analysis time with a routine spectrometer (9.4 Tesla) is not prohibitive since a single analysis requires three hours. Therefore, this procedure could be applied for quantification of squalene in olive oils of Corsican origin.

2.3. Quantification of Squalene in Various Olive Oil Samples of Corsican Origin

Twenty-four olive oil samples from various localities in Corsica and from various olive varieties have been analyzed using ^{13}C NMR, according to the experimental procedure previously described. In the ^{13}C NMR spectrum of olive oil (Figure 2), eight out of 12 of the protonated carbons of squalene were observed. All of these resonances were perfectly resolved and did not overlap with resonances of other components of olive oil, and their relaxation times were between 0.7 s and 4.5 s. The mass of squalene in every olive oil sample has been calculated using Formula (2), taking into account the mean areas of these resonances. Then, the mass percentages of squalene have been calculated using Formula (3), which are reported in the Table 3.

$$\%C = \frac{m_{SQ}}{m} \times 100 \tag{3}$$

%C: percentage of squalene; m_{SQ}: calculated mass (mg) of squalene; m: mass of the olive oil sample.

Table 3. Quantification of squalene in olive oils from Corsica using ^{13}C NMR.

Sample	Olive Variety	Squalene (%)*
1		0.35
2	Zinzala	0.37
3		0.41
4		0.35
5	Sabine	0.35
6		0.40
7		0.42
8	Picholine	0.38
9		0.40
10		0.43
11		0.44
12	Germaine	0.49
13		0.51
14		0.51
15		0.83
16	Cortenaise	0.42
17		0.47
18	Capannacce	0.52
19	Germaine/Picholine	0.37
20	Germaine/Capanacce	0.44
21		0.67
22	Germaine/Sabine	0.47
23		0.49
24	Sabine/Picholine	0.46

*: percentages calculated using formula (3).

Among the 24 olive oil samples, 18 samples were obtained from olive of a single variety, the last six samples coming from olives of two varieties. From Table 3, it is observed that Corsican olive oils contained appreciable amount of squalene comprised between 0.35% and 0.52% for 22 samples out of 24. The two last samples exhibited higher contents (0.67% and 0.83%). These results are in agreement with those reported in the literature (0.3–0.7%) [12].

Although the number of samples from every locality and from every olive variety is limited, it seems that there is no direct relation between the content of squalene in a given olive oil sample and the variety of the olive. However, it could be observed that zinzala, sabine, and picholine olives produced an oil containing 0.35–0.42% of squalene. The olive oil from Germaine, Cortenaise and Capanacce varieties exhibited a slightly higher content of squalene (0.40–0.83%). Finally, olive oil coming from two varieties of fruits contained 0.37–0.67% of squalene.

3. Materials and Methods

3.1. Chemicals

Squalene, triolein and di-2-methoxyethyloxide (diglyme) were obtained from Sigma-Aldrich (St-Louis, MO, USA), Acros Organics (Geel, Belgium), and Jansen Chimica (Geel, Belgium), respectively. Olive oil samples were supplied by Mrs. Henneman (Chambre d'Agriculture de la Haute Corse, Bastia, Corsica, France).

3.2. NMR Experiments

3.2.1. Quantitative ^{13}C NMR Spectra

Quantitative ^{13}C NMR spectra were recorded on a Bruker (Wissembourg, France) AVANCE 400 Fourier Transform spectrometer operating at 100.13 MHz for ^{13}C, equipped with a 5 mm probe, in CDCl$_3$ with all shifts referred to internal TMS. ^{13}C NMR spectra were recorded with the following parameters: inverse gated decoupling, flip angle 30°, acquisition time = 2.7 s for 128 K data table with a spectral width of 24,000 Hz (240 ppm), a relaxation delay D_1 = 1.0 s, composite pulse decoupling of the proton band, and a digital resolution of 0.366 Hz/pt. The internal reference used was diglyme. The number of accumulated scans was 3000 for each sample. Exponential line broadening multiplication (1 Hz) of the free induction decay was applied before Fourier transformation.

3.2.2. T_1 Measurements

The longitudinal relaxation times of the ^{13}C nuclei (T_1 values) were determined by the inversion-recovery method, using the standard sequence: 180°–τ–90°–D_1, with an acquisition time of 0.68 s (for 32 K data table with a spectral width of 25,000 Hz) and a relaxation delay D_1 of 20 s. Each delay of inversion (τ) was thus taken into account for the computation of the corresponding T_1 using the function $I_p = I_0 + pe^{-\tau/T_1}$ (Bruker microprogram; I_p and I_0 are populations of nuclear spins; p is a constant of integration).

3.2.3. Calibration Line

A weighted amount of 0.37–1.66 mg of squalene was diluted in 0.5 mL of CDCl$_3$ containing 1.49 mg of diglyme.

3.2.4. Quantification of Squalene in Olive Oils

A weighted amount of 140–150 mg of olive oil was diluted in 0.5 ml of CDCl$_3$ containing 1.53 mg of diglyme.

4. Conclusions

An experimental procedure, based on ^{13}C NMR spectroscopic analysis, was developed and allowed for the quantification of squalene in olive oil samples. An optimized pulse sequence (flip angle α = 30°, inverse gated decoupling, total recycling time 3.7 s) was checked and led to reliable quantitative determination of squalene in olive oil samples from Corsica with an analysis time of less than three hours using a medium field NMR spectrometer (9.4 T). In the 24 olive oil samples investigated, squalene accounted for 0.35–0.83% of the whole composition.

Acknowledgments: The authors are indebted to the Collectivité Territoriale de Corse for a research grant (A.-M.N.) and for financial support (FEDER AGRIEX). Thanks to Mrs. Henneman (Chambre d'Agriculture de la Haute Corse) for olive oil samples.

Author Contributions: A.B., F.T. and J.C. conceived and designed the experiments; A.-M.N. and M.P. performed the experiments; A.B. and F.T. analyzed the data; J.C. and M.P. wrote the paper.

Conflicts of Interest: The authors declare no conflict of interest.

References

1. Psomiadou, E.; Tsimidou, M. On the role of squalene in olive oil stability. *J. Agric. Food Chem.* **1999**, *47*, 4025–4032. [CrossRef] [PubMed]
2. Smith, T.J.; Yang, G.Y.; Seril, D.N.; Liao, J.; Kim, S. Inhibition of 4-(methylnitrosamino)-1-(3-pyridyl)-1-butanone-induced lung tumorigenesis by dietary olive oil and squalene. *Carcinogenesis* **1998**, *19*, 703–706. [CrossRef] [PubMed]
3. Newmark, H.L. Squalene, olive oil, and cancer risk: A review and hypothesis. *Cancer Epidemiol. Biomark. Prev.* **1997**, *6*, 1101–1103. [CrossRef]
4. Rao, C.V.; Newmark, H.L.; Reddy, B.S. Chemopreventive effect of squalene on colon cancer. *Carcinogenesis* **1998**, *19*, 287–290. [CrossRef] [PubMed]
5. Reddy, L.H.; Couvreur, P. Squalene: A natural triterpene for use in disease management and therapy. *Adv. Drug Deliv. Rev.* **2009**, *61*, 1412–1426. [CrossRef] [PubMed]

6. Fox, C.B. Squalene emulsions for parenteral vaccine and drug delivery. *Molecules* **2009**, *14*, 3286–3312. [CrossRef] [PubMed]

7. Auffray, B. Protection against singlet oxygen, the main actor of sebum squalene peroxidation during sun exposure, using *Commiphora myrrha* essential oil. *Int. J. Cosmet. Sci.* **2007**, *29*, 23–29. [CrossRef] [PubMed]

8. Jame, P.; Casabianca, H.; Batteau, M.; Goetinck, P.; Salomon, V. Differentiation of the origin of squalene and squalane using stable isotopes ratio analysis. *SOFW J.* **2010**, *136*, 2–7.

9. Deprez, P.; Volkman, J.; Davenport, S. Squalene content and neutral lipids composition of livers from deep-sea sharks caught in Tasmanian waters. *Mar. Freshw. Res.* **1990**, *41*, 375–387. [CrossRef]

10. Wetherbee, B.M.; Nichols, P.D. Lipid composition of the liver oil of deep-sea sharks from the chatham rise, New Zealand. *Comp. Biochem. Physiol. B Biochem. Mol. Biol. B* **2000**, *125*, 511–521. [CrossRef]

11. Ryan, E.; Galvin, K.; O'Connor, T.P.; Maguire, A.R.; O'Brien, N.M. Phytosterol, squalene, tocopherol content and fatty acid profile of selected seeds, grains, and legumes. *Plant Foods Hum. Nutr.* **2007**, *62*, 85–91. [CrossRef] [PubMed]

12. Liu, G.C.; Ahrens, E.H.; Schreibman, P.H.; Crouse, J.R. Measurement of squalene in human tissues and plasma: Validation and application. *J. Lipid Res.* **1976**, *17*, 38–45. [PubMed]

13. Grigoriadou, D.; Androulaki, A.; Psomiadou, E.; Tsimidou, M.Z. Solid phase extraction in the analysis of squalene and tocopherols in olive oil. *Food Chem.* **2007**, *105*, 675–680. [CrossRef]

14. AOAC (Association of Official Analytical Chemists). *Squalene in Oils and Fats. Titrimetric Method, Official Method AOAC*; AOAC: Rockville, MD, USA, 1999.

15. Popa, O.; Băbeanu, N.E.; Popa, I.; Niță, S.; Dinu-Pârvu, C.E. Methods for Obtaining and Determination of Squalene from Natural Sources. *BioMed Res. Int.* **2015**. [CrossRef] [PubMed]

16. Bondioli, P.; Mariani, C.; Lanzani, A.; Fedeli, E.; Muller, A. Squalene recovery from olive oil deodorizer distillates. *J. Am. Oil Chem. Soc.* **1993**, *70*, 763–766. [CrossRef]

17. De Leonardis, A.; Macciola, V.; De Felice, M. Rapid determination of squalene in virgin olive oils using gas-liquid chromatography. *Ital. J. Food Sci.* **1998**, *10*, 75–80.

18. Giacometti, J. Determination of aliphatic alcohols, squalene, α-tocopherol and sterols in olive oils: Direct method involving gas chromatography of the unsaponifiable fraction following silylation. *Analyst* **2001**, *126*, 472–475. [CrossRef] [PubMed]

19. He, H.P.; Cai, Y.; Sun, M.; Corke, H. Extraction and purification of squalene from *Amaranthus* grain. *J. Agric. Food Chem.* **2002**, *50*, 368–372. [CrossRef] [PubMed]

20. Nenadis, N.; Tsimidou, M. Determination of squalene in olive oil using fractional crystallization for sample preparation. *J. Am. Oil Chem. Soc.* **2002**, *79*, 257–259. [CrossRef]

21. Oueslati, I.; Anniva, C.; Daoud, D.; Tsimidou, M.Z.; Zarrouk, M. Virgin olive oil (VOO) production in Tunisia: The commercial potential of the major olive varieties from the arid Tataouine zone. *Food Chem.* **2009**, *112*, 733–741. [CrossRef]

22. Owen, R.W.; Giacosa, A.; Hull, W.E.; Haubner, R.; Wurtele, G.; Spiegelhalder, B.; Bartsch, H. Olive-oil consumption and health: The possible role of antioxidants. *Lancet Oncol.* **2000**, *1*, 107–112. [CrossRef]

23. Villén, J.; Blanch, G.P.; Ruiz del Castillo, M.L.; Herraiz, M. Rapid and simultaneous analysis of free sterols, tocopherols, and squalene in edible oils by coupled Reversed-Phase Liquid Chromatography-gas chromatography. *J. Agric. Food Chem.* **1998**, *46*, 1419–1422. [CrossRef]

24. Russo, A.; Muzzalupo, I.; Perri, E.; Sindona, G. A new method for detection of squalene in olive oils by mass spectrometry. *J. Biotechnol.* **2010**, *150*, 296–297. [CrossRef]

25. Mannina, L.; Sobolev, A.P. High resolution NMR characterization of olive oils in terms of quality, authenticity and geographical origin. *Magn. Reson. Chem.* **2011**, *49*, 3–11. [CrossRef] [PubMed]

26. Dais, P.; Hatzakis, E. Quality assessment and authentication of virgin olive oil by NMR spectroscopy: A critical review. *Anal. Chim. Acta* **2013**, *765*, 1–27. [CrossRef] [PubMed]

27. Alexandri, E.; Ahmed, R.; Siddiqui, H.; Choudhary, M.I.; Tsiafoulis, C.G.; Gerothanassis, I.P. High resolution NMR spectroscopy as structural and analytical tool for unsaturated lipids in solution. *Molecules* **2017**, *22*, 1663. [CrossRef] [PubMed]

28. Alonso-Salces, R.M.; Héberger, K.; Holland, M.V.; Moreno-Rojas, J.M.; Mariani, C.; Bellan, G.; Reniero, F.; Guillou, C. Multivariate analysis of NMR fingerprint of the unsaponifiable fraction of virgin olive oils for authentication purposes. *Food Chem.* **2010**, *118*, 956–965. [CrossRef]

29. Mannina, L.; D'Imperio, M.; Capitani, D.; Rezzi, S.; Guillou, C.; Mavromoustakos, T.; Vilchez, M.D.; Fernández, A.; Thomas, F.; Aparicio, R. ^{1}H NMR-based protocol for the detection of adulterations of refined olive oil with refined Hazelnut oil. *J. Agric. Food Chem.* **2009**, *57*, 11550–11556. [CrossRef] [PubMed]

30. Robosky, L.C.; Wade, K.; Woolson, D.; Baker, J.D.; Manning, M.L.; Gage, D.A.; Reily, M.D. Quantitative evaluation of sebum lipid components with nuclear magnetic resonance. *J. Lipid Res.* **2008**, *49*, 686–692. [CrossRef] [PubMed]

31. Kontogianni, V.G.; Exarchou, V.; Troganis, A.; Gerothanassis, I.P. Rapid and novel discrimination and quantification of oleanolic and ursolic acids in complex plant extracts using two-dimensional nuclear magnetic resonance spectroscopy—Comparison with HPLC methods. *Anal. Chim. Acta* **2009**, *635*, 188–195. [CrossRef] [PubMed]

32. Tsiafoulis, C.G.; Skarlas, T.; Tzamaloukas, O.; Miltiadou, D.; Gerothanassis, I.P. Direct nuclear magnetic resonance identification and quantification of geometric isomers of conjugated linoleic acid in milk fraction without derivatization steps: Overcoming sensitivity and resolution barriers. *Anal. Chim. Acta* **2014**, *821*, 62–71. [CrossRef] [PubMed]

33. Blanc, M.C.; Bradesi, P.; Casanova, J. Identification and Quantitative Determination of Eudesman-Type Acids from *Dittrichia viscosa* sp. *viscosa* Essential Oil using ^{13}C-NMR Spectroscopy. *Phytochem. Anal.* **2005**, *16*, 150–154. [CrossRef] [PubMed]

34. Ferrari, B.; Castilho, P.; Tomi, F.; Rodrigues, A.I.; Costa, M.C.; Casanova, J. Direct Identification and Quantitative Determination of Costunolide and Dehydrocostuslactone in *Laurus novocanariensis* Fixed Oil using ^{13}C-NMR Spectroscopy. *Phytochem. Anal.* **2005**, *16*, 104–107. [CrossRef] [PubMed]

35. Nam, A.M.; Paoli, M.; Castola, V.; Casanova, J.; Bighelli, A. Identification and Quantitative Determination of Lignans in *Cedrus atlantica* Resins using 13C NMR Spectroscopy. *Nat. Prod. Commun.* **2011**, *6*, 379–385. [PubMed]

36. Castola, V.; Bighelli, A.; Casanova, J. Direct Qualitative and Quantitative Analysis of Triterpenes Using ^{13}C NMR Spectroscopy Exemplified by Dichloromethanic Extracts of Cork. *Appl. Spectrosc.* **1999**, *53*, 344–350. [CrossRef]

37. Duquesnoy, E.; Castola, V.; Casanova, J. Identification and quantitative determination of triterpenes in the hexane extract of *Olea europaea* L. leaves using ^{13}C NMR spectroscopy. *Phytochem. Anal.* **2007**, *18*, 347–353. [CrossRef] [PubMed]

38. Pogliani, L.; Ceruti, M.; Ricchiardi, G.; Viterbo, D. An NMR and molecular mechanics study of squalene and squalene derivatives. *Chem. Phys. Lipids* **1994**, *70*, 21–34. [CrossRef]

39. Bronzini de Caraffa, V.; Gambotti, C.; Giannettini, J.; Maury, J.; Berti, L.; Gandemer, G. Using lipid profiles and genotypes for the characterization of Corsican olive oils. *Eur. J. Lipid Sci. Technol.* **2008**, *110*, 40–47. [CrossRef]

40. Duquesnoy, E.; Castola, V.; Casanova, J. Identification and quantitative determination of carbohydrates in ethanol extracts of two conifers using ^{13}C NMR spectroscopy. *Carbohydr. Res.* **2008**, *343*, 893–902. [CrossRef] [PubMed]

41. Paoli, M.; Bighelli, A.; Castola, V.; Tomi, F.; Casanova, J. Quantification of taxanes in a leaf and twig extract from *Taxus baccata* L. using ^{13}C NMR spectroscopy. *Magn. Reson. Chem.* **2013**, *51*, 756–761. [CrossRef] [PubMed]

42. Günther, H. *La Spectroscopie de RMN. Principes de Base, Concepts et Applications de la Spectroscopie de Résonance Magnétique Nucléaire du Proton et du Carbone-13 en Chimie*; Masson: Paris, France, 1994.

43. Becker, E.D.; Ferretti, J.A.; Gambhir, P.N. Selection of optimum parameters for pulse Fourier transform nuclear magnetic resonance. *Anal. Chem.* **1979**, *51*, 1413–1420. [CrossRef]

magnetochemistry

MDPI

Review

Nuclear Magnetic Resonance, a Powerful Tool in Cultural Heritage

Noemi Proietti, Donatella Capitani * and Valeria Di Tullio

Laboratorio di Risonanza Magnetica "Annalaura Segre", Istituto di Metodologie Chimiche, CNR Area Della Ricerca di Roma 1, Via Salaria km. 29,300, 00015 Monterotondo, Italy; noemi.proietti@cnr.it (N.P.); valeria.ditullio@cnr.it (V.D.T.)
* Correspondence: donatella.capitani@cnr.it; Tel.: +39-06-90672700

Received: 28 December 2017; Accepted: 13 January 2018; Published: 17 January 2018

Abstract: In this paper five case studies illustrating applications of NMR (Nuclear Magnetic Resonance) in the field of cultural heritage, are reported. Different issues were afforded, namely the investigation of advanced cleaning systems, the quantitative mapping of moisture in historic walls, the investigation and evaluation of restoration treatments on porous stones, the stratigraphy of wall paintings, and the detection of CO_2 in lapis lazuli. Four of these case studies deal with the use of portable NMR sensors which allow non-destructive and non-invasive investigation in situ. The diversity among cases reported demonstrates that NMR can be extensively applied in the field of cultural heritage.

Keywords: portable NMR; NMR stratigraphy; porous stone; wall paintings; cultural heritage

1. Introduction

Nuclear Magnetic Resonance (NMR) is a powerful tool in research and diagnostics in many different fields. Despite its versatility, NMR is inherently insensitive compared with other analytical techniques. To overcome this problem, new NMR techniques and probeheads are continuously being developed and improved.

High resolution NMR spectroscopy allows the investigation of the structure of liquid [1] and solid [2,3] samples. New probeheads such as microprobes and cryoprobes [4,5] have enabled the NMR analysis of liquid samples down to the μg scale. In the case of cryogenically cooled NMR probeheads the detection coil and the preamplifier are cooled at low temperature in order to achieve a net increase in sensitivity. Microprobes allow the sample to be confined in a small volume. In the early times of solid state NMR about 400–500 mg of sample were required for performing the analysis. Nowadays the use of new microprobes and technologies to enhance the sensitivity of rare nuclei has allowed the amount of sample to be reduced down to a few mg, making NMR a suitable technique for analyzing solid samples available only in a small amount. High Resolution Magic Angle Spinning (HRMAS) [6] allows the analysis of a very low amount of soft matter.

Monitoring and diagnosis of artworks enable the prevention or delay of their degradation. The knowledge of the state of degradation of artworks, the characterization of constitutive materials, and the development of new methods and materials aimed at lengthening the life time of artworks, are important achievements in the correct safeguard of Cultural Heritage. The amount of samples obtained from precious artifacts must be reduced to a minimum, therefore micro-destructive, non-destructive, and possibly non-invasive analytical techniques are advisable. Another problem is that often objects to be analyzed cannot be brought to the laboratory because they are too precious or unmovable. These difficulties have been overcome with the use of portable NMR sensors which allow the study of arbitrarily sized objects in a non-destructive and non-invasive way [7]. The magnetic field generated by these sensors is external to the sensor and allows measurements to be performed inside the object under investigation [8]. Because the magnetic field generated by these sensors is inhomogeneous, the signal (Free Induction Decay-FID) decays too quickly to be observed, hence it must be detected as an echo [9]. Note that the inhomogeneous field is a source of relaxation which makes the transverse relaxation time shorter than that which would be measured in a homogeneous field. Nevertheless, these sensors allow the measurements of important NMR parameters such as the proton population, the self-diffusion coefficient D, relaxation times T_1 and T_2, and even allow the acquisition of correlation 2D-maps such as (T_1, T_2) and (T_2, D).

In recent years several issues regarding cultural heritage have been afforded by NMR [10–12]. Among these are wall paintings and oil paintings [13–19], paper [20,21], moisture in historical walls [22–24], lead soaps [25–28], pigments [29,30], organic paint binders [31,32], modern art materials [33–35], archaeological and fossil wood [36–39], ancient leather [40,41], amber [42,43], ancient pottery [44–46], porous stones [47–51], and advanced cleaning and conservation systems for cultural heritage [52–55].

In this paper some examples of the use of NMR in affording issues involving cultural heritage, are reported.

2. Results and Discussion

2.1. Cleaning Process of Cultural Heritage Studied by Single-Sided NMR

The cleaning of cultural heritage is carried out to remove layers that may damage artefacts such as salt efflorescences, metal stains, crusts, or layers formed by natural ageing [56–58]. The cleaning procedure is a delicate and potentially damaging operation to the artefact [59]. To improve the procedure highly performing cleaning systems are required [52,60,61].

An important improvement to classical cleaning procedures has been the formulation of solutions confined into host systems like physical and chemical gels. The use of thickeners and gels in chemical cleaning treatments helps to reduce solvent permeation into the porous microstructure and to control the effect of the cleaning action on the surface of artefacts. Thickeners and gels can be used with water as the principal cleaning method or in combination with other methods. In the latest case, they may be modified by adding compounds able to target a specific type of stain or surface coating. These materials are purposely designed to remove layers and contaminants that are insoluble in water and/or have penetrated deep into the surface of the artefact. The most common materials used to obtain thickeners include:

1. Natural and synthetic silicates, i.e., clays such as sepiolite and attapulgite.
2. Cellulosic and wooden materials, i.e., cotton fibers, paper pulp, wooden pulp.
3. Gelling materials, i.e., natural polysaccharide-based gels such as agarose and agaropectine, Gellan gum, Xanthan gum, etc.

Sepiolite is a fine, grey-white naturally occurring clay containing hydrated magnesium trisilicate. It consists of extremely small laths, or needle-shaped particles each with twenty four to twenty seven channels running lengthwise which help to create a system of pores. Pores are responsible for the ability of the material to absorb an amount of solvent suitable to form a thick paste. Some drawbacks must be taken into account. For example the complete removal of particles from an irregular and damaged surface is sometimes difficult, and the impurities present in the clay may cause additional problems. In the case of sepiolite the skin should be protected as it has been shown to act as an irritant.

Basically, synthetic silicates are related to the natural clay mineral $[Si_8Mg_{5.34} Li_{0.66} (Ca, Na)0.66]$, a tri-octahedral sheet silicate. The free-flowing powder forms a thixotropic gel when mixed with liquid, and stays in place when applied. Furthermore the risk of skin irritation after prolonged contact is low.

Gellan gum is an anionic polysaccharide produced by the bacterium *Sphingomonas elodea*. At room temperature the gel is rigid due to its macro-reticulate structure [61].

Carbogel is made of a neutralized polyacrylic acid, which allows a gel to be prepared by simply adding water, whose viscosity can be varied at will. For example, an amount of Carbogel between

0.5% and 4% by weight in aqueous solution is sufficient to yield a highly viscous gel. Carbogel has also a high water retention capacity, and water evaporates over a very long time.

Due to the absence of chemical aggressiveness and toxicity, water based cleaning systems are preferred with respect to traditional systems. In recent years, the use of *Desulfovibrio vulgaris* bacteria confined in hydrogels has become a quoted method to remove sulfates from stone material belonging to cultural heritage. Thanks to their metabolism, sulfate-reducing bacteria *D. Vulgaris* are able to reduce SO_4^{2-} ions to H_2S. Compared with other traditional methods, this procedure results in a more homogeneous removal of the surface deposits and preserves the layers under salt efflorescences and black crusts. In the case of wall paintings this method is used by applying sulfate-reducing bacteria entrapped in hydrogel poultices.

In this study hydrogels combined with sulfate-reducing bacteria formulations were applied on wall painting specimens affected by sulfate efflorescence. After the application of gel formulations, specimens were investigated by single-sided NMR. These measurements allowed detailed information to be obtained on the interaction between water molecules, the gel network, and the porous matrix.

Water transport in Carbogel, Sepiolite, Laponite RD and Gellan gels was investigated by measuring the longitudinal relaxation time (T_1), the effective transverse relaxation time (T_{2eff}), and the self-diffusion coefficient (D).

Relaxation times of polymer gels depend on the molecular weight, degree of branching, crosslink density, size of the side groups, and temperature [62]. Relaxation measurements are affected by both rotational and translational molecular motions, whereas diffusion measurements are related only to the translational molecular displacement. T_1 and T_2 values along with the corresponding populations found in Carbogel, Sepiolite, Laponite RD, and Gellan gels are reported in Table 1.

Table 1. T_1 and T_2 relaxation times and relative populations, and the self-diffusion coefficient measured in Carbogel 0.5%, Laponite RD 5%, Sepiolite 6%, and Gellan 2%.

Sample	T_1 (s)	W_A (%)	T_{2A} (ms)	W_B (%)	T_{2B} (ms)	$D \times 10^{-9}$ m^2/s
Carbogel 0.5%	1.7 ± 0.1	10	6.4 ± 0.8	90	52.2 ± 0.4	2.05 ± 0.05
Laponite RD 5%	1.2 ± 0.1	9	4.7 ± 0.4	91	41.9 ± 0.1	1.89 ± 0.02
Sepiolite 6%	0.03 ± 0.01	29	3.4 ± 0.4	71	11.2 ± 0.1	1.75 ± 0.03
Gellan 2%	0.7 ± 0.1	5	5.6 ± 0.5	95	39.2 ± 0.1	1.97 ± 0.02

The very short T_1 value measured in sepiolite is likely to be due to the presence of paramagnetic impurities, such as iron, which are present in the natural clay. These impurities should be absent in laponite which is a synthetic clay.

As reported in the literature [63], the transverse relaxation time of a nucleus is sensitive to slow and fast motions, whereas the longitudinal relaxation time is sensitive to fast motions. The observation that T_2 was found to be much shorter than T_1 indicated that water in gels had a slowed motion due to the presence of the network of the gel. The presence of two transverse relaxation times was ascribed to two proton species, the shortest one, T_{2A}, related to 'bound' water, and the longest one, T_{2B}, related to 'free' water. Proton nuclei in the 'bound' water domain are affected by hydrogen bonding interactions and/or chemical exchange with the gel network.

In order to study the translational motion of water molecules in the gel system, the self-diffusion coefficient was measured. This parameter may give information about molecular dynamics and sample microstructure [64,65].

The self-diffusion coefficients measured in Carbogel, Gellan gum, Sepiolite, and Laponite gels are reported in Table 1. All self-diffusion coefficients were found to be slower than that of bulk water (2.1×10^{-9} m^2/s at room temperature) indicating that the translational motion of water molecules is more hindered in these gels than in bulk water. Specifically, the water self-diffusion coefficient in silicates was found to be slower than that measured in Carbogel and Gellan. This is likely to be due to the restricted diffusion caused by the basic structure of phyllosilicates constituted of interconnected six member rings of SiO_4^{4-} tetrahedra extending outward in infinite sheets packaged into layer networks which impair free water diffusion.

The mean square displacement $\langle r^2 \rangle$ of water molecules with a self-diffusion coefficient D during a diffusion time Δ can be calculated from the relationship:

$$\langle r^2 \rangle = 2D\Delta$$

In a diffusion time Δ of 10 ms, bulk water molecules should undergo a mean square displacement of about 6.48 μm (diffusion length). During this diffusion time, the translational motion of water molecules in gels was increasingly hindered as the average probability of collision between water molecules and the gel network increased. The mean square displacement slightly decreased to 6.33 μm in Carbogel, and to 6.29 in Gellan gum, whereas a more marked shortening was obtained in Laponite and Sepiolite with a mean square displacement of 6.15 and 5.92 μm, respectively.

^{1}H NMR depth profiles were collected to evaluate the amount of water absorbed by the wall painting specimens and the depth of penetration of water after the application of the cleaning system. The amplitude of the ^{1}H depth profile is proportional to the amount of water absorbed by the specimen as a function of the depth scanned. Therefore, each experimental point represents the amount of water at the corresponding depth of measurement. The trend shown by depth profiles depended on the capability of the cleaning system to release water into the specimen. Figure 1a reports the depth profiles collected after applying Carbogel, Sepiolite, Laponite, and Gellan gum/bacteria cleaning systems for 1 h to wall painting specimens. Carbogel released the largest amount of water. Sepiolite and Laponite RD exhibited a very similar trend. The Gellan gum/bacteria system exhibited the lowest water release capability. The same trend was found after a 4-h application, see Figure 1b.

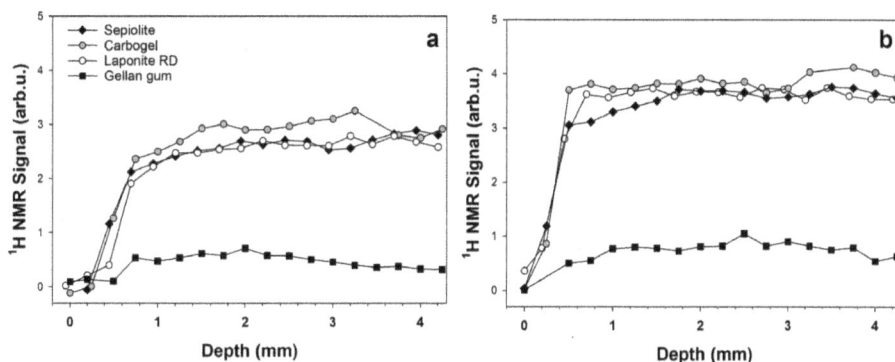

Figure 1. ^{1}H NMR depth profiles of water released into wall painting specimens after (**a**) 1-h and (**b**) 4-h application of Carbogel, Sepiolite, Laponite, and Gellan gum/bacteria cleaning systems.

To summarize, by single-sided ^{1}H NMR we investigated, in a fully non-invasive and non-destructive way, different types of gel formulations to be used as cleaning systems for wall painting. Transverse relaxation times measured in gel formulations indicated the presence of water in two environments. Bound water exhibiting the shortest relaxation time was involved in hydrogen bonding interactions and/or proton exchange with gel macromolecules, whereas the water molecules with the longest relaxation time were free to move through the network, although self-diffusion measurements demonstrated that their translational motion was increasingly hindered as the average probability of collisions with the gel network increased. The release and diffusion of water at the gel-specimen interface were also investigated by ^{1}H NMR depth profiles. The profiles indicated that the Gellan gum/bacteria system is the one showing the lowest amount of water release. All systems exhibited a penetration depth of the released water deeper than 4 mm, see Figure 1.

Water absorption, distribution and penetration in the support depend not only on the physical-chemical and rheological properties of the gel but also on the properties of the support to which the gel is applied. This study is part of a more extensive research effort aimed at obtaining an analytical protocol for establishing in situ the performance of cleaning systems.

^{1}H NMR depth profiles allow some questions to be answered such as the thickness of an unwanted organic layer to be removed from the surface of a wall painting, its distribution in the wall painting, and the presence of residues on the wall painting surface after performing a cleaning treatment.

For example, shellac is a natural organic resin of animal origin that was widely used in the past as a varnish for wall paintings. When shellac ages, it becomes insoluble, changes color, and becomes very hard to remove. Many old paintings became yellowish due to the shellac coating.

Single-sided NMR was applied before and after performing a cleaning procedure to remove a layer of degraded shellac [55]. Figure 2 shows the ^{1}H depth profiles collected before and after the cleaning. Basically, the profiles exhibited two regions. At depths deeper than about 0.6 mm the amplitude of the profile was due to the NMR signal of the moisture in the porous structure. Between 0.2 and 0.6 mm, the profile collected before the cleaning procedure showed a rather intense signal of the hydrogen atoms of shellac, this region was about 300 μm thick. After the cleaning, due to a decrease in the hydrogen level of shellac, the intensity of the profile was found to be definitely weaker than the profile before cleaning, and the thickness of the degraded organic layer reduced to about 100 μm. Actually, the intensity of the profiles between 0.2 and 0.6 mm indicated the efficiency of the cleaning treatment, the more effective the treatment, the lower the level of hydrogens encoded by the profile after the cleaning.

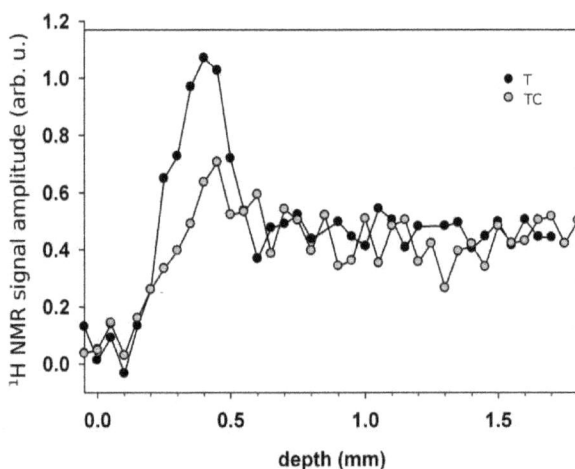

Figure 2. ^{1}H NMR depth profiles in the presence of a degraded organic layer of shellac (T), and after applying a cleaning procedure to remove shellac (TC). Adapted from [55]. Copyright Nova Science Publishers, Inc., 2013.

This technique may be applied in situ in a non-destructive and non-invasive way to evaluate the efficiency of cleaning methods aimed at removing degraded organic layers from artworks.

2.2. Detection of Moisture in Historical Walls and Wall Paintings

Water plays a fundamental role in the degradation of porous materials such as historical walls, wall paintings, stones, mortar, concrete etc. [66,67]. In many cases water may affect the material causing fractures, stress, and expansion. Condensation/evaporation cycles facilitate the migration of solubilized salts which may crystallize on the surface of the porous material causing efflorescence, and sometimes crumbling and fretting of the surface. Capillary rise of water and collected rainfall are among the main causes of the presence of dampness. Wetting/drying cycles cause solubilization and recrystallization of salts inside the porous structure inducing the breaking down of the structure itself. Chemical reactions may occur between pollutants deposited on the surface and the porous material. The presence of an anomalous amount of moisture may also cause the formation of crusts on monuments. Specifically, runoff is associated with the presence of white areas due to reprecipitated crystals of calcite formed when water evaporates. Condensation is associated with the presence of grey regions where the porous material has been previously covered by a layer of dust and particles [66]. In general, many degradation processes such as leaching, transport, accumulation, solute precipitation and fractionation, as well as biological colonization are caused by the moisture content of the masonry. Moreover the moisture content is also affected by seasonal variation and environmental conditions.

Despite its importance, the quantitative detection of water distribution in precious artefacts such as historical walls or wall paintings, is difficult to achieve. Methods used to obtain information on the presence of moisture are gravimetric tests, IR thermography (IRT), and electrical conductivity. However these methods show some drawbacks. In fact gravimetric tests require sampling, IRT does not allow a quantitative evaluation of moisture, and electrical conductivity is affected by the presence of salts.

Portable NMR sensors have permitted the mapping of water distribution in historical walls and wall paintings [23,24]. Figure 3 shows a portable NMR sensor detecting moisture in an ancient wall painting affected by capillary rise of water from the ground.

Figure 3. Single-sided NMR measuring moisture in a wall painting dating back to the XV century located in Nostra Signora del Sacro Cuore Church, Rome, Italy.

Water distribution was quantitatively determined in two regions A and B of a tuff wall affected by capillary rise of water from the ground. Measurements were carried out choosing a matrix of 21 points in region A and 21 points in region B. Each experimental point covered an area of 2×5 cm^2 corresponding to the area of the probehead, see Figure 4a,b. Experimental data were processed to obtain a contour plot [68,69]. As previously reported [23], a contour plot is a graphical way of obtaining a 2D representation of a 3D surface, where x and y are the coordinates of a strip of the region of the tuff wall, and z is the integral of the NMR signal. In the contour plots obtained the difference in the moisture level is shown as a gradient of color, red corresponds to the lowest water content, while dark blue corresponds to the highest water content. In both maps the distribution of the moisture gave a clear image of the front of the rising damp. According to a suitable calibration procedure [23], in map A the maximum value of the intensity of the NMR signal corresponded to 11% of the moisture content, whereas in map B it corresponded to 10%. Because in the fully saturated water specimens used for calibration the moisture content was found to be 25%, these results indicated a rather high level of moisture in the regions of the tuff wall affected by the capillary rise.

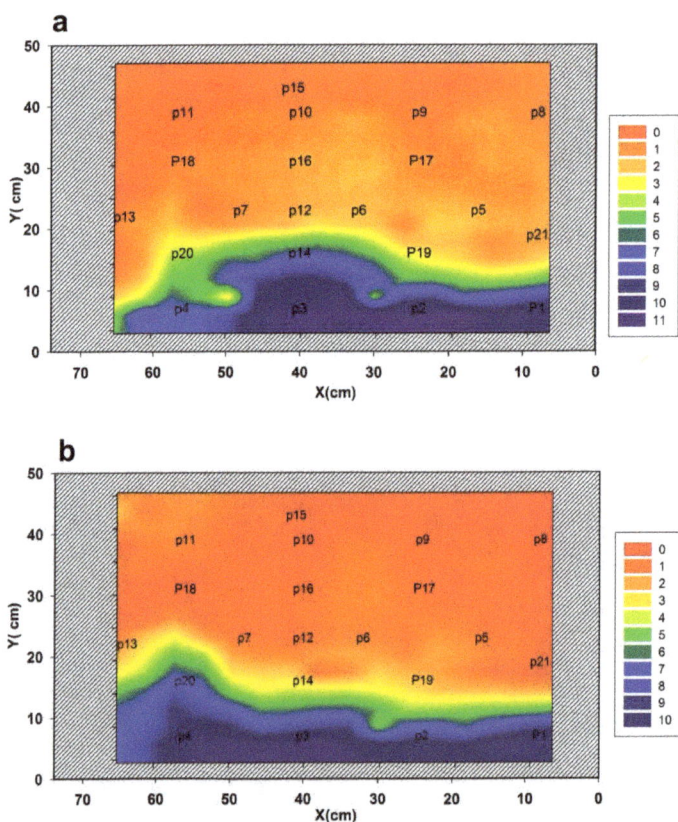

Figure 4. In (a,b) the moisture distribution maps of regions A and B of the tuff wall affected by capillary rise of water, are reported. The moisture level is shown as a gradient of color, red corresponds to the lowest water content, while dark blue corresponds to the highest water content.

In region A transverse relaxation time measurements were performed as a function of the height of the wall. After applying an inverse Laplace transformation to the magnetization decays measured with the Carr Purcell Meiboom Gill CPMG sequence, the distributions of transverse relaxation times were obtained, see Figure 5. In this representation, peaks represent the most probable T_2 values, whereas peak areas represent the population of each component. The distributions reported in Figure 5 indicate the presence of water confined in pores with various sizes. In fact relaxation times of a fluid confined in a porous structure are strictly related to its geometry, as water confined in small pores relaxes faster than water confined in large pores [70]. Therefore at the lowest height of 7 cm (P3), where the highest amount of water was measured (about 11%), up to four peaks are observed indicating the heterogeneous nature of tuff with water confined in pores having a different size. At a height of 16 cm (P14) water of type 4 confined in the largest pores disappeared and only three types of water were observed. At a height of 23 cm (P12) the amount of water of type 3 decreased, and at 31 cm (P16) it decreased further. Water of type 3 fully disappeared at a height of 39 cm (P10). Basically, by increasing the height of measurement, the decrease of the total amount of water indicated by the moisture distribution map reported in Figure 4a was accompanied by a progressive decrease of the amount of water confined in large pores. Values of transverse relaxation times and the corresponding populations are reported in Table 2.

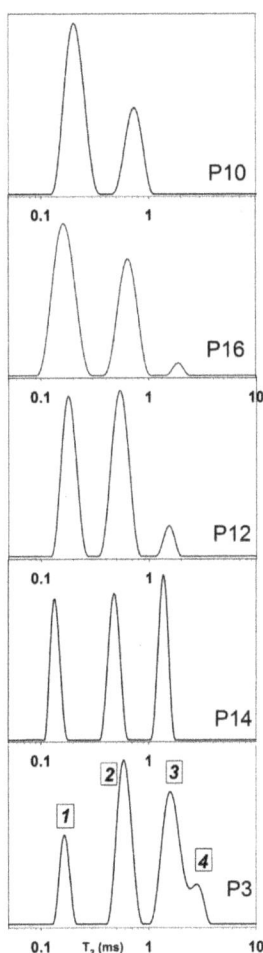

Figure 5. Transverse relaxation time distributions measured at different heights in region A of the tuff wall.

Table 2. T_2 relaxation times and relative populations obtained at different heights in region A of the tuff wall. The error on the parameters is less than 10% of the nominal value.

(x,y) (cm)	W_a (%)	T_{2a} (ms)	W_b (%)	T_{2b} (ms)	W_c (%)	T_{2c} (ms)	W_d (%)	T_{2d} (ms)
P3 (40,7)	14	0.17	38	0.60	40	1.63	8	2.85
P14 (40,16)	29	0.15	36	0.40	34	1.57	–	-
P12 (40,23)	43	0.18	52	0.54	5	1.58	–	-
P16 (40,31)	58	0.17	40	0.65	2	1.7	–	-
P10 (40,39)	69	0.20	21	0.80	–	–	–	-

To summarize, using single-sided NMR a detailed map of the moisture distribution in two regions of the tuff wall affected by the capillary rise of water from the ground, was obtained. A proper calibration of the intensity of the NMR signal allowed a quantitative determination of the moisture level. Measurement of relaxation times evidenced the presence of water confined in pores of various sizes.

2.3. Protective and Consolidating Treatment on Lapideous Material Studied by Single-Sided NMR

2.3.1. Sandstone Specimens Treated with Hydrophobic Treatments

As previously mentioned, water plays a fundamental role in the degradation of porous stones. A protective treatment is carried out to prevent the penetration of water and decay agents in the porous structure and, therefore, to delay degradation processes [66,67]. A suitable protective treatment should also permit vapor leakage and avoid alteration in the optical properties of the treated stone.

The choice of the treatment and its performances depend on many chemico-physical parameters such as the porosity and permeability of the material. Furthermore, information about the penetration depth of the treatment and the suitable time of application to the selected type of stone is important for choosing and optimizing the treatment.

Another important requirement in stone conservation is that the compound absorbed in the stone should not accumulate in some regions into the stone. In fact, in this case, chemical and physical inhomogeneity may occur between the impregnated layers of the stone and the layers underneath. These layers might differently respond to changes in thermo-hygrometric conditions causing mechanical damage to the porous structure.

Figure 6a reports the depth profile collected on water-saturated untreated sandstone (UT), whereas Figure 6b reports depth profiles collected on water-saturated sandstone specimens treated with dimethylsiloxane (DMS) for 5 (T5), 600 (T600), and 1800 s (T1800). The amplitude of these profiles depends on the level of absorbed water. Hence, by comparing the amplitude of the depth profiles of untreated and treated specimens, information on the hydrophobic action of the treatment is obtained.

Figure 6b shows that in treated specimens the amplitude of profiles was reduced as a function of the time of application of the hydrophobic treatment. Specifically, treatments for 600 and 1800 s halved the amount of absorbed water with respect to the amount absorbed in the untreated sandstone. At a depth deeper than 3 mm, the amplitude of profiles of treated specimens was well comparable with that of the untreated specimen, indicating that the hydrophobic action was no longer effective. The profiles were fit to Equation (4) with $k = 2$, see the solid lines through the experimental points.

Table 3 reports the parameters obtained from the best-fit. Note that x_1 and b_1 values, which encoded the fast rising initial part of profiles, were affected by surface effects, and, therefore, were not further considered in the following discussion. It is worth noting that the second inflection point x_2 observed in the profiles of all treated specimens depended on the duration of the hydrophobic treatment, see Table 3. However, at depths greater than x_2, the amplitude of profiles increased to reach the values measured in the untreated specimen, indicating that x_2 was the deepest depth at which the treatment was still capable of exerting a hydrophobic action. Table 3 also reports slopes at inflection points. Any change in slope in the profile indicates a variation of the amount of the absorbed product inside the stone, and, as a consequence, it is possible to evaluate the extent of inhomogeneity in the treated stone from the value of the slope at inflection points. For example, according to data reported in Table 3, the inhomogeneity is more pronounced in the specimen treated for 1800 s ($b_2 = 2.703$) than in that treated for 600 s ($b_2 = 2.174$). As a role, the greatest $|b_k|$ the more pronounced the inhomogeneity.

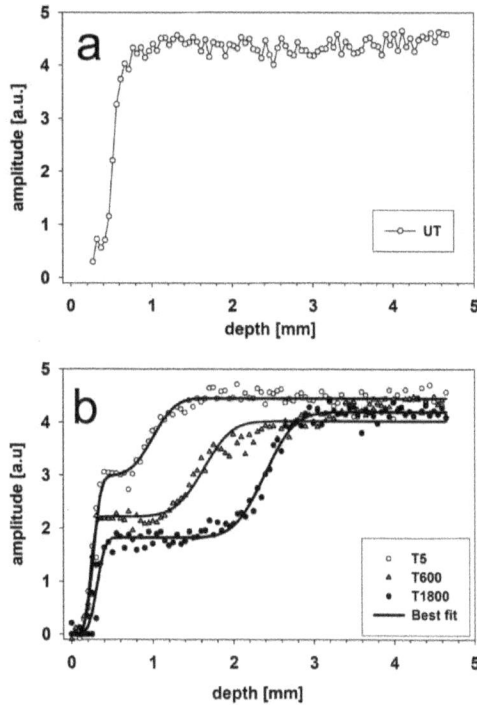

Figure 6. (a) ^{1}H NMR depth profile collected on a water-saturated untreated sandstone specimen (UT); (b) Profiles collected on water-saturated sandstone specimens treated with dimethylsiloxane (DMS) for 5 s (T5), 600 s (T600), and 1800 s (T1800). Adapted from [50].

Table 3. Parameters obtained fitting depth profiles to Equation (4) and slope at inflection points obtained from Equation (5).

Specimen	x_1 (mm)	Δ_1 (mm)	w_1 (arb.u.)	x_2 (mm)	Δ_2 (mm)	w_2 (arb.u.)	R^2	b_1	b_2
T5	0.27 ± 0.01	0.073 ± 0.008	2.97 ± 0.09	1.00 ± 0.03	0.22 ± 0.04	1.48 ± 0.09	0.99	14.29	2.325
T600	0.23 ± 0.01	0.05 ± 0.01	2.22 ± 0.06	1.63 ± 0.03	0.28 ± 0.04	1.80 ± 0.06	0.96	14.29	2.174
T1800	0.31 ± 0.01	0.08 ± 0.01	1.82 ± 0.03	2.39 ± 0.02	0.30 ± 0.02	2.37 ± 0.04	0.99	8.33	2.703

As is well-known, the relaxation times of fluids confined in porous media are strictly related to the geometry of the structure [70]. Whereas differences in longitudinal relaxation times (T_1) may be not sufficient to clearly identify water in different compartments, differences in transverse relaxation times (T_2) are usually more pronounced. Therefore transverse relaxation times permit the detection of changes of the open porosity which may occur after performing a treatment on a porous stone.

With the aim of investigating differences in the open porosity of sandstone before and after treatments, we measured transverse relaxation times. All CPMG decays exhibited a multi-exponential behavior indicating the presence of water distributed in pores with various sizes. Transverse relaxation times and the corresponding spin populations obtained fitting the decays to Equation (2), are reported in Table 4.

Table 4. T_2 relaxation times and relative populations measured in sandstone specimens. The error on the parameters is less than 10% of the nominal value.

Specimen	W_a (%)	T_{2a} (ms)	W_b (%)	T_{2b} (ms)	W_c (%)	T_{2c} (ms)
UT	40	0.18	40	0.9	20	4.4
T5	42	0.17	42	1.1	16	4.7
T600	38	0.20	44	0.71	16	4.3
T1800	31	0.20	44	0.93	25	4.1

All CPMG decays measured at a depth of 5 mm overlapped, see Figure 7, indicating that, at this depth, the hydrophobic treatment was no longer effective. This result agrees well with the result obtained from depth profiles reported in Figure 6b.

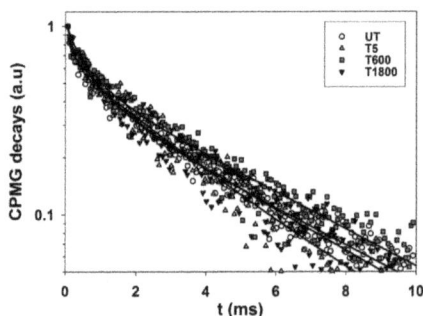

Figure 7. Carr Purcell Meiboom Gill CPMG magnetization decays measured at a depth of 5 mm on water-saturated sandstone specimens. Adapted from [50]. Copyright Spriger Nature, 2011.

For example, the effect of the hydrophobic treatment may be evidenced by comparing the distribution of transverse relaxation times measured at depths of 1 and 4 mm on the sandstone specimen treated for 1800 s, see Figure 8. In fact, the distribution at a depth of 4 mm showed three T_2 peaks centered at about 20, 6.5, and 1.7 ms corresponding to water in large, medium, and small pores, with a relative spin population of 6%, 42%, and 52%, respectively. Instead, the distribution at a depth of 1 mm showed only two peaks centered at about 7 and 1.8 ms with spin population of 15% and 85%, indicating that the largest amount of water was confined in small pores, whereas the peak corresponding to water confined in large pores was lacking. Again this result was in accordance with the result obtained from depth profiles reported in Figure 6b.

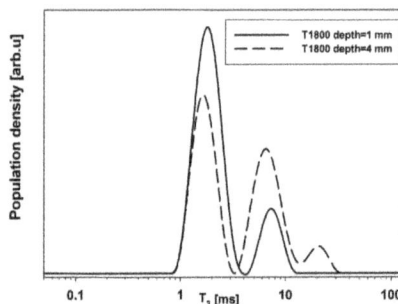

Figure 8. T_2 relaxation time distributions in sandstone treated with DMS for 1800 s (T1800) at a depth of 1 mm (solid line), and 4 mm (dashed line). Adapted from [50]. Copyright Spriger Nature, 2011.

2.3.2. An Unsuitable Consolidative Treatment on Calcarenite Studied by Single-Sided NMR

Figure 9 reports depth profiles collected on water-saturated calcarenite specimens untreated and consolidated with a solution of tetraethoxysilane (TEOS) in ethyl alcohol for 5 s (T5), 600 s (T600), and 1800 s (T1800). The amplitude of the profile of the specimen treated for 5 s overlapped with that of the untreated specimen indicating that the treatment for 5 s did not show any consolidative action. On the contrary, profiles of specimens treated for 600 and 1800 s clearly exhibited a very low amplitude with respect to the amplitude measured in the untreated specimen, indicating that water absorption was impaired by the treatment.

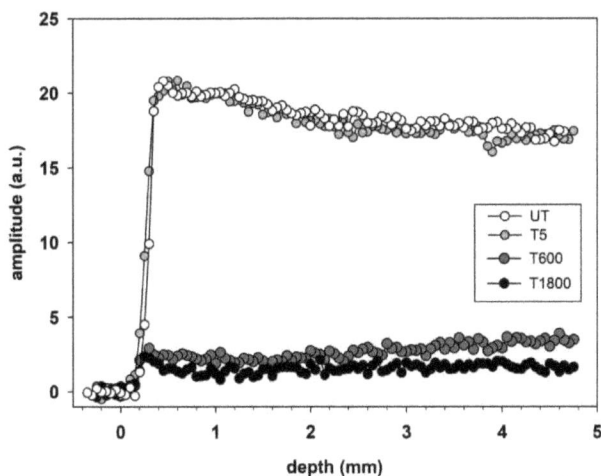

Figure 9. ^{1}H NMR depth profile collected on water-saturated untreated calcarenite (UT), and calcarenite treated with tetraethoxysilane (TEOS) for 5 s (T5), 600 s (T600), and 1800 s (T1800).

The amount of water in untreated calcarenite and in calcarenite treated for 600 s was obtained by numerical integration of the corresponding profiles applying the Newton–Cotes quadrature rule, see Figure 9. The data obtained indicated that the amount of water absorbed by consolidated calcarenite was reduced by about 80% with respect to the amount of water absorbed by untreated calcarenite. Because of the absence of water repellency of the TEOS treatment, the reduction of the amplitude of the profiles may be ascribed either to the pores coating that strongly reduced the pores volume available to water, or to the presence of a film formed on the surface of the specimens which hindered water penetration inside the stone. In the former case the consolidative treatment would have obstructed many pores deeply upsetting the porosity of the treated specimen along the full thickness. In the latter case the presence of a TEOS film would have strongly impaired the correct exchange of moisture between the porous structure and the environment. Both cases should be avoided when performing a consolidative treatment. In fact a suitable consolidative treatment should not markedly affect the water vapor permeability, assuring the correct stone breathing [71].

The effect of the consolidating treatment can be also rationalized by measuring T_2 values and the relative populations. As an example Table 5 reports the values measured at different depths in water-saturated untreated calcarenite and calcarenite, treated for 600 s. Figure 10 reports the corresponding transverse relaxation time distributions. Peaks of the distributions indicate the presence of water in pores with a different size in the analyzed sandstone. In untreated calcarenite the major amount of water was confined in large pores, with a percentage between 70% and 90% of the total amount of water, whereas in the treated calcarenite, the amount of water confined in large pores was strongly reduced. This effect is observed at any depth measured, namely 250, 800, 700, and 1000 μm, see Figure 10. To summarize, after the treatment the total amount of absorbed water obtained by integrating the profiles was reduced from 100% to 20%, and water was mostly distributed in small pores. It is evident that the treatment with the TEOS solution not only drastically reduced the amount of water absorbed by the specimen, but also made large pores almost completely unavailable to water along the full thickness of the treated specimen.

Table 5. T_2 relaxation times and relative populations measured in untreated calcarenite and calcarenite treated with TEOS for 600 s. Measurements were carried out at different depths in the specimens, namely 250, 500, 700, and 1000 μm. The error on the parameters is less than 10% of the nominal value.

Depth (μm)	Sample	W_a	T_{2a} (ms)	W_b	T_{2b} (ms)	W_c	T_{2c} (ms)	W_d	T_{2d} (ms)	W_e	T_{2e} (ms)
250	UT	3%	0.070	2%	0.52	-	-	14%	12.172	81%	40.1
	T600	14%	0.099	46%	0.94	35%	3.32	-	-	5%	106.0
500	UT	-	-	4%	0.81	17%	6.91	-	-	89%	33.00
	T600	6%	0.090	64%	1.25	24%	5.15	-	-	6%	122.6
700	UT	-	-	4%	0.85	10%	4.91	-	-	86%	22.3
	T600	16%	0.100	27%	0.43	48%	2.27	5%	13.1	4%	115.2
1000	UT	9%	0.083	4%	1.18	-	-	21%	9.8	66%	38.0
	T600	28%	0.126	60%	2.06	7%	6.84	-	-	5%	108.0

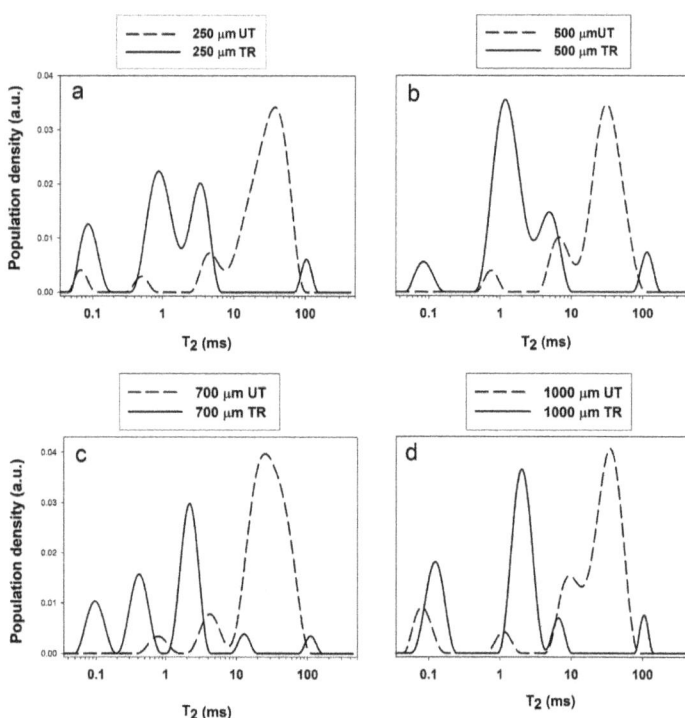

Figure 10. T_2 relaxation time distributions in untreated calcarenite (dashed lines) and calcarenite treated with TEOS for 600 s (solid lines) at 250 μm (**a**); 500 μm (**b**); 700 μm (**c**); and 1000 μm (**d**) of depth.

2.4. NMR Stratigraphy of Paintings

NMR stratigraphy is able to reveal non-invasively and in situ the different layers of a painting. The first NMR stratigraphy was published few years ago by Presciutti et al. [16]. The stratigraphy encodes the amplitude of the ^{1}H-NMR signal as a function of the depth scanned. The intensity of the NMR signal enables one layer to be differentiated from another one according to the hydrogen content.

As an example we report the stratigraphies collected on a Byzantine icon [19]. Figure 11a shows a portable NMR sensor while scanning the icon, while in b and c details of two regions measured by NMR are reported.

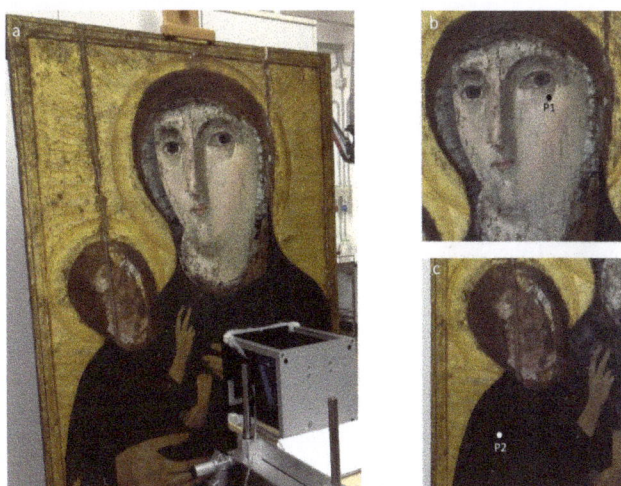

Figure 11. (**a**) Single-sided NMR collecting a stratigraphy of the Madonna Hodigitria icon. (**b,c**) Details of two regions of the icon measured by NMR. Adapted from [19].

In Figure 12a the stratigraphy of the Virgin's face (region P1), is shown. The stratigraphy enabled the detection of three layers, the first one, 0.5 mm thick was ascribed to the pictorial layer, the second one, 0.3 mm thick, was ascribed to the incamottatura (canvas + glue), and the third one was constituted by the wood of the panel.

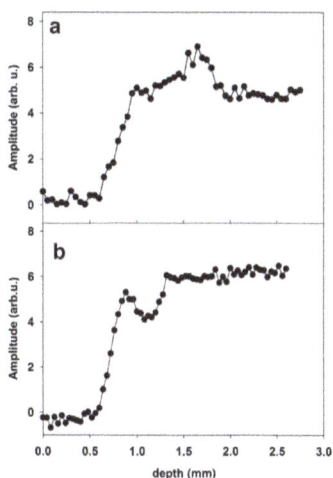

Figure 12. ^{1}H NMR stratigraphy of (**a**) region P1 (Virgin's face), and (**b**) region P2 (Virgin's mantle). Adapted from [19].

Figure 12b shows the stratigraphy of the Virgin's mantle (region P2). Three layers were observed, the first one, 0.4 mm thick, was ascribed to the pictorial layer, the second one, 0.5 mm thick, was ascribed to the primer, and the third one was the wood of the panel.

2.5. Detection of CO_2 in Lapis Lazuli by ^{13}C MAS NMR

The ultramarine pigment is obtained by purification of lapis lazuli, a semi-precious stone. It has been one of the most valued pigments in Europe since the 13th century. It was typically used to paint the robes and mantel of Christ and the Virgin. Ultramarine pigments are feldspathoids of sodalite structure. Natural and synthetic feldspathoids are able to absorb a variety of molecular species. Miliani et al. [72] applied Fourier Transform Infrared Spectroscopy (FTIR) to investigate the presence of adsorbed species on synthetic and natural ultramarine pigments. Their data indicated the presence of CO_2 in natural Afghan ultramarine, and suggested that CO_2 molecules, although not free to rotate, were loosely physisorbed to the sodalite cage [73].

The presence of CO_2 in the sodalite cage of natural Afghan ultramarine was confirmed by ^{13}C MAS NMR spectroscopy.

It has been shown that ^{13}C NMR may be successfully used to characterize the motion of molecules adsorbed on molecular sieves [74]. In particular, the ^{13}C NMR spectrum has been successfully used to investigate the local structure and dynamics of absorbed CO_2 in porous materials such as zeolites with micro- or mesopores [73]. We exploited ^{13}C MAS NMR spectroscopy to ascertain the presence of CO_2 in natural abundance included into the structure of a sample of Afghan lapis lazuli. In Figure 13 the ^{13}C MAS NMR spectra of a sample of Afghan lapis lazuli (bottom) and a sample of synthetic ultramarine (top) are compared. In the spectrum of Afghan lapis lazuli a sharp resonance at 125 ppm is observed and assigned to $^{13}CO_2$ while in the spectrum of synthetic ultramarine, the peak was not present. No other signals related to other carbon based materials (such as $CaCO_3$ or CO) were observed in the ^{13}C MAS NMR spectrum. Because the sample of Afghan lapis lazuli was finely ground the presence of CO_2 was not related to microfluid inclusions.

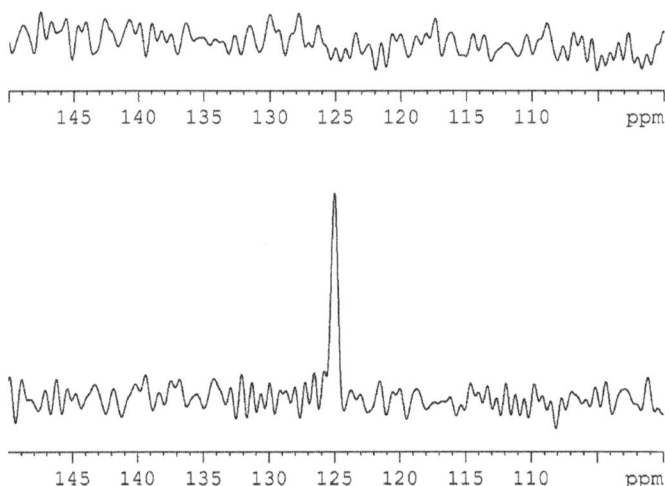

Figure 13. ^{13}C MAS NMR spectra of a sample of Afghan lapis lazuli (**bottom**), and a sample of synthetic ultramarine (**top**).

3. Materials and Methods

3.1. Gel Preparation and Application to Wall Painting Specimens

Carbogel, Gellan gum, Sepiolite, and Laponite RD gels were prepared at different concentrations. A 2% Gellan gum gel (Kelkogel, CTS srl, Altavilla Vicentina, (VI), Italy), a 0.5% Carbogel (CTS srl) and a 5% Laponite RD and a 6% Sepiolite clay (Bresciani srl, Milano, Italy), were prepared by dissolving the powder in water.

The gellan gum formulation (1 g gellan gum in 50 mL H_2O) was heated in a microwave oven at 750 W for 3 min and cooled to room temperature to obtain a rigid gel about 5 cm thick.

The cleaning formulations were prepared by dissolving 2.5 g lyophilized bacteria in 2 L water; 50 mL of this suspension was added in 100 mL gel. In the case of Gellan gum, the suspension was added during the gel cooling at about 40 °C.

Tests were performed on wall painting specimens (30 × 30 × 2 cm^3) prepared ad hoc with 8 mm arriccio (1 lime putty + 3 sand, grain size 0.1–1 mm) and 4 mm fine plaster (1 lime putty + 2 sand, grain size 0.5 mm max.). After a 3-h drying time, pigments were applied using the fresco technique. Gels were applied on the surface of specimens for 1 and 4 h. Analyses by portable NMR were performed on the hydrogel/bacteria systems and on wall painting specimens before and after the treatment.

3.2. Measurements on Hydrogel-Based Cleaning Systems

Non-invasive, non-destructive measurements were performed at 13.62 MHz with a portable NMR instrument from Bruker Biospin (Ettlingen, Germany) interfaced with a purposely built single-sided sensor by RWTH Aachen University, Aachen, Germany [75].

Longitudinal relaxation times T_1 were measured with the Saturation Recovery pulse sequence followed by a CPMG train in the detection period to increase the sensitivity [76]. Effective transverse relaxation times T_{2eff} [77,78] were measured with the CPMG pulse sequence, 4096 echoes were measured using an echo time 2τ of 71.2 μs. Due to the dead time of the probe (about 40 μs) and the strong steady magnetic field gradient, the transverse relaxation of non-exchangeable protons of the gel network which usually ranges between 10 and 20 μs, was too short to be detected [79]. In fact, in the presence of a magnetic field gradient, the transverse relaxation time does not depend only on the spin-spin interaction but it is also controlled by molecular diffusion in the magnetic field gradient that in the case of portable NMR is strong (14.28 T/m).

The longitudinal relaxation time was obtained fitting the magnetization decay to the following equation:

$$M(t) = M_0\left[1 - e^{\left[\frac{t}{T_1}\right]}\right]$$

(1)

where M_0 is the magnetization at the equilibrium.

Effective transverse relaxation times were obtained fitting the magnetization decay to the following equation:

$$Y(\tau) = \sum_{i=1}^{n} W_i e^{\frac{-2\tau}{T_{2i}}}$$

(2)

where n is the number of components, and W_i and T_{2i} are the weight and the transverse relaxation time of the ith component, respectively. Before fitting, the sum of weights was normalized to 100%.

^{1}H NMR depth profiles collected on specimens after the application of gels were obtained by applying the CPMG pulse sequence with an echo time of 86 μs and a nominal resolution of 23 μm. Profiles were acquired by repositioning the single-sided sensor in steps of 250 μm to scan the desired spatial range, from the surface of the specimen to a depth of about 5 mm.

Self-diffusion coefficients D were measured at 0.36 T (13.62 MHz) with a steady gradient (SG) of 14.28 T/m using a stimulated echo (SGSTE) pulse sequence followed by a CPMG echo train to improve the signal/noise. The attenuation of the spin echo resulting from the dephasing of nuclear spins was used to measure the molecular displacement according to the following equation:

$$\ln\left(\frac{A}{A_0}\right) = -\gamma^2 G^2 \tau_1^2 \left(\Delta + \frac{2}{3}\tau_1\right) D - \frac{2\tau_1}{T_2} - \frac{\Delta}{T_1}$$

where A_0 is the amplitude of the echo at the shortest time τ_1, D is the self-diffusion coefficient, Δ is the diffusion time, G is the steady magnetic field gradient, and γ is the ^{1}H gyromagnetic ratio (2.6752×10^8 s^{-1} rad T^{-1}). For large D values, strong magnetic field gradient, and provided that $\tau_1 \ll T_2$, and $\Delta \ll T_1$, diffusion terms dominate over relaxation terms and the self-diffusion coefficient may be obtained by the following equation:

$$\ln\left(\frac{A}{A_0}\right) = -\gamma^2 G^2 \tau_1^2 \left(\Delta + \frac{2}{3}\tau_1\right) D \tag{3}$$

The self-diffusion coefficient D of water molecules is obtained from the slope of the plot of $\ln\left(\frac{A_1}{A_0}\right)$ vs. $\gamma^2 G^2 \tau_1^2 \left(\Delta + \frac{2}{3}\tau_1\right)$.

The uncertainty associated with T_1, T_{2eff}, and D was obtained by repeating the measurement three times on each sample.

3.3. Moisture Measurements

Measurements were carried out in situ at the proton frequency of 16 MHz by a single-sided NMR instrument from Bruker Biospin. The probehead used detected the hydrogen signal from the moisture at 5 mm of depth inside the wall. The pulse width corresponding to the $\pi/2$ pulse was 10.4 μs, and the dead time was 15 μs. Because the inhomogeneous magnetic field generated by single-sided NMR makes the signal decay very quickly, the signal must be recovered stroboscopically as a Hahn echo [9]. The echo time (2τ) was set as short as possible, i.e., 20 μs.

3.4. Protective and Consolidating Treatment on Lapideous Material

3.4.1. Materials

Sandstone had a granularity between 200 and 800 μm. The finer class was made up of quartz, feldspars, and mica flakes and the coarser class by mica schists and rare bioclastic grains. The bounding material was mainly made of a clay matrix. The total porosity was found to be 10%, and the porosity distribution determined by mercury intrusion porosimetry (MIP) was found to be unimodal with a radius between 0.064 and 1 μm [50].

Calcarenite had a medium grain size less than 100 μm. It was an organogenic rock, mostly made up of micro-fossils and bearing a micritic matrix. The insoluble residue was constituted of clay minerals, silicate minerals, phosphates, and a small amount of ferrous components. The total porosity was found to be 35%, and the porosity distribution determined by MIP was found to be unimodal with a radius between 0.256 and 4 μm [50].

Stones were cut to obtain specimens with a size of $5 \times 5 \times 2$ cm^3.

3.4.2. Treatments Application

Before treating, all specimens were dried up to constant mass in a ventilated oven at a temperature of 60 ± 2 °C. Then specimens were kept for 24 h at 23 ± 2 °C and $50\% \pm 5\%$ relative humidity.

Sandstone specimens were treated with a commercially available protective and hydrophobic solution of dimethylsiloxane (DMS) in white spirit, whereas calcarenite specimens were treated with a commercially available consolidating solution of tetraethoxysilane (TEOS) in ethyl alcohol.

Treatments were absorbed by capillarity letting the specimens come in contact with the solutions for 5, 600, and 1800 s, respectively. Before and after treating, specimens were weighed taking care to ensure the complete evaporation of the solvent when the specimen weight reached a constant value. NMR measurements were carried out on water saturated specimens making the specimens absorb water by capillarity.

3.5. Depth Profiles and Transverse Relaxation Times Collected on Sandstone and Calcarenite

Depth profiles of specimens were obtained with an echo time of 94 μs and a resolution of 20 μm, the single-sided sensor was repositioned in steps of 60 μm to scan from the surface of the specimen to a depth of 10 mm.

Transverse relaxation times T_2 were measured with the CPMG sequence, and 2048 echoes were recorded with an echo time 2τ of 50 μs. T_2 values were obtained by fitting the magnetization decays to Equation (2).

Processing of Depth Profiles

Depth profiles were fit to the following equation:

$$f(x) = \sum_{k=1}^{N} \frac{W_k}{2} erf\left[\frac{x_k - x}{\Delta_k \sqrt{2}}\right] + q_0 \tag{4}$$

where N is the number of transitions of the amplitude in the depth profile, x_k is the penetration depth of the kth component at the inflection point x_k, Δ_k is the half width of the transition of the amplitude from low to high value of the spin population, w_k is the spin population of kth component, and q_0 is the lowest spin population.

The slope at inflection point b_k was obtained from the first derivate of $f(x)$ calculated at $x = x_k$ using the parameters obtained from the best fit of depth profiles:

$$b_k = \frac{W_k}{2\Delta_k} \tag{5}$$

3.6. ^{1}H NMR Stratigraphy of an Ancient Icon

NMR stratigraphies were collected by repositioning the sensor in steps of 50 μm to scan from the outermost surface of the painting to a depth of 0.25 cm with a resolution of 92 μm or to a depth of 0.45 cm with a resolution of 57 μm.

3.7. Measurements on Lapis Lazuli

^{13}C MAS spectra were carried out at 100.61 MHz on a Bruker Avance 400 spectrometer (Bruker, Bremen, Germany). Samples were packed into 4 mm zirconia rotors and sealed with Kel-F caps. The spin-rate was 8 KHz. The 90° pulse width was 3.5 μs, the relaxation delay was 10 s, 4000 scans were collected. ^{13}C chemical shifts were reported in ppm with respect to tetramethylsilane.

4. Conclusions

In this paper case studies regarding the use of NMR techniques for characterizing and monitoring artefacts were reported. These cases, though not exhaustive, indicate that NMR can be extensively applied to cultural heritage.

Because new NMR techniques and methods are continuously being developed and improved to enhance the sensitivity of this technique, in the next years NMR will be probably increasingly employed in the field of cultural heritage. In fact NMR is a key tool for the chemical characterization of soluble and insoluble materials constituting artefacts and may help to shed light on the techniques used by the artists. NMR is also an important tool to understand transformations and structural modifications caused by degradation processes occurring in artefacts. The development of portable NMR sensors suitable for non-destructive and non-invasive analysis in situ has made possible the investigation of precious and unmovable artefacts, and their monitoring over time. The use of portable NMR combined with the use of laboratory NMR techniques that require ever smaller amounts of sample will probably make NMR more and more competitive with other analytical techniques traditionally applied in the field of cultural heritage.

Conflicts of Interest: The authors declare no conflict of interest.

References

1. Ernst, R.R.; Bodenhausen, G.; Wokaun, A. *Principles of Nuclear Magnetic Resonance in One and Two Dimensions*; Clarendon Press: Oxford, UK, 1987.
2. Schmidt-Rohr, K.; Spiess, H.W. *Multidimensional Solid-State NMR and Polymers*; Academic Press: London, UK, 1994.
3. MacKenzie, K.J.D.; Smith, M.E. *Multinuclear Solid-State NMR of Inorganic Materials*; Pergamon Materials Series 6; Elsevier Science Ltd.: Amsterdam, The Netherlands, 2002.
4. Spraul, M.; Feund, A.S.; Nast, R.E.; Withers, R.S.; Maas, W.E.; Corcoran, O. Advancing NMR sensitivity for LC-NMR-MS using a cryoflow probe: Application to the analysis of acetaminophen in urine. *Anal. Chem.* **2003**, *75*, 1546–1551. [CrossRef]
5. Schlotterbeck, G.; Ross, A.; Hochstrasser, R.; Senn, H.; Kühn, T.; Marek, D.; Schett, O. High-resolution capillary tube NMR. A miniaturized 5-μL highsensitivity TXI Probe for mass-limited samples, off-line LC NMR, and HT NMR. *Anal. Chem.* **2002**, *74*, 4464–4471. [CrossRef] [PubMed]
6. Alam, T.M.; Jenkins, J.E. HR-MAS NMR Spectroscopy in Material Science. In *Advanced Aspect of Spectroscopy*; InTech: Rijeka, Croatia, 2012; pp. 279–306.
7. Blümich, B.; Perlo, J.; Casanova, F. Mobile single-sided NMR. *Prog. Nucl. Magn. Reson. Spectrosc.* **2008**, *52*, 197–269. [CrossRef]
8. Blümich, B.; Casanova, F.; Perlo, J.; Anferova, S.; Anferov, V.; Kremer, K.; Goga, N.; Kupferschläger, K.; Adams, M. Advances of unilateral mobile NMR in nondestructive materials testing. *Magn. Reson. Imaging* **2005**, *23*, 197–201. [CrossRef] [PubMed]
9. Blümich, B.; Anferova, S.; Kremer, K.; Sharma, S.; Herrmann, V.; Segre, A.L. Unilateral NMR for Quality Control: The NMR-MOUSE®. *Spectroscopy* **2003**, *18*, 18–32.
10. Capitani, D.; Di Tullio, V.; Proietti, N. Nuclear Magnetic Resonance to characterize and monitor Cultural Heritage. *Prog. Nucl. Magn. Reson. Spectrosc.* **2012**, *64*, 29–69. [CrossRef] [PubMed]
11. Blümich, B.; Casanova, F.; Perlo, J.; Presciutti, F.; Anselmi, C.; Doherty, B. Noninvasive testing of art and cultural heritage by mobile NMR. *Acc. Chem. Res.* **2010**, *43*, 761–770. [CrossRef] [PubMed]
12. Baias, M. Mobile NMR: An essential tool for protecting our cultural heritage. *Magn. Reson. Chem.* **2017**, *55*, 33–37. [CrossRef] [PubMed]
13. Haber, A.; Blümich, B.; Souvorova, D.; Del Federico, E. Ancient Roman walls mapped non-destructively by portable NMR. *Anal. Bioanal. Chem.* **2011**, *401*, 1441–1452. [CrossRef] [PubMed]
14. Blümich, B.; Casanova, F.; Harber, A.; Del Federico, E.; Boardman, V.; Stiliano, A.; Wald, G.; Isolani, L. Non-Invasive Depth-Profiling of Walls by Portable Nuclear Magnetic Resonance. *Anal. Bioanal. Chem.* **2010**, *397*, 3117–3125. [CrossRef] [PubMed]
15. Fife, G.R.; Stabik, B.; Kelley, A.E.; King, J.N.; Blümich, B.; Hoppenbrouwers, R.; Meldrum, T. Characterization of aging and solvent treatments of painted surfaces using single-sided NMR. *Magn. Reson. Chem.* **2015**, *53*, 58–63. [CrossRef] [PubMed]
16. Presciutti, F.; Perlo, J.; Casanova, F.; Glöggler, S.; Miliani, C.; Blümich, B.; Brunetti, B.G.; Sgamellotti, A. Noninvasive nuclear magnetic resonance profiling of painting layers. *Appl. Phys. Lett.* **2008**, *93*, 033505-1–033505-3. [CrossRef]
17. Proietti, N.; Di Tullio, V.; Presciutti, F.; Gentile, G.; Brunetti, B.G.; Capitani, D. A Multi-Analytical study of ancient nubian detached mural paintings. *Microchem. J.* **2016**, *125*, 719–725. [CrossRef]
18. Di Tullio, V.; Capitani, D.; Atrei, A.; Benetti, F.; Perra, G.; Presciutti, F.; Proietti, N.; Marchettini, N. Advanced NMR methodologies and micro-analytical techniques to investigate the stratigraphy and materials of 14th century Sienese Wooden Paintings. *Microchem. J.* **2016**, *125*, 208–218. [CrossRef]

19. Proietti, N.; Capitani, D.; Di Tullio, V. Applications of Nuclear Magnetic Resonance Sensors to Cultural Heritage. *Sensor* **2014**, *14*, 6977–6997. [CrossRef] [PubMed]

20. Del Federico, E.; Centeno, S.; Kehlet, C.; Currier, P.; Jerschow, A. Unilateral NMR applied to the conservation of works of art. *Anal. Bioanal. Chem.* **2010**, *396*, 213–222. [CrossRef] [PubMed]

21. Lepore, A.; Baccaro, S.; Casieri, C.; Cemmi, A.; De Luca, F. Role of water in the ageing mechanism of paper. *Chem. Phys. Lett.* **2012**, *531*, 206–209. [CrossRef]

22. Oligschläger, D.; Waldow, S.; Haber, A.; Zia, W.; Blümich, B. Moisture dynamics in wall paintings monitored by single-sided NMR. *Magn. Reson. Chem.* **2015**, *53*, 48–57. [CrossRef] [PubMed]

23. Di Tullio, V.; Proietti, N.; Gobbino, M.; Capitani, D.; Olmi, R.; Priori, S.; Riminesi, C.; Giani, E. Non-destructive mapping of dampness and salts in degraded wall paintings in hypogeous building: The case of St. Clement at mass fresco in St. Clement Basilica, Rome. *Anal. Bioanal. Chem.* **2010**, *396*, 1885–1896. [CrossRef] [PubMed]

24. Capitani, D.; Proietti, N.; Gobbino, M.; Soroldoni, L.; Casellato, U.; Valentini, M.; Rosina, E. An integrated study for mapping the moisture distribution in an ancient damaged wall painting. *Anal. Bioanal. Chem.* **2009**, *395*, 2245–2253. [CrossRef] [PubMed]

25. Catalano, J.; Yao, Y.; Murphy, A.; Zumbulyadis, N.; Centeno, S.A.; Dybowski, C. Nuclear magnetic resonance spectra and 207Pb chemical-shift tensors of lead carboxylates relevant to soap formation in oil paintings. *Appl. Spectrosc.* **2014**, *68*, 280–286. [CrossRef] [PubMed]

26. Catalano, J.; Murphy, A.; Yao, Y.; Alkan, F.; Zumbulyadis, N.; Centeno, S.A.; Dybowski, C. 207Pb and 119Sn Solid-State NMR and Relativistic Density Functional Theory Studies of the Historic Pigment Lead–Tin Yellow Type I and Its Reactivity in Oil Paintings. *J. Phys. Chem. A* **2014**, *18*, 7952–7958. [CrossRef] [PubMed]

27. Del Federico, E.; Centeno, S.A.; Kehlet, C.; Ulrich, K.; Yamazaki-Kleps, A.; Jerschow, A. In Situ Unilateral 1H-NMR Studies of the Interaction Between Lead White Pigments and Collagen-Based Binders. *Appl. Magn. Reson.* **2012**, *42*, 363–376. [CrossRef]

28. Kobayashi, T.; Perras, F.A.; Murphy, A.; Yao, Y.; Catalano, J.; Centeno, S.A.; Dybowski, C.; Zumbulyadis, N.; Pruski, M. DNP-enhanced ultrawideline 207Pb solid-state NMR spectroscopy: An application to cultural heritage science. *Dalton Trans.* **2017**, *46*, 3535–3540. [CrossRef] [PubMed]

29. Del Federico, E.; Schoefberger, W.; Jerschow, A.; Tyne, L.; Schelvis, J.; Kapetanaki, S. Insight into Framework Destruction in Ultramarine Pigments. *Inorg. Chem.* **2006**, *45*, 1270–1276. [CrossRef] [PubMed]

30. Holmes, S.T.; Dybowski, C. Carbon-13 chemical-shift tensors in indigo: A two-dimensional NMR-ROCSA and DFT Study. *Solid State Nucl. Magn. Reson.* **2015**, *7*, 90–95. [CrossRef] [PubMed]

31. Spyros, A.; Anglos, D. Studies of organic paint binders by NMR spectroscopy. *Appl. Phys. A* **2006**, *83*, 705–708. [CrossRef]

32. Sfakianaki, S.; Kouloumpi, E.; Anglos, D.; Spyros, A. Egg yolk identification and aging in mixed paint binding media by NMR spectroscopy. *Magn. Reson. Chem.* **2015**, *53*, 22–26. [CrossRef] [PubMed]

33. Stamatakis, G.; Knuutinen, U.; Laitinen, K.; Spyros, A. Analysis and aging of unsaturated polyester resins in contemporary art installations by NMR spectroscopy. *Anal. Bioanal. Chem.* **2010**, *398*, 3203–3214. [CrossRef] [PubMed]

34. Proietti, N.; Di Tullio, V.; Capitani, D.; Tomassini, R.; Guiso, M. Nuclear Magnetic Resonance in contemporary art: The case of Moon Surface by Turcato. *Appl. Phys. A Mater. Sci. Process.* **2013**, *113*, 1009–1017. [CrossRef]

35. Ulrich, K.; Centeno, S.A.; Arslanoglu, J.; Del Federico, E. Absorption and diffusion measurements of water in acrylic paint films by single-sided NMR. *Prog. Organ. Coat.* **2011**, *71*, 283–289. [CrossRef]

36. Bardet, M.; Pournou, A. Fossil wood from the Miocene and Oligocene epoch: Chemistry and morphology. *Magn. Reson. Chem.* **2015**, *53*, 9–14. [CrossRef] [PubMed]

37. Bardet, M.; Pournou, A. NMR Studies of Fossilized Wood. *Ann. Rep. NMR Spectrosc.* **2017**, *90*, 41–83.

38. Bardet, M.; Gerbaud, G.; Doan, C.; Giffard, M.; Hediger, S.; De Paëpe, G.; Trân, Q.K. Dynamics property recovery of archaeological-wood fibers treated with polyethylene glycol demonstrated by high-resolution solid-sate NMR. *Cellulose* **2012**, *19*, 1537–1545. [CrossRef]

39. Bardet, M.; Gerbaud, G.; Giffard, M.; Doan, C.; Hediger, S.; Le Pape, L. 13C high-resolution solid-sate NMR for structural elucidation of archaeological woods. *Prog. Nucl. Magn. Reson. Spectrosc.* **2009**, *55*, 199–224. [CrossRef]

40. Bardet, M.; Gerbaud, G.; Le Pape, L.; Hediger, S.; Trân, Q.K.; Boumlil, N. NMR and EPR as Analytical Tools to Investigate Structural Features of Archaeological Leathers. *Anal. Chem.* **2009**, *81*, 1505–1511. [CrossRef] [PubMed]

41. Badea, E.; Şendrea, C.; Carşote, C.; Adams, A.; Blümich, B.; Iovu, H. Unilateral NMR and thermal microscopy studies of vegetable tanned leather exposed to dehydrothermal treatment and light irradiation. *Microchem. J.* **2016**, *129*, 158–165. [CrossRef]

42. Lambert, J.B.; Santiago-Blay, J.A.; Wu, Y.; Allison, J. Levy Examination of amber and related materials by NMR spectroscopy. *Magn. Reson. Chem.* **2015**, *53*, 2–8. [CrossRef] [PubMed]

43. Lambert, J.B.; Tsai, C.Y.H.; Shah, M.C.; Hurtley, A.E.; Santiago-Blay, J.A. Distinguishing Amber And Copal Classes By Proton Magnetic Resonance Spectroscopy. *Archaeometry* **2012**, *54*, 332–348. [CrossRef]

44. Casieri, C.; De Luca, F.; Nodari, L.; Russo, U.; Terenzi, C.; Tudisca, V. Effects of time and temperature of firing on Fe-rich ceramics studied by Mossbauer spectroscopy and two-dimensional H-1-nuclear magnetic resonance relaxometry. *J. Appl. Phys.* **2012**, *112*. [CrossRef]

45. Casieri, C.; Terenzi, C.; De Luca, F. Two-dimensional longitudinal and transverse relaxation time correlation as a low-resolution nuclear magnetic resonance characterization of ancient ceramics. *J. Appl. Phys.* **2009**, *105*, 034901. [CrossRef]

46. Terenzi, C.; Casieri, C.; De Luca, F.; Quaresima, R.; Quarta, G.; Tudisca, V. Firing-Induced Microstructural Properties of Quasi-Diamagnetic Carbonate-Based Porous Ceramics: A 1H NMR Relaxation Correlation Study. *Appl. Magn. Reson.* **2015**, *46*, 1159–1178. [CrossRef]

47. Camaiti, M.; Fantazzini, P.; Bortolotti, V. Stone Porosity, wettability changes and other features detected by MRI and NMR relaxometry: A more than 15-year study. *Magn. Reson. Chem.* **2015**, *53*, 34–47. [CrossRef] [PubMed]

48. Di Tullio, V.; Cocca, M.; Avolio, R.; Gentile, G.; Proietti, N.; Ragni, P.; Errico, M.E.; Capitani, D.; Avella, M. Unilateral NMR investigation of multifunctional treatments on stones based on colloidal inorganic and organic nanoparticles. *Magn. Reson. Chem.* **2015**, *53*, 64–77. [CrossRef] [PubMed]

49. Brai, M.; Casieri, C.; De Luca, F.; Fantazzini, P.; Gombia, M.; Terenzi, C. Validity of NMR pore-size analysis of cultural heritage ancient building materials containing magnetic impurities. *Solid State Nucl. Magn. Reson.* **2007**, *25*, 461–465. [CrossRef] [PubMed]

50. Di Tullio, V.; Proietti, N.; Capitani, D.; Nicolini, I.; Mecchi, A.M. NMR depth profiles as a non-invasive analytical tool to probe the penetration depth of hydrophobic treatments and inhomogeneities in treated porous stones. *Anal. Bioanal. Chem.* **2011**, *400*, 3151–3164. [CrossRef] [PubMed]

51. Presciutti, F.; Doherty, B.; Anselmi, C.; Brunetti, B.G.; Sgamellotti, A.; Miliani, C. A non-invasive NMR relaxometric characterization of the cyclododecane–solvent system insideporous substrates. *Magn. Reson. Chem.* **2015**, *53*, 27–33. [CrossRef] [PubMed]

52. Canevali, C.; Fasoli, M.; Bertasa, M.; Botteon, A.; Colombo, A.; Di Tullio, V.; Capitani, D.; Proietti, N.; Scalarone, D.; Sansonetti, A. A multi-analytical approach for the study of copper stain removal by agar gels. *Microchem. J.* **2016**, *129*, 249–258. [CrossRef]

53. Di Tullio, V.; Capitani, D.; Proietti, N. Unilateral NMR to study water diffusion and absorption in stone-hydrogel systems. *Microporous Mesoporous Mater.* **2017**. [CrossRef]

54. Angelova, L.; Ormsby, B.; Richardson, E. Diffusion of water from a range of conservation treatment gels into paint films studied by unilateral NMR. Part I. Acrylic emulsion paint. *Microchem. J.* **2016**, *124*, 311–320. [CrossRef]

55. Capitani, D.; Di Tullio, V.; Proietti, N. Advanced Nuclear Magnetic Resonance Methodologies in Cultural Heritage. In *Cultural Heritage Protection, Developments and International Perspectives*; Frediani, P., Frediani, M., Rosi, L., Eds.; Nova Science Publishers, Inc.: New York, NY, USA, 2013.

56. Troiano, F.; Gulotta, D.; Balloi, A.; Polo, A.; Toniolo, L.; Lombardi, E.; Daffonchio, D.; Sorlin, C.; Cappitelli, F. Successful combination of chemical and biological treatments for the cleaning of stone artworks. *Int. Biodeterior. Biodegrad.* **2013**, *85*, 294–304. [CrossRef]

57. Burnstock, A.; White, R. Cleaning gels: Further studies. In *Conservation Science in the UK*; Tennent, N.H., Ed.; James & James: London, UK, 1993; pp. 36–39.

58. Gioventù, E.; Lorenzi, P.F.; Villa, F.; Sorlini, C.; Rizzi, M.; Cagnini, A.; Griffo, A.; Cappitelli, F. Comparing the bioremoval of black crusts on colored artistic lithotypes of the Cathedral of Florence with chemical and laser treatment. *Int. Biodeterior. Biodegrad.* **2011**, *65*, 832–839. [CrossRef]

59. Gauri, K.L.; Bandyopadhyay, J.K. *Carbonate Stone: Chemical Behaviour, Durability and Conservation*; John Wiley & Sons: New York, NY, UK, 1999.

60. Baglioni, M.; Giorgi, R.; Berti, D.; Baglioni, P. Smart cleaning of cultural heritage: A new challenge for soft nanoscience. *Nanoscale* **2012**, *4*, 42–53. [CrossRef] [PubMed]

61. Lu, Y.; Sathasivam, S.; Song, J.; Crick, C.R.; Carmalt, C.J.; Parkin, I.P. Robust self-cleaning surfaces that function when exposed to either air or oil. *Science* **2015**, *347*, 1132–1135. [CrossRef] [PubMed]

62. Challagan, P.T. *Principles of Nuclear Magnetic Resonance Microscopy*; Clarendon Press: Oxford, UK, 1991.

63. Homans, S.W. *A Dictionary of Concepts in NMR*; Clarendon Press: Oxford, UK; New York, NY, USA, 1992.

64. Kimmich, R. *NMR: Tomography, Diffusiometry, Relaxometry*; Springer: Berlin, Germany, 1979.

65. Rata, D.G.; Casanova, F.; Perlo, J.; Demco, D.E.; Blümich, B. Self-diffusion measurements by a mobile single-sided NMR sensor with improved magnetic field gradient. *J. Magn. Reson.* **2006**, *180*, 229–235. [CrossRef] [PubMed]

66. Amoroso, V.G.G.; Fassina, V. *Stone Decay and Conservation*; Elsevier: Lausanne, Switzerland, 1983.

67. Camuffo, D. Physical weathering of stones. *Sci. Total Environ.* **1995**, *167*, 1–14. [CrossRef]

68. Di Zenzo, S. A Note on the Gradient of a Multi-Image. *Comput. Vis. Graph. Image Process.* **1986**, *33*, 116–125. [CrossRef]

69. Couran, R.; Herbert, R.; Ian, S. *What is Mathematics? An Elementary Approach to Ideas and Methods*; Oxford University Press: New York, NY, USA, 1996; p. 344.

70. Watson, A.T.; Chang, C.T. Characterizing porous media with NMR methods. *Prog. Nucl. Magn. Reson. Spectrosc.* **1997**, *31*, 343–386. [CrossRef]

71. Ferreira Pinto, A.P.; Delgado Rodrigues, J. Stone consolidation: The role of treatment procedures. *J. Cult. Heritage* **2008**, *9*, 38–53. [CrossRef]

72. Miliani, C.; Daveri, A.; Brunetti, B.G.; Sgamellotti, A. CO_2 entrapment in natural ultramarine blue. *Chem. Phys. Lett.* **2008**, *466*, 148–151. [CrossRef]

73. Omi, H.; Ueda, T.; Miyakubo, K.; Eguchi, T. Dynamics of CO_2 molecules confined in the micropores of solids as studied by ^{13}C NMR. *Appl. Surf. Sci.* **2005**, *252*, 660–667. [CrossRef]

74. SefciK, M.D.; Schaefer, J.; Stejskal, E.O. *Molecular Sieves II*; Katzer, J.R., Ed.; ACS Symposium Series; American Chemical Society: Washington, DC, USA, 1977; Volume 40.

75. Perlo, J.; Casanova, F.; Blümich, B. Profiles with microscopic resolution by single-sided NMR. *J. Magn. Reson.* **2005**, *176*, 64–70. [CrossRef] [PubMed]

76. Perlo, J.; Casanova, F.; Blümich, B. *Single-Sided NMR*; Springer-Verlag: Berlin, Germany, 2011.

77. Goelman, G.; Prammer, M.G. The CPMG pulse sequence in strong magnetic field gradients with applications to oil-well logging. *J. Magn. Reson.* **1995**, *113*, 11–18. [CrossRef]

78. Hurlimann, M.D.; Griffin, D.D. Spin dynamics of Carr-Purcell-Meiboom-Gill-like sequences in grossly inhomogeneous B(0) and B(1) fields and application to NMR well logging. *J. Magn. Reson.* **2000**, *143*, 120–125. [CrossRef] [PubMed]

79. Belton, P.S. NMR and the mobility of water in polysaccharide gels. *Int. J. Biol. Macromol.* **1997**, *21*, 81–88. [CrossRef]

magnetochemistry

MDPI

Communication

Nuclear Magnetic Resonance Spectroscopy Investigations of Naphthalene-Based 1,2,3-Triazole Systems for Anion Sensing

Karelle Aiken [1], Jessica Bunn [1], Steven Sutton [1], Matthew Christianson [1], Domonique Winder [1], Christian Freeman [1], Clifford Padgett [2], Colin McMillen [3], Debanjana Ghosh [1,*] and Shainaz Landge [1,*]

[1] Department of Chemistry and Biochemistry, Georgia Southern University, 521 College of Education Drive, Statesboro, GA 30460-8064, USA; kaiken@georgiasouthern.edu (K.A.); jlbunn@email.meredith.edu (J.B.); ss06616@georgiasouthern.edu (S.S.); mc04317@georgiasouthern.edu (M.C.); dw00648@georgiasouthern.edu (D.W.); cf03292@georgiasouthern.edu (C.F.)

[2] Department of Chemistry and Physics, Armstrong State University, 11935 Abercorn Street, Savannah, GA 31419, USA; clifford.padgett@armstrong.edu

[3] Department of Chemistry, Clemson University, Clemson, SC 29634-1905, USA; cmcmill@clemson.edu

* Correspondence: dghosh@georgiasouthern.edu (D.G.) slandge@georgiasouthern.edu (S.L.); Tel.: +1-912-478-1883 (S.L.)

Received: 3 November 2017; Accepted: 13 December 2017; Published: 6 February 2018

Abstract: Detailed Nuclear Magnetic Resonance (NMR) spectroscopy investigations on a novel naphthalene-substituted 1,2,3-triazole-based fluorescence sensor provided evidence for the *"turn-on"* detection of anions. The one-step, facile synthesis of the sensors was implemented using the "Click chemistry" approach in good yield. When investigated for selectivity and sensitivity against a series of anions (F^-, Cl^-, Br^-, I^-, $H_2PO_4^-$, ClO_4^-, OAc^-, and BF_4^-), the sensor displayed the strongest fluorometric response for the fluoride anion. NMR and fluorescence spectroscopic studies validate a 1:1 binding stoichiometry between the sensor and the fluoride anion. Single crystal X-ray diffraction evidence revealed the structure of the sensor in the solid state.

Keywords: Nuclear Magnetic Resonance spectroscopy; naphthalene; 1,2,3-triazole; Click chemistry; fluorometric; *turn-on*; fluoride; anion

1. Introduction

Small, inexpensive organic molecular sensors are making an impact in the field of molecular recognition and as a result, have captured the attention of chemists [1–3]. When chemosensors respond to external stimuli, distinct and significant changes can be observed—for example, in color or fluorescence [4–6]. Anions are crucial in biological and environmental systems; optimum concentrations are critical to proper functioning, as an excess or diminished amount of anions can prove fatal in both systems [7].

The ability to detect fluoride is important for the environment, industry, biological systems, and the military [6,8]. Developing cost-effective, high-performance, easily portable methods for the detection of this anion is highly beneficial to society [8]. The fluoride (F^-) anion, in particular, has a significant impact on health. With groundwater concentrations of 0.5 to 48 ppm [9], this anion is important for healthy dental and bone development [10,11]. However, overexposure causes fluorosis [12] and high levels in utero can impede children's cognitive development [13]. In military operations, the detection of fluoride could be quite useful in the uranium enrichment process for nuclear power and weaponry development [14,15]. In chemical warfare, quick measurement of fluoride levels can expedite tracking of harmful phosphorofluoridate nerve agents such as sarin that hydrolyze to release the anion upon contact with the atmosphere [16,17].

A number of colorimetric and fluorometric sensors have been developed to detect anions through a Brønsted acid–base reaction and/or hydrogen bond formation at the N–H and O–H moieties [18]. For ion recognition, complicated molecules with multistep syntheses are used to create highly conjugated systems with common scaffolds such as ureas, amides, and/or phenolic groups [19,20]. Photoinduced electron transfer (PET) [21,22], metal-to-ligand charge transfer (MLCT) [22,23]; excimer/exciplex formation [23], intramolecular charge transfer (ICT) [24]; and excited state intra/intermolecular proton transfer (ESPT) [25,26] are some of the signaling mechanisms by which the anions are detected.

One of the greatest challenges for chemists is to create chemosensors that are stable, fast, sensitive at the parts per million (ppm) level, and efficient [27]. Organosensors, through reversible interactions, present an avenue into various applications such as resettable logic gate systems [28–31], molecular security devices [32,33], micellar devices [34,35], dual sensors [36,37], corrosion inhibitors [38], fabrication of materials and polymers, etc. [39,40]. The optically and chemically stable naphthalene substituted-1,2,3-triazole molecule, **NpTP** ((2-(4-(naphthalen-2-yl)-1*H*-1,2,3-triazol-1-yl)phenol, Scheme 1) described herein is produced with straightforward synthesis, targeted design, and sensitive as well as selective ion-recognition properties.

Scheme 1. Synthesis of 2-(4-(naphthalen-2-yl)-1*H*-1,2,3-triazol-1-yl)phenol (**NpTP**).

Bypassing complicated synthetic steps, the 1,2,3-triazole chemosensors is accessed in one step with an azide–alkyne cycloaddition utilizing the most well-known "Click" reaction [41–46]. With this approach, the recognition core can be easily retained while the signaling units are readily modified using commercially available precursors. Furthermore, unlike other naphthyl-based fluoride sensors reported in the literature [47,48] that function through the interaction of the anion with N-H groups in cage-like bisurea systems, **NpTP** utilizes a much simpler phenolic–triazole binding core.

The triazole units serve distinct roles in sensing. They can be a part of the response group, participate in cation and anion chelation, or function as a ligation unit that links one part of a sensor to another [42,44]. These scaffolds are N-donors via one of the *sp*2-nitrogens during cation binding, and H-donors at the C*sp*2-H proton in anion binding [41,44]. In the case of **NpTP**, the triazole serves in three capacities: ligation, signaling, and recognition.

Previous work by our group investigated the sensing properties of **PTP** (2-(4-phenyl-1-H-1,2,3-triazol-1-yl)phenol) [49]. This molecule exhibited a blue "*turn-on*" fluorescent response to fluoride (F−), acetate (OAc−) and dihydrogen phosphate (H$_2$PO$_4$−). **PTP** was equally responsive to OAc− and H$_2$PO$_4$−, and three times more sensitive to F− compared to the other two anions. **NpTP** presented herein illustrates a red-shift effect on the signal output, a yellow "*turn-on*" fluorescence upon interacting with fluoride, acetate, and dihydrogen phosphate due to increased conjugation length in the sensor, i.e., replacing the phenyl group in **PTP** with naphthyl in **NpTP**. Additionally, our investigations have revealed that the incorporation of the naphthyl unit significantly improved the ion selectivity and fluoride sensitivity of the sensor relative to **PTP**.

The results presented in this study show a strong fluoride response with **NpTP**. The sensor "*turns-on*" in the presence of a fluoride ion upon irradiation at 300 nm. The binding pocket is created by the phenolic –OH and triazole C*sp*2-H site (Scheme 2). Detailed NMR investigations (1) showed the structural skeleton of **NpTP**; (2) displayed the binding interaction with F⁻; and (3) revealed the stoichiometry between the sensor and the analyte after titrating with varied concentrations of fluoride anion with **NpTP**. NMR, Ultra violet-visible (UV-Vis) and fluorescence spectroscopy studies detailed the molecule's response to host anions as their tetrabutylammonium salts (F⁻, Cl⁻, Br⁻, I⁻, $H_2PO_4^-$, ClO_4^-, OAc⁻, BF_4^-).

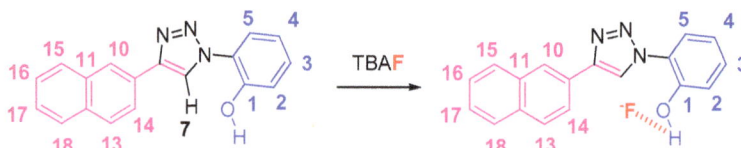

Scheme 2. The proposed binding mode of **NpTP** and fluoride anion.

2. Results and Discussion

The detailed characterization of **NpTP** was carried out by NMR (1D and 2D) and single crystal X-ray analysis. NMR studies revealed the structure, the anion-binding site and the stoichiometry between the sensor and the fluoride anion. The photophysical properties of **NpTP** with and without the anions (F⁻, Cl⁻, Br⁻, I⁻, $H_2PO_4^-$, ClO_4^-, OAc⁻, BF_4^-) were investigated by steady state absorption and fluorescence spectroscopy.

2.1. Nuclear Magnetic Resonance Spectroscopic Studies

The **NpTP** structure was characterized by ¹H-NMR in various deuterated solvents such as acetonitrile-d_3 (CD₃CN), dimethylsulfoxide-d_6 ([(CD₃)₂SO]), and acetone-d_6 [(CD₃)₂CO]. Figure 1 shows the comparison between the three solvents. The aromatic peaks in these solvents appear in the range of δ 7.00 to 9.05 ppm with anticipated coupling patterns. The triazole C$sp2$–H (H7) proton singlet, as expected, is highly deshielded and hence is a reference peak in many studies (δ 9.05 ppm in DMSO; δ 9.00 ppm in acetone, and δ 8.78 ppm in acetonitrile) [41,50]. The H5 proton has noticeably shifted its position in all three solvents, with acetone being highly downfield (δ 7.82 ppm as doublet of doublets in acetone-d_6). The naphthyl core has seven resonance signals; the singlet at δ 8.52 in DMSO-d_6 is easily identified as the H10 proton. The H14 and H13 proton split as a doublet (δ 8.11 ppm and δ 8.00 ppm, respectively). The phenolic –OH proton is visible as a broad singlet in DMSO-d_6 at δ 10.60 ppm (Figure 2 and Figure S1). The correlational spectroscopy (g-COSY) studies helped in assigning all the aromatic protons. A strong *meta* coupling (⁴*J*) cross peak of the H10 and H14 proton is visible and is marked below in Figure 3.

The ¹³C-NMR signals (Figure 4, Figures S2 and S3) for the aromatic region ranged from δ 117.5 to 150.5 ppm in DMSO-d_6. With the help of 1D DEPT 90 (Figure S4) and 2D HSQC (Figure 5), all the single bonded carbon hydrogen correlations were marked. The HMBC studies (Figure S5), aided in allocating the quaternary carbons (1, 6, 8, 9, 11, and 12). The strong peaks for example seen for C1 carbon are for H5 and H3 protons through three bond correlations and the weak peak is seen for H2 proton via two bond correlation. Both the 1D and 2D studies guided in assigning the ¹H and ¹³C resonances for the **NpTP** molecule.

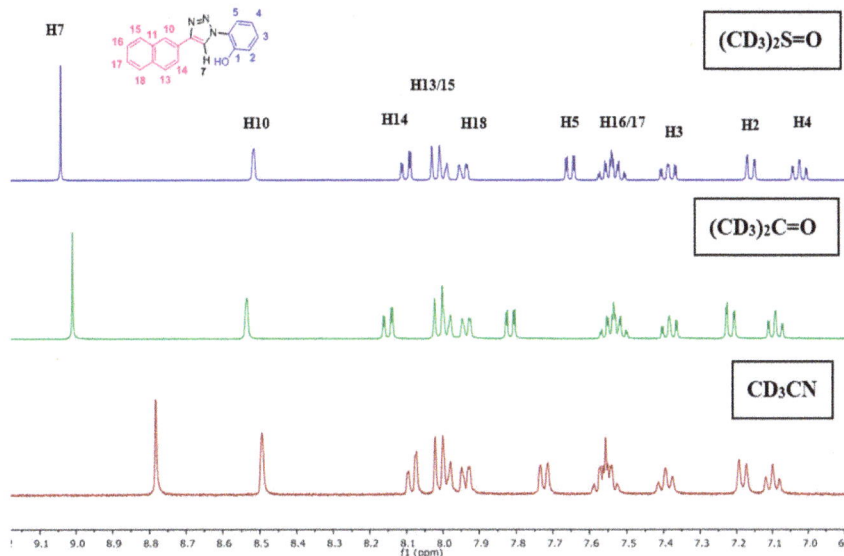

Figure 1. ^{1}H-NMR spectra of **NpTP** molecule in various deuterated solvents showing the expansion of the aromatic region from δ 6.80–9.2 ppm of **NpTP**.

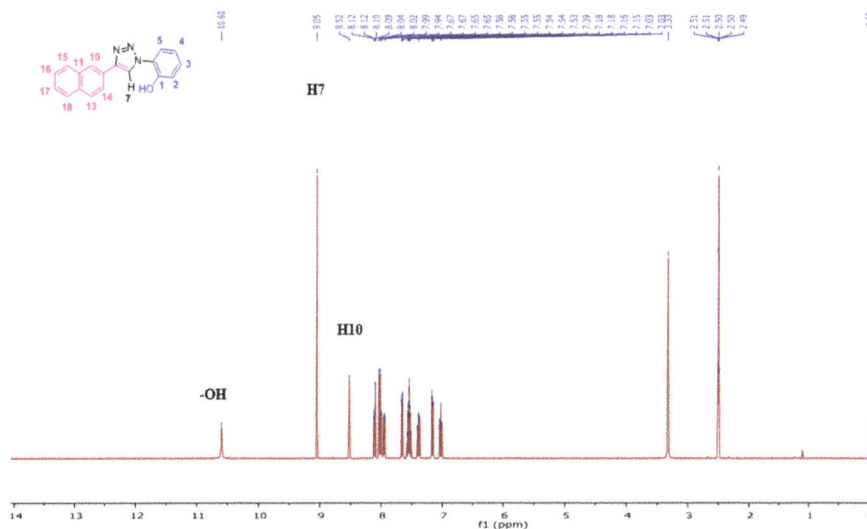

Figure 2. Full ^{1}H-NMR spectrum (400 MHz, (CD$_3$)$_2$S=O, RT) of **NpTP**; selected few peaks are assigned (for the expanded version of the aromatic region, see Figure 1).

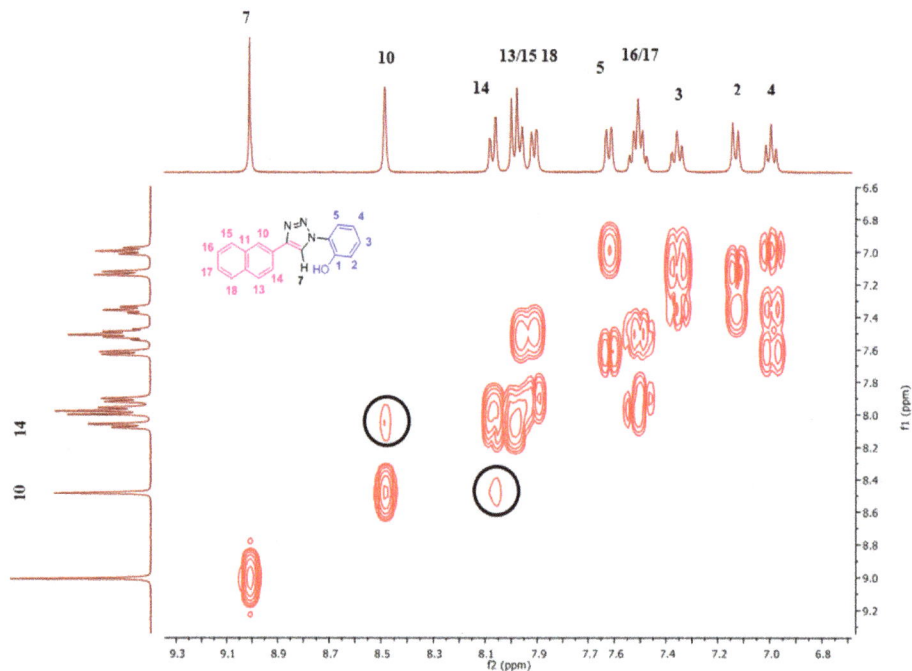

Figure 3. The 2D-COSY spectrum (400 MHz, (CD$_3$)$_2$S=O, RT) of **NpTP** molecule, the correlation with all aromatic protons is seen.

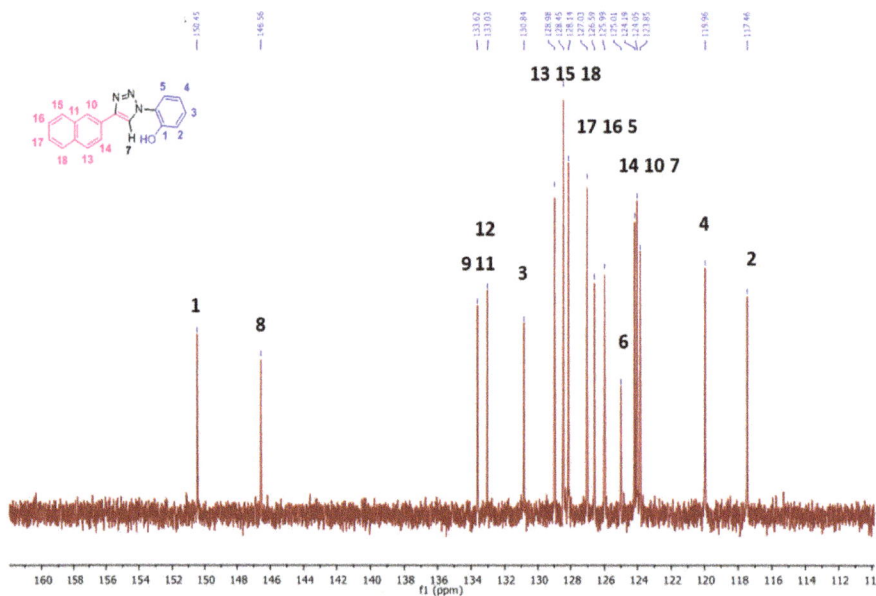

Figure 4. The ^{13}C-NMR spectrum (100 MHz, (CD$_3$)$_2$S=O, RT) showing the expansion from δ 110–162 ppm of **NpTP**. (For the full version, see Figure S2.)

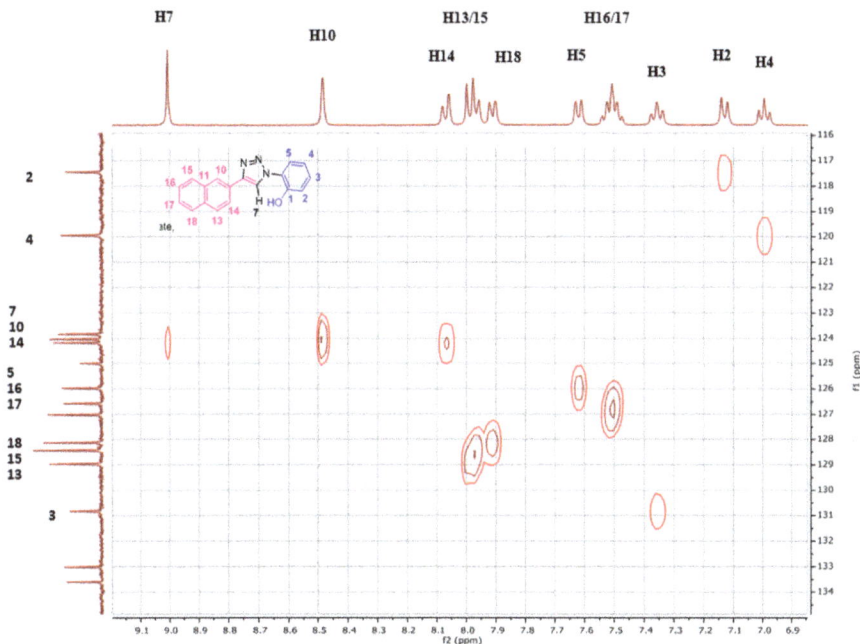

Figure 5. The 2D HSQC spectrum of **NpTP**, showing single bond carbon hydrogen correlation (all the C–H bonds are marked).

Fluoride ion interaction with **NpTP** was investigated by evaluating the binding mechanism of the anion with the sensor. For this purpose, the Csp2-H triazole proton signal was used to verify the site of interaction between the **NpTP** molecule and the fluoride anion. The ^{1}H-NMR titration experiments were carried out with the sensor by gradual addition of 0 to 4.0 equivalents of tetrabutylammonium fluoride (TBAF) in CD$_3$CN (Figure 6). The protons on the phenyl ring (H2–H4) and the triazole proton were greatly affected. The H4 proton, which is in the *para* position to the –OH group, is shielded from 7.10 to 6.24 ($\Delta\delta$ = 0.84) with increased concentration of TBAF from 0 to 4.0 equivalents. The protons that are *ortho* and *meta* to –OH (H2 and H3) initially unite to form the broadened peak at ~δ 7.27 after the addition of 0.2 equivalents of TBAF. At a higher concentration of approximately one equivalent, it splits again into two distinct resonances. These results are in accordance with the previously reported **PTP** sensor from our group [49]. The initial downfield and later upfield shifts of the H2 phenyl proton showed the impact of fluoride binding on the ring. The proton *meta* (H3) to the –OH group was moderately affected at higher concentrations of F$^-$ ion. This confirmed our hypothesis of the increased electron density in the phenyl ring displaying a significant impact on *ortho* and *para* position of phenyl protons with a through-bond propagation [51,52].

Substantial change was observed in the chemical shift of the Csp2-H triazole proton (δ 8.78 in CD$_3$CN) before and after its binding with fluoride. The change in delta value is significant from 8.78 to 9.58 ($\Delta\delta$ = 0.80) with increasing equivalents of TBAF. This strong deshielding effect is attributed to the fact that the triazole proton is in the vicinity of the anion through possible hydrogen bonding-like interaction [53]. The H5 proton on the phenyl ring, in comparison to other protons' chemical shifts, is minimally affected throughout the course of the titration, providing additional evidence for a binding site with fluoride. The above results support the fact that the triazole proton and the phenolic proton (–OH) are part of the binding pocket. The naphthyl protons H10 and H13–H18 are not influenced by fluoride binding [49].

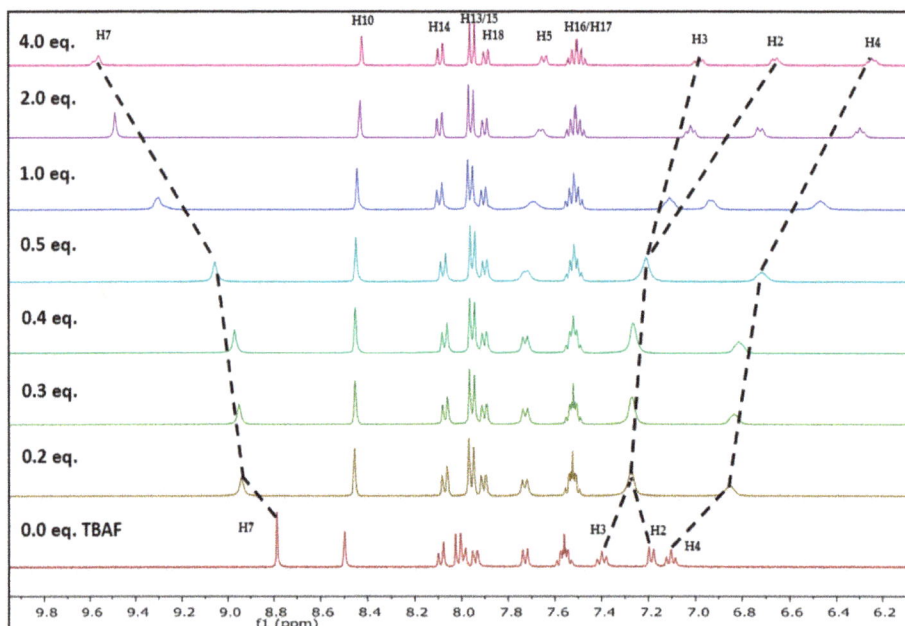

Figure 6. Changes in partial ^{1}H-NMR (400 MHz) spectra of **NpTP** (6.2×10^{-2} M) in CD_3CN upon increasing equivalents of TBAF (0 to 4.0 eq.).

The titration experiments also helped to find the binding stoichiometry of the sensor with the F^- ions through Job's plot [54,55]. The change in the delta value of the triazole proton H7 ($\Delta\delta = \delta x - \delta o$) at δ 8.78 [55] was plotted against the mole fraction of the sensor **[NpTP]**/(**[NpTP]** + TBAF). The plot was fitted using the non-linear curve fit parameters of the ORIGIN 8.0 software. It showed a maximum intensity at 0.45 mole fraction, revealing the binding stoichiometry of F^- ions to **NpTP** as 1:1 (Figure 7).

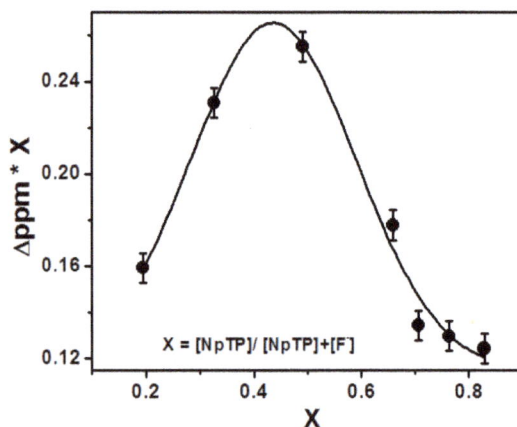

Figure 7. Job's plot through NMR titration of **NpTP** upon the addition of TBAF in CD_3CN.

To understand the structural conformations of **NpTP** in solution, Nuclear Overhauser Effect (NOE) experiments were carried out (Figures S6 and S7). We did not find any correlation between the triazole C*sp*2-H proton (H7) and the phenyl H5 proton, indicating that the phenolic –OH is in close proximity to the H7 proton. A strong NOE correlation is observed between the a) triazole proton H7 and the H10 and H14 protons and b) H10 and H15 protons on the naphthyl ring (Figure S6). The NOE correlation is also seen for the –OH proton and the H2 proton, suggesting the position of the phenolic group towards phenyl proton rather than the triazole proton (Figure S7). NOE studies of **NpTP** in presence of four equivalents of TBAF (Figure S8) showed the correlation between the H7, H10, and H14 protons, indicating that the naphthyl core is unaffected. There is no conformational change before and after fluoride binding. It also suggests that the fluoride anion is in close vicinity to the triazole proton and the –OH proton, shown in Scheme 2. The single X-ray crystal structure for **NpTP** and our previous studies on cation sensor (**BPT**) [56] also support this hypothesis. The crystal structure substantiated that there is no intramolecular hydrogen bonding between the phenolic oxygen (–OH) and the triazole proton (see Section 2.4).

Since the phenolic (–OH) group was part of the binding pocket, the behavior of the –OH group was studied in DMSO-d_6 by a titration study with **NpTP** and TBAF. In this polar aprotic solvent, the phenolic –OH resonates distinctly at δ 10.60 ppm. Upon addition of 0.5 equivalents of fluoride anion, the –OH proton signal completely disappears indicating a hydrogen bond between F- and the phenolic –OH. Higher equivalents of TBAF (2.0 equivalents) generated a triplet at 16.1 ppm, which intensified with 4.0 equivalents of TBAF. The highly deshielded triplet peak is an indication of the stable hydrogen bonded HF_2^- ion. This provided evidence to a deprotonation pathway in the ion-recognition process (Figures S9 and S10) [57,58].

In ^{1}H-NMR, the H7 proton in deuterated DMSO (Figure S9) at 1 equivalent of TBAF started splitting. Our understanding is that the fluoride anion is in close proximity to the triazole proton and hence is severely affected. To confirm this hypothesis, we monitored the interaction through ^{19}F-NMR titration in DMSO-d_6, where the singlet for pure TBAF appears at δ −106 ppm, and HF_2^- can be seen at −144 ppm (weak signal). At one equivalent of TBAF, the −106 ppm peak completely disappears indicating the formation of a complex between the **NpTP** molecule and the F$^-$ ion. At higher concentrations (2–4 equivalents), the doublet for HF_2^- ion at −144 ppm was clearly observed (Figure S11) [59,60].

In comparison to the previously reported sensor, **PTP** [49], from our group, the ^{1}H-NMR studies with **NpTP** showed similar observations. The phenolic –OH proton was significantly affected but the aryl core (phenyl in **PTP** and the naphthyl in **NpTP**) was not. The binding site for fluoride in both sensors consisted of the triazole proton and the phenolic proton. The triazole proton in **PTP** appeared at δ 8.65 ppm and in **NpTP** at δ 8.78 ppm in CD$_3$CN. For both sensors, the ^{1}H-NMR titration experiments conducted in CD$_3$CN at four equivalents of TBAF resulted in considerable deshielding of the triazole proton and shielding of the proton *para* to the phenolic –OH. The change in the triazole proton's chemical shift for **PTP** was 0.70 ppm and 0.8 ppm for **NpTP**.

2.2. Absorption and Fluorescence Studies

With the **NpTP** molecule, anion recognition was investigated using steady state absorption and fluorescence experiments. This was carried out by screening the molecule with the tetrabutylammonium salts of various anions: F$^-$, Cl$^-$, Br$^-$, I$^-$, $H_2PO_4^-$, ClO_4^-, OAc$^-$, and BF_4^-. Significant spectral changes for **NpTP** in both absorption and fluorescence spectra were noted in the presence of F$^-$, OAc$^-$, and $H_2PO_4^-$ ions (Figure 8). However, fluoride showed the most significant response compared to the other two anions. In acetonitrile, **NpTP** showed the lowest energy absorption band in the range of 275–310 nm, peaking at 290 nm. The structured absorption is characteristic of the $\pi-\pi^*$ transition in the polyaromatic ring system [61–64]. Development of a new absorption peak around 355 nm at the cost of the pre-existing **NpTP** original band indicates effective interaction of these ions with **NpTP** leading to the formation of a new complex (Figure 8).

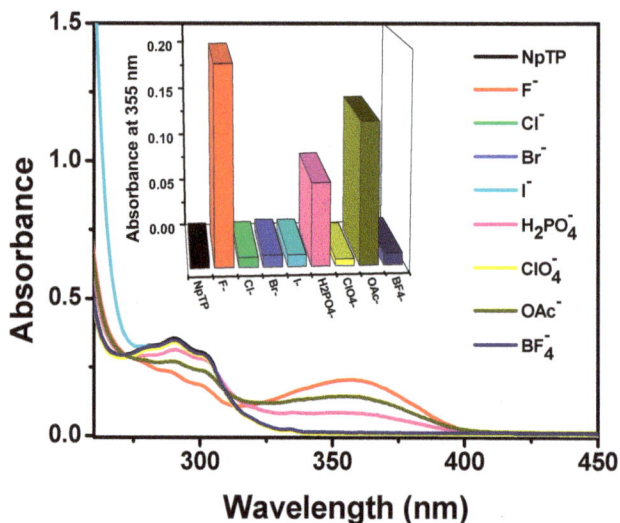

Figure 8. Absorption spectra and bars (in the inset) representing the spectral response of **NpTP** ($\sim 1 \times 10^{-5}$ M) upon the addition of 4×10^{-4} M of various anions in acetonitrile. The bars in the inset are plotted by monitoring the absorbance of **NpTP** at 355 nm in presence of anions.

Modulation in the fluorescence color change of the molecule, as observed in the presence of a UV lamp (Section 2.3) under the influence of fluoride, acetate, and dihydrogen phosphate anions, steered us to investigate the emission properties of **NpTP** in the presence of these ions. An emission spectrum of **NpTP** molecule was obtained upon exciting the molecule at 300 nm (around λ_{max} absorption). The structured emission band spanning between 345 nm and 380 nm is due to the naphthalene moiety [65,66], whereas the band at 330 nm is a signature of the phenol triazole group [49]. With the addition of a series of tetrabutylammonium salts of ions, emission spectra of **NpTP** in presence of the ions (Figure 9) resulted in a similar spectral changes as the absorption profiles (Figure 8). With the addition of F$^-$, OAc$^-$, and H$_2$PO$_4^-$ ions, the **NpTP** emission from the naphthalene moiety was quenched and two new bands developed—one around 410 nm and the other at 530 nm. The band at 530 nm revealed a low quantum yield with respect to the one at 410 nm. Also, in the presence of fluoride ions, the 410 nm band had comparatively higher fluorescence than the other two ions. Fluorescence color study (see Section 2.3, Figure 12) under a UV lamp (long wavelength ~365 nm) resulted in yellow fluorescence for **NpTP** in the presence of F$^-$, OAc$^-$, and H$_2$PO$_4^-$ ions, which validated the existence of a 530 nm emission band (yellow region in the color spectrum) in the presence of these three ions.

To further study the interaction of fluoride with **NpTP**, TBAF was progressively added to the molecule in acetonitrile. The fluorescence intensity of the 330 nm band of the sensor gradually decreased with a concomitant development of the bands at 410 nm and 530 nm (Figure 10). To unveil the characteristics of each emission band, excitation scans were collected by monitoring the emission wavelengths at 530 nm (Figure 11a) and 410 nm (Figure 11b). The spectrum obtained at λ_{em} 530 nm resembled the absorption spectrum when fluoride was added to the molecule, with the peak maximizing at 350 nm. To substantiate our result, the molecule was excited at 350 nm and a broad emission band peaking at 545 nm correlates to the emission band due to fluoride ion interaction with **NpTP** (Figure 11a).

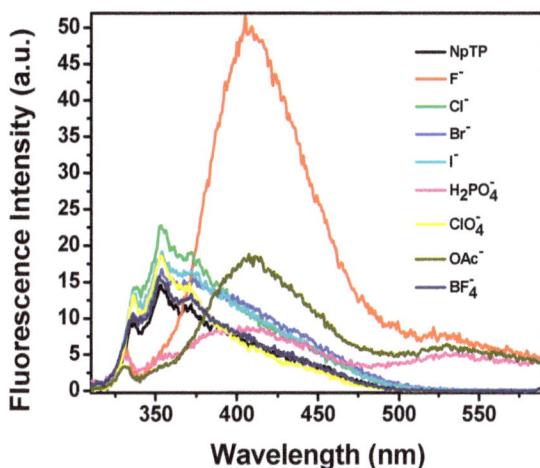

Figure 9. Fluorescence spectra of **NpTP** (~1 × 10^{-6} M) upon the addition of 4 × 10^{-4} M of various anions in acetonitrile. λ_{exc} = 300 nm.

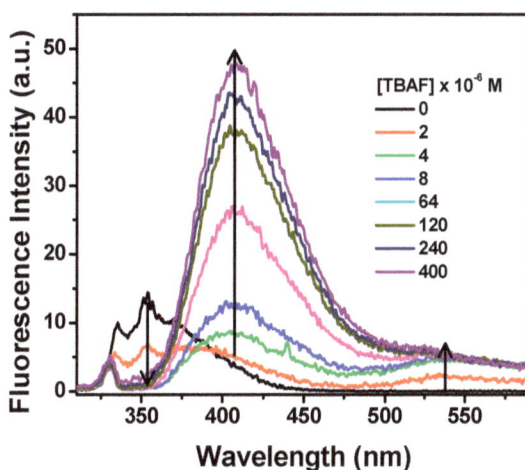

Figure 10. Fluorescence spectra of **NpTP** (~1 × 10^{-6} M) upon the addition of TBAF in acetonitrile. Concentrations of TBAF are provided on the legends. λ_{exc} = 300 nm.

The spectrum obtained at λ_{em} 410 nm (Figure 11b) is rather interesting as it showed the excitation band specific to the naphthalene triazole itself. This hinted at the fact that fluoride addition resulted in a notable change in the photophysical property of the **NpTP** molecule. The presence of an electron withdrawing group in the form of F^- ions dramatically influenced the excited state dynamics of the molecule through an inductive effect caused after fluoride is bound to **NpTP**. This allowed an electron flow throughout the aromatic rings, causing further conjugation in the system and, as a result, a red shifted emission band of **NpTP** appeared at 410 nm (Figure 11b). This also demonstrated the excited state proton transfer (ESPT) process occurring in this molecule when fluoride abstracts the phenolic proton [26]. The appearance of the 410 nm emission band is attributed to the formation of an anion of **NpTP**. Deprotonation occurs during the excited state lifetime of the molecule. This phenomenon correlates to the widely studied photophysics of 2-Naphthol [25,67–71] where the molecule undergoes the ESPT process at high pH conditions, revealing emission of the Naptholate anion around 410 nm [67]. Observation of the emissions from both the deprotonated **NpTP** and **NpTP**-F$^-$ hinted at the fact that the excited state reaction is partially completed during the excited state lifetime [67]. Similar observations in the fluorescence spectral patterns of OAc$^-$ and H$_2$PO$_4^-$ ions indicated that deprotonation and anion binding are simultaneously taking place with anions that have higher basicity. Though acetate is considered more basic than fluoride, the fluorescence response for fluoride ions is relatively higher than for acetate (Figure 8). Here, the size of the anion played an important part in its binding with the molecule. Fluoride, being smaller in size than acetate, has better proximity to fit in the triazole pocket and bind with the –OH proton.

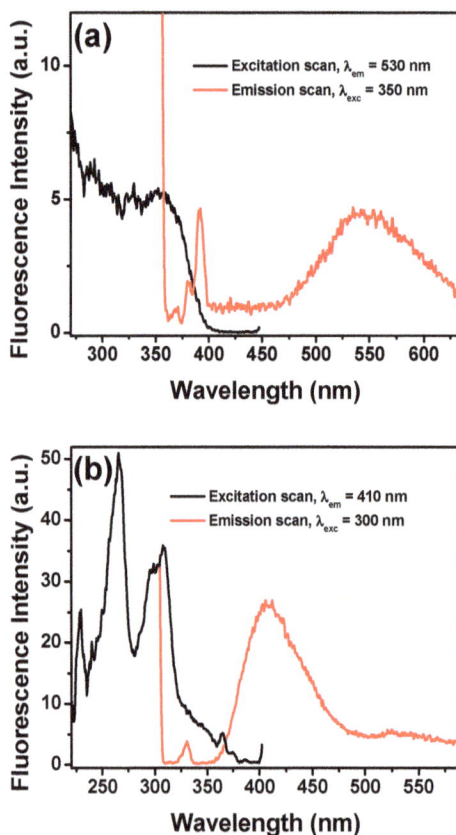

Figure 11. Emission (red) and fluorescence excitation (black) spectra of **NpTP** (~1 × 10^{-6} M) in acetonitrile for (**a**) λ_{exc} is 350 nm to obtain emission spectrum and λ_{em} is monitored at 530 nm for excitation scan; (**b**) λ_{exc} is 300 nm to obtain emission spectrum and λ_{em} is monitored at 410 nm for excitation scan.

Overall, relative to our previous study with a phenyl-based sensor, **PTP**, the spectroscopic investigation with **NpTP** revealed increased ion-selectivity with the replacement of the phenyl unit by the naphthyl group. While fluoride induces the strongest response in both molecules, with **PTP** the intensity of the fluorescence with OAc$^-$ and H$_2$PO$_4^-$ was on par [49]. For **NpTP**, the improvement in ion selectivity is verified by a fluorescence output with H$_2$PO$_4^-$ that is significantly lower than that for OAc$^-$ (Figure 9), a clear and marked distinction occurring between F$^-$ versus OAc$^-$ versus H$_2$PO$_4^-$.

The stoichiometry of the sensor with F^- was determined from the modified Benesi–Hildebrand equation (Equation (1)) [72]. The graph was plotted with $1/\Delta F$ against $1/[L]$ (Figure S12):

$$1/\Delta F = 1/\Delta F_{max} + 1/K. \, 1/\Delta F_{max}. \, 1/[L], \qquad (1)$$

where $\Delta F = F_x - F_0$ and $\Delta F_{max} = F_\infty - F_0$;

F_0, F_x, and F_∞ are the fluorescence intensities of the **NpTP** molecule considered in the absence, at an intermediate concentration, and at a concentration of complete interaction of the anion, respectively. K is the binding constant and $[L]$ is the concentration of the fluoride anion. The fluorescence was monitored at 530 nm. **NpTP** showed a linear variation upon addition of fluoride, justifying the validity of the above equation and confirming the 1:1 interaction between the sensor and the anion. The binding constant, K, determined from the slope to be $2.8 \times 10^4 \, M^{-1}$ for **NpTP**–fluoride binding, demonstrated higher sensitivity towards fluoride compared to our previously studied triazole molecule, **PTP** ($K = 9.0 \times 10^3 \, M^{-1}$) [49].

2.3. Color Studies

The color study showed (Figure 12) a *"turn-on"* yellow fluorescence enhancement of **NpTP**, under a UV lamp of wavelength 365 nm, with tetrabutylammonium salts of fluoride. Salts of $H_2PO_4^-$ and OAc^- ions also showed the *"turn-on"* fluorescence but the color intensity was low compared to the F^- anion. The results corroborated the absorption and fluorescence spectroscopy experiments. The observations from spectroscopic and color studies of **NpTP** upon addition of F^- were attributed to hydrogen bonding, which is consistent with previous studies on hydrogen bonding interactions between the sensor and the analyte [24,49,73,74].

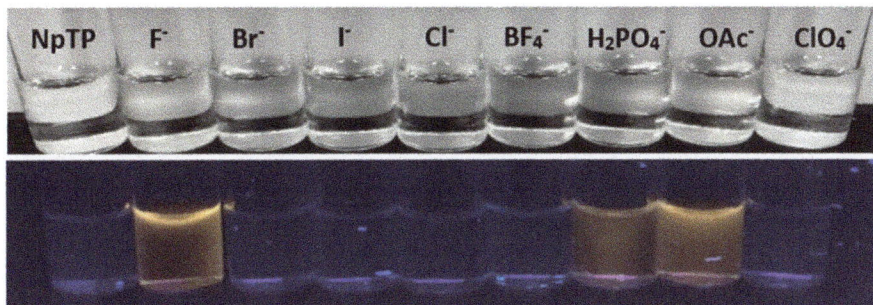

Figure 12. Color changes of **NpTP** under a UV lamp of long wavelength (365 nm) upon addition of $\sim 2 \times 10^{-2}$ M of anions to $\sim 1 \times 10^{-3}$ M sensor in acetonitrile.

2.4. Single Crystal X-ray Crystallography Studies

The single X-ray crystal of **NpTP** (15 mg) was obtained by slow evaporation of a mixed solvent system (0.5 mL methanol + 0.2 mL acetonitrile + 2–3 drops of DMSO). A colorless pyramidal crystal of $C_{18}H_{13}N_3O$ having approximate dimensions of 0.067 mm $\times$ 0.068 mm $\times$ 0.071 mm was used for the X-ray crystallographic analysis. Crystal data, data collection, and structure refinement details are summarized in Table 1. **NpTP** crystallized in the tetragonal space group $P4_3$ (#78) with the unit cell parameters a = 7.3806(16) Å, b = 7.3806(16) Å, c = 50.665(11) Å, volume = 2759.9(13) Å^3. The structure was collected at 140 K and had an unweighted r factor of 4.88% (R1). The thermal ellipsoid of the single-crystal structure of **NpTP** is shown in Figure 13.

The structure has two independent **NpTP** molecules in the asymmetric unit that only differ in the orientation of the naphthyl ring to the triazole. The structure is held together by two independent hydrogen bonding chains with H-bond between the phenolic O–H and the triazole nitrogen of neighboring molecules. The first chain runs parallel to the *a*-axis (Figure S15), the second runs parallel to the *b*-axis (Figure S16). Combined, these form a network of hydrogen bonds that holds the overall structure together. The detail reports on NpTP crystal structure is provided in Table S1–S7.

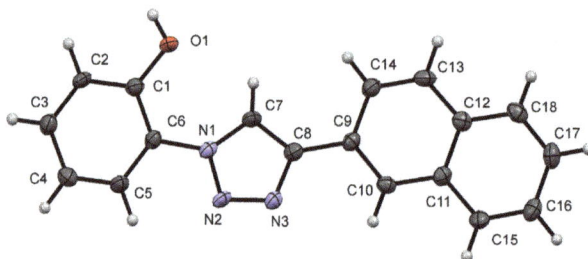

Figure 13. A view of the molecular structure of NpTP, showing the atom labeling. Displacement ellipsoids are drawn at the 50% probability level.

Table 1. Experimental details for **NpTP** molecule for single X-ray crystal studies.

Crystal Data for NpTP	
Chemical formula	$C_{18}H_{13}N_3O$
M_r	287.31
Crystal system, space group	Tetragonal, $P4_3$
Temperature (K)	173
a, b, c (Å)	7.3806 (16), 7.3806 (16), 50.665 (16)
V (Å^3)	2759.9(13)
Z	8
Radiation type	Mo $K\alpha$
μ (mm^{-1})	0.09
Crystal size (mm)	$0.07 \times 0.07 \times 0.07$
Data collection for NpTP	
Diffractometer	Bruker D8 Venture Photon 100 diffractometer
Absorption correction	Multi-scan SADABS, Bruker
T_{min}, T_{max}	0.883, 1.00
No. of measured, independent and observed [$F^2 > 2.0\sigma(F^2)$] reflections	17713, 5409, 4263
R_{int}	0.045
Refinement	
$R[F^2 > 2\sigma(F^2)]$, $wR(F^2)$, S	0.049, 0.096, 1.02
No. of reflections	5409
No. of parameters	398
H-atom treatment	H-atom parameters constrained
$\Delta\rho_{max}$, $\Delta\rho_{min}$ (e Å^{-3})	0.16, −0.20
Special treatment	Refined as a 2-component inversion twin

Computer programs: Data collection: Bruker *APEX3*; cell refinement: Bruker *SAINT*; data reduction: Bruker *SAINT*; program(s) used to solve structure: SHELXT-2014 (Sheldrick 2014); program(s) used to refine structure: *SHELXL2014* (Sheldrick 2014).

3. Materials and Methods

All chemicals and reactants for **NpTP** synthesis and the tetrabutylammonium salts of anions were obtained from commercial sources (Sigma-Aldrich (St. Louis, MO, USA) and Acros Organics (Pittsburgh, PA, USA)) and used without further purification. Column chromatography was performed with Selecto Scientific Si-gel (particle size 100–200 microns). All reactions were monitored by thin-layer chromatography (TLC) using Agela Technologies silica gel plates 60 F$_{254}$ (Wilmington, DE, USA). Visualization was accomplished with UV light and/or staining with appropriate stains (iodine, vanillin). Melting points were measured with a Vernier Melt Station (Beaverton, OR, USA) using a Vernier LabQuest 2 and are uncorrected. NMR spectra were recorded on an Agilent MR400DD2 spectrometer (Santa Clara, CA, USA), with a multinuclear probe with two RF channels and variable temperature capability (^{1}H-NMR: 400 MHz, ^{13}C-NMR: 100 MHz). The solvent used was CD_3CN, [(CD_3)_2SO], [(CD_3)_2CO] purchased from Sigma-Aldrich (St. Louis, MO, USA) and Acros Organics (Pittsburgh, PA, USA) with TMS as an internal standard set at 0 ppm in both ^{1}H-NMR and ^{13}C-NMR spectra. The NMR signals are reported in parts per million (ppm) relative to the residual in the solvent. Signals are described with multiplicity, singlet (*s*), doublet (*d*), triplet (*t*), triplet of doublet (*td*), quartet (*q*) and multiplet (*m*); coupling constants (*J; Hz*) and integration. High Resolution MS (HRMS) analyses were performed using MALDI, Q-TOF micro, 3200API, LCMS, GCMS EI (DI) (Figure S13). The Electrospray Ionization Mass Spectrometry (ESI-MS) was conducted using a Shimadzu LCMS-2020 Single Quad (Korneuburg, Austria) (Figure S14).

Room-temperature absorption and steady-state fluorescence measurements were performed using a Shimadzu UV-2450 spectrophotometer and PerkinElmer LS55 (Waltham, MA, USA) with a well plate reader fluorimeter, respectively. Concentration of **NpTP** was kept at ~1.0×10^{-6} M in acetonitrile to avoid any possible intermolecular effect. Stock concentrations of ~1.0×10^{-2} M for the tetrabutylammonium salts of anions were also prepared in acetonitrile. The solvents used are of HPLC grade and all the experiments were performed at ambient temperature (27 °C) with air-equilibrated solutions.

The single X-ray crystal structure measurements were made on a Bruker D8 Venture Photon 100 diffractometer using Mo-Kα radiation (Madison, WI, USA).

General procedure for synthesis of sensor 2-(4-(naphthalen-2-yl)-1*H*-1,2,3-triazol-1-yl)phenol *(*NpTP*)*:
2-Azidophenol (225.4 mg, 1.67 mmol) [75] and naphthylene-2-acetylene (253.9 mg, 1.67 mmol) were suspended in tert-butanol/water (33.4 mL; 1:1, *v/v*) in a round bottomed flask. In order to dissolve the solids completely, the mixture was warmed slightly above the room temperature. An aqueous solution of copper(II) sulfate pentahydrate (8.10 mg, 0.03 mmol in 2 mL of water) was then added dropwise, followed by sodium ascorbate (64.0 mg, 0.32 mmol in 2 mL of water). The reaction was stirred vigorously while refluxing for 24 h. Upon cooling to room temperature, the resulting mixture was placed in an ice bath and diluted with water (~20 mL) to induce precipitation. The crude, solid product was collected with vacuum filtration and purified by flash column chromatography (10% ethyl acetate in hexanes followed by 40% ethyl acetate in hexanes) to provide a beige powder, 329.6 mg (69%).

Melting point: 225.9–226.5 °C.
[1]**H-NMR** (400 MHz, DMSO-d_6) δ 10.60 (brs, 1H, OH), 9.05 (s, 1H), 8.52 (s, 1H), 8.11 (dd, 1H, 1.6, 8.0 Hz), 7.98–8.04 (m, 2H), 7.95 (dd, 1H, 1.4, 8.0 Hz), 7.66 (dd, 1.68, 7.92), 7.50–7.60 (m, 2H), 7.39 (ddd, 1H, 0.8, 1.76, 7.48 Hz), 7.17 (dd, 1H, 1.28, 8.3 Hz), 7.03 (td, 1H, 1.2, 7.6 Hz).
[13]**C-NMR** (100 MHz, DMSO-d_6) δ 150.0, 146.1, 133.2, 132.6, 130.4, 128.5, 128.0, 127.7, 126.6, 126.1, 125.5, 124.6, 123.7, 123.6, 123.4, 119.5, 117.0.
ESI-MS *m/z* 288.0 [M + H]$^+$; calculated value for $C_{18}H_{13}N_3O$ = 287.0, found from experiment 288.0.
HRMS (ESI/QTOF) *m/z*: [M + H]$^+$ calculated for $C_{18}H_{14}N_3O$ 288.1131; Found 288.1119.

4. Conclusions

A new, simple, efficient synthesis of a naphthalene-based -1,2,3-triazole fluorescent sensor that showed yellow *"turn-on"* fluorescence response in the presence of fluoride ions has been developed. The single crystal and NMR studies confirmed the skeletal structure of **NpTP**. The binding interaction of **NpTP** with the fluoride anion through the phenolic group and the triazole proton of the sensor was confirmed from the upfield shift of the phenolic protons and the downfield shift of the triazole proton. Job's plot using NMR studies revealed 1:1 binding between the **NpTP** molecule and the anion. Steady state studies of UV-Vis and fluorescence supported the formation of the new species after the interaction of **NpTP** and F$^-$. The presence of fluoride ions demonstrated an ESPT process in the molecule. The recognition behavior of the **NpTP** molecule towards anions can find applications in logic gate systems, molecular switches, dual detection systems, and in a biological environment. The described sensing system can also open up avenues in "structure–signal" (structure–property) investigations for developing a predictive model for tuning the signal-output of 1,2,3-triazole sensors for efficient and selective sensing.

Supplementary Materials: The following are available online at http://www.mdpi.com/2312-7481/4/1/15/s1. Figure S1. ^{1}H-NMR spectrum of **NpTP**. Figure S2. ^{13}C-NMR spectrum of **NpTP** in DMSO-d_6. Figure S3. ^{13}C-NMR spectrum of **NpTP** in Acetone-d_6. Figure S4. 1D DEPT90 spectrum of **NpTP**. Figure S5. 2D HMBC spectrum of **NpTP**. Figure S6. The 2D NOESY spectrum of **NpTP** in ((CD$_3$)$_2$S=O, RT) showing the correlation between the H7 proton and H10 and H14 protons and in between H10 and H15 proton. Figure S7. The 2D NOESY spectrum of **NpTP** in ((CD$_3$)$_2$S=O, RT) showing the correlation between the -OH proton and H2 proton. Figure S8. 2D NOESY spectrum of **NpTP** + 4 equivalence of TBAF. Figure S9. ^{1}H-NMR Titration experiments of **NpTP** with TBAF in DMSO-d_6. Figure S10. ^{1}H-NMR Titration experiments of **NpTP** with TBAF in DMSO-d_6, region expanded from 13.5 to 18.0 ppm. Figure S11. ^{19}F-NMR Titration experiments of **NpTP** with TBAF in DMSO-d_6. Figure S12. Benesi–Hildebrand plot of **NpTP**. Figure S13. HRMS of **NpTP**. Figure S14. ESI of **NpTP**. Figures S15 and S16. Single Crystal X-ray spectroscopic study. Tables S1–S7. Detail reports on **NpTP** crystal structure.

Acknowledgments: The authors acknowledge the Department of Chemistry and Biochemistry, Georgia Southern University, College of Science and Mathematics and the College Office of Undergraduate Research (COUR) for the financial assistance. The authors are also thankful to Jeffrey Orvis for the intellectual and helpful discussion. Jessica Bunn, ShainazLandge, and KarelleAiken acknowledge the support of the National Science Foundation (Award Number: NSF-CHE (REU) 1359229).

Author Contributions: Georgia Southern University is a Primarily Undergraduate Institution and hence multiple undergraduates have been involved in this project. Jessica Bunn (summer REU scholar), Steven Sutton, Matthew Christianson, and Christian Freeman are all undergraduate scholars who have contributed towards this project with synthesis and analysis. Domonique Winder was a Master's student who was involved in the initial synthesis of **NpTP**. Clifford Padgett and Colin McMillen analyzed and solved the crystal structure. Shainaz Landge, Karelle Aiken, and Debanjana Ghosh conceived, designed, and analyzed the experiments. Karelle Aiken is primarily responsible for the synthesis of the sensors. Debanjana Ghosh performed the steady-state studies and Shainaz Landge performed the NMR and color studies. Shainaz Landge, Karelle Aiken, and Debanjana Ghosh all are equally responsible for writing this manuscript.

Conflicts of Interest: The authors declare no conflict of interest.

References

1. Lehn, J.-M. Molecular and Supramolecular Devices. In *Supramolecular Chemistry*; Wiley-VCH Verlag GmbH & Co. KGaA: Weinheim, Germany, 1995.

2. Gopel, W.; Hesse, J.; Zemel, J.N. *Sensors: A Comprehensive Survey*; Wiley: Hoboken, NJ, USA, 1989.

3. Zaccheroni, N.; Palomba, F.; Rampazzo, E. Luminescent Chemosensors: From Molecules to Nanostructures. In *Applied Photochemistry: When Light Meets Molecules*; Bergamini, G., Silvi, S., Eds.; Springer International Publishing: Cham, Switzerland, 2016.

4. Kubo, Y.; Yamamoto, M.; Ikeda, M.; Takeuchi, M.; Shinkai, S.; Yamaguchi, S.; Tamao, K. A Colorimetric and Ratiometric Fluorescent Chemosensor with Three Emission Changes: Fluoride Ion Sensing by a Triarylborane– Porphyrin Conjugate. *Angew. Chem. Int. Ed.* **2003**, *42*, 2036–2040. [CrossRef] [PubMed]

5. Udhayakumari, D.; Naha, S.; Velmathi, S. Colorimetric and fluorescent chemosensors for Cu^{2+}. A comprehensive review from the years 2013–15. *Anal. Methods* **2017**, *9*, 552–578. [CrossRef]

6. Zhou, Y.; Zhang, J.F.; Yoon, J. Fluorescence and Colorimetric Chemosensors for Fluoride-Ion Detection. *Chem. Rev.* **2014**, *114*, 5511–5571. [CrossRef] [PubMed]

7. Du, J.; Hu, M.; Fan, J.; Peng, X. Fluorescent chemodosimeters using "mild" chemical events for the detection of small anions and cations in biological and environmental media. *Chem. Soc. Rev.* **2012**, *41*, 4511–4535. [CrossRef] [PubMed]

8. Hussain, I.; Ahamad, K.U.; Nath, P. Low-Cost, Robust, and Field Portable Smartphone Platform Photometric Sensor for Fluoride Level Detection in Drinking Water. *Anal. Chem.* **2017**, *89*, 767–775. [CrossRef] [PubMed]

9. Ayoob, S.; Gupta, A.K. Fluoride in Drinking Water: A Review on the Status and Stress Effects. *Crit. Rev. Environ. Sci. Technol.* **2006**, *36*, 433–487. [CrossRef]

10. Carey, C.M. Focus on Fluorides: Update on the Use of Fluoride for the Prevention of Dental Caries. *J. Evid. Based Dent. Pract.* **2014**, *14*, 95–102. [CrossRef] [PubMed]

11. Everett, E.T. Fluoride's Effects on the Formation of Teeth and Bones, and the Influence of Genetics. *J. Dent. Res.* **2010**, *90*, 552–560. [CrossRef] [PubMed]

12. Jha, S.K.; Singh, R.K.; Damodaran, T.; Mishra, V.K.; Sharma, D.K.; Rai, D. Fluoride in Groundwater: Toxicological Exposure and Remedies. *J. Toxicol. Environ. Health* **2013**, *16*, 52–66. [CrossRef] [PubMed]

13. Valdez Jiménez, L.; López Guzmán, O.D.; Cervantes Flores, M.; Costilla-Salazar, R.; Calderón Hernández, J.; Alcaraz Contreras, Y.; Rocha-Amador, D.O. In utero exposure to fluoride and cognitive development delay in infants. *Neurotoxicology* **2017**, *59*, 65–70. [CrossRef] [PubMed]

14. Van Veen, E.H.; De Loos-Vollebregt, M.T.C.; Wassink, A.P.; Kalter, H. Determination of trace elements in uranium by inductively coupled plasma-atomic emission spectrometry using Kalman filtering. *Anal. Chem.* **1992**, *64*, 1643–1649. [CrossRef]

15. Orlov, A.A.; Tsimbalyuk, A.F.; Malyugin, R.V. Desublimation for Purification and Transporting UF6: Process Description and Modeling. *Sep. Purif. Rev.* **2017**, *46*, 81–89. [CrossRef]

16. El Sayed, S.; Pascual, L.; Agostini, A.; Martínez-Máñez, R.; Sancenón, F.; Costero, A.M.; Parra, M.; Gil, S. A Chromogenic Probe for the Selective Recognition of Sarin and Soman Mimic DFP. *ChemistryOpen* **2014**, *3*, 142–145. [CrossRef] [PubMed]

17. Jang, Y.J.; Kim, K.; Tsay, O.G.; Atwood, D.A.; Churchill, D.G. Update 1 of: Destruction and Detection of Chemical Warfare Agents. *Chem. Rev.* **2015**, *115*, PR1–PR76. [CrossRef] [PubMed]

18. Duke, R.M.; Veale, E.B.; Pfeffer, F.M.; Kruger, P.E.; Gunnlaugsson, T. Colorimetric and fluorescent anion sensors: an overview of recent developments in the use of 1,8-naphthalimide-based chemosensors. *Chem. Soc. Rev.* **2010**, *39*, 3936–3953. [CrossRef] [PubMed]

19. Amendola, V.; Fabbrizzi, L.; Mosca, L.; Schmidtchen, F.-P. Urea-, Squaramide-, and Sulfonamide-Based Anion Receptors: A Thermodynamic Study. *Chemistry* **2011**, *17*, 5972–5981. [CrossRef] [PubMed]

20. Perez-Ruiz, R.; Diaz, Y.; Goldfuss, B.; Hertel, D.; Meerholz, K.; Griesbeck, A.G. Fluoride recognition by a chiral urea receptor linked to a phthalimide chromophore. *Org. Biomol. Chem.* **2009**, *7*, 3499–3504. [CrossRef] [PubMed]

21. Gunnlaugsson, T.; Davis, A.P.; Glynn, M. Fluorescent photoinduced electron transfer (PET) sensing of anions using charge neutral chemosensors. *Chem. Commun.* **2001**, *24*, 2556–2557. [CrossRef]

22. Beer, P.D. Transition-Metal Receptor Systems for the Selective Recognition and Sensing of Anionic Guest Species. *Acc. Chem. Res.* **1998**, *31*, 71–80. [CrossRef]

23. Ghosh, D.; Nandi, N.; Chattopadhyay, N. Differential Förster Resonance Energy Transfer from the Excimers of Poly(*N*-vinylcarbazole) to Coumarin 153. *J. Phys. Chem. B* **2012**, *116*, 4693–4701. [CrossRef] [PubMed]

24. Ghosh, D.; Sarkar, D.; Chattopadhyay, N. Intramolecular charge transfer promoted fluorescence transfer: A demonstration of re-absorption of the donor fluorescence by the acceptor. *J. Mol. Liq.* **2010**, *156*, 131–136. [CrossRef]

25. Loken, M.R.; Hayes, J.W.; Gohlke, J.R.; Brand, L. Excited-state proton transfer as a biological probe. Determination of rate constants by means of nanosecond fluorometry. *Biochemistry* **1972**, *11*, 4779–4786. [CrossRef] [PubMed]

26. Zhang, X.; Guo, L.; Wu, F.-Y.; Jiang, Y.-B. Development of Fluorescent Sensing of Anions under Excited-State Intermolecular Proton Transfer Signaling Mechanism. *Org. Lett.* **2003**, *5*, 2667–2670. [CrossRef] [PubMed]

27. Kompa, K.L.; Levine, R.D. A molecular logic gate. *Proc. Natl. Acad. Sci. USA* **2001**, *98*, 410–414. [CrossRef] [PubMed]

28. Bojinov, V.; Georgiev, N. Molecular sensors and molecular logic gates (review). *J. Univ. Chem. Technol. Metall.* **2011**, *46*, 3–26.

29. Borghetti, J.; Snider, G.S.; Kuekes, P.J.; Yang, J.J.; Stewart, D.R.; Williams, R.S. /'Memristive/' switches enable /'stateful/' logic operations via material implication. *Nature* **2010**, *464*, 873–876. [CrossRef] [PubMed]

30. De Silva, A.P. Molecular Logic Gates. In *Supramolecular Chemistry*; John Wiley & Sons, Ltd.: Hoboken, NJ, USA, 2012.

31. Landge, S.M.; Aprahamian, I. A pH Activated Configurational Rotary Switch: Controlling the E/Z Isomerization in Hydrazones. *J. Am. Chem. Soc.* **2009**, *131*, 18269–18271. [CrossRef] [PubMed]

32. Kumar, M.; Dhir, A.; Bhalla, V. A Molecular Keypad Lock Based on the Thiacalix[4]arene of 1,3-Alternate Conformation. *Org. Lett.* **2009**, *11*, 2567–2570. [CrossRef] [PubMed]

33. Rezaeian, K.; Khanmohammadi, H. Molecular logic circuits and a security keypad lock based on a novel colorimetric azo receptor with dual detection ability for copper(II) and fluoride ions. *Supramol. Chem.* **2016**, *28*, 256–266. [CrossRef]

34. Ahmad, Z.; Shah, A.; Siddiq, M.; Kraatz, H.-B. Polymeric micelles as drug delivery vehicles. *RSC Adv.* **2014**, *4*, 17028–17038. [CrossRef]

35. Petcu, A.R.; Rogozea, E.A.; Lazar, C.A.; Olteanu, N.L.; Meghea, A.; Mihaly, M. Specific interactions within micelle microenvironment in different charged dye/surfactant systems. *Arab. J. Chem.* **2016**, *9*, 9–17. [CrossRef]

36. Park, G.J.; Jo, H.Y.; Ryu, K.Y.; Kim, C. A new coumarin-based chromogenic chemosensor for the detection of dual analytes Al^{3+} and F. *RSC Adv.* **2014**, *4*, 63882–63890. [CrossRef]

37. Tang, S.; Meng, Q.; Sun, H.; Su, J.; Yin, Q.; Zhang, Z.; Yu, H.; Chen, L.; Gu, W.; Li, Y. Dual pH-sensitive micelles with charge-switch for controlling cellular uptake and drug release to treat metastatic breast cancer. *Biomaterials* **2017**, *114*, 44–53. [CrossRef] [PubMed]

38. Lebrini, M.; Traisnel, M.; Lagrenée, M.; Mernari, B.; Bentiss, F. Inhibitive properties, adsorption and a theoretical study of 3,5-bis(*n*-pyridyl)-4-amino-1,2,4-triazoles as corrosion inhibitors for mild steel in perchloric acid. *Corros. Sci.* **2008**, *50*, 473–479. [CrossRef]

39. Lu, W.; Chen, D.; Jiang, H.; Jiang, L.; Shen, Z. Polymer-based fluoride-selective chemosensor: Synthesis, sensing property, and its use for the design of molecular-scale logic devices. *J. Polym. Sci. Part A* **2012**, *50*, 590–598. [CrossRef]

40. Wagner, D.B. The Use of Coumarins as Environmentally-Sensitive Fluorescent Probes of Heterogeneous Inclusion Systems. *Molecules* **2009**, *14*, 210–237. [CrossRef] [PubMed]

41. Hua, Y.; Flood, A.H. Click chemistry generates privileged CH hydrogen-bonding triazoles: The latest addition to anion supramolecular chemistry. *Chem. Soc. Rev.* **2010**, *39*, 1262–1271. [CrossRef] [PubMed]

42. Lau, Y.H.; Rutledge, P.J.; Watkinson, M.; Todd, M.H. Chemical sensors that incorporate click-derived triazoles. *Chem. Soc. Rev.* **2011**, *40*, 2848–2866. [CrossRef] [PubMed]

43. Wang, C.; Wang, D.; Yu, S.; Cornilleau, T.; Ruiz, J.; Salmon, L.; Astruc, D. Design and Applications of an Efficient Amphiphilic "Click" CuI Catalyst in Water. *ACS Catal.* **2016**, *6*, 5424–5431. [CrossRef]

44. Watkinson, M. Click Triazoles as Chemosensors. In *Click Triazoles*; Košmrlj, J., Ed.; Springer: Berlin/Heidelberg, Germany, 2012.

45. Zhang, F.; Moses, J.E. Benzyne Click Chemistry with in Situ Generated Aromatic Azides. *Org. Lett.* **2009**, *11*, 1587–1590. [CrossRef] [PubMed]

46. Rostovtsev, V.V.; Green, L.G.; Fokin, V.V.; Sharpless, K.B. A Stepwise Huisgen Cycloaddition Process: Copper(I)-Catalyzed Regioselective "Ligation" of Azides and Terminal Alkynes. *Angew. Chem. Int. Ed.* **2002**, *41*, 2596–2599. [CrossRef]

47. Cho, E.J.; Ryu, B.J.; Lee, Y.J.; Nam, K.C. Visible Colorimetric Fluoride Ion Sensors. *Org. Lett.* **2005**, *7*, 2607–2609. [CrossRef] [PubMed]

48. Cho, E.J.; Moon, J.W.; Ko, S.W.; Lee, J.Y.; Kim, S.K.; Yoon, J.; Nam, K.C. A New Fluoride Selective Fluorescent as Well as Chromogenic Chemosensor Containing a Naphthalene Urea Derivative. *J. Am. Chem. Soc.* **2003**, *125*, 12376–12377. [CrossRef] [PubMed]

49. Ghosh, D.; Rhodes, S.; Hawkins, K.; Winder, D.; Atkinson, A.; Ming, W.; Padgett, C.; Orvis, J.; Aiken, K.; Landge, S. A simple and effective 1,2,3-triazole based "turn-on" fluorescence sensor for the detection of anions. *New J. Chem.* **2015**, *39*, 295–303. [CrossRef]

50. Govan, R.D. *Approaches Toward Novel 1,2,3-Triazole Sensors for the Detection of Anions and Heavy Metal Cations, Electronic Theses & Dissertations*; Georgia Southern University: Statesboro, GA, USA, 2017; Available online: http://digitalcommons.georgiasouthern.edu/etd/1604 (accessed on 15 December 2017).

51. Boiocchi, M.; Del Boca, L.; Gómez, D.E.; Fabbrizzi, L.; Licchelli, M.; Monzani, E. Nature of Urea–Fluoride Interaction: Incipient and Definitive Proton Transfer. *J. Am. Chem. Soc.* **2004**, *126*, 16507–16514. [CrossRef] [PubMed]

52. Peng, X.; Wu, Y.; Fan, J.; Tian, M.; Han, K. Colorimetric and Ratiometric Fluorescence Sensing of Fluoride: Tuning Selectivity in Proton Transfer. *J. Org. Chem.* **2005**, *70*, 10524–10531. [CrossRef] [PubMed]

53. Juwarker, H.; Lenhardt, J.M.; Castillo, J.C.; Zhao, E.; Krishnamurthy, S.; Jamiolkowski, R.M.; Kim, K.-H.; Craig, S.L. Anion Binding of Short, Flexible Aryl Triazole Oligomers. *J. Org. Chem.* **2009**, *74*, 8924–8934. [CrossRef] [PubMed]

54. Job, P. *Spectrochemical Methods of Analysis*; Wiley Interscience: New York, NY, USA, 1971.

55. Choi, K.; Hamilton, A.D. Selective Anion Binding by a Macrocycle with Convergent Hydrogen Bonding Functionality. *J. Am. Chem. Soc.* **2001**, *123*, 2456–2457. [CrossRef] [PubMed]
56. Ghosh, D.; Rhodes, S.; Winder, D.; Atkinson, A.; Gibson, J.; Ming, W.; Padgett, C.; Landge, S.; Aiken, K. Spectroscopic investigation of bis-appended 1,2,3-triazole probe for the detection of Cu(II) ion. *J. Mol. Struct.* **2017**, *1134*, 638–648. [CrossRef]
57. Charisiadis, P.; Exarchou, V.; Troganis, A.N.; Gerothanassis, I.P. Exploring the "forgotten" -OH NMR spectral region in natural products. *Chem. Commun.* **2010**, *46*, 3589–3591. [CrossRef] [PubMed]
58. Siskos, M.G.; Kontogianni, V.G.; Tsiafoulis, C.G.; Tzakos, A.G.; Gerothanassis, I.P. Investigation of solute-solvent interactions in phenol compounds: Accurate ab initio calculations of solvent effects on ^{1}H NMR chemical shifts. *Org. Biomol. Chem.* **2013**, *11*, 7400–7411. [CrossRef] [PubMed]
59. Jain, A.; Gupta, R.; Agarwal, M. Benzimidazole scaffold as dipodal molecular cleft for swift and efficient naked eye fluoride ion recognition via preorganized N-H and aromatic C-H in aqueous media. *Indian J. Chem.* **2017**, *56*, 513–518.
60. Li, J.; Xu, X.; Shao, X.; Li, Z. A novel colorimetric fluoride sensor based on a semi-rigid chromophore controlled by hydrogen bonding. *Luminescence* **2015**, *30*, 1285–1289. [CrossRef] [PubMed]
61. Jones, R.N. The Ultraviolet Absorption Spectra of Aromatic Hydrocarbons. *Chem. Rev.* **1943**, *32*, 1–46. [CrossRef]
62. Mondal, K.; Bhattacharyya, S.; Sharma, A. Photocatalytic Degradation of Naphthalene by Electrospun Mesoporous Carbon-Doped Anatase TiO_2 Nanofiber Mats. *Ind. Eng. Chem. Res.* **2014**, *53*, 18900–18909. [CrossRef]
63. Silva, A.F.; Fiedler, H.D.; Nome, F. Ionic Quenching of Naphthalene Fluorescence in Sodium Dodecyl Sulfate Micelles. *J. Phys. Chem. A* **2011**, *115*, 2509–2514. [CrossRef] [PubMed]
64. Maeda, H.; Maeda, T.; Mizuno, K. Absorption and Fluorescence Spectroscopic Properties of 1- and 1,4-Silyl-Substituted Naphthalene Derivatives. *Molecules* **2012**, *17*, 5108–5125. [CrossRef] [PubMed]
65. Zhang, Y.; Ye, X.; Petersen, J.L.; Li, M.; Shi, X. Synthesis and Characterization of Bis-*N*-2-Aryl Triazole as a Fluorophore. *J. Org. Chem.* **2015**, *80*, 3664–3669. [CrossRef] [PubMed]
66. Rasouli, M.; Tavassoli, S.H.; Mousavi, S.J.; Darbani, S.M.R. Measuring of naphthalene fluorescence emission in the water with nanosecond time delay laser induced fluorescence spectroscopy method. *Opt. Int. J. Light Electron Opt.* **2016**, *127*, 6218–6223. [CrossRef]
67. Lakowicz, J.R. *Principles of Fluorescence Spectroscopy*; Springer: New York, NY, USA, 2007.
68. Harris, C.M.; Selinger, B.K. Acid-base properties of 1-naphthol. Proton-induced fluorescence quenching. *J. Phys. Chem.* **1980**, *84*, 1366–1371. [CrossRef]
69. Boyer, R.; Deckey, G.; Marzzacco, C.; Mulvaney, M.; Schwab, C.; Halpern, A.M. The photophysical properties of 2-naphthol: A physical chemistry experiment. *J. Chem. Ed.* **1985**, *62*, 630. [CrossRef]
70. Laws, W.R.; Brand, L. Analysis of two-state excited-state reactions. The fluorescence decay of 2-naphthol. *J. Phys. Chem.* **1979**, *83*, 795–802. [CrossRef]
71. Marciniak, B.; Kozubek, H.; Paszyc, S. Estimation of pKa* in the first excited singlet state. A physical chemistry experiment that explores acid-base properties in the excited state. *J. Chem. Ed.* **1992**, *69*, 247. [CrossRef]
72. Benesi, H.A.; Hildebrand, J.H. A Spectrophotometric Investigation of the Interaction of Iodine with Aromatic Hydrocarbons. *J. Am. Chem. Soc.* **1949**, *71*, 2703–2707. [CrossRef]
73. Bhosale, S.V.; Bhosale, S.V.; Kalyankar, M.B.; Langford, S.J. A Core-Substituted Naphthalene Diimide Fluoride Sensor. *Org. Lett.* **2009**, *11*, 5418–5421. [CrossRef] [PubMed]
74. He, X.; Hu, S.; Liu, K.; Guo, Y.; Xu, J.; Shao, S. Oxidized Bis(indolyl)methane: A Simple and Efficient Chromogenic-Sensing Molecule Based on the Proton Transfer Signaling Mode. *Org. Lett.* **2006**, *8*, 333–336. [CrossRef] [PubMed]
75. Pirali, T.; Gatti, S.; DiBrisco, R.; Tacchi, S.; Zaninetti, R.; Brunelli, E.; Massarotti, A.; Sorba, G.; Canonico, P.L.; Moro, L.; et al. Estrogenic Analogues Synthesized by Click Chemistry. *ChemMedChem* **2007**, *2*, 437–440. [CrossRef] [PubMed]

magnetochemistry

MDPI

Article

Towards a Microscopic Theory of the Knight Shift in an Anisotropic, Multiband Type-II Superconductor

Richard A. Klemm

Department of Physics, 4111 Libra Drive, University of Central FLlorida, Orlando, FL 32816-2385, USA;
richard.klemm@ucf.edu; Tel.: 1-407-823-1543

Received: 24 November 2017; Accepted: 4 January 2018; Published: 29 January 2018

Abstract: A method is proposed to extend the zero-temperature Hall-Klemm microscopic theory of the Knight shift K in an anisotropic and correlated, multi-band metal to calculate $K(T)$ at finite temperatures T both above and into its superconducting state. The transverse part of the magnetic induction $\mathbf{B}(t) = \mathbf{B}_0 + \mathbf{B}_1(t)$ causes adiabatic changes suitable for treatment with the Keldysh contour formalism and analytic continuation onto the real axis. We propose that the Keldysh-modified version of the Gor'kov method can be used to evaluate $K(T)$ at high $\mathbf{B}_0$ both in the normal state, and by quantizing the conduction electrons or holes with Landau orbits arising from $\mathbf{B}_0$, also in the entire superconducting regime for an anisotropic, multiband Type-II BCS superconductor. Although the details have not yet been calculated in detail, it appears that this approach could lead to the simple result $K_S(T) \approx a(\mathbf{B}_0) - b(\mathbf{B}_0)|\Delta(\mathbf{B}_0, T)|^2$, where $2|\Delta(\mathbf{B}_0, T)|$ is the effective superconducting gap. More generally, this approach can lead to analytic expressions for $K_S(T)$ for anisotropic, multiband Type-II superconductors of various orbital symmetries that could aid in the interpretation of experimental data on unconventional superconductors.

Keywords: Knight shift; Type-II superconductor; nuclear magnetic resonance; anisotropy; electron correlations

---

1. Introduction

In nuclear magnetic resonance (NMR) measurements of a nuclear spin, there is a difference between the resonance frequency of that spin when it is in a metal from when it is in vacuum or in an insulator. This is known as the Knight shift [1]. Although the temperature T dependence of the Knight shift in a superconductor $K_S(T)$ has long been considered to be a probe of the spin state of the paired electrons [2–4], the only theoretical basis for that assumption was the 1958 Yosida model that assumed the probed nuclear spins could be entirely neglected [5], and that the only quantity of interest was the T dependence of the zero-magnetic-field H limit of the electron spin susceptibility of an isotropic and uncorrelated Type-I superconductor [5]. This led for a BCS singlet-pair-spin superconductor to $K_S(T) \propto x/(1+x)$, where $x = (\beta/\Delta)\frac{d\Delta}{d\beta}$, $\beta = 1/T$, $\Delta(T)$ is one-half the BCS energy gap, and we set $k_B = 1$, so that $K_S(T) \to 0$ as $T \to 0$, unlike most experimental results [5].

For isotropic Type-I superconductors in the Meissner state, crushing the sample to a powder of crystallites the cross-sections of which were less than the magnetic penetration depth was usually found to provide a reasonable method for that conventional theory to be applicable [2–4]. In the first years following the BCS theory, a few elemental transition metal superconductors were found to behave somewhat differently, as $K_S(T)$ did not vanish as $T \to 0$ [6], and it was thought that surface impurity spin-orbit scattering could explain the near-cancelation of $K_S(0)$ in transition metals [6,7]. However, surface impurity spin-orbit scattering could not explain the observed non-vanishing $K_S(0)$ results observed in clean materials. It is now understood that there is also a component to the Knight shift due to the orbital motion of the electrons in a superconductor, and for an anisotropic superconductor, this orbital contribution to the Knight shift depends upon the magnetic induction B direction.

There have since been many examples of unexplained behaviors of the Knight shift in exotic superconductors. Since one possibility of a T-independent Knight shift result would be a parallel-spin, triplet pair-spin superconducting state, the use of the Knight shift has been considered to be a principal tool for the identification of a triplet pair-spin state. Some examples of triplet-pair-spin or some other types of exotic behavior claimed to exist in unusual materials based upon the unconventional Knight shift T-dependence are listed in the bibliography [8–14]. However, one of those materials was a quasi-one-dimensional organic superconductor [10,11], some examples of which often exhibit spin-density waves [15], and another was the very dirty sodium cobaltate hydrate material [12–14]. In the latter example, the upper critical field parallel to that layered compound is Pauli-limited, which normally only occurs when the magnetic field breaks the oppositely-oriented pair spins [16,17]. Since dirt drastically suppresses p-wave superconductivity [18], the sodium cobaltate hydrate Knight shift results, if correct, are likely to arise from some other mechanism.

Moreover, in highly anisotropic Type-II superconductors, such as the cuprates and heavy fermion materials, other significant breakdowns in the Yosida theory have been found to exist. In the first Knight shift measurements on the cuprate $YBa_2Cu_3O_{7-\delta}$, Bennett et al. found that the Yosida theory appeared to work for the ^{63}Cu spins in the CuO chains for all applied H directions, although the orbital contributions are different for each of those three orthogonal H directions, and it also appeared to work well for the ^{63}Cu spins in the CuO_2 planes when the strong constant H was applied parallel to the CuO_2 layers. However, when H was applied normal to the CuO_2 layers, no T dependence to the Knight shift was observed in that cuprate [19]. This result was later described by Slichter as possibly being due to a "fortuitous" cancelation of the effect from an isolated planar ^{63}Cu spin by its interaction with its near-neighbor planar ^{63}Cu spins [20]. Subsequently, in a number of layered correlated superconductors, the T dependence of the Knight shift probes of the nuclear spins in the layers with the field applied normal to the layers has been observed to vary strongly with field strength, approaching a constant $K_S(T)$ in the large normal field strength limit, as first observed by Bennett et al. [19–24].

Especially in the case of Sr_2RuO_4, numerous Knight shift measurements of the ^{17}O, ^{99}Ru, ^{101}Ru, and ^{87}Sr have all led to temperature-independent Knight shift measurements [25–27], as did polarized neutron scattering experiments [28]. These experiments were all interpreted as evidence for a parallel-spin pair state in that material. However, several upper critical field measurements with the field parallel to the layers showed strong Pauli limiting effects [29,30], which is inconsistent with a parallel-spin pair state[31,32]. In addition, the fact that T-independent Knight shift measurements were obtained for the field both parallel and perpendicular to the RuO_2 layers is incompatible with any of the crystal point-group-compatible p-wave states. Thus, the only way for a T-independent Knight shift to legitimately arise from a parallel-pair-spin state in both field directions is for the d-vector (the vector describing the components of the three triplet spin states) to rotate with the magnetic field [33,34]. This argument was used to show that while the upper critical field of Sr_2RuO_4 is strongly Pauli limited for the field applied parallel to the layers, it could possibly be consistent with one or more p-wave helical states, provided that the d-vector is allowed to rotate freely with the magnetic field direction [32]. This means that spin-orbit coupling with the lattice would have to be negligible. However, there is strong evidence that spin-orbit coupling in Sr_2RuO_4 is very strong at some points on the Fermi surface, ruling out such d-vector rotation possibilities[35]. More worrisome for the Knight shift measurement results is the fact that carefully performed scanning tunneling measurements of the electronic density of states provided very strong evidence of a nodeless superconducting order parameter orbital symmetry in Sr_2RuO_4 [17,36], consistent with a nearly isotropic gap function that is essentially identical on all three of its Fermi surfaces. Since the theories behind the Pauli limiting effects and the BCS gap density of states are very well established, but the Knight shift measurement interpretations rely entirely on the complete neglect of the probed nuclear spins, the development of a microscopic theory of the T dependence of the Knight shift in anisotropic and correlated Type-II superconductors is sorely needed.

We further note that the time dependence of a spin-1/2 particle in a classic magnetic resonance experiment is now a textbook example of an exactly soluble first quantization quantum mechanics problem giving rise to a Berry phase [37,38]. In that case, the Berry, or geometric, phase is a combination of the resonance profile with the frequency of the oscillatory transverse applied magnetic field. In higher spin I systems, there are $2I$ combinations of those two quantities, giving rise to a multiplet of Berry phases, as discussed in the following. Note that the probed nuclear spins of Sr_2RuO_4 are either 5/2 or 9/2. Since nothing was known about the Berry phase in 1958, its possible implications for the interpretation of Knight shift measurements have been generally and perhaps completely ignored in the literature.

In fairness to the pioneering work of Yosida [5], there have been a few cases in which a complete lack of any T-dependence to the Knight shift has been confirmed by other experiments consistent with a parallel-pair-spin superconducting state [39–41]. These are for the uranium-based compounds UCoGe and UPt$_3$, for which the T-independent Knight shift in UCoGe is in agreement with the general assessment of the upper critical field and muon depolarization experiments [18,40]. In UPt$_3$, the seeming incompatibility of the Knight shift and the upper critical field appears to have been resolved by polarized neutron diffraction experiments [41], favoring a parallel-spin pair state in all three superconducting phases. In the ferromagnetic superconductors UGe$_2$, UCoGe, and URhGe, the weak Ising-like ferromagnetism appears to allow for a parallel-spin, p-wave superconducting order parameter in the plane perpendicular to the ferromagnetism, but the Knight shift measurements have not yet been made on URhGe and UGe$_2$, the latter of which is only superconducting under pressure. In these three ferromagnetic superconductors, there is at least a plausible mechanism for a parallel-spin pair superconducting state, and in URhGe the upper critical field fits the predictions for all three crystal axis directions of a parallel-spin p-wave polar state fixed to the crystal a-axis direction normal to the c-axis Ising ferromagnetic order [40,42], and there is a reentrant, high field phase that violates the Pauli limit by a factor of 20 [40]. In order to obtain further evidence that the classic Yosida interpretation of a T-independent $K_S(T)$ can correctly imply a parallel-spin superconducting state, we urge that ^{73}Ge, with a strong nuclear moment, (or possibly ^{103}Rh, with a much weaker nuclear moment) $K_S(T)$ measurements on URhGe be carefully performed in the low-field superconducting phase.

2. The Model

The first microscopic model of the Knight shift at $T = 0$ in anisotropic and correlated metals was recently presented by Hall and Klemm [43]. This model assumed that the applied magnetic fields probe the nuclear spins, and the spins of the electrons orbiting the nucleus interact with the nucleus via the hyperfine interaction in the form of a diagonal D tensor with two distinct components $D_x = D_y \neq D_z$. The assumption $D_x = D_y$ was made to simplify the calculations, as discussed in more detail in the following. After interacting with the nuclear spins, the orbital electrons can be excited into one of multiple bands, each of which was assumed to have an ellipsoidal Fermi surface of arbitrary anisotropy and shape. The orbital motion of the electrons in each of these bands was constrained by the strong, time-independent part B_0 of the magnetic induction $B(t)$ to be in Landau levels, and the electron spins also could interact weakly with B_0. It was found that the self-energy due to D_z led to the Knight shift, and that due to $D_x = D_y$ led to the first formulas for the linewidth changes associated with the Knight shift at $T = 0$. However, since those calculations were made at $T = 0$, they could not be used to probe the superconducting state. In the following, a method is proposed to do so.

Following Haug and Jauho [44], we write the Hamiltonian as $\mathcal{H} = \mathcal{H}_0 + \mathcal{H}_{int} + \mathcal{H}'(t)$, where $\mathcal{H}_0 + \mathcal{H}_{int}$ is the time-independent part and $\mathcal{H}'(t)$ is the time-dependent part due to the oscillatory (or pulsed) magnetic field transverse to the constant applied magnetic field H_0, and the time-independent part consists of the simple (or exactly soluble) part $\mathcal{H}_0$ and the interaction part $\mathcal{H}_{int}$ that involves the interactions between the particles that must be treated perturbatively. In the case at hand, there are four types of particles: (1) the nuclear spins probed in the NMR experiment, which are assumed to have the general spin $I \neq 0$ with $(2I + 1)$ substates denoted m_I; (2) the local orbital electrons surrounding each of the nuclei probed in the NMR experiment; (3) the conduction electrons or holes that propagate from the local nuclei throughout the metal/superconductor; and (4) the superconducting Cooper pairs of electrons or holes. We note that complicated materials such as Sr_2RuO_4 contain multiple Fermi surfaces, which can be a mix of electron and hole Fermi surfaces. In this model, we do not account for competing ferromagnetism or charge-density wave (CDW) or spin-density wave (SDW) formation, at least one of which is normally present in the transition metal dichalcogenides, the organic layered superconductors, the cuprates, the iron pnictides, and the ferromagnetic superconductors. Such competing effects will be the subjects of future studies.

2.1. The Simple Hamiltonian $\mathcal{H}_0$

Since in an NMR experiment, the applied magnetic field can be applied in any direction, we assume the resulting constant magnetic induction $B_0 = B_0(\sin\theta\cos\phi, \sin\theta\sin\phi, \cos\theta) = B_0\hat{r}$ with respect to the crystalline Cartesian x, y, z axes. We then quantize the spins along B_0. We thus write

$$\mathcal{H}_0 = \mathcal{H}_{n,0} + \mathcal{H}_{e,0} + \mathcal{H}_{cond,0} \quad, \tag{1}$$

where

$$\mathcal{H}_{n,0} = -\omega_n \sum_{i,m_I} m_I a^\dagger_{i,m_I} a_{i,m_I} \quad, \tag{2}$$

$$\mathcal{H}_{e,0} = \sum_{i,q,\sigma} [\epsilon_q - \sigma\omega_e/2] b^\dagger_{i,q,\sigma} b_{i,q,\sigma} \quad, \tag{3}$$

$$\mathcal{H}_{cond,0} = \sum_{j,\sigma} \int d^3r_j \psi^\dagger_{j,\sigma}(r_j) \left(\sum_{\nu=1}^{3} \frac{1}{2m_{j,\nu}} [\nabla_{j,\nu}/i - eA_{j,\nu}(r_j)]^2 - \sigma\omega'_{j,e}/2 \right) \psi_{j,\sigma}(r_j) \quad, \tag{4}$$

where e is the electronic charge, $a^\dagger_{i,m_I}$ creates a nucleus of spin I in the subspin state $m_I = -I, -I + 1, \ldots, I - 1, I$ at the atomic position i, $b^\dagger_{i,q,\sigma}$ creates an electron with energy ϵ_q and spin-1/2 eigenstates indexed by $\sigma = \pm 1$ orbiting that nucleus at site i, where $q \in (n, \ell, m)$ is nominally its weak-spin-orbit local electron orbital quantum number set [or its fully relativistic set (n, j)], $\psi^\dagger_{j,\sigma}(r_j)$ creates an electron or hole with spin eigenstate $\sigma = \pm 1$ at position r_j in the jth conduction band, $\omega_n = \mu_n \cdot B_0$, $\omega_e = \mu_e \cdot B_0$, $\omega'_{j,e} = \mu_e \cdot g_j \cdot B_0$ are respectively the Zeeman energies for the probed nucleus, local orbital electrons, and conduction electrons, respectively, μ_n is the nuclear magneton for the probed nucleus (the value of which can be positive or negative), $|\mu_e| = \mu_B$ is the Bohr magneton, $g_j \cdot B_0$ defines the quantization axis direction for the anisotropic but assumed diagonal g_j tensor in the jth of the N_b conduction bands with effective mass $m_{j,\nu}$ in the νth spatial direction, $A_{j,\nu}(r_j)$ is the νth component of the magnetic vector potential at the position r_j of the conduction electron in the jth band, $B_0 = \nabla_{j,\nu} \times A_{j,\nu}$ is the magnetic induction that is independent of j, ν, and the time t, $i = \sqrt{-1}$, and we set $\hbar = 1$. Here we use the previous notation [43], but rearrange the terms in the overall Hamiltonian in order to properly take account of both the t and T dependencies essential for probing the superconducting state. We note that for integer or half-integer I, the nuclei would normally be expected to obey Bose-Einstein or Fermi-Dirac statistics, but since different nuclei correspond to different atoms and do not come in contact with one another, that statistics is not expected to be an important feature of the Knight shift. Equation (1) is the extension to arbitrary nuclear spin I of the bare Hamiltonian studied previously, except that $\mathcal{H}_{cond,0}$ was the time independent part of $H_{A,2}$ [43]. We note that for a diagonal g_j tensor,

$$\omega'_{j,e} = \mu_B B_0 [g_{j,xx} \sin^2\theta \cos^2\phi + g_{j,yy} \sin^2\theta \sin^2\phi + g_{j,zz} \cos^2\theta]^{1/2}. \tag{5}$$

As a starting point, we assume B_0 is uniform in the probed material, but when the material goes into the superconducting state, and B_0 is in an arbitrary direction with respect to the crystal axes, this is only true at the upper critical field H_{c2} above which the superconductor becomes a normal metal [17,45]. However, in the mixed state for which the time-independent part of the applied magnetic field H_0 satisfies $H_{c1} < H_0 < H_{c2}$, if H_0 is along a crystal axis, the direction of B_0 is the same as the direction of H_0 [17,46].

2.2. The Time-Independent Interaction Hamiltonian $\mathcal{H}_{int}$

We write the time-independent interaction part $\mathcal{H}_i$ of the Hamiltonian as

$$\mathcal{H}_{int} = \mathcal{H}_{hf} + \mathcal{H}_{e,int} + \mathcal{H}_{e,cond} + \mathcal{H}_{sc} \quad , \tag{6}$$

where

$$\mathcal{H}_{hf} = -\frac{D_z}{4} \sum_{i,q,\sigma,m_I} m_I \sigma a^\dagger_{i,m_I} a_{i,m_I} b^\dagger_{i,q,\sigma} b_{i,q,\sigma} - \frac{D_x}{2} \sum_{i,q,\sigma,m_I} A^\sigma_{I,m_I} a^\dagger_{i,m_I+\sigma} a_{i,m_I} b^\dagger_{i,q,-\sigma} b_{i,q,\sigma} \quad , \tag{7}$$

$$\mathcal{H}_{e,int} = \frac{1}{2} \sum_{i,q,\sigma} U_q \hat{n}_{i,q,\sigma} \hat{n}_{i,q,-\sigma} = \sum_{i,q} U_q b^\dagger_{i,q,\uparrow} b_{i,q,\uparrow} b^\dagger_{i,q,\downarrow} b_{i,q,\downarrow}, \tag{8}$$

$$\mathcal{H}_{e,cond} = \sum_{i,q,j,\sigma} \int d^3 r_j \left(v_{i,q,j} \psi^\dagger_{j,\sigma}(r_j) b_{i,q,\sigma} + H.c. \right) \delta^{(3)}(r_j - r_i) \quad , \tag{9}$$

where $A^\sigma_{I,m_I} = \sqrt{I(I+1) - m_I(m_I + \sigma)}$, and depending upon what is calculated, the superconducting pairing interaction may be written either in real space as

$$\mathcal{H}^{pos}_{sc} = \frac{1}{2} \sum_{j,j',\sigma,\sigma'} \int d^3 r_j \int d^3 r'_{j'} \psi^\dagger_{j,\sigma}(r_j) \psi^\dagger_{j',\sigma'}(r'_{j'}) V_{j,j';\sigma,\sigma'}(r_j - r'_{j'}) \psi_{j',\sigma'}(r'_{j'}) \psi_{j,\sigma}(r_j), \tag{10}$$

or in momentum space as

$$\mathcal{H}^{mom}_{sc} = \frac{1}{2} \sum_{j,j',\sigma,\sigma'} \int \frac{d^3 k_j}{(2\pi)^3} \int \frac{d^3 k'_{j'}}{(2\pi)^3} \psi^\dagger_{j,\sigma}(k_j) \psi^\dagger_{j',\sigma'}(k'_{j'}) V_{j,j';\sigma,\sigma'}(k_j - k'_{j'}) \psi_{j',\sigma'}(k'_{j'}) \psi_{j,\sigma}(k_j). \tag{11}$$

Although it appears at first sight to be easier to extend the calculation of the Knight shift into the BCS superconducting state by using $\mathcal{H}^{pos}_{sc}$ in order to include the Zeeman terms, we have included the momentum-space pairing interaction $\mathcal{H}^{mom}_{sc}$ for p-wave superconductors in magnetic fields [18,42], for which the simplest single-band parallel-spin pairing interaction $V_{j,j';\sigma,\sigma'}(k_j - k'_{j'}) = -V_0 \delta_{j,j'} \delta_{j,1} \delta_{\sigma,\sigma'} k_1 \cdot k'_1$ [18], and a modification of $\mathcal{H}^{mom}_{sc}$ more naturally treats the pairing of conduction electrons (or holes) in the presence of a strong B_0.

We note that $\sigma = \pm 1$ present in A^σ_{I,m_I} corresponds to the correct matrix elements for raising and lowering the m_I value and also corresponds to our description of the spin-1/2 electron spins [43]. Of course, m_I and σ are restricted by $-I \leq m_I, m_I + \sigma \leq I$. The first three of these terms were presented previously [43], except for a slightly different normalization factor proportional to N_b, and respectively represent the hyperfine interaction between the nuclear and surrounding orbital electrons, the effective local electron correlation interaction, and the effective Anderson interaction that allows an orbital electron to leave a local atomic site and jump into a conduction band [47]. The last term $\mathcal{H}_{sc}$ is responsible for superconducting pairing, and in the form presented allows for pairing between electrons or holes in different bands and with either the same ($\sigma' = \sigma$) or different ($\sigma' = -\sigma$) spins. In most superconductors, interband pairing is generally considered to be less important than is intraband pairing, but such complications might be important in cases such as Sr_2RuO_4, for which two of the bands are nearly identical. For standard BCS pairing, we would have $V_{\sigma,\sigma'}(r_j - r'_{j'}) \to -V_0 \delta_{\sigma',-\sigma} \delta_{j,j'} \delta^{(3)}(r_j - r'_{j'})$, at least in the standard approximation. For parallel-spin p-wave superconductors, one cannot assume the paired electrons are at the same location, but different approximations have been found to give reliable results for the upper critical induction in ferromagnetic superconductors [18,32,42,48,49].

2.3. The Time-Dependent Hamiltonian $\mathcal{H}'(t)$

The crucial part of a magnetic resonance experiment arises from the time-dependent field transverse to the stronger constant magnetic field. In a conventional NMR experiment, the time-dependent induction $B_1(t)$ oscillates in the plane normal to the strong, constant magnetic induction B_0. For $B_0 = B_0(\sin\theta\cos\phi, \sin\theta\sin\phi, \cos\theta) = B_0\hat{r}$, one may then write $B_1(t) = B_1\{\cos[\omega_0(t-t_0)]\hat{\theta} - \sin[\omega_0(t-t_0)]\hat{\phi}\}$, where $\hat{\theta} = (\cos\theta\cos\phi, \cos\theta\sin\phi, -\sin\theta)$ and $\hat{\phi} = (-\sin\phi, \cos\phi, 0)$ in the same Cartesian coordinates, and in order not to get confused with the time contours, we may choose $B_1(t_0)$ to be along $\hat{\theta}$. This is the classic way to obtain a resonance in the power spectrum associated with flipping an electron or proton spin from up to down, or in a spin I nucleus, to obtain a regular pattern of resonance frequencies associated with changes in the multiple Zeeman-like nuclear spin levels. Since one generally takes $B_1 \ll B_0$, this classic case is generally adiabatic [37,38], and is the simplest case to treat analytically.

For the above classic NMR case of a single angular frequency ω_0 in $B_1(t)$, we then have

$$\mathcal{H}'(t) = \mathcal{H}'_n(t) + \mathcal{H}'_e(t) + \mathcal{H}'_{cond}(t) \; , \tag{12}$$

$$\mathcal{H}'_n(t) = -\frac{\Omega_n}{2}\sum_{i,m_I,\sigma} e^{i\sigma\omega_0(t-t_0)} A^\sigma_{I,m_I} a^\dagger_{i,m_I+\sigma} a_{i,m_I} \; , \tag{13}$$

$$\mathcal{H}'_e(t) = -\frac{\Omega_e}{2}\sum_{i,\sigma} e^{i\sigma\omega_0(t-t_0)} b^\dagger_{i,\sigma} b_{i,-\sigma} \; , \tag{14}$$

$$\mathcal{H}'_{cond}(t) = -\frac{1}{2}\sum_{j,\sigma} e^{i\sigma\omega_0(t-t_0)} \int d^3 r_j \psi^\dagger_{j,\sigma}(r_j)\Omega'_{j,e}\psi_{j,-\sigma}(r_j) \; , \tag{15}$$

where $\Omega_n = \mu_n B_1$, $\Omega_e = \mu_e B_1$, and $\Omega'_{j,e} = \mu_e \cdot g_j \cdot B_1$, and A^σ_{I,m_I} is given by Equation (11) [43].

3. The Keldysh Contours

Following Haug and Jauho [44] and with regard to the contours, Rammer and Smith [50], we may treat the time and temperature dependence of the particles together in the same formulas, as long as we properly order the time integrations around the appropriate contours. When there is only one type of particle, which we take to be a fermion, the fields at the three-dimensional positions r_1 and $r_{1'}$ evolve in time according to the simple Hamiltonian H_0,

$$\psi_{\mathcal{H}_0}(r_1, t_1) \equiv \psi_{\mathcal{H}_0}(1) = e^{i\mathcal{H}_0 t_1}\psi(r_1)e^{-i\mathcal{H}_0 t_1}, \tag{16}$$

$$\psi^\dagger_{\mathcal{H}_0}(r_{1'}, t_{1'}) \equiv \psi^\dagger_{\mathcal{H}_0}(1') = e^{i\mathcal{H}_0 t_{1'}}\psi^\dagger(r_{1'})e^{-i\mathcal{H}_0 t_{1'}}, \tag{17}$$

with its density matrix also involving only $\mathcal{H}_0$ (and the number operator $\hat{N}$ in the grand canonical ensemble statistics),

$$\hat{\rho}_0 = \frac{e^{-\beta(\mathcal{H}_0 - \mu\hat{N})}}{\text{Tr}[e^{-\beta(\mathcal{H}_0 - \mu\hat{N})}]}, \tag{18}$$

and the Green function is given by the two contour integration paths C and C_{int} sketched in Figure 1,

$$G(1, 1') = -i\frac{\text{Tr}\left\{\hat{\rho}_0 T_C\left[S_{C_{int}} S'_C \psi_{\mathcal{H}_0}(1)\psi^\dagger_{\mathcal{H}_0}(1')\right]\right\}}{\text{Tr}[T_C(S_{C_{int}} S'_C)]}, \tag{19}$$

$$S'_C = \exp\left[-i\int_C d\tau \mathcal{H}'_{\mathcal{H}_0}(\tau)\right], \tag{20}$$

$$S_{C_{int}} = \exp\left[-i\int_{C_{int}} d\tau \mathcal{H}_{int,\mathcal{H}_0}(\tau)\right], \tag{21}$$

where the operators in $\mathcal{H}'_{\mathcal{H}_0}(\tau)$ and $\mathcal{H}_{int,\mathcal{H}_0}(\tau)$ evolve in time via the easily soluble Hamiltonian $\mathcal{H}_0$. The Greek letter τ implies that one needs to consider it as being just above or just below the real axis until the contours merge into one. Roman lettering (t) indicates the integrals are on the real axis.

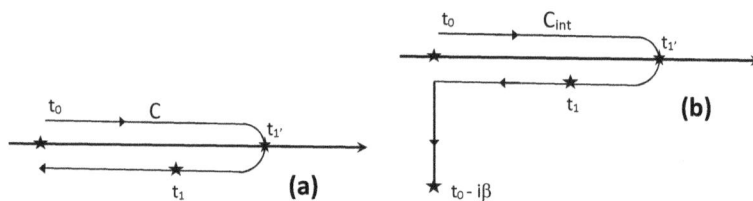

Figure 1. (**a**) Sketch of the "closed path" contour C; (**b**) Sketch of the "interaction" contour C_{int}.

For the case of the time-dependent Hamiltonian $\mathcal{H}'(t)$ making adiabatic changes, as in the case considered here, the two contours C and C_{int} respectively shown in Figure 1a,b merge into contour C shown in Figure 1a. We use the standard short-hand notation

$$G(1,1') = -i\langle T_C[\psi_{\mathcal{H}}(1)\psi^\dagger_{\mathcal{H}}(1')]\rangle, \qquad (22)$$

where the particle type, its position, and its energy are still undefined. In order to treat the various time orderings on the contour C, we define the following in the Heisenberg representation for the full Hamiltonian $\mathcal{H}$,

$$G^>(1,1') = -i\langle\psi_{\mathcal{H}}(1)\psi^\dagger_{\mathcal{H}}(1')\rangle, \qquad (23)$$

$$G^<(1,1') = +i\langle\psi^\dagger_{\mathcal{H}}(1')\psi_{\mathcal{H}}(1)\rangle, \qquad (24)$$

$$G^C(1,1') = -i\langle T[\psi_{\mathcal{H}}(1)\psi^\dagger_{\mathcal{H}}(1')]\rangle = \Theta(t_1 - t_{1'})G^>(1,1') + \Theta(t_{1'} - t_1)G^<(1,1'), \qquad (25)$$

$$G^{\tilde{C}}(1,1') = -i\langle\tilde{T}[\psi_{\mathcal{H}}(1)\psi^\dagger_{\mathcal{H}}(1')]\rangle = \Theta(t_1 - t_{1'})G^<(1,1') + \Theta(t_{1'} - t_1)G^>(1,1'), \qquad (26)$$

where the ordinary time-ordering operator T and inverse-time-ordering operator $\tilde{T}$ describe opposite directions in time, as sketched by lines C_1 and C_2 in Figure 2. We note that $G^C(1,1') + G^{\tilde{C}}(1,1) = G^<(1,1') + G^>(1,1')$, so only three of these Green functions are linearly independent.

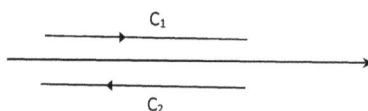

Figure 2. Sketch of the Keldysh contour C_K.

Here we need to describe three particles, all of which are effectively fermions.

3.1. Bare Nuclear Contour Green Functions

We first consider the nuclei, which are assumed not to interact with one another, as they are fixed in the crystalline locations, which if there is more than one isotope of a particular type with spin I, may be at a random selection of crystalline sites. In the presence of the constant magnetic induction B_0, it can be in any one of the $2I + 1$ manifold of nuclear Zeeman states, but because each of these local states at the probed nuclear site i can be either unoccupied or singly occupied, this manifold of local nuclear spin states is precisely that of a fermion with $(2I + 1)$ states. Its occupancy in the local state m_I on site i in the grand canonical ensemble is therefore easily seen to be

$$\langle \hat{n}^n_{i,m_I} \rangle = \frac{1}{e^{\beta(\epsilon_{m_I} - \mu_{ncp})} + 1}, \tag{27}$$

where $\epsilon_{m_I} = -\omega_n m_I$ and μ_{ncp} is the nuclear chemical potential. We then have for the bare nuclear Green functions with $\mathcal{H} = \mathcal{H}_{0,n}$,

$$G^{(0,n),<}_{i,i';m_I,m'_I}(1,1') = +i\delta_{i,i'}\delta_{m_I,m'_I} e^{i\epsilon_{m_I}(t_1 - t_{1'})} \langle \hat{n}^n_{i,m_I} \rangle, \tag{28}$$

$$G^{(0,n),>}_{i,i';m_I,m'_I}(1,1') = -i\delta_{i,i'}\delta_{m_I,m'_I} e^{i\epsilon_{m_I}(t_1 - t_{1'})} [1 - \langle \hat{n}^n_{i,m_I} \rangle], \tag{29}$$

and $G^{(0,n),C}_{i,i';m_I,m'_I}(1,1'), G^{(0,n),\tilde{C}}_{i,i';m_I,m'_I}(1,1')$ are constructed from these according to Equations (25) and (26). There are only three distinct bare neutron Green functions. This is also true when interactions are included [44]. Although it is somewhat surprising that the nuclear occupation density has the Fermi function form even for integral spin I, this is due to the nuclear Zeeman magnetic level occupancy being either 0 or 1 for each level on a given probed nuclear site.

3.2. Bare Orbital Electron Contour Green Functions

For the surrounding orbital electrons, we assume that the magnetic induction $B_0 + B_1(t)$ is sufficiently weak that it does not change the electronic structure of the orbital electrons or lead to transitions between the orbital electron states and energy levels. Thus, we assume that it only interacts with the orbital electron spins. We note that this is expected to be a good approximation, as the total charge of the nucleus plus its orbital electrons is on the order of one electron charge (for an ion), and the mass of the ion is so large that any Landau levels describing the orbital electrons and their central nucleus is completely negligible in comparison with the Landau levels of the conduction electrons. The only point then to consider for the interaction of B_0 with the orbital electrons is that there can be either 0, 1, or 2 electrons in a given orbital energy ϵ_q, and two possible magnetic energies for up and down spins. Hence, it is elementary to show that the average orbital electron occupation number in the grand canonical ensemble is

$$\langle \hat{n}^e_{i,q,\sigma} \rangle = \frac{1}{e^{\beta(\epsilon_q - \sigma\omega_e/2 - \mu_{ecp})} + 1}, \tag{30}$$

where μ_{ecp} is the orbital electron chemical potential. It is then easy to show that the bare orbital electron Green functions are

$$G^{(0,e),<}_{i,i';q,q'}(1,1') = +i\delta_{i,i'}\delta_{q,q'}\delta_{\sigma,\sigma'} e^{i(\epsilon_q - \sigma\omega_e/2)(t_1 - t_{1'})} \langle \hat{n}^e_{i,q,\sigma} \rangle, \tag{31}$$

$$G^{(0,e),>}_{i,i';q,q'}(1,1') = -i\delta_{i,i'}\delta_{q,q'}\delta_{\sigma,\sigma'} e^{i(\epsilon_q - \sigma\omega_e/2)(t_1 - t_{1'})} [1 - \langle \hat{n}^e_{i,q,\sigma} \rangle], \tag{32}$$

and the time-ordered and inverse-time-ordered bare orbital electron Green functions are obtained analogously using Equations (25) and (26). Only three of the bare orbital electron Green functions are linearly independent. This is also true when interactions are included [44].

3.3. Bare Conduction Electron Contour Green Functions

In a normal metal (above the superconducting transition of all superconductors including the cuprates and the record high transition temperature superconductor hydrogen sulfide, which probably transforms to H_3S under the 155 GPa pressure that causes it to become superconducting at 203 K [51]), the conduction electrons or holes propagate throughout the metal with wave vectors on or nearly on one or more Fermi surfaces. Both the spins and the charges of the conduction electrons interact with B_0, the spins via the Zeeman interaction and the charges couple to the magnetic vector potential, leading to Landau orbits. Here we assume each of these potentially multiple Fermi surfaces has an ellipsoidal shape, but the shapes and orientations of each of the Fermi surfaces can be different from one another.

We first use the Klemm-Clem transformations to transform each of the ellipsoidal conduction electron band dispersions into spherical forms [43,45]. For each ellipsoidal band, the anisotropic scale transformation that preserves the Maxwell equation $\nabla \cdot B_0 = 0$ transforms the elliptical Fermi surface into a spherical one, but rotates the transformed induction differently in each band. Then, one rotates these bands so that the rotated induction is along the z direction in each band [43]. In the jth band, the conduction electrons behave as free particles with wave vector $k_{j,\parallel}$ along the transformed $\hat{z}$ direction, but propagate in Landau orbits indexed by the harmonic oscillator quantum number n_j. Thus, we need to requantize the conduction electron fields as $\tilde{\psi}_{j,n_j,\sigma}(k_{j,\parallel})$.

We therefore rewrite the transformed $\tilde{\mathcal{H}}_{cond,0}$ as

$$\tilde{\mathcal{H}}_{cond,0} = \sum_{j,\sigma} g_{L,j} \sum_{n_j=0}^{\infty} \int \frac{dk_{j,\parallel}}{2\pi} \tilde{\psi}_{j,n_j,\sigma}^{\dagger}(k_{j,\parallel}) [\varepsilon_j(n_j,k_{j,\parallel}) - \sigma\tilde{\omega}_{j,e}'/2] \tilde{\psi}_{j,n_j,\sigma}(k_{j,\parallel}), \tag{33}$$

$$\varepsilon_j(n_j,\tilde{k}_{j,\parallel}) = \frac{\tilde{k}_{j,\parallel}^2}{2m_j\alpha_j^2} + \frac{(n_j+1/2)eB_0\alpha_j}{m_j}, \tag{34}$$

$$\tilde{\omega}_{j,e}' = \mu_B B_0 \beta_j(\theta,\phi), \tag{35}$$

where $n_j = 0,1,2,\ldots$ are the two-dimensional simple harmonic oscillator quantum numbers of the Landau orbits for band j, $k_{j,\parallel}$ are the free-particle dispersions along the transformed induction direction,

$$\alpha_j(\theta,\phi) = [\overline{m}_{j,1}\sin^2\theta\cos^2\phi + \overline{m}_{j,2}\sin^2\theta\sin^2\phi + \overline{m}_{j,3}\cos^2\theta]^{1/2}, \tag{36}$$

$$m_j = (m_{j,1}m_{j,2}m_{j,3})^{1/3}, \tag{37}$$

$$\overline{m}_{j,v} = \frac{m_{j,v}}{m_j}, \tag{38}$$

$$\beta_j(\theta,\phi) = [g_{j,xx}^2\overline{m}_{j,1}\sin^2\theta\cos^2\phi + g_{j,yy}^2\overline{m}_{j,2}\sin^2\theta\sin^2\phi + g_{j,zz}^2\overline{m}_{j,3}\cos^2\theta]^{1/2}, \tag{39}$$

for a diagonal g_j tensor describing the spins of the jth conduction band, and

$$g_{L,j} = \frac{eB_0\alpha_j}{2\pi} \tag{40}$$

is the spatially-transformed Landau degeneracy for a single electron in the jth band. We may then write the conduction electron occupation number as

$$\langle \hat{n}_{j,k_{j,\parallel},n_j,\sigma}^{cond} \rangle = \frac{1}{e^{\beta[\varepsilon_j(n_j,\tilde{k}_{j,\parallel}) - \sigma\tilde{\omega}_{j,e}'/2 - \mu_{cond,cp}]} + 1}, \tag{41}$$

where $\mu_{cond,cp}$ is the chemical potential of the conduction electrons. We note that all of the bands that cross this conduction electron chemical potential make important contributions to the Knight shift.

The bare conduction electron Green functions can then be found to be

$$G^{(0,\text{cond}),<}_{\substack{j,j',k_{j,\parallel},k'_{j',\parallel} \\ n_j,n'_{j'};\sigma,\sigma'}}(1,1') = +i\delta_{j,j'}\delta_{k_{j,\parallel},k'_{j',\parallel}}\delta_{n_j,n'_{j'}}\delta_{\sigma,\sigma'}g_{L,j}e^{i[\varepsilon_j(\tilde{k}_{j,\parallel},n_j)-\sigma\tilde{\omega}'_e/2](t_1-t_{1'})}\langle\hat{n}^{\text{cond}}_{i,\tilde{k}_{j,\parallel},n_j,\sigma}\rangle, \tag{42}$$

$$G^{(0,\text{cond}),>}_{\substack{j,j',k_{j,\parallel},k'_{j',\parallel} \\ n_j,n'_{j'};\sigma,\sigma'}}(1,1') = -i\delta_{j,j'}\delta_{k_{j,\parallel},k'_{j',\parallel}}\delta_{n_j,n'_{j'}}\delta_{\sigma,\sigma'}g_{L,j}e^{i[\varepsilon_j(\tilde{k}_{j,\parallel},n_j)-\sigma\tilde{\omega}'_e/2](t_1-t_{1'})}[1-\langle\hat{n}^{\text{cond}}_{i,\tilde{k}_{j,\parallel},n_j,\sigma}\rangle], \tag{43}$$

and the contour-ordered and inverse-contour-ordered bare conduction electron Green functions are obtained as in Equations (25) and (26), so that there are only three independent bare conduction electron Green functions.

Furthermore, due to the strong B_0, we also need to spatially transform all of the other terms in the Hamiltonian that contain the conduction electrons. Thus, we have [43]

$$\mathcal{H}'_{\text{cond}}(t) = -\frac{1}{2}\sum_{j,\sigma}g_{L,j}e^{i\sigma w_0(t-t_0)}\sum_{n_j=0}^{\infty}\int\frac{dk_{j,\parallel}}{2\pi}\tilde{\psi}^{\dagger}_{j,n_j,\sigma}(k_{j,\parallel})\tilde{\Omega}'_{j,e}\tilde{\psi}_{j,n_j,-\sigma}(k_{j,\parallel}), \tag{44}$$

$$\tilde{\Omega}'_{j,e} \approx \mu_B B_1\gamma_j(\theta,\phi), \tag{45}$$

$$\gamma_j(\theta,\phi) = [g^2_{j,xx}(\overline{m}_{j,2}\sin^2\theta\sin^2\phi+\overline{m}_{j,3}\cos^2\theta)$$
$$+g^2_{j,yy}(\overline{m}_{j,1}\sin^2\theta\cos^2\phi+\overline{m}_{j,3}\cos^2\theta)+g^2_{j,zz}(\overline{m}_{j,1}\cos^2\phi+\overline{m}_{j,2}\sin^2\phi)]^{1/2}. \tag{46}$$

4. Transformations in Time of the Operators with the Bare Hamiltonian

In order to proceed with the perturbation expansions, we first need to transform the nuclear, orbital electronic, and conduction electronic operators in real time, using the bare Hamiltonian in Equation (1). For the nuclear and orbital electronic operators, this is elementary. We have

$$a_{i,m_I}(t) = e^{i\mathcal{H}_{n,0}t}a_{i,m_I}e^{-i\mathcal{H}_{n,0}t}$$
$$\frac{da_{i,m_I}(t)}{dt} = i\left[\mathcal{H}_{n,0},a_{i,m_I}(t)\right]$$
$$= +i\omega_n m_I a_{i,m_I}(t), \tag{47}$$

and integrating the elementary differential equation, we immediately find

$$a_{i,m_I}(t) = e^{i\omega_n m_I t}a_{i,m_I}(0) = e^{i\omega_n m_I(t-t_0)}a_{i,m_I}(t_0), \tag{48}$$

in order to use this in Equation (16). The quantity $a^{\dagger}_{i,m_I+\sigma}(t)$ is instantly obtained from the Hermitian conjugate of Equation (52) and letting $m_I\rightarrow m_I+\sigma$, and hence $a^{\dagger}_{i,m_I+\sigma}(t)=e^{-i\omega_n(m_I+\sigma)(t-t_0)}a^{\dagger}_{i,m_I+\sigma}(t_0)$, so that the time-transformed Equation (16) becomes

$$\mathcal{H}'_{n,\mathcal{H}_{n,0}}(t) = -\frac{\Omega_n}{2}\sum_{i,m_I,\sigma}e^{i\sigma(w_0-w_n)(t-t_0)}A^{\sigma}_{I,m_I}a^{\dagger}_{i,m_I,\sigma}(t_0)a_{i,m_I}(t_0). \tag{49}$$

Similarly, for the local orbital electron operators, we have

$$b_{i,q,\sigma}(t) = e^{i\mathcal{H}_{e,0}(t-t_0)}b_{i,q,\sigma}(t_0)e^{-i\mathcal{H}_{e,0}(t-t_0)} \tag{50}$$
$$= e^{-i(\varepsilon_q-\sigma w_e/2)(t-t_0)}b_{i,q,\sigma}(t_0), \tag{51}$$

and

$$\mathcal{H}'_{e,\mathcal{H}_{e,0}}(t) = -\frac{\Omega_e}{2}\sum_{i,q,\sigma}e^{i\sigma(w_0-w_e)(t-t_0)}b^{\dagger}_{i,q,\sigma}(t_0)b_{i,q,-\sigma}(t_0), \tag{52}$$

For the spatially-transformed conduction electron operators,

$$\tilde{\psi}_{j,n_j,\sigma}(k_{j,||},t) = e^{i\tilde{\mathcal{H}}_{\text{cond},0}(t-t_0)}\tilde{\psi}_{j,n_j,\sigma}(k_{j,||},t_0)e^{-i\tilde{\mathcal{H}}_{\text{cond},0}(t-t_0)} \tag{53}$$

$$= e^{-i[\varepsilon_j(n_j,k_{j,||})-\sigma\tilde{\omega}'_{j,e}/2](t-t_0)}\tilde{\psi}_{j,n_j,\sigma}(k_{j,||},t_0), \tag{54}$$

so that the time-transformed Equation (44). becomes

$$\tilde{\mathcal{H}}'_{\text{cond},\tilde{\mathcal{H}}_{\text{cond},0}}(t) = -\frac{1}{2}\sum_{j,\sigma}g_{\text{L},j}\sum_{n_j=0}^{\infty}\int\frac{dk_{j,||}}{2\pi}e^{i\sigma(\omega_0-\tilde{\omega}'_{j,e})(t-t_0)}\tilde{\psi}^{\dagger}_{j,n_j,\sigma}(k_{j,||},t_0)\tilde{\Omega}'_{j,e}\tilde{\psi}_{j,n_j,-\sigma}(k_{j,||},t_0). \tag{55}$$

We note that all three of these transformed Hamiltonians correspond to spin-dependent external field interactions, where the fields are

$$U_{n,i,i';m_I,m'_I}(t) = -\frac{\Omega_n}{2}\delta_{i,i'}\sum_{\sigma=\pm1}\delta_{m'_I,m_I+\sigma}A^{\sigma}_{I,m_I}e^{i\sigma(\omega_0-\omega_n)(t-t_0)}, \tag{56}$$

$$U_{e,i,i';q,q'}(t) = -\frac{\Omega_e}{2}\delta_{i,i'}\delta_{q,q'}\delta_{\sigma',-\sigma}e^{i\sigma(\omega_0-\omega_e)(t-t_0)}, \tag{57}$$

$$U_{\text{cond},j,j';n_j,n'_j}(t) = -\frac{\tilde{\Omega}'_{j,e}}{2}\delta_{j,j'}\delta_{n_j,n'_j}\delta_{k_{j,||},k'_{j,||}}\delta_{\sigma',-\sigma}e^{i\sigma(\omega_0-\tilde{\omega}'_{j,e})(t-t_0)}, \tag{58}$$

Then, we time transform the difficult (interaction) parts of the full Hamiltonian. The hyperfine and local electron-electron interactions are elementary to transform. We obtain

$$\mathcal{H}_{hf,\mathcal{H}_{n,0}+\mathcal{H}_{e,0}}(t) = -\frac{D_z}{4}\sum_{i,q,\sigma,m_I}m_I\sigma a^{\dagger}_{i,m_I}(t_0)a_{i,m_I}(t_0)b^{\dagger}_{i,q,\sigma}(t_0)b_{i,q,\sigma}(t_0)$$

$$-\frac{D_x}{2}\sum_{i,q,\sigma,m_I}A^{\sigma}_{I,m_I}e^{i\sigma(\omega_e-\omega_n)(t-t_0)}a^{\dagger}_{i,m_I+\sigma}(t_0)a_{i,m_I}(t_0)b^{\dagger}_{i,q,-\sigma}(t_0)b_{i,q,\sigma}(t_0), \tag{59}$$

$$\mathcal{H}_{e,\text{int},\mathcal{H}_{e,0}}(t) = \frac{1}{2}\sum_{i,q,\sigma}U_q\hat{n}_{i,q,\sigma}(t_0)\hat{n}_{i,q,-\sigma}(t_0), \tag{60}$$

Of these, only the transverse (D_x) part of the hyperfine interaction picks up a time dependence. Before we time transform the remaining two interaction Hamiltonians, we first spatially transform the conduction electron operators in the presence of the magnetic field necessary for the NMR experiment. Then we rewrite $\mathcal{H}_{e,\text{cond}}$ in terms of the spatially-transformed conduction electron fields,

$$\mathcal{H}_{e,\text{cond}} \to \tilde{\mathcal{H}}_{e,\text{cond}} = \sum_{i,q,j}g_{\text{L},j}\sum_{n_j=0}^{\infty}\int\frac{dk_{j,||}}{2\pi}\left(v_{i,q,j}\tilde{\psi}^{\dagger}_{j,n_j,\sigma}(k_{j,||})b_{i,q,\sigma}+H.c.\right), \tag{61}$$

which after time-transformation with respect to $\mathcal{H}_{e,0}$ and $\tilde{\mathcal{H}}_{\text{cond},0}$ becomes

$$\tilde{\mathcal{H}}_{e,\text{cond},\mathcal{H}_{e,0}+\tilde{\mathcal{H}}_{\text{cond},0}}(t) = \sum_{i,q,j}g_{\text{L},j}\sum_{n_j=0}^{\infty}\int\frac{dk_{j,||}}{2\pi}\left(v_{i,q,j}e^{i[\varepsilon_j(n_j,k_{j,||})-\varepsilon_q+\sigma(\omega_e-\tilde{\omega}'_{j,e})/2](t-t_0)}\right.$$

$$\left.\tilde{\psi}^{\dagger}_{j,n_j,\sigma}(k_{j,||},t_0)b_{i,q,\sigma}(t_0) + H.c.\right) \tag{62}$$

The most important Hamiltonian for the Knight shift in a superconductor is the pairing interaction $\mathcal{H}_{sc}$, which in position space was written in Equation (10). Since in a Knight shift measurement, the experimenter first measures the Knight shift in the applied field $H(t)$ and hence the induction

$B(t) = \mu_0 H(t)$ while the superconductor is in its normal (metallic) state, and then cools the material through its superconducting transition at $T_c(H)$, it is clear that the correct formulation for the superconducting pairing interaction must be in momentum space, and more precisely, to account for the pairing of the electrons (or holes) while they are in Landau orbits in the normal state. We therefore first rewrite $\mathcal{H}_{sc}$ in a fully spatially-transformed magnetic-induction-quantized form that allows for different pairing interactions, such as those giving rise to various types of spin-singlet and spin-triplet superconductors arising from a multiple-band metal. As a start to understand the orbital motion of the paired superconducting electrons (or holes), we first assume the standard approximation for the evaluation of the upper critical field H_{c2} that the paired particles of combined charge $2e$ move together in Landau levels [18,42,52,53]. For a BCS superconductor for which $V_{j,j';\sigma,\sigma'}(k_j - k'_{j'}) = -V_0 \delta_{j,j'} \delta_{\sigma,-\sigma'}$, there is no need to transform the wave vector dependence of the pairing interaction due to the Landau orbits formed by the strong applied field [18]. Such pairing interactions will be considered elsewhere. Thus, we begin by considering only the simplest case of isotropic intraband pairing of equivalent strength in all of the bands, which after spatial transformation due to the magnetic induction may be written as

$$\tilde{H}_{sc,0} = -V_0 \sum_\sigma \sum_j g^2_{L,j} \sum_{n_j,n'_j=0}^\infty \int \frac{dk_{j,||}}{2\pi} \int \frac{dk'_{j,||}}{2\pi} \tilde{\psi}^\dagger_{j,n_j,\sigma}(k_{j,||}) \tilde{\psi}^\dagger_{j,n_j,-\sigma}(-k_{j,||}) \tilde{\psi}_{j,n'_j,-\sigma}(-k'_{j,||}) \tilde{\psi}_{j,n'_j,\sigma}(k'_{j,||}),$$

(63)

and transforming this in time using $\mathcal{H}_{cond,0}$, we have

$$\tilde{H}_{sc,0,\mathcal{H}_{cond,0}}(t) = -V_0 \sum_\sigma \sum_j g^2_{L,j} \sum_{n_j,n'_j=0}^\infty \int \frac{dk_{j,||}}{2\pi} \int \frac{dk'_{j,||}}{2\pi} \tilde{\psi}^\dagger_{j,n_j,\sigma}(k_{j,||},t_0) \tilde{\psi}^\dagger_{j,n_j,-\sigma}(-k_{j,||},t_0)$$
$$\times \tilde{\psi}_{j,n'_j,-\sigma}(-k'_{j,||},t_0) \tilde{\psi}_{j,n'_j,\sigma}(k'_{j,||},t_0),$$

(64)

which is independent of t.

5. Dyson's Equations for the Green Functions

For a system with continuous position variables r, the contour C Dyson equation for an adiabatic time-dependent interaction for which $C_{int} \rightarrow C$ can be written as [44]

$$G(1,1') = G_0(1,1') + \int d^3r_2 \int_C d\tau_2 G_0(1,2) U(2) G(2,1')$$
$$+ \int d^3r_2 \int d^3r_3 \int_C d\tau_2 \int_C d\tau_3 G_0(1,2) \Sigma(2,3) G(3,1'),$$

(65)

where Σ is the self-energy and U is an external field. By carefully keeping the order of the times in going about the contour C, one can analytically continue the integrals off the real axis to the real axis. We first need to define the retarded and advanced Green functions, which are

$$G^r(1,1') = \Theta(t_1 - t_{1'})[G^>(1,1') - G^<(1,1')],$$

(66)

$$G^a(1,1') = \Theta(t_{1'} - t_1)[G^<(1,1') - G^>(1,1')],$$

(67)

Then, letting $\int_C d\tau_2 G_0(1,2) G(2,1')$ be represented by $C = AB$, one can analytically continue the appropriate contour-ordered Green function components on the real axis, so that

$$C^< = A^r B^< + A^< B^a,$$

(68)

$$C^> = A^r B^> + A^> B^a,$$

(69)

where the integration $\int_C d\tau_2 \rightarrow \int_{-\infty}^{\infty} dt_2$. Similarly, by representing the double contour integral $\int_C d\tau_2 \int_C d\tau_3 G_0(1,2)\Sigma(2,3)G(3,1')$ by $D = ABC$, one can analytically continue these contour integration paths to the real axis, obtaining [44]

$$D^< = A^r B^r C^< + A^r B^< C^a + A^< B^a C^a, \tag{70}$$

$$D^> = A^r B^r C^> + A^r B^> C^a + A^> B^a C^a, \tag{71}$$

We then implement the three Dyson equations for the nuclear, local orbital electron, and conduction electron Green functions. We first consider the Dyson equation for the nuclear Green function. In this case, there are two terms to consider: the external field U_n given by Equation (56), and the hyperfine interaction given by Equation (59). However, the hyperfine interaction does not involve two different times, as in the self-energy, which is analogous to the exchange interaction in the electron gas with electron-electron Coulomb interactions. It is instead analogous to the direct interaction with a fermion loop, but in this case, the fermion loop is for the local orbital electrons. As first shown by Hall and Klemm, the leading self-energy diagrams for the Knight shift and the linewidth changes in a metal at $T = 0$ are shown in Figure 3.

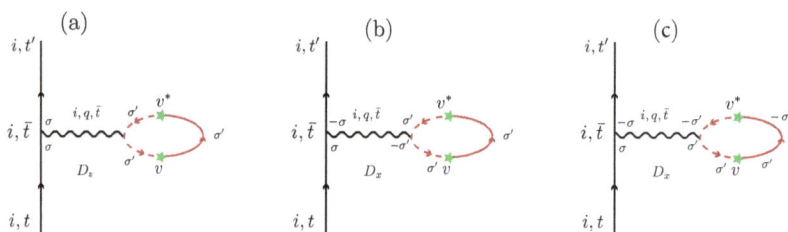

Figure 3. Hall-Klemm diagrams for the Knight shift linewidth changes at $T = 0$. The vertical solid lines on the left are the nuclear $G_n(1,1')$, the wiggly horizontal line represents the hyperfine interaction, the dashed curves represent G_e, the stars represent the excitation from the local orbitals to the conduction band, and the solid counterclockwise arrowed curves represent the conduction electron G_{cond}. (**a**) represents the leading Knight shift contribution arising from D_z; (**b,c**) represent the two leading contributions to the linewidth changes arising from D_x [43]. Reprinted with permission of B. E. Hall and R. A. Klemm. Microscopic model of the Knight shift in anisotropic and correlated metals. *J. Phys. Condens. Matter* **2016**, *28*, 03LT01. Copyright©2016 Institute of Physics.

6. Proposed Calculation of $K(T)$ in the Normal and Superconducting States

6.1. Gor'kov's Derivation of the Ginzburg-Landau Equations

Since the upper critical field has been obtained for anisotropic superconductors with a variety of pairing interactions [18,42,52,53] and also that the most rapid temperature variation, a discontinuity in slope, of the conventional Knight shift in superconductors, occurs just at the superconducting transition, it is evident that an extension of those upper critical field calculations to the Ginzburg-Landau regime just below $T_{c2}(B_0)$ can provide the crucial information for $K_S(T)$ in the superconducting state. We propose to extend the Hall-Klemm $T = 0$ Knight shift calculation in the presence of a strong magnetic induction $B(t)$ into the superconducting state using an extension of the microscopic derivation of the Ginzburg-Landau expression for the gap function as pioneered by Gor'kov [54,55] to include the time-dependent applied field. That work was generalized for a general $V(r - r')$

single-band pairing by Scharnberg and Klemm [18]. In the superconducting state, we require the regular (conduction) and anomalous Green functions

$$G_{j,j';\sigma,\sigma'}(1,1') = -i\langle T_C[\psi_{j,\sigma,\mathcal{H}}(1)\psi^\dagger_{j',\sigma',\mathcal{H}}(1')]\rangle, \tag{72}$$

$$F_{j,j';\sigma,\sigma'}(1,1') = \langle T_C[\psi_{j,\sigma,\mathcal{H}}(1)\psi_{j',\sigma',\mathcal{H}}(1')]\rangle, \tag{73}$$

$$F^\dagger_{j,j'\sigma,\sigma'}(1,1') = \langle T_C[\psi^\dagger_{j,\sigma,\mathcal{H}}(1)\psi^\dagger_{j',\sigma',\mathcal{H}}(1')]\rangle, \tag{74}$$

and the gap function, which in real space for intraband pairing only is

$$\Delta_{j;\sigma,\sigma'}(r_j,r'_j) = V_j(r_j - r'_j)\delta_{j,j'}F_{j,j';\sigma,\sigma'}(1,1')|_{t_1-t_{1'}=0+}, \tag{75}$$

As discussed in the next subsection, in order to include the temperature dependence of the normal state of the nuclei, the local orbital electrons, and the conduction electrons, we need to quantize the conduction electrons in momentum space and Landau orbits, as was done for the bare conduction electron Green functions in Equations (42) and (43). Although this was never done in upper critical field calculations [17,18,32,42,48,49,52,53], the reasons given for not doing it were that impurities would broaden the levels, smearing out the Landau level spacings [18,52]. However, with the present quality of some materials, that argument should be reexamined. More important, in order to calculate the upper critical induction B_{c2}, one requires the paired electrons (or holes) to be in Landau levels [17,18,32,42,48,49,52,53]. However, as discussed in the next section, it is not clear that electrons (or holes) will only pair with other electrons (or holes) in the same Landau orbit. With multiple bands, an electron in one single-particle Landau level corresponding to one conduction band could in principle pair with another electron in a different Landau level corresponding to another band. So one will have to make some assumptions about the pairing processes to simplify the calculations. However, to get a preliminary microscopic idea of how $K_S(T)$ picks up its T dependence below T_c, we will first revisit the Gor'kov procedure for deriving the Ginzburg-Landau equations in real space.

In the standard real-space finite temperature formalism, the Gor'kov equations of motion generalized to include multiple ellipsoidally anisotropic bands and their Zeeman energies without interband pairing are [18]

$$\left(i\omega_n - \sum_{\nu=1}^{3}\frac{1}{2m_{j,\nu}}[\nabla_{j,\nu}/i - eA_{j,\nu}(r_j)]^2 + \mu - \sigma\omega'_{j,e}/2\right)G_{j,j';\sigma,\sigma'}(r_j,r'_{j'},\omega_n)$$

$$+ \sum_\rho \int d^3r''_j \Delta_{j;\sigma,\rho}(r_j,r''_j)F^\dagger_{j,j';\rho,\sigma'}(r''_j,r'_{j'},\omega_n) = \delta_{\sigma,\sigma'}\delta_{j,j'}\delta^{(3)}(r_j - r'_j), \tag{76}$$

$$\left(-i\omega_n - \sum_{\nu=1}^{3}\frac{1}{2m_{j,\nu}}[\nabla_{j,\nu}/i + eA_{j,\nu}(r_j)]^2 + \mu - \sigma\omega'_{j,e}/2\right)F^\dagger_{j,j';\sigma,\sigma'}(r_j,r'_{j'},\omega_n)$$

$$+ \sum_\rho \int d^3r''_j \Delta^*_{j;\sigma,\rho}(r_j,r''_j)G_{j,j';\rho,\sigma'}(r''_j,r'_{j'},\omega_n) = 0, \tag{77}$$

where $\omega_n = (2n+1)\pi/\beta$ is the fermion Matsubara frequency. Letting $G^{(0)}_{j,j';\sigma,\sigma'}(r_j,r'_{j'},\omega_n)$ be the solution for the G function in the normal state with $\Delta = 0$, one can rewrite the above equations for G and $F^\dagger$ in the finite temperature formalism as

$$G_{j,j';\sigma,\sigma'}(r_j,r'_{j'},\omega_n) = G^{(0)}_{j,j;\sigma,\sigma'}(r_j,r'_{j'},\omega_n)\delta_{j,j'}\delta_{\sigma,\sigma'}$$

$$- \sum_\rho \int d^3r''_j \int d^3r'''_j G^{(0)}_{j,j;\sigma,\sigma}(r_j,r''_j,\omega_n)\Delta_{j,\rho,\sigma}(r''_j,r'''_j)$$

$$\times \int d^3\xi_j \int d^3\xi'_j \sum_{\rho'} G^{(0)}_{j,j;\sigma,\sigma'}(r'_j,\xi_j,-\omega_n)\Delta^*_{j,\rho',\sigma'}(\xi_j,\xi'_j)G^{(0)}_{j,\rho,\rho'}(r'''_j,\xi_j,\omega_n), \tag{78}$$

to order Δ^2. One can then substitute this into the equation for $F^\dagger$, multiply by the pairing interaction, and obtain a self-consistent equation for Δ to order Δ^3 [54,55]. The coefficients of the two terms proportional to Δ define the upper critical induction B_{c2} [18,32,42,48,49,52,53]. Functionally integrating the cubic equation with respect to $\Delta_{j,\sigma,\sigma'}(r_j, r'_j)$ and by neglecting the field dependence of the resulting term proportional to $|\Delta|^4$, one can obtain the generalized Ginzburg-Landau free energy.

We note that even if one completely neglects the field dependence of the term of order Δ^3 in the Gor'kov expansion of $F^\dagger$ for $\Delta(r)$, it is easy to see that this procedure will lead to the following phenomenological general result:

$$K_S(T) \;=\; a(B_0) - b(B_0)|\Delta(B_0, T)|^2, \tag{79}$$

where a and b strongly depend upon the magnitude and direction of B_0, but not much upon T, and $2|\Delta(B_0, T)|$ is the effective superconducting gap in the Ginzburg-Landau regime. This simple result includes the pairing in all of the bands, which couple together to give one effective $T_{c2}(B_0)$, below which $|\Delta(B_0, T)|^2 \propto [T_{c2}(B_0) - T]$. It remains to be seen if this form could be generalized to the full BCS superconducting gap $|\Delta(B_0, T)|$ temperature dependence, which saturates at low T values. If so, it could lead to a quantitative theory of the Knight shift that would be valid for essentially any type of superconductor involving Cooper pairing. Hence, a proper calculation of $a(B_0)$ and $b(B_0)$ can provide a microscopic understanding of the behavior for the $^{63}CuO_2$ $K_S(T)$ for B_0 parallel and normal to the layers of YBa$_2$Cu$_3$O$_{7-\delta}$, which was described by Slichter as "fortuitous" [20]. It could in principle explain the small or vanishing b term in Sr$_2$RuO$_4$, at least for the field normal to the layers, for which Landau level formation would be highly restricted on two of the Fermi surfaces.

6.2. High-Field Solution for an Anisotropic, Multiband Type-II BCS Superconductor

More important, we note that a major simplification of the Keldysh contour procedure can be made by first taking the mean-field approximation of the BCS pairing interaction represented in momentum space by Equation (63). We write the mean-field gap (or isotropic order parameter) for singlet pairing in band j as

$$\Delta_{j,-\sigma,\sigma} \;=\; V_0 g_{L,j} \sum_{n_j=0}^{\infty} \int \frac{dk_{j,\parallel}}{2\pi} \langle \tilde{\psi}_{j,n_j,-\sigma}(-k_{j,\parallel}) \tilde{\psi}_{j,n_j,\sigma}(k_{j,\parallel}) \rangle, \tag{80}$$

where the expectation value is in the grand canonical ensemble, so that the mean-field effective Hamiltonian for the conduction electrons in the superconducting state becomes

$$\tilde{\mathcal{H}}_{sc,cond} \;=\; \sum_{j,\sigma} g_{L,j} \sum_{n_j=0}^{\infty} \int \frac{dk_{j,\parallel}}{2\pi} \left(\tilde{\psi}_{j,n_j,\sigma}^\dagger(k_{j,\parallel})[\varepsilon_j(n_j, k_{j,\parallel}) - \mu_{cond,cp} - \sigma \tilde{\omega}'_{j,e}/2] \tilde{\psi}_{j,n_j,\sigma}(k_{j,\parallel}) \right.$$
$$\left. + \left[\tilde{\psi}_{j,n_j,\sigma}^\dagger(k_{j,\parallel}) \tilde{\psi}_{j,n_j,-\sigma}^\dagger(-k_{j,\parallel}) \Delta_{j,-\sigma,\sigma} + H.c. \right] \right) \tag{81}$$

where we have included the chemical potential of the conduction electrons. Note that we assume the total momentum of the paired electrons (or holes) is zero, as both are assumed to be on opposite sides of the same Landau orbit, and have opposite momenta in the direction normal to the plane of the Landau orbits. This effective quadratic Hamiltonian can then be diagonalized by a standard Bogoliubov-Valatin transformation [56], letting

$$\tilde{\psi}_{j,n_j,\uparrow}(k_{j,\parallel}) \;=\; u_{j,n_j,k_{j,\parallel}} \gamma_{j,n_j,\uparrow}(k_{j,\parallel}) + v_{j,n_j,k_{j,\parallel}} \gamma_{j,n_j,\downarrow}^\dagger(k_{j,\parallel}), \tag{82}$$

$$\tilde{\psi}_{j,n_j,\downarrow}^\dagger(-k_{j,\parallel}) \;=\; -v_{j,n_j,k_{j,\parallel}}^* \gamma_{j,n_j,\uparrow}(k_{j,\parallel}) + u_{j,n_j,k_{j,\parallel}}^* \gamma_{j,n_j,\downarrow}^\dagger(k_{j,\parallel}), \tag{83}$$

where we require the γ operators to obey independent fermion statistics. Using the standard transformation procedure to eliminate the off-diagonal terms, we then obtain

$$|u_{j,n_j,k_{j,\|}}|^2 \;=\; \frac{1}{2}\left[1 + \left(\frac{\varepsilon_j(n_j,k_{j,\|}) - \mu_{\text{cond,cp}}}{E_j(n_j,k_{j,\|})}\right)\right], \tag{84}$$

$$|v_{j,n_j,k_{j,\|}}|^2 \;=\; \frac{1}{2}\left[1 - \left(\frac{\varepsilon_j(n_j,k_{j,\|}) - \mu_{\text{cond,cp}}}{E_j(n_j,k_{j,\|})}\right)\right], \tag{85}$$

and the diagonalized superconducting Hamiltonian becomes

$$\tilde{\mathcal{H}}_{\text{sc,cond}} \;\rightarrow\; \sum_{j,\sigma} g_{L,j} \sum_{n_j=0}^{\infty} \int \frac{dk_{j,\|}}{2\pi} \gamma_{j,n_j,\sigma}^\dagger(k_{j,\|})\gamma_{j,n_j,\sigma}(k_{j,\|})\left[E_j(n_j,k_{j,\|}) + \sigma\tilde{\omega}'_{j,e}/2\right], \tag{86}$$

$$E_j(n_j,k_{j,\|}) \;=\; \sqrt{[\varepsilon_j(n_j,k_{j,\|}) - \mu_{\text{cond,cp}}]^2 + |\Delta_j|^2}, \tag{87}$$

where $|\Delta_j|^2 = \Delta_{j,-\sigma,\sigma}\Delta_{j,-\sigma,\sigma}^\dagger$ is positive definite for each j value. We note that the quasiparticle dispersions in $\tilde{\mathcal{H}}_{\text{sc,cond}}$ are nearly identical to the BCS quasiparticle dispersions, as they do indeed have a real energy gap $2|\Delta|$, but there is in addition an effective Zeeman term arising from the difference in the spin up and spin down quasiparticle energies, leading to a magnetic gap function. Thus, the self-consistent expression from Equation (80) for $\Delta_{j,-\sigma,\sigma}$ becomes

$$\Delta_{j,\downarrow,\uparrow} \;=\; -V_0 g_{L,j} \sum_{\sigma=\pm} \sum_{n_j=0}^{\infty} \int \frac{dk_{j,\|}}{2\pi} \frac{\Delta_{j,\downarrow,\uparrow}}{E_j(n_j,k_{j,\|})}\left(\frac{1}{e^{\beta[E_j(n_j,k_{j,\|})+\sigma\tilde{\omega}'_{j,e}/2]} + 1} - \frac{1}{2}\right), \tag{88}$$

which, combined with Equation (86), explicitly demonstrates the presence of the superconducting gap Δ_j in each band that is involved in $K_S(T)$ in the superconducting state. Thus, it is clear that the effective or phenomenological Equation (79) mentioned in the abstract for $K_S(T)$ applies in the mixed state of a type-II superconductor, not just in the Ginzburg-Landau region. However, by quantizing the superconducting order parameter at a finite induction strength B_0, both the Landau orbits and the Zeeman interaction can greatly affect its B_0 dependence, and the Landau orbits in particular can be distinctly different for layered compounds with B_0 parallel or perpendicular to the layers, especially at large induction strengths, as first noted in experiments on $YBa_2Cu_3O_{7-\delta}$ [19,20].

The road ahead to construct the first microscopic theory of the Knight shift in a superconductor of any type is now clear. The conduction electrons must be quantized in Landau orbits, and this can be done for any number of ellipsoidally anisotropic electron or hole bands, as outlined above. The procedure will be extended for our model of multiple ellipsoidal bands with the Zeeman couplings and time-dependent Zeeman couplings in each band to construct the Bogoliubov-Valatin transformed contour G functions. To do this properly, one needs to apply those transformations presented in Equations (82) and (83) to the time-dependent Zeeman interaction on the conduction electrons in Equation (44) and also to the Anderson-like interaction in Equation (61) that removes a local orbital electron and places it in the superconducting state and *vice versa*. This will cause Equations (44), (55), (61) and (62) to be rewritten in terms of the quasiparticle operators $\gamma_{j,n_j,\sigma}(k_{j,\|})$ and $\gamma_{j,n_j,\sigma}^\dagger(k_{j,\|})$, and will modify Equation (58). Then, the Keldysh contour method can be used to perform a microscopic theory of $K_S(T)$ in the mixed superconducting state of an anisotropic, multiband, type-II BCS superconductor. Since the conduction electrons are transformed into non-interacting quasiparticles in the superconducting state, the self-energy $\Sigma(2,3)$ in Dyson's equation will only apply to the orbital electrons via the Hubbard interaction U_q. All other interactions reduce to effective external fields. After a detailed microscopic evaluation of $K_S(T)$ using the contour-extended version of the diagram pictured in Figure 3a, special attention will be directed at the conditions for a near vanishing of $b(B_0)$, which could lead to a T-independent $K_S(T)$, even for a "conventional" superconductor. The linewidth changes can be evaluated in the superconducting state from the contour-extended versions of the diagrams pictured in Figure 3b,c. Eventually, other superconducting pairing symmetries could also be studied with this technique, although the pairing interaction would have to be transformed as above, including the Landau orbits. Eventually, this could be done for charge-density and spin-density wave systems, for which no theory of the Knight shift is presently available. We note that $2H$-TaS_2 has a nodal charge-density wave below 75 K, with a presumably s-wave superconducting state entering below 0.6 K [17,57], which is very similar to the complex situation in the high-temperature superconductor $Bi_2Sr_2CaCu_2O_{8+\delta}$, in which the nodal pseudogap (probably charge-density wave) regions and isotropic s-wave superconducting regions break up into spatial domains [58]. These results are consistent with previous c-axis twist Josephson junction experiments on that material [59]. Although the NMR linewidths in that material are too broad to perform Knight shift measurements, they could be done on other materials, such as the dichalcogenides, and also in improved $YBa_2Cu_3O_{7-\delta}$ samples.

7. Discussion and Conclusions

We have outlined a procedure to obtain a microscopic theory of the Knight shift in an anisotropic Type-II superconductor. This was based upon the Hall-Klemm microscopic model of the effect at $T = 0$ [43], for which multiple anisotropic conduction bands of ellipsoidal shapes were included. We considered the simplest magnetic resonance case of $B(t) = B_0 + B_1(t)$ with $|B_1| \ll |B_0|$ and $B_1 \cdot B_0 = 0$ with $B_1(t)$ oscillating at a single frequency ω_0. For this simple case, the time changes to the system are adiabatic, so that the interaction Keldysh contour C_{int} shown in Figure 1b effectively coincides with contour C depicted in Figure 1a, and the integrations can be analytically continued onto the real axis. The procedure can effectively treat any nuclear spin value I. The conduction electrons (or holes) were quantized in Landau orbits in the applied field in the normal state, and the Hamiltonian for a generalized anisotropic, multiband BCS type-II superconductor was diagonalized, allowing for a full treatment of the superconducting state.

We emphasize that by quantizing the superconducting order parameter in the presence of a strong time-independent magnetic induction B_0, the energy spacings of the Landau orbits can depend strongly upon the direction of B_0. At very weak B_0 values, the Landau levels primarily give rise to overall anisotropic constant backgrounds of $K(T)$ and $K_S(0)$, with $K_S(T)$ being predominantly governed by the anisotropic Zeeman interactions and D_z. However, for sufficiently strong B_0 values in anisotropic materials with layered or quasi-two-dimensional anisotropy, the spacings between the Landau energy levels depends strongly upon the direction of B_0, so that $K_S(T)$ could become independent of T for $T \leq T_c$, as first observed for $B_0 || \hat{c}$ in $YBa_2Cu_3O_{7-\delta}$ [19,20]. Such behavior could also arise for quasi-one-dimensional materials in all B_0 directions, although to different degrees for B_0 parallel and perpendicular to the most conducting crystal direction.

Since the crucial interaction for the Knight shift is the hyperfine interaction between the probed nuclei and their surrounding orbital electrons, the symmetry of this interaction can be very important. Generally, the hyperfine interaction can arise from the electrons in any of the orbital levels. For s-orbitals, the Fermi contact term is important, but the induced-dipole induced-dipole interactions can arise from the nucleus of any spin for any spin $I \geq 1/2$ and its surrounding electrons in any orbital, and induced-quadrupole induced-quadrupole and higher order interactions can also occur for certain orbitals and nuclear spin values [60,61]. In the Hall-Klemm model [43], the hyperfine interaction crucial for the Knight shift was taken to be diagonal in the spin representations of a lattice with tetragonal symmetry $D_x = D_y \neq D_z$. In that simple model, the $T = 0$ results indicated that the Knight shift arose from D_z, and the line width was modified by $D_x = D_y$. In more realistic examples of correlated and anisotropic materials, the hyperfine interaction would be represented by a symmetric matrix unless time-reversal symmetry-breaking interactions were present. Such matrices can be diagonalized by a set of rotations, but in complicated cases the quantization axes would not necessarily be the same as for the overall crystal structure. Such complications would mix the Knight shift and its linewidth, depending upon the direction of B_0.

As noted previously, in first quantization, an isolated nuclear spin wave function in an NMR experiment was found to have the form

$$|I, m_I\rangle(t) = e^{im_I\omega_0 t} \sum_{m_I'=-I}^{I} C_{m_I'}^{m_I} e^{im_I'\Gamma_n t},\qquad(89)$$

where $\Gamma_n = [(\omega_0 - \omega_n)^2 + \Omega_n^2]^{1/2}$ is the nuclear resonance function and the constants $C_{m_I'}^{m_I}$ depend upon the initial conditions [43]. Those authors found this form to hold for $I = 1/2, 1, 3/2$, and in second quantization, up to $I = 2$, so it is likely to hold for arbitrary I. In the adiabatic regime, we have $\omega_0 \ll \omega_n$ [37,38], so that there will be a manifold of geometrical phases that will arise with higher I values.

We remark that it is possible to generalize this treatment to more complicated $B_1(t)$ functions, such as a periodic function of square-wave or triangle-wave shape. This can be represented as a Fourier series, but if the primary angular frequency is ω_0, terms of higher multiples n of ω_0 can be present, some of which would violate the adiabatic requirement that they be much smaller than the Zeeman energy spacings. Hence, this experiment would make some amount of non-adiabatic changes that could drive the system out of thermal equilibrium, and the two contours C and C_{int} shown in Figure 1 and discussed above would not coincide, greatly complicating the analysis.

Conflicts of Interest: The authors declare no conflict of interest.

References

1. Knight, W.D. Nuclear magnetic resonance shift in metals. *Phys. Rev.* **1949**, *76*, 1259–1260.
2. Knight, W.D.; Androes, G.M.; Hammond, R.H. Nuclear magnetic resonance in a superconductor. *Phys. Rev.* **1956**, *104*, 852–853.
3. Reif, F. Observation of nuclear magnetic resonance in superconducting mercury. *Phys. Rev.* **1956**, *102*, 1417–1418.
4. Reif, F. The study of superconducting Hg by nuclear magnetic resonance techniques. *Phys. Rev.* **1957**, *106*, 208–221.
5. Yosida, K. Paramagnetic susceptibility in superconductors. *Phys. Rev.* **1958**, *110*, 769–770.
6. Gladstone, G.; Jensen, M.A.; Schrieffer, J.R. Superconductivity in the transition metals. In *Superconductivity*; Parks, R.D., Eds.; Marcel Dekker, Inc.: New York, NY, USA, 1969; pp. 801–803.
7. Abrikosov, A.A.; Gor'kov, L.P. Spin-orbit interaction and the Knight shift in superconductors. *Sov. Phys. JETP* **1962**, *15*, 752–757.
8. Baek, S.-H.; Harnegea, L.; Wurmehl, S.; Büchner, B.; Grafe, H.-J. Anomalous superconducting state in LiFeAs implied by the ^{75}As Knight shift measurement. *J. Phys. Condens. Matter* **2013**, *25*, 162204, doi:10.1088/0953-8984/25/16/162204.
9. Kohori, Y.; Yamato, Y.; Iwamoto, Y.; Kohara, T.; Bauer, E.D.; Maple, M.B.; Sarrao, J.L. NMR and NQR studies of the heavy fermion superconductors CeTIn$_5$ (T = Co and Ir). *Phys. Rev. B* **2001**, *64*, 134526, doi:10.1103/Phys RevB.64.134526.
10. Lee, I.J.; Brown, S.E.; Clark, W.G.; Strouse, M.J.; Naughton, M.J.; Kang, W.; Chaikin, P.M. Triplet superconductivity in an organic superconductor probed by NMR Knight shift. *Phys. Rev. Lett.* **2002**, *88*, 17004, doi:10.1103/PhysRevLett.88.017004.
11. Lee, I.J.; Brown, S.E.; Clark, W.G.; Strouse, M.J.; Naughton, M.J.; Chaikin, P.M.; Brown, S.E. Evidence from ^{77}Se Knight shifts for triplet superconductivity in (TMTSF)$_2$PF$_6$. *Phys. Rev. B* **2003**, *68*, 92510, doi:10.1103/PhysRevB.68.092510.
12. Michioka, C.; Ohta, H.; Itoh, Y.; Yoshimura, K.; Kato, M.; Sakurai, H.; Takayama-Muromachi, E.; Takada, K.; Sasaki, T. Knight shift of triangular lattice superconductor Na$_{0.35}$CoO$_2$·1.3H$_2$O. *Phys. B* **2006**, *628–629*, doi:10.1016/j.physb.2006.01.181.

13. Kato, M.; Michioka, C.; Waki, T.; Itoh, Y.; Yoshimura, K.; Ishida, K.; Sakurai, H.; Takayama-Muromachi, E.; Takada, K.; Sasaki, T. Possible spin triplet superconductivity in $Na_xCoO_2 \cdot yH_2O$: ^{59}Co NMR Studies. *J. Phys. Condens. Matter* **2006**, *18*, 669–682.

14. Sakurai, H.; Ihara, Y.; Takada, K. Superconductivity of cobalt oxide hydrate $Na_x(H_3O)_zCoO_2 \cdot yH_2O$. *Phys. C* **2015**, 378–387, doi:10.1016/j.physc.2015.02.010.

15. Klemm, R.A. Striking similarities between the pseudogap phenomenon in the cuprates and in layered organic and dichalcogenide superconductors. *Physica C* **2000**, *341–348*, 839–842.

16. Chou, F.C.; Cho, J.H.; Lee, P.A.; Abel, E.T.; Matan, K.; Lee, Y.S. Thermodynamic and transport measurements of superconducting $Na_{0.3}CoO_2 \cdot 1.3H_2O$ single crystals prepared by electrochemical deintercalation. *Phys. Rev. Lett.* **2004**, *92*, 157004, doi:10.1103/PhysRevLett.92.157004.

17. Klemm, R.A. *Layered Superconductors Volume 1*; Oxford University Press: Oxford, UK, 2012.

18. Scharnberg, K.; Klemm, R.A. *P*-Wave superconductors in magnetic fields. *Phys. Rev. B* **1980**, *22*, 5233–5244.

19. Barrett, S.E.; Durand, D.J.; Pennington, C.H.; Slichter, C.P.; Friedmann, T.A.; Rice, J.P.; Ginsberg, D.M. ^{63}Cu Knight shifts in the superconducting state of $YBa_2Cu_3O_{7-\delta}$ (T_c = 90 K). *Phys. Rev. B* **1990**, *41*, 6283–6296.

20. Slichter, C.P. The Knight shift — A powerful probe of condensed-matter systems. *Philos. Mag. Part B* **1999**, *79*, 1253–1261.

21. Fujiwara, K.; Kitaoka, Y.; Ishida, K.; Asayama, K.; Shimakawa, Y.; Manako, T.; Kubo, Y. NMR and NQR studies of superconductivity in heavily doped $Tl_2Ba_2CuO_{6+y}$ with a Single CuO_2 plane. *Physica C* **1991**, *184*, 207–219.

22. Zheng, G.Q.; Kitaoka, Y.; Asayama, K.; Hamada, K.; Yamauchi, H.; Tanaka, S. NMR study of local hole distribution, spin fluctuation, and superconductivity in $Tl_2Ba_2Ca_2Cu_3O_{10}$. *Physica C* **1996**, *260*, 197–210.

23. Zheng, G.Q.; Sato, T.; Kitaoka, Y.; Fujita, M.; Yamada, K. Fermi-liquid ground state in the *n*-Type $Pr_{0.91}LaCe_{0.09}CuO_{4-y}$ copper-oxide superconductor. *Phys. Rev. Lett.* **2003**, *90*, 197005, doi:10.1103/PhysRevLett.90.197005.

24. Kotegawa, H.; Masaki, S.; Awai, Y.; Tou, H.; Mizuguchi, Y.; Takano, Y. Evidence for unconventional superconductivity in arsenic-free iron-based superconductor FeSe: A ^{77}Se-NMR study. *J. Phys. Soc. Jpn.* **2008**, *77*, 113703, doi:10.1143/JPSJ.77.113703.

25. Ishida, K.; Mukuda, H.; Kitaoka, Y.; Asayama, K.; Mao, Z.Q.; Mori, Y.; Maeno, Y. Spin-triplet superconductivity in Sr_2RuO_4 identified by ^{17}O Knight shift. *Nature* **1998**, *396*, 658–660.

26. Ishida, K.; Mukuda, H.; Kitaoka, Y.; Mao, Z.Q.; Fukazawa, H.; Maeno, Y. Ru NMR probe of the spin susectibility in the superconducting state of Sr_2RuO_4. *Phys. Rev. B* **2001**, *63*, 60507, doi:10.1103/PhysRevB.63.060507.

27. Mackenzie, A.P.; Maeno, Y. The superconductivity of Sr_2RuO_4 and the physics of spin-triplet pairing. *Rev. Mod. Phys.* **2003**, *25*, 657–712.

28. Duffy, J.A.; Hayden, S.M.; Maeno, Y.; Mao, Z.; Kulda, J.; McIntyre, G.J. Polarized-neutron scattering study of the Cooper-pair moment in Sr_2RuO_4. *Phys. Rev. Lett.* **2000**, *85*, 5412–5415.

29. Deguchi, K.; Tanatar, M.A.; Mao, Z.Q.; Ishiguro, T.; Maeno, Y. Superconducting double transition and the upper critical field limit of Sr_2RuO_4 in parallel magnetic fields. *J. Phys. Soc. Jpn.* **2002**, *71*, 2839–2842.

30. Kittaka, S.; Nakamura, T.; Aono, Y.; Yonezawa, S.; Ishida, K.; Maeno, Y. Angular dependence of the upper critical field of Sr_2RuO_4. *Phys. Rev. B* **2009**, *80*, 174514, doi:10.1103/PhysRevB.80.174514.

31. Machida, K.; Ichioka, M. Magnetic field dependence of low-temperature specific heat in Sr_2RuO_4. *Phys. Rev. B* **2008**, *77*, 184515, doi:10.1103/PhysRevB.77.184515.

32. Zhang, J.; Lörscher, C.; Gu, Q.; Klemm, R.A. Is the anisotropy of the upper critical field of Sr_2RuO_4 consistent with a helical *p*-wave state? *J. Phys. Condens. Matter* **2014**, *26*, 252201, doi:10.1088/0953-8984/26/25/252201.

33. Annett, J.F.; Györffy, B.L.; Litak, G.; Wysokiński, K.I. Magnetic field induced rotation of the *d*-vector in Sr_2RuO_4. *Physica C* **2007**, *460–462*, 995–996.

34. Leggett, A.J. A theoretical description of the new phases of liquid ^{3}He. *Rev. Mod. Phys.* **1975**, *47*, 331–414.

35. Rozbicki, E.J.; Annett, J.F.; Souquet, J.-R.; Mackenzie, A.P. Spin-orbit coupling and *k*-dependent Zeeman splitting in strontium ruthenate. *J. Phys. Condens. Matter* **2011**, *23*, 94201, doi:10.1088/0953-8984/23/9/094201.

36. Suderow, H.; Crespo, V.; Guillamon, I.; Vieira, S.; Servant, F.; Lejay, P.; Brison, J.P.; Flouquet, J. A nodeless superconducting gap in Sr_2RuO_4 from tunneling spectroscopy. *New J. Phys.* **2009**, *11*, 93004, doi:10.1088/1367-2630/11/9/093004.

37. Berry, M.V. Quantal phase factors accompanying adiabatic changes. *Proc. R. Soc. Lond. A* **1984**, *392*, 45–57.

38. Griffiths, D.J. *Introduction to Quantum Mechanics*, 2nd ed.; Pearson: Upper Saddle River, NJ, USA, 2005.

39. Hattori, T.; Karube, K.; Ihara, Y.; Ishida, K.; Deguchi, K.; Sato, N.K.; Yamamura, T. Spin susceptibility in the superconducting state of the ferromagnetic superconductor UCoGe. *Phys. Rev. B* **2013**, *88*, 85127, doi:10.1103/PhysRevB.88.085127.

40. Aoki, D.; Flouquet, J. Ferromagnetism and superconductivity in uranium compounds. *J. Phys. Soc. Jpn.* **2012**, *81*, 11003, doi:10.1143/JPSJ.81.011003.

41. Gannon, W.J.; Halperin, W.P.; Rastovski, C.; Eskildsen, M.R.; Dai, P.C.; Stunault, A. Magnetism in the superconducting state of UPt_3 from polarized neutron diffraction. *Phys. Rev. B* **2012**, *86*, 104510, doi:10.1103/PhysRevB.86.104510.

42. Scharnberg, K.; Klemm, R.A. Upper critical field in *p*-wave superconductors with broken symmetry. *Phys. Rev. Lett.* **1985**, *54*, 2445–2448.

43. Hall, B.E.; Klemm, R.A. Microscopic model of the Knight shift in anisotropic and correlated metals. *J. Phys. Condens. Matter* **2016**, *28*, 03LT01, doi:10.1088/0953-8984/28/3/03LT01.

44. Haug, H.J.W.; Jauho, A.-P. *Quantum Kinetics in Transport and Optics of Semiconductors*; Springer: Berlin, Germany, 2008.

45. Klemm, R.A.; Clem, J.R. Lower critical field of an anisotropic Type-II superconductor. *Phys. Rev. B* **1980**, *21*, 1868–1875.

46. Klemm, R.A. Lower critical field of a superconductor with uniaxial anisotropy. *Phys. Rev. B* **1993**, *47*, 14630, doi:10.1103/PhysRevB.47.14630.

47. Anderson, P.W. Localized magnetic states in metals. *Phys. Rev.* **1961**, *124*, 41–53.

48. Lörscher, C.; Zhang, J.; Gu, Q.; Klemm, R.A. Anomalous angular dependence of the upper critical induction of orthorhombic ferromagnetic superconductors with completely broken *p*-wave symmetry. *Phys. Rev. B* **2013**, *88*, 24504, doi:10.1103/PhysRevB.88.024504.

49. Zhang, J.; Lörscher, C.; Gu, Q.; Klemm, R.A. First-order chiral to non-chiral transition in the angular dependence of the upper critical induction of the Scharnberg-Klemm *p*-wave pair state. *J. Phys. Condens. Matter* **2014**, *26*, 252202, doi:10.1088-8984/26/25/252202.

50. Rammer, J.; Smith, H. Quantum field-theoretical methods in transport theory of metals. *Rev. Mod. Phys.* **1986**, *58*, 323–359.

51. Drozdov, A.P.; Eremets, M.I.; Troyan, I.A.; Ksenofontov, V.; Shylin, S.I. Conventional superconductivity at 203 K at high pressures in the sulfur hydride system. *Nature* **2015**, *525*, 73–77.

52. Werthamer, N.R.; Helfand, E.; Hohenberg, P.C. Temperature and purity dependence of the superconducting critical field H_{c2}. III. spin and spin-orbit effects. *Phys. Rev.* **1966**, *147*, 295–302.

53. Klemm, R.A.; Luther, A.; Beasley, M.R. Theory of the upper critical field in layered superconductors. *Phys. Rev. B* **1975**, *12*, 877–891.

54. Abrikosov, A.A.; Gor'kov, L.P.; Dzyaloshinskii, I.E. *Methods of Quantum Field Theory in Statistical Physics*; Prentice-Hall: Englewood Cliffs, NJ, USA, 1963.

55. Gor'kov, L.P. Microscopic derivation of the Ginzburg-Landau equations in the theory of superconductivity. *Sov. Phys. JETP* **1959**, *9*, 1364–1367.

56. Rickayzen, G. The theory of Bardeen, Cooper, and Schrieffer. In *Superconductivity*; Parks, R.D., Eds.; Marcel Dekker, Inc.: New York, NY, USA, 1969; pp. 51–115.

57. Klemm, R.A. Pristine and intercalated transition metal dichalcogenide superconductors. *Physica C* **2015**, *514*, 86–94.

58. Zhong, Y.; Wang, Y.; Han, S.; Lv, Y.-F.; Wang, W.-L.; Zhang, D.; Ding, H.; Zhang, Y.-M.; Wang, L.; He, K.; et al. Nodeless pairing in superconducting copper-oxide monolayer films on $Bi_2Sr_2CaCu_2O_{8+\delta}$. *Sci. Bull.* **2016**, *61*, 1239–1247.

59. Li, Q.; Tsay, Y.N.; Suenaga, M.; Klemm, R.A.; Gu, G.D.; Koshizuka, N. $Bi_2Sr_2CaCu_2O_{8+\delta}$ Bicrystal *c*-axis twist Josephson junctions: A new phase-sensitive test of order parameter symmetry. *Phys. Rev. Lett.* **1999**, *83*, 4160–4163.

60. Schaefer, H.F., III; Klemm, R.A., Harris, F.E. Atomic hyperfine structure. II. First-order wavefunctions for the ground states of B, C, N, O, and F. *Phys. Rev.* **1969**, *181*, 137–143.

61. Armstrong, L. Jr. *Theory of the Hyperfine Structure of Free Atoms*; Wiley Interscience: New York, NY, USA, 1971.

magnetochemistry

MDPI

Article

The Radiofrequency NMR Spectra of Lithium Salts in Water; Reevaluation of Nuclear Magnetic Moments for ^{6}Li and ^{7}Li Nuclei

Włodzimierz Makulski

Faculty of Chemistry, University of Warsaw, L. Pasteura 1, 02-093 Warsaw, Poland; wmakul@chem.uw.edu.pl; Tel.: +48-22-552-6346

Received: 2 December 2017; Accepted: 4 January 2018; Published: 10 January 2018

Abstract: LiCl and LiNO$_3$ water solutions in the presence of small amounts of 3-helium have been investigated by means of multinuclear resonance spectroscopy. The resulting concentration dependences of the ^{3}He, ^{6,7}Li$^+$, ^{14}NO$_3$$^-$ and ^{35}Cl$^-$ resonance radiofrequencies are reported in the infinite limit. This data along with new theoretical corrections of shielding lithium ions was analyzed by a known NMR relationship method. Consequently, the nuclear magnetic moments of ^{6}Li and ^{7}Li were established against that of the helium-3 dipole moment: $\mu(^6$Li$) = +0.8220457(50)\mu_N$ and $\mu(^7$Li$) = +3.256418(20)\mu_N$. The new results were shown to be very close to the previously obtained values of the (ABMR) atomic beam magnetic resonance method. This experiment proves that our helium method is well suited for establishing dipole moments from NMR measurements performed in water solutions. This technique is especially valuable when gaseous substances of the needed element are not available. All shielding constants of species present in water solutions are consistent with new nuclear magnetic moments and these taken as a reference. Both techniques—NMR and ABMR—give practically the same results provided that all shielding corrections are properly made.

Keywords: ^{6}Li and ^{7}Li nuclear magnetic moments; NMR liquid-phase studies; nuclear magnetic shielding constants

1. Introduction

The electromagnetic moments of nuclei, dipole and quadrupole, have great significance for theory of nuclear structure. The magnetic moments are of prime importance for all nuclei with spin number $I \geq 1/2$. They were established for the first time in the famous molecular beam experiments carried out by Rabi (1939) [1] and, afterwards, improved values were experimentally determined by means of NMR bulk experiments e.g., by Walchli (1954), for the sequence of nuclear moments from lithium up to thallium [2]. The method relies on the accurate measurements of two frequencies for different nuclei placed in one sample at the same magnetic field. One of these frequencies should belong to the nucleus with a well-known magnetic moment and can be taken as a reference. The main problem with this procedure lies in ensuring that the shielding effects of nuclei in the particular experimental conditions are known with enough accuracy. The spectacular growth of quantum theoretical methods in this field provided new impetus for improving existing data. Several such works were performed in the Laboratory of NMR Spectroscopy at the University of Warsaw. We utilize the gas phase conditions as a rule, because of the importance of the shielding results for the isolated molecules when extrapolation to the zero-pressure limit is possible [3,4]. Unfortunately, we do not have any stable gaseous substances at normal conditions available for several elements (e.g., Li, Be, Na, K, Sc). Instead of gaseous species, the liquid solutions should be used in these cases. In this work, water solutions of common salts of lithium were applied—LiCl and LiNO$_3$ in the presence of dissolved ^{3}He atoms. This procedure has several advantages: very narrow NMR signals, good sensitivity and well-known shielding parameters of different ions in liquid samples.

Without a doubt, lithium nuclei are of great account from the point of view of nuclear physics. Accurate and precise experimental values of nuclear properties are of prime importance in this case. There are eight lithium isotopes ranging from ^{4}Li up to ^{11}Li; only two of them are stable: ^{6}Li (7.59(4)%) and ^{7}Li (92.41(4)%) [5]. Both these nuclei possess different moments, electric quadrupole and dipole magnetic, connected with magnetic numbers $I^\pi = 1^+$ (with three neutrons) and $I^\pi = 3/2^-$ (with four neutrons), respectively. Since the two isotopes vary by a single spin-1/2 neutron, they exhibit different quantum statistics: ^{6}Li is a composite fermion while a ^{7}Li nucleus is a composite boson particle. In these circumstances, they represent one of the smallest objects, whose nuclear parameters could be precisely calculated in the near future. Interestingly, in spite of different mass numbers, the charge radius in ^{7}Li is smaller, which indicates the valuable differences in the magnetic distribution inside both nuclei [6].

The first hints about the ^{7}Li nuclear magnetic moment were made by Goudsmit and Young [7] and soon after deduced by Granath [8] as the nuclear spin 3/2 and magnetic moment possess 3.29 times the theoretical magnetic moment of the proton ($\mu_N = e\hbar/2m_p$, where e is the elementary charge and m_p is the proton's mass). A further investigation into the magnetic properties of lithium isotopes was carried out by Rabi's molecular/atomic beam MR experiments in the resonance absorption method. The determination of the nuclear spin and magnetic moment of lithium isotopes was obtained for LiCl, LiF and Li$_2$ molecules [9,10]. Next, more precise results were received by NMR measurements performed in water solutions of lithium salts and calculated against the deuterium NMR reference [11,12]. Soon after, precise lithium nuclei dipole moments were measured by the atomic beam magnetic resonance method [13]. These last results were cited later in the most pronounced tabulated compilations of magnetic moments for stable nuclei [14–16]. All of the remaining lithium nuclei are radioactive and have very short half-lives (^{4}Li-4.9–8.9 $\times$ 10^{-23} s, ^{5}Li-5.4 $\times$ 10^{-22} s, ^{8}Li-0.84 s, ^{9}Li-0.178 s,^{10}Li-5.5 $\times$ 10^{-22}–5.5 $\times$ 10^{-21} s and ^{11}Li-0.0087 s) [5].

The aim of this work is twofold. Firstly, precise NMR measurements of frequencies for LiCl and LiNO$_3$ in water solutions were performed and analysis of new ^{6}Li/^{7}Li NMR data collected for water solutions at low concentrations was performed and compared to the results for ^{3}He dissolved in the same samples. Up to now, the addition of helium ingredients has only been carried out in our lab only in the gas phase. We are now trying to extend our method to the liquid samples. As a second step, the nuclear magnetic moments of ^{6}Li and ^{7}Li nuclei were recalculated using new shielding constants of lithium cations solvated in water solutions [17]. New magnetic moments measured in our work were compared with these established before by the atomic beam method. It is obvious that accurate values of the nuclear ground-state properties of isotopes, such as the magnetic dipole and electric quadrupole moments, are ideal tools for testing the validity of nuclear structure models. Subsequently a comparison of different experimental and purely theoretical results was made.

2. Results and Discussion

2.1. NMR Experiments in Water Solutions

Lithium has NMR spectroscopy based on two different nuclei. Both are quadrupolar, then the interaction with the electric field gradient at the nucleus is important by definition. It is worth noting anomalous, very small quadrupolar moment of ^{6}Li (0.00082(2) barn, 1 barn = m^2) [15] (contrary to that of ^{7}Li-0.0406(8) barn), which as a consequence yields rather sharp resonance signals. The chemical shift range of both nuclides is small and reaches only ~30 ppm. Fortunately, lithium cation shows a high symmetry structure characterized by a small electric field gradient and its line width for reference solution (9.7 M LiCl in D$_2$O) not even achieving ~0.1 Hz. For this reason, water solutions of lithium salts seem to be ideal for precise measurements.

For the derivation of the lithium nuclear magnetic moments we have used the usual form of equation, which connects two observed frequencies at the zero concentration of lithium salts and nuclear dipole moments. They should be corrected for shielding values of Li$^+$ and ^{3}He measured in aqueous solutions:

$$\Delta\mu_{Li}^z = \frac{\nu_{Li}}{\nu_{He}} \cdot \frac{(1-\sigma_{He})}{(1-\sigma_{Li})} \cdot \frac{I_{Li}}{I_{He}} \Delta\mu_{He}^z, \tag{1}$$

where ν_{Li} and ν_{He} mean appropriate radiofrequencies extrapolated to the infinite diluted solutions. I_x are magnetic quantum numbers of measured nuclei, and $\sigma_{He,Li}$ are also shielding corrections for nuclei in the experimental conditions. The above equation makes it possible to calculate the magnetic moment μ_{Li} when all other quantities are known. The experimental results of NMR measurements are shown in Table 1. The suitable concentration dependencies of specific extrapolations are illustrated in Figures 1 and 2. In general, the concentration dependences of chemical shifts/shielding for cations or anions should not be linear, particularly at higher concentrations. For uniformity, all analyses were done by single-variable quadratic functions. It is known that virial expansions can be used for models of aqueous ionic solutions [18]. All coefficients are shown in Table 1 as δ [ppm], δ_1 [ppm $\times$ mL $\times$ mol^{-1}] and δ_2 [ppm $\times$ mL $\times$ mol^{-2}]. The course of the functions (Figures 1 and 2) reflects the magnetic susceptibility effect of solutions and a complex intermolecular forces arising during rapidly equilibration of solvent-separated cations and anions.

A crucial role in the estimations of lithium nuclear magnetic moments has been played by knowledge of the diamagnetic corrections for helium atoms and lithium cations. As the reference point of helium measurements was chosen very precise shielding value of single atom nuclei $\sigma_0(^3\text{He}) = 59.96743(10)$ ppm calculated with relativistic corrections, QED (Quantum Electrodynamics) corrections and nuclear mass effects [19]. At the beginning, we measured the ^{3}He NMR signal against that of gaseous systems; the difference is 2.7675(25) ppm in the chemical shift category, independently on the concentration of helium in water. It corresponds to the 0.2384(5) ppm deshielding effect when going from isolated molecule in gaseous state to the liquid water solution. This value was used to correct the helium frequency by electron screening. For comparison, the chemical shift corrected for the susceptibility of ^{3}He in water solution against that of gaseous sample (1-atm gas sample used for the gas reference) was measured previously by Jokisaari [20] $\Delta\delta = 0.297(39)$ ppm.

Table 1. NMR parameters measured in LiCl and LiNO$_3$ water solutions*. Shielding results of ^{6}Li$^+$ and ^{7}Li$^+$ cations calculated for: (a) six-coordinated and (b) four-coordinated water complexes.

Water Solution	Nuclide	ν_0 (Radiofrq.) MHz	δ/ppm	δ_1/ppm mL mol^{-1} δ_2/ppm mL mol^{-2}	σ/ppm	Reference
LiCl						
(^{6}Li$^+$)$_{aq.}$	73.6695828(2)	−0.1472	−0.0632 0.0148	90.89(300) (a) 91.69(300) (b)	[17]
	(^{7}Li$^+$)$_{aq.}$	194.5544573(2)	−0.1469	−0.0632 0.0148	90.89(300) (a) 91.69(300) (b)	[17]
	^{35}Cl$^-$	49.0491386(10)	4.7125	0.9358 −0.0461	998.28(500)	[21] (This work)
	^{3}He	381.3564690(5)	−2.7675	−0.0478 0.0102	59.729(1)	[19] (This work)
LiNO$_3$						
	(^{6}Li$^+$)$_{aq.}$	73.6695829(2)	−0.147	−0.003 −0.0059	90.89(300) (a) 91.69(300) (b)	[17]
	(^{7}Li$^+$)$_{aq.}$	194.5544571(2)	−0.147	−0.003 0.0059	90.89(300) (a) 91.69(69) (b)	[17]
	^{14}NO$_3^-$	36.1752096(10)	−5.595	−0.107 0.0165	−132.14	[4] (This work)
	^{3}He	381.3564691(5)	−2.7676	−0.0045 −0.004	59.729(1)	(This work)

* Lock system tuned to $\nu(\text{D}_2\text{O}) = 76.8464$ MHz

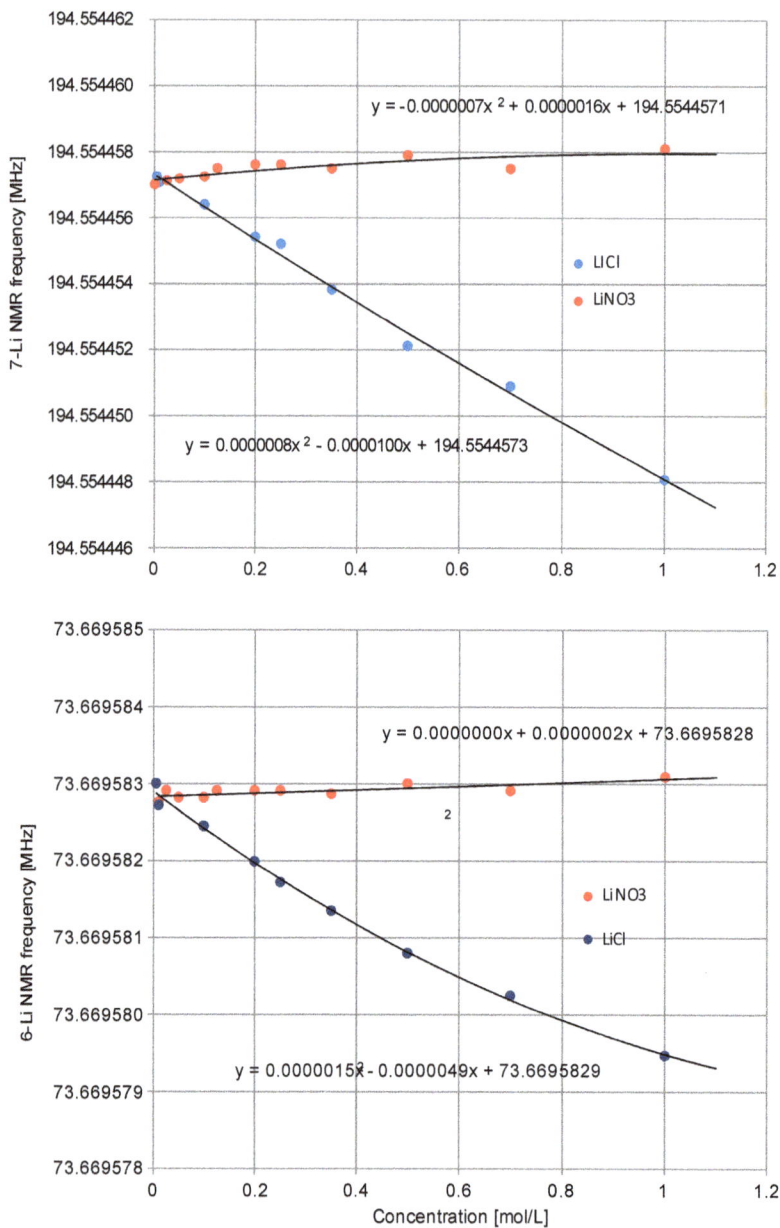

Figure 1. ^{6}Li and ^{7}Li NMR frequencies versus concentration of LiCl and LiNO$_3$ in water solutions.

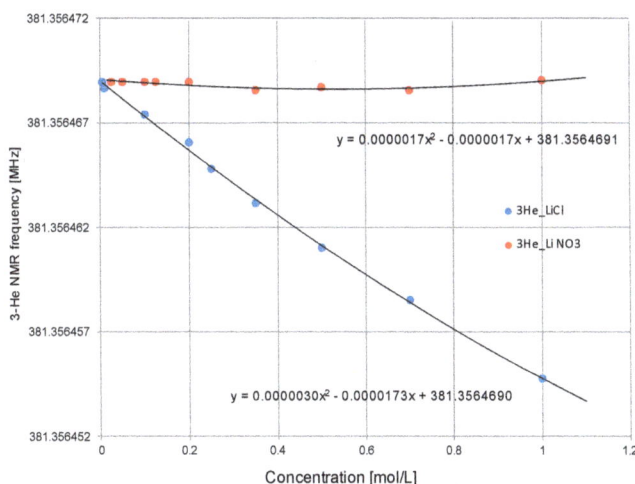

Figure 2. The ^{3}He NMR frequencies of helium atoms dissolved in water solutions of LiCl (lover curve) and LiNO$_3$ (upper curve) against salt concentration. Intersection of the axis represents frequency value at infinite dilution.

More significant correction is needed in the case of lithium nuclei. The ^{6,7}Li$^+$ cation's solvation properties in water solutions were actively studied in many theoretical simulations [22,23] and experimental research used different spectroscopy techniques [24–26]. The structure of the water complex is the subject of many controversies. The Li$^+$ cation in water solution has the smallest ionic radius of 90 pm (as 4-coordinated) and 76 pm (as-6-coordinated), and the highest positive charge density compared to other alkali metals. The stability of four, five or six water molecules in the inner shell of Li$^+$ ion is still under consideration. Most of this data refers to strong solutions in which there are very few water molecules that are not in the primary hydration spheres of the lithium cation, which may account for some of the solvation number variations with solute concentration. In the lithium aqueous ions have been found to have the solvation numbers of 3–6 and solvation numbers less than 4 can be suitable when the formation of contact ion pairs is possible. In the infinite dilution, we can exclude the possibility of interaction between a solvated cation and an anion and forming an ion pair. It is clear that the measured solvation number is a time-averaged value in the water solutions. The primary solvation number seen is fractional; there are two or more species with integral solvation numbers present in equilibrium with each other:

$$[Li(H_2O)_6]^+ \rightleftharpoons [Li(H_2O)_5]^+ + H_2O \rightleftharpoons [Li(H_2O)_4]^+ + 2H_2O, \qquad (2)$$

The higher solvation numbers may be interpreted in terms of water molecules in a tetrahedron coordination [Li(H$_2$O)$_4$]$^+$ or even higher coordinated complexes e.g., an octahedral aqua ions which are revealed by molecular dynamic simulations. The final suggestion of P.E. Mason et al. [25] shows that an infinitely diluted water solution at room temperature is mainly composed of 4 coordinated lithium complexes of great stability. Without pre-empting composition at the infinite dilution we decided to calculate lithium moments when tetrahedral or/and octahedral coordination take place. If the coordination number of central lithium cation varies, its shielding values change, starting from 95.30–95.41 ppm for an isolated ion up to 90.18 ppm in the hexacoordinated complex [17]. In the last case the small correction of 0.8 ppm for 2 water molecules, which distorts the first tetrahedral solvation shell of lithium ion, was applied [27]. The final shielding effect, with the small relativistic term 0.08 ppm calculated by the CCSD/utA,tz (Coupled Cluster) quantum method, was then 90.89 ppm. If four coordinating lithium cations are present then shielding constant 91.69 ppm should be valid [17].

Taking into account of the ν_{Li}/ν_{He} frequency ratio (see Table 1) and both shielding corrections for ^{3}He and ^{6,7}Li nuclei we can deduce the nuclear magnetic dipole moments of ^{6}Li and ^{7}Li nuclei (see Table 2). Two values in the table were quoted for different shielding corrections for the lithium nucleus (90.89 and 91.69 ppm) as the lower and upper limit for the magnetic moment. It is worth noting that both results are in good agreement with previously results used in establishing the absolute lithium shielding scale by J.Mason [28], see also [13]: 90.0(8) ppm (^{6}Li) and 90.4(7) ppm (^{7}Li). In any case, the effect is small and will be used as reference against ABMR results (see Table 3).

Table 2. $^{6/7}$Li nuclear magnetic shielding values calculated from Equation (3) and selected nuclear magnetic moments. Theoretical results for hexa- and tetra-coordinated water complex.

$\mu(^7Li)/\mu_N$	Method/Reference	Nucleus	$\sigma(^7Li^+)_{aq.}$/ppm
	Theory/[17]		90.89 ÷ 91.69(300)
3.2564169(98)	NMR/(This work)	$^{35}Cl^-$	91.16
		$^{14}NO_3^-$	90.36
3.2564195(98)	NMR/(This work)	$^{35}Cl^-$	91.53
		$^{14}NO_3^-$	90.73
3.2564157(30)	ABMR/[13], (This work)	$^{35}Cl^-$	90.76
		$^{14}NO_3^-$	89.96
3.2564625(4)	NMR/[15]	$^{35}Cl^-$	105.13
		$^{14}NO_3^-$	104.33
$\mu(^6Li)/\mu_N$			$\sigma(^6Li^+)_{aq.}$/ppm
	Theory/[17]		90.89 ÷ 91.69(300)
0.8220453(25)	NMR/(This work)	$^{35}Cl^-$	91.09
		$^{14}NO_3^-$	90.30
0.8220459(25)	NMR/(This work)	$^{35}Cl^-$	91.82
		$^{14}NO_3^-$	91.03
0.8220445(10)	ABMR*/[13], (This work)	$^{35}Cl^-$	90.12
		$^{14}NO_3^-$	89.32
0.822567(3)	NMR/[15]	$^{35}Cl^-$	725.27
		$^{14}NO_3^-$	724.47

2.2. ABMR Experiments for Atoms

An extensive ABMR (atomic beam magnetic resonance) experiment was carried out to examine ^{6,7}Li nuclear magnetic moments [13]. Several improvements to the original technique were made to avoid all systematic errors involved in this approach. The method of separated oscillatory fields with triple resonance technique and special calibration of the magnetic field offered very precise final results. For a proper comparison of our results with ABMR values, several new corrections were applied to the original quantities, i.e., proton-to-electron mass ratio $m_p/m_e = 1836.15267389(17)$ [29] and diamagnetic correction factor in Li atom $(1-\sigma_{Li})^{-1} = 1.0000101472$ [30]. This last value is very consistent with previous received theoretical results −101.4 and 101.45 ppm [31,32]. The g_J factor for the $2^2S_{1/2}$ state was taken from the original work-2.002301100(64), which agrees very closely with the purely theoretical data, 2.00230101 [33]. The final, corrected magnetic moments established by Beckmann et al. [13] are shown in Table 2 as ABMR* results. The differences between nuclear magnetic moments measured in our NMR investigation and the ABMR method are then of the order 0.8–1.5 × 10^{-4}%. Remarkably, our refine results are much closer to the ABMR results than those cited in several current specifications [14–16] received from previous NMR measurements performed in aqueous solutions. It is certainly not without significance that the final results are more closely related to the ABMR results when shielding lithium cations were used for the strictly hexacoordinated water complex.

Table 3. Electromagnetic properties of lithium, chlorine, nitrogen, helium and deuterium nuclei.

Nuclide	I^π	Q Barn	Abundance %	μ/μ_N	Diamagnetic Correction	g_I Factor	$\gamma_I \times 10^7$	Reference
^{6}Li	1^+	0.00082(2)	7.59(4)	0.8220453(25)	1.00009089	0.822045(3)	3.93712(1)	(This work)
				0.8220459(25)	1.00009169			
				0.8220445(10)	1.000101472			[13]
				0.832; 0.835				[34]
				0.843(5);				
				0.843(2)				
^{7}Li	$3/2^-$	0.0406(8)	92.41(4)	3.2564169(98)	1.00009089	2.170945(7)	10.39756(3)	(This work)
				3.2564195(98)	1.00009169			
				3.2564157(2)	1.000101472			[13]
				2.993; 3.036				[34]
				3.01(2);				
				3,02(2)				
^{35}Cl	$3/2^+$	0.0850(11)	75.78(4)	0.821721(5)		0.547814(3)	2.62371(1)	[21]
^{14}N	1^+	0.02001(10)	99.632(7)	0.4035729(45)		0.403573(5)	1.93288(2)	[4]
^{3}He	$1/2^+$		0.000137	2.127625308(25)	1.00005973	4.25525061(5)	20.3801680(2)	[29]
^{2}H(D)	1^+	0.00286(2)	0.0156	0.8574382311(48)		0.857438231(5)	4.1066289(1)	[29]

2.3. Shielding Factors

The new nuclear magnetic moments from NMR and ABMR experiments (Table 2) can certainly be tested, because a few shielding constants of different additional nuclei present in the solution are known with great precision. The concentration dependencies for $^{35}Cl^-$ and $^{14}NO_3^-$ anions are shown in Figure 3.

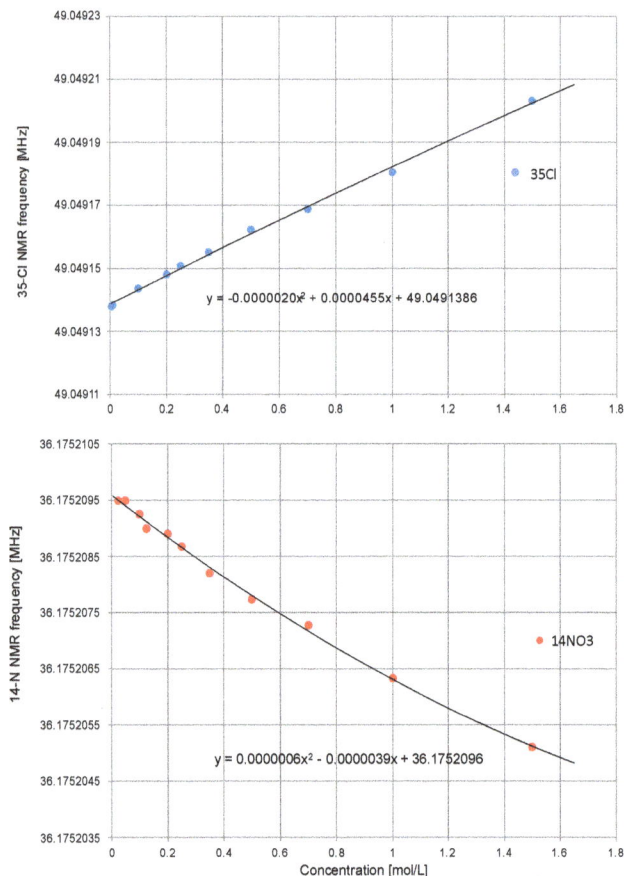

Figure 3. The ^{14}N and ^{35}Cl NMR frequencies of Cl^- and NO_3^- anions as function of LiCl and LiNO₃ concentration in water solutions. Intersection of the axis represents frequency value at infinite dilution.

In order to verify the conformity of the nuclear shielding values of lithium nuclei in water solution a different form of Equation (1) was used:

$$\sigma_X = 1 - \frac{\nu_X}{\nu_Y} \cdot \frac{\Delta\mu_Y}{\Delta\mu_X} \cdot \frac{I_X}{I_Y}(1 - \sigma_Y), \tag{3}$$

Formula (3) was carried out for each pair of nuclei: $^{6,7}Li/^{14}N$ and $^{6,7}Li/^{35}Cl$ present in our samples of H_2O solutions. ^{14}N nuclear shielding in the NO_3^- anion at infinite dilution was calculated from nuclear magnetic shielding of liquid CH_3NO_2, which is equal to -132.14 ppm [4]. ^{35}Cl nuclear shielding in the Cl^- anion was calculated against shielding value in 0.1 M NaCl/D_2O solution, which is equal to 998.28 ppm [21]. From the results collected in Table 2, it is clear that only our new $^{6,7}Li$ nuclear magnetic moments are consistent with shielding calculations against D_2O, $^{35}Cl^-$ and $^{14}NO_3^-$ species accordingly to Equation (3).

Subsequently, the uncertainty error of lithium shielding is much less then suggested by theoretical predictions (± 3 ppm) [17] and possibly remains ± 1.5 ppm an order of magnitude.

It is worth noting that measurements of lithium dipole moments, contrary to many heavier isotopes, depend on diamagnetic corrections of NMR frequencies only in limited degree. This is a consequence of the relatively narrow spectral ranges of all nuclei in magnetic resonance studies ($^{6,7}Li$, ^2H, ^3He) and the small screening factors. It means that $^{6,7}Li$ magnetic moments belong to the class of most precise and accurately known dipole moments for all elements in the whole periodic table.

The lithium nuclei are very promising objects in the theoretical quantum calculation field. It is known that pure theoretical methods are still a long way from the precision of resonance experiments. Formerly performed calculations are valid to the three or four digit numbers, i.e., $\mu(^6Li) = 0.832\mu_N$ and $\mu(^7Li) = 2.993\mu_N$ or $\mu(^6Li) = 0.835\mu_N$ and $\mu(^7Li) = 3.036\mu_N$ [34]. On the other hand the lithium magnetic moments of another isotopes are still a subject of great interest. New developments have also involved short living isotopes: $^{8,9,11}Li$ nuclei. The investigation into magnetic moments for stable isotopes forms only a part of the studies, which include the short living isotopes at different excitation levels. The nuclear moments of $^8Li(1.653560(18))\mu_N$, $^9Li(3.43682(5))\mu_N$ and $^{11}Li(3.6712(5))\mu_N$ were measured by β-NMR experiments with major precision [35].

3. Materials and Methods

$LiNO_3$ (Sigma-Aldrich, Saint Louis, MO, USA, 99.99%) and LiCl (Sigma-Aldrich, 99.998%, anhydrous) were used for preparing water solutions at total densities in the range 0.25–1.2 mol/L. Samples of 0.3 mL in Pyrex tubes (4 mm o.d. and 56 mm long) were frozen in liquid nitrogen and pumped to a pressure of ~10^{-3} mmHg. Small amounts of ^3He (Chemgas, Boulogne, France, 99.9%) $\leq 3.0 \times 10^{-3}$ mol/L were then added before sealing the ampoules by torch. Only a small amount of helium can be dissolved in water solutions (~0.0015 g/kg in pure H_2O at room temperature). These ampoules were fitted into standard 5 mm o.d. NMR test tubes (Wilmad-LabGlass Co., Vineland, NJ, USA) 548-PP or 10 mm tubes with liquid D_2O in the annular space. The reference samples were 0.1 M NaCl in D_2O for $^{35,37}Cl$ NMR spectra ($\Delta_{1/2} = 0.38$ Hz) and 0.1 M LiCl for $^{6,7}Li$ NMR spectra. The lock system, operated at 76.8464 MHz, allows the same magnetic field $B_0 = 11.7570$ T to be preserved. All measurements were performed at a constant temperature of 300 K. The small isotope effect when H_2O was changed by D_2O was equal to 0.02 ppm in 1 M lithium chloride solution. The rise of temperature causes deshielding effect of the lithium-7 signal by 0.0076 ppm/deg in the range 288.8–328.8 K.

High resolution $^{6,7}Li$, ^{35}Cl and ^{14}N NMR spectra were recorded on a Varian-INOVA 500 spectrometer (Varian Inc., Palo-Alto, CA, USA) equipped with sw5 (switchable) and BB10 (broad band) probes operating at 194.5544 MHz, 73.6695 MHz, 49.0491 MHz and 36.1752 MHz, respectively. For the enhancement of ^6Li signals, the $^2H(D)$ filter was omitted in the detection circuit. The primary reference solutions—$^{6,7}LiCl$ (9.7 M in D_2O), $Na^{35}Cl$ (0.1 M in D_2O), $CH_3^{14}NO_2$ (liquid) were used for standardization of lithium, chlorine and nitrogen spectra. The ^3He NMR spectra in liquid water solutions were measured by a special, homemade (Helium) probe, relative to the gas phase result, received from the extrapolation of helium shielding in gaseous mixtures CF_4-^3He and C_2F_6-^3He to the zero-point density.

The observed line width of an NMR signal at half-height was different for particular nuclei: $\Delta_{1/2}(^7Li) = 0.30 \div 0.45$ Hz with digital resolution (d.r.) = 0.18 Hz, $\Delta_{1/2}(^6Li) = 0.18 \div 0.35$ Hz with (d.r.) = 0.18 Hz, $\Delta_{1/2}(^3He) = 0.55 \div 1.15$ Hz with (d.r.) = 0.38 Hz, $\Delta_{1/2}(^{14}N) = 2.24 \div 6.45$ Hz with (d.r.) = 0.23 Hz, $\Delta_{1/2}(^{35}Cl) = 8.0 \div 9.75$ Hz with (d.r.) = 0.61 Hz. All spectra were subjected to line broadening (l.b. = 0.1) and zero-filling procedures to improve of spectral quality.

The shielding susceptibility effect for water (3.006 ppm) was calculated treating the formula $\sigma_{lb} = -4\pi/3\chi_v$ and $\chi_v = \chi_M \cdot M_p/\varrho$ where $\chi_M = -12.97$, $M_p = 18.0002$ and $\varrho = 0.999865$ g/cm^3 [36].

4. Conclusions

The nuclear magnetic moment is a very important basic parameter of each nuclide, which is a fundamental measure of nucleus magnetic structure. The lithium isotopes belong to the most investigated nuclei of the past eight decades. NMR measurements offer the highest precision in relative measurements. In this work the dipole moments of ^{6}Li and ^{7}Li were found to be $\mu(^6\text{Li}) = +0.8220453(25) \div +0.8220459(25)$ and $\mu(^7\text{Li}) = +3.2564169(98) \div +3.2564195(98)$ in nuclear magnetons (μ_N). Our new results are more valuable than those previously established by NMR spectroscopy of lithium salts in water solvents. The results are very close to the earlier given numbers measured by the ABMR method: $\mu(^6\text{Li}) = +0.8220445(10)\mu_N$ and $\mu(^7\text{Li}) = +3.2564157(30)\mu_N$. Because both lithium nuclei differ by one only neutron this indicates significant differences in the magnetic distribution in ^{6}Li and ^{7}Li nuclei, which is confirmed by the nuclear theory.

The shielding constants received from theoretical calculations were verified by our experimental investigations against other shielding constants measured simultaneously in solutions. Both kinds of procedures lead to general agreement what means that nuclear shielding and magnetic moments built the orderly set of compatible data. This provided a very important check of the consistency and reliability of the magnetic properties of lithium nuclei. The limiting factor of the nuclear magnetic moments values is therefore diamagnetic corrections.

The applicability of the dissolved helium as a shielding reference in salt water solutions is then proved. Our new measurements did not solve the problem of the different kinds of lithium water complex ions present in solutions. Further investigations into these questions are strongly recommended. Nevertheless, our experimental findings can give new input towards the understanding of subnucleonic effects in magnetic moments when compared to new theoretical calculations involving higher-order corrections. I hope that the new "helium-3" method can be easily expanded to other alkali and alkaline earth metals to find their nuclear properties. The first attempts in this field are in progress.

Conflicts of Interest: The author declares no conflict of interest. The founding sponsors had no role in the design of the study; in the collection, analyses, or interpretation of data; in the writing of the manuscript, and in the decision to publish the results.

References

1. Rabi, I.I.; Zacharias, J.R.; Millman, S.; Kusch, P. A New Method of Measuring Nuclear Magnetic Moment. *Phys. Rev.* **1938**, *53*, 318. [CrossRef]
2. Walhli, H.E. *Some Improved Measurements of Nuclear Magnetic Dipole Moments by Means of Nuclear Magnetic Resonance*; Spectroscopy Research Laboratory—Union Carbide and Carbon Corporation: Oak Ridge, TN, USA, 1954.
3. Antušek, A.; Jackowski, K.; Jaszuński, M.; Makulski, W.; Wilczek, M. Nuclear magnetic dipole moments from NMR spectra. *Chem. Phys. Lett.* **2005**, *411*, 111–116. [CrossRef]
4. Jaszuński, M.; Antušek, A.; Garbacz, P.; Jackowski, K.; Makulski, W.; Wilczek, M. The determination of accurate nuclear magnetic dipole moments and direct measurement of NMR shielding constants. *Prog. Nucl. Magn. Reson. Spectrosc.* **2012**, *67*, 49–63. [CrossRef] [PubMed]
5. Baum, E.M.; Ernesti, M.C.; Knox, H.D.; Miller, T.R.; Watson, A.M. *Nuclides and Isotopes. Chart of the Nuclides*, 17th ed.; Bechtel: San Francisco, CA, USA, 2010.
6. Puchalski, M.; Pachucki, K. Ground state hyperfine splitting in ^{6,7}Li atoms and the nuclear structure. *Phys. Rev. Lett.* **2013**, *111*, 243001. [CrossRef] [PubMed]
7. Goudsmit, S.; Young, L.A. The Nuclear Moment of Lithium. *Nature* **1930**, *125*, 461–462. [CrossRef]
8. Granath, L.D. The Nuclear Spin and Magnetic Moment of Li^7. *Phys. Rev.* **1932**, *42*, 44. [CrossRef]
9. Rabi, I.I.; Millman, S.; Kush, P.; Zacharias, J.R. The Magnetic Moments of Li^6, Li^7 and F^{19}. *Phys. Rev.* **1938**, *53*, 495. [CrossRef]

10. Rabi, I.I.; Millman, S.; Kusch, P.; Zacharias, J.R. The Molecular Beam Resonance Method for Measuring Nuclear Magnetic Moments. *Phys. Rev.* **1939**, *55*, 526–535. [CrossRef]

11. Lutz, O. The g_I-factors and the magnetic moments of alkali nuclei and the shielding of Rb^+ by water. *Phys. Lett. A* **1967**, *25*, 440–441. [CrossRef]

12. Lutz, O. Untersuchungen über die magnetische Kernresonanz von Alkalikernen in wäßriger Lösung. *Z. Naturforsch. A* **1968**, *23*, 1202–1209. [CrossRef]

13. Beckmann, A.; Böklen, K.D.; Elke, D. Precision Measurements of the Nuclear Magnetic Dipole Moments of ^{6}Li, ^{7}Li, ^{23}Na, ^{39}K and ^{41}K. *Z. Phys.* **1974**, *270*, 173–186. [CrossRef]

14. Raghavan, P. Table of nuclear moments. *Atomic Data Nucl. Data Tables* **1989**, *42*, 189–291. [CrossRef]

15. Stone, N.J. Table of nuclear magnetic dipole and electric quadrupole moments. *Atomic Data Nucl. Data Tables* **2005**, *90*, 75–176. [CrossRef]

16. Stone, N.J. *Nuclear Data Section*; IAEA, Vienna International Centre: Vienna, Austria, 2014.

17. Antušek, A.; Kędziera, D.; Kaczmarek-Kędziera, A.; Jaszuński, M. Coupled cluster study of NMR shielding of alkali metal ions in water complexes and magnetic moments of alkali metal nuclei. *Chem. Phys. Lett.* **2012**, *532*, 1–8. [CrossRef]

18. Friedman, H.L. *Ionic Solution Theory: Based on Cluster Expansion Methods*; Interscience Pub.: Miami, FL, USA, 1962; ISBN-13: 978-1124075259.

19. Rudziński, A.; Puchalski, M.; Pachucki, K. Relativistic, QED, and nuclear mass effects in the magnetic shielding of ^{3}He. *J. Chem. Phys.* **2009**, *130*, 244102. [CrossRef] [PubMed]

20. Seydoux, R.; Diehl, P.; Mazitov, R.K.; Jokisaari, J. Chemical Shifts in Magnetic Resonance of the ^{3}He Nucleus in Liquid Solvents and Comparison with Other Noble Gases. *J. Magn. Reson. A* **1993**, *101*, 78–83. [CrossRef]

21. Jaszuński, M.; Repisky, M.; Demissie, T.B.; Komorovsky, S.; Malkin, E.; Ruud, K.; Garbacz, P.; Jackowski, K.; Makulski, W. Spin-rotation and NMR shielding constantsin HCl. *J. Chem. Phys.* **2013**, *139*, 234302. [CrossRef] [PubMed]

22. Rao, J.S.; Dinadayalane, T.C.; Leszczynski, J.; Sastry, G.N. Comprehensive Study on the Solvation of Mono- and Divalent Metal Cations: Li^+, Na^+, K^+, Be^{2+}, Mg^{2+} and Ca^{2+}. *J. Phys. Chem. A* **2008**, *112*, 12944–12953. [CrossRef] [PubMed]

23. Llanio-Trujillo, J.L.; Marques, J.M.C.; Pereira, F.B. New insights on lithium-cation microsolvation by solvents forming hydrogen-bonds: Water versus methanol. *Comput. Theor. Chem.* **2013**, *1021*, 124–134. [CrossRef]

24. Rodriguez, O., Jr.; Lisy, J.M. Infrared spectroscopy of $Li^+(CH_4)_n$, $n = 1–9$, clusters. *Chem. Phys. Lett.* **2011**, *502*, 145–149. [CrossRef]

25. Mason, P.E.; Ansell, S.; Neilson, G.W.; Rempe, S.B. Neutron Scattering Studies of the Hydration Structure of Li^+. *J. Phys. Chem. B* **2015**, *119*, 2003–2009. [CrossRef] [PubMed]

26. Zeng, Y.; Wang, C.; Zhang, X.; Ju, S. Solvation structure and dynamic of Li^+ ion in liquid water, methanol and ethanol: A comparison study. *Chem. Phys.* **2014**, *433*, 89–97. [CrossRef]

27. Alam, T.M.; Hart, D.; Rempe, S.L.B. Computing the ^{7}Li NMR chemical shielding of hydrated Li^+ using cluster calculations and time-averaged configurations from ab initio molecular dynamics simulations. *Phys. Chem. Chem. Phys.* **2011**, *13*, 13629–13637. [CrossRef] [PubMed]

28. Mason, J. (Ed.) *Multinuclear NMR*; Plenum Press: New York, NY, USA, 1987; p. 56, ISBN 978-1-4613-1783-8.

29. Mohr, P.J.; Newell, D.B.; Taylor, B.N. CODATA recommended values of the fundamental physical constants: 2014. *Rev. Mod. Phys.* **2016**, *88*, 035009. [CrossRef]

30. AL-Khafiji, K.S.; Selman, A.M.; Al-Shebly, S.A.K. Calculation of the Standard Deviation and Nuclear Magnetic Shielding Constant for Lithium Atom. *J. Kerbala Univ.* **2008**, *6*, 107–110.

31. Ormand, F.T.; Matsen, F.A. Nuclear Magnetic Shielding Constants for Several 2-, 3-, and 4-Electron Atoms and Ions. *J. Chem. Phys.* **1959**, *30*, 368–371. [CrossRef]

32. Malli, G.; Froese, C. Nuclear Magnetic Shielding Constants Calculated from Numerical Hartree-Fock Wave Functions. *Int. J. Quantum Chem.* **1967**, *1*, 95–98. [CrossRef]

33. Guan, X.-X.; Wang, Z.-W. Calculation of the Zeeman effect in the $^2S_{1/2}$, $n^2P_{1/2}$, and $n^2P_{3/2}$ ($n = 2, 3, 4,$ and 5) states of lithium atom. *Phys. Lett. A* **1998**, *244*, 120–126. [CrossRef]

34. Cockrell, R.C. Ab Initio Nuclear Structure Calculations for Light Nuclei, 2012. Ph.D Thesis, Iowa State University, Ames, IA, USA, 30 January 2012.

35. Borremans, D.; Balabanski, D.L.; Blaum, K.; Geithner, W.; Gheysen, S.; Himpe, P.; Kowalska, M.; Lassen, J.; Lievens, P.; Mallion, S.; et al. New measurement and reevaluation of the nuclear magnetic and quadrupole moments of ^{8}Li and ^{9}Li. *Phys. Rev. C* **2005**, 044309. [CrossRef]
36. Haynes, W.M. (Ed.) *CRC Handbook of Chemistry and Physics*, 96th ed.; CRC Press: Boca Raton, FL, USA, 2015; ISBN 978-1482260960.

magnetochemistry

MDPI

Review

Characterization of Halogen Bonded Adducts in Solution by Advanced NMR Techniques

Gianluca Ciancaleoni

Dipartimento di Chimica e Chimica Industriale, Università degli Studi di Pisa, via Giuseppe Moruzzi, 13, 56124 Pisa, Italy; gianluca.ciancaleoni@unipi.it; Tel.: +39-050-221-9233

Received: 7 September 2017; Accepted: 20 September 2017; Published: 25 September 2017

Abstract: In the last 20 years, a huge volume of experimental work into halogen bonding (XB) has been produced. Most of the systems have been characterized by solid state X-ray crystallography, whereas in solution the only routine technique is titration (by using ^{1}H and ^{19}F nuclear magnetic resonance (NMR), infrared (IR), ultraviolet–visible (UV–Vis) or Raman spectroscopies, depending on the nature of the system), with the aim of characterizing the strength of the XB interaction. Unfortunately, titration techniques have many intrinsic limitations and they should be coupled with other, more sophisticated techniques to provide an accurate and detailed description of the geometry and stoichiometry of the XB adduct in solution. This review will show how crucial information about XB adducts can be obtained by advanced NMR techniques, nuclear Overhauser effect-based spectroscopies (NOESY, ROESY, HOESY . . .) and diffusion NMR techniques (PGSE or DOSY).

Keywords: halogen bonding; supramolecular chemistry; NMR spectroscopy

1. Introduction

The attractive interaction between halogen atoms and nucleophilic species (hereafter called "halogen bond" or "XB") has drawn the attention of chemists since 1954, when Hassel discovered that the Br–O distance in the 1:1 adduct between Br_2 and dioxane was only 2.71 Å, smaller than the sum of the corresponding van der Waals radii (3.35 Å) [1]. The adduct was immediately recognized as a charge-transfer pair [2], but the details of the interactions were largely unknown. To date, a huge body of experimental and theoretical data [3–7] has been collected, producing a detailed knowledge of the XB interaction and making it a routinely used tool in many fields of chemistry.

In 2013, the International Union of Pure and Applied Chemistry (IUPAC) released a definition for XB, stating that: "A halogen bond occurs when there is evidence of a net attractive interaction between an electrophilic region associated with a halogen atom in a molecular entity and a nucleophilic region in another, or the same, molecular entity" [8]. The easiest way to evidence a net attractive interaction is to measure the interatomic distance, as Hassel did in 1954, between the halogen and the nucleophile in the solid-state. For this reason, X-ray crystallography is the main technique used to characterize halogen-bonded adducts. The large amount of structures produced allows many interesting contributions based on the analysis of structural databases [9–13]. And, indeed, most of the applications of XB are in the materials science: porous systems [14–16], liquid crystals [17], light-emitting materials [18,19] and magnetic materials [20,21] are only some applicative fields fruitfully explored with XB-based materials.

More recently, XB found applications also in solution, mostly for anion recognition [22–24] and catalysis [25–27], with the difference that structural characterization in solution is much more complicated than in the solid state, since most of the experimental techniques are less "direct" than X-ray crystallography. The general method, in the case of an intermolecular interaction, is to monitor a property of the system, which often (but not necessarily) is a spectroscopic observable, in the absence and the presence of that interaction. In most cases, information is derived under the hypothesis that the entire effect on the observable is due to the interaction under examination.

For example, the most used technique for the characterization of XB adducts in solution is titration [28]. It can be performed by using any NMR-active nucleus of the pair (generally ^{1}H or ^{19}F for their high sensitivity and favorable isotope abundance), even if, depending on the nature of the system, Raman, IR and UV–V is spectroscopies [29,30] can also be used [31]. But the underlying assumption is the same: the effect of the increasing concentration of one component on the chosen property (nuclear magnetic resonance (NMR) chemical shift, infrared (IR) absorption frequency, or whatever) of the second component is supposed to be entirely due to XB. But if two different kinds of adducts were present in solution, one held by XB and the other by a different weak interaction, the disentanglement of the two effects in the experimental data would not be straightforward.

Another critical issue of titration is the stoichiometry of the adduct: each kind of stoichiometry (1:1, 2:1 and so on) requires a different equation for the fitting of titration data, but it can happen that more than one equation satisfactorily fits experimental data, leaving the question to the discretion of the user. A partial solution to the problem is the use of Job's plot, but it is not entirely reliable [28,32].

In both cases, for the structure and the stoichiometry of the adduct, assumptions are generally made on the basis of common sense, or, if the solid-state structure is available, it is just assumed that in solution the adducts have the same structure/stoichiometry. Given the importance of a correct, detailed and accurate characterization of XB adducts, especially when applications in solution are involved, more sophisticated tools should be employed and coupled with classical titrations for a thorough description of the system. Supramolecular chemistry often took advantage of advanced NMR techniques [33–37], especially the nuclear Overhauser effect (NOE)-based NMR spectroscopies [38] and diffusion NMR techniques [39–41]. The former allows the researcher (i) to verify the presence of an adduct (or a particular conformation for intra-molecular interactions) just by detecting a NOE between the nuclei of one fragment with the nuclei of the other; and (ii) to gain information on the relative orientation of the two fragments in the adduct. In a complementary way, the latter allows the direct measurement of the hydrodynamic volume (V_H) of the species in solution, thereby revealing if and how much a single species is involved in the formation of supramolecular adducts. Such advanced NMR techniques are, surprisingly, not routinely used in the characterization of XB adducts, but when they are employed crucial information on the behavior of the XB donors and acceptors in solution can be derived, allowing for the correct interpretation of other experimental data (titrations, for example).

In this review, after some examples on the application of 1D NMR techniques, the basic principles of NOE-based spectroscopies and diffusion techniques will be briefly presented. The paper will then show how informative and useful they can be, through a critical discussion of a selection of recently published papers.

2. 1D Nuclear Magnetic Resonance (NMR) Techniques: Applications

Most of the XB systems studied by solution NMR merely take advantage of standard 1D NMR techniques, as the measurement of the chemical shift in different solvents or in the presence of increasing concentration of another component (titration, or Job's plot). Indeed, the information that can be extracted by this reliable, fast and simple technique is impressive.

Already by 1979, Bertrán and Rodríguez had used ^{1}H NMR spectra to demonstrate the presence of XB in solution [42], using the difference of δ_{CH} in cyclohexane (a very weakly interacting solvent) and in the solvent of interest. Interestingly, very good correlations arose from the results of iodo- and bromoform, whereas the correlations between the results of iodo- and chloroform were poor. This was likely due to the weakness of the chloroform/solvent XB and to the strength of chloroform/solvent hydrogen bonding (HB).

Similarly, Metrangolo and Resnati compared the XB interaction between halogenated hydrocarbons and different solvents by means of ^{19}F NMR spectroscopy [43], measuring for each solvent the value of $\Delta\delta_{CF2X}$, which is the difference between the chemical shift of the fluorine in the -CF_2X moiety in pentane and in the solvent of choice. Results showed that the interaction depends on the nature of X ($\Delta\delta_{CF2I} > \Delta\delta_{CF2Br} > \Delta\delta_{CF2Cl}$) and the solvent; that $\Delta\delta_{CF2I}$ decreases passing from primary to tertiary amines; and that, regarding pyridine derivatives, the methyl groups in positions 2 and 6 decreases $\Delta\delta_{CF2I}$ with respect to the unsubstituted pyridine, likely due to the steric hindrance, whereas electron-donating (withdrawing) groups in position 4 tend to increase (decrease) $\Delta\delta_{CF2I}$.

Clearly, other nuclei can also provide useful information, especially in those cases in which there are no hydrogen or fluorine nuclei close to the interaction site, or their response is too slight to be accurate. Erdelyi and co-workers demonstrated the applicability of ^{15}N NMR spectroscopy, which is useful for medium/strong XBs [44,45], but is not accurate enough for weak ones (in particular, pyridine and para-substituted halobenzenes) [46]. On the same topic, Philp and co-workers demonstrated that in the case of a iodotriazole having a pending pentafluorophenyl moiety, the ^{19}F δ is almost insensitive to the addition of a pyridine, making titration unsuccessful; whereas monitoring the chemical shift of the nitrogen in the pyridine at increasing concentrations of the iodotriazole (through the ^1H-^{15}N heteronuclear multiple-bond correlation spectroscopy (HMBC) 2D NMR technique to shorten the acquisition time) led to the determination of the association constant (K_a) [47]. On the other hand, Goroff and co-workers demonstrated the usefulness of ^{13}C NMR spectroscopy and, analyzing the spectra of two iodoalkynes [48], showed that the frequency of the α-carbon is not strictly correlated to the polarity of the solvent, but to the solvent basicity.

The analysis of ^{13}C NMR spectra led also to another interesting result: in a paper by Wang and co-workers, published in 2012, the authors analyzed the trend of ^{13}C NMR δ of C_6F_5X (X = Cl, Br) with the concentration in different solvents, concluding that in some cases XB is not the only weak interaction active in solution [49]. Indeed, depending on the nature of X and of the solvent, a competitive lone pair-π interaction is also possible, favored by the electronic depauperation of the aromatic ring due to the presence of the fluorine atoms [50,51].

It is also noteworthy that nuclei different to ^1H are of fundamental importance in solid-state NMR studies on XB [52–54].

For quantitative information, as mentioned in the Introduction, ^1H and ^{19}F NMR titrations are generally very useful. For example, Cabot and Hunter published a systematic study on the values of K_a for many IC_6F_{13} (**11**)-B XB adducts [55], where B is a Lewis base, such as *tri-n*-butylphosphine oxide, an amine or pyridine, in three representative solvents: benzene, CCl_4 and chloroform. Notably, in the latter, log K_a is positive only when B = quinuclidine, 1,4-diazabicyclo[2.2.2]octane (**DABCO**) or piperidine, an indication that the XB between neutral species is generally measurable only in apolar solvents. In the case of anionic XB acceptors, K_a can also be large in polar solvents [56,57]. The implicit approximation of all the titrations is that the entire effect is due to XB while, as mentioned, other non-XB adducts can be present. Obviously, this is especially important for weaker XBs, for which the approximation is less acceptable.

Another 1D NMR technique that is used in the characterization of XB in solution is Job's plot. Such a technique is often used in supramolecular chemistry to elucidate the stoichiometry of an adduct, and is based on the concept that the concentration of a D_mA_n complex is at maximum when the [D]/[A] ratio is equal to m/n. Unfortunately, this method is reliable under two conditions: (i) D_mA_n is the only adduct present in solution; and (ii) the two components do not self-aggregate [28,32]. Sometimes, if the X-ray structure of the adduct is available, it can be assumed that the stoichiometry in solution is the same, but this assumption is not always safe (see later). Clearly, an accurate determination of the stoichiometry is always desirable, since the equation used to fit the experimental data of a titration depends on the stoichiometry of the adduct. In principle, titration data could be fitted with many equations and the stoichiometry could be decided on the basis of the goodness of the fitting results but, especially for many-body adducts, it is possible that more than one equation satisfactorily fit the data, leaving the final decision to the user. In some cases, it is difficult to find a binding model that fits the experimental data, because of a non-conventional trend in the data, as in the case of sulfate in reference [58].

Many of the potential problems exposed up to now can be solved by combining 1D NMR techniques with advanced NMR ones, as the next sections will elucidate. Anyway, it is important to underline that the verb "combine" has been used, and not "substitute", as 1D NMR techniques are rapid, reliable, do not require a special technical training and, in many cases, more advanced techniques merely corroborate the results obtained with 1D methods.

3. Advanced NMR Techniques: Theory

3.1. Nuclear Overhauser Effect

The nuclear Overhauser effect (NOE) arises from the dipole–dipole interaction between NMR-active nuclei, and depends on the competition between multiple- and zero-quantum relaxation mechanisms [59,60]. If we consider two not scalarly-coupled spins, namely I and S, separated by the distance r_{IS}, they have four energy levels, according to the spin states of the two spins ($\alpha\alpha$, $\alpha\beta$, $\beta\alpha$ and $\beta\beta$). The rate constant for the transition between the $\alpha\alpha$ and the $\beta\beta$ states is denoted W_{2IS} (the "2" indicates it is a double-quantum transition), whereas the rate constant for the transition between the $\alpha\beta$ and the $\beta\alpha$ states is denoted W_{0IS} (the "0" indicates it is a zero-quantum transition). The difference between W_{2IS} and W_{0IS} is called the cross-relaxation rate constant and it is generally abbreviated with the symbol σ_{IS}. The basis of NOE is given by the Solomon equation (Equation (1)).

$$\frac{d(I - I^0)}{dt} = -R_I\left(I - I^0\right) - \sigma_{IS}\left(S - S^0\right)$$ (1)

where the apex "0" indicates the equilibrium state; R_I is the self-relaxation rate constant of the spin I (the sum of all the possible rate constants W); and the transition between I (or S) and $2IS$ are neglected (formally, $\Delta_I = \Delta_S = 0$). Equation (1) says that when S spin magnetization deviates from the equilibrium, the I spin magnetization will change proportionally to σ_{IS} and to the extent of the deviation of the S spin from the equilibrium. Clearly, the cross-relaxation term must be different to zero and, since the two spins are not scalarly coupled, this happens only when there is a dipolar relaxation between I and S. In the steady state NOE, and for a system isotropically tumbling in solution, Equation (1) can be written as in Equations (2)–(4).

$$NOE_I\{S\} = \frac{I - I^0}{I^0} = \frac{\gamma_S}{\gamma_I}\frac{\sigma_{IS}}{\rho_{IS}}$$ (2)

$$\sigma_{IS} = \left(\frac{\mu_0}{4\pi}\right)^2\frac{\hbar^2\gamma_I^2\gamma_S^2}{10}\left(\frac{6\tau_c}{1 + (\omega_I + \omega_S)^2\tau_c^2} - \frac{\tau_c}{1 + (\omega_I - \omega_S)^2\tau_c^2}\right)r_{IS}^{-6}$$ (3)

$$\rho_{IS} = W_{2IS} + 2W_{1I} + W_{0IS}$$
$$= \left(\frac{\mu_0}{4\pi}\right)^2\frac{\hbar^2\gamma_I^2\gamma_S^2}{10}\left(\frac{6\tau_c}{1 + (\omega_I + \omega_S)^2\tau_c^2} + \frac{3\tau_c}{1 + \omega_I^2\tau_c^2}\right.$$
$$\left. - \frac{\tau_c}{1 + (\omega_I - \omega_S)^2\tau_c^2}\right)r_{IS}^{-6}$$ (4)

where μ_0 is the permeability constant in a vacuum; $\hbar$ is the Planck's constant divided by 2π; τ_c is the rotational correlation time; ω_I and ω_S are the resonance frequencies of I and S nuclei, respectively; and ρ_{IS} is the dipolar longitudinal relaxation rate constant. The dependence of both σ_{IS} and ρ_{IS} on r^{-6} implies that the steady-state NOE cannot be directly related to the internuclear distances. For an estimation of the latter, the measurement of the kinetics of the NOE buildup, i.e., the measurement of NOE at different values of mixing time (τ_m), is needed (Equation (5))

$$NOE_I\{S\}(\tau_m) = e^{-(R - \sigma_{IS})\tau_m}\left(1 - e^{-2\sigma_{IS}\tau_m}\right)$$ (5)

where R represents the total longitudinal relaxation rate constants of both I and S spins, assumed to be equal. If quantitative information on r_{IS} is needed, the experimental data should be collected in an extensive range of τ_m and fitted using Equation (5) (Figure 1); or, if only small values of τ_m are used (linear buildup), the data can be fitted with a straight line, whose slope will be equal to $2\sigma_{IS}$.

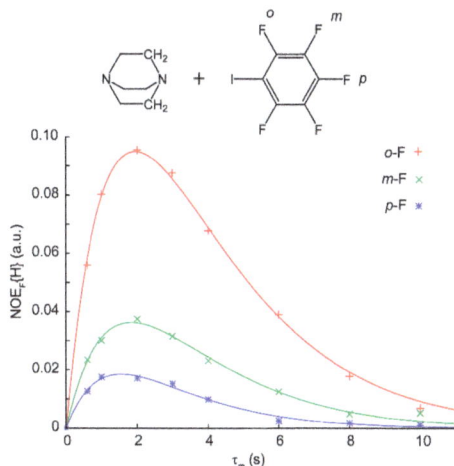

Figure 1. Adapted with permission from reference [61]. Experimental data relative to the intensity of nuclear Overhauser effect (NOE) on the fluorine nuclei of iodopentafluorobenzene as a function of τ_m after the irradiation of CH_2 protons of **DABCO**. The straight lines are the best fitting functions using Equation (5). Copyright 2015 Wiley-VCH Verlag GmbH & Co. KGaA.

Now, r_{IS} can be evaluated after an evaluation of τ_c, the other variable present in Equation (3), or by comparing σ_{IS} with σ_{AB}, where A and B are a couple of spins of the same nature, whose distance is known and with the same rotational correlation time [62,63]. Under these circumstances, Equation (6) allows the measurement of r_{IS}.

$$\frac{\sigma_{IS}}{\sigma_{AB}} = \left(\frac{r_{IS}}{r_{AB}}\right)^{-6} \tag{6}$$

The simplest pulse sequence to perform 2D NOESY is shown in Figure 2. The first 90° rf pulse rotates the magnetization of the spin I on the xy plane, where it can evolve according to its frequency ω_I and the second 90° rf pulse turns a part of the magnetization back on the z-axis. The size of this magnetization depends on ω_I and t1 and, therefore, it is said to be a frequency-labelled magnetization. During the mixing time (τ_m), this magnetization can be partly transferred to the spin S by NOE or chemical exchange, keeping its dependence from ω_I in the process. The last 90° rf pulse rotates this z-magnetization back onto the y-axis, where it can be read. The part of the magnetization that did not undergo NOE or exchange will evolve again according to its ω_I, giving a signal at $[\omega_I, \omega_I]$ (diagonal peak), whereas the part that underwent NOE or exchange to spin S will evolve according to ω_S, giving a signal at $[\omega_I, \omega_S]$ (off diagonal peak). Such a simple pulse sequence is not used anymore and many other, more complex sequences have been developed.

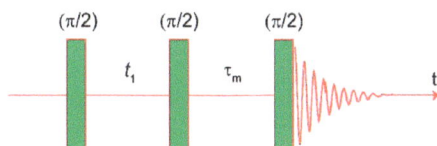

Figure 2. Pulse sequence for the NOE experiment (basic version).

The NOE-based techniques (NOESY, its heteronuclear version HOESY or the experiment performed under spin-locked conditions, ROESY) are therefore of primary importance in the structural elucidation of an adduct, either qualitatively, since the simple detection of a NOE between the nuclei of two molecular entities is already enough to demonstrate the presence of an intermolecular adducts in solution [64], or quantitatively, since the quantification of different NOEs can give precious information on the internal structure and the geometry of the adduct [65,66]. In Section 4, we will see some examples of how NOE spectroscopies can be practically applied in this sense.

Obviously, the potential of the NOE technique goes far beyond the examples here reported, and its possibilities and limitations can also be effectively explored by coupling experimental data with model theories [66–68].

3.2. Diffusion NMR

Comparing the intensity of a NMR signal in the absence and presence of a gradient of the magnetic field along the z axis $G(z)$, the former is always more intense than the latter. What is responsible for this attenuation is the translational self-diffusion [69,70], that is, the net result of the thermal motion induced by the random Brownian motion experienced by particles or molecules in solution. Starting from this, it can be understood that by performing a series of spectra at different values of G, the translational self-diffusion coefficient (D_t) can be directly measured plotting the signal attenuation as a function of G.

In more detail, the basis of diffusion NMR techniques rely on the fact that the Larmor frequency (ω) depends on the strength of the magnetic field and on the gyromagnetic ratio (γ) of the nucleus of choice. In the presence of a homogenous magnetic field B_0, ω has the same value at every position of the sample (Equation (7)).

$$\omega = \gamma B_0 \tag{7}$$

If a second magnetic field, whose intensity linearly depends on z ($G(z)$, [T m^{-1}]) is added to B_0, homogeneity is lost and Equation (7) can be written as a function of the z coordinate (Equation (8)).

$$\omega = \gamma(B_0 + G(z)\cdot z) \tag{8}$$

Now, ω depends on the position of the nucleus, but how this labeling can be used to measure D_t requires a short discussion of the actual pulse sequence.

The simplest pulse sequence is a modification of the spin-echo sequence published by Hahn in 1950 [71], proposed by Stejskal and Tanner in 1965 (Figure 3) [72].

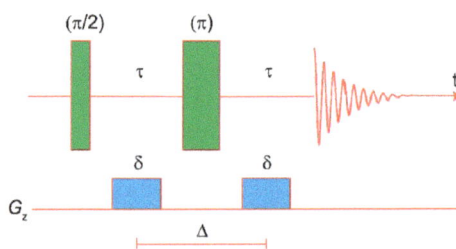

Figure 3. Pulse sequence for the pulsed-field gradient spin echo (PGSE) experiment (basic version).

The first act is a 90° rf pulse that rotates the magnetization on the xy plane, where it undergoes many dephasing phenomena: chemical shift, hetero- and homonuclear J-coupling evolution, and spin-spin transverse relaxation (T_2). Furthermore, the presence of a gradient introduces an additional dephasing component. At $t = \tau$, a 180° rf pulse is applied. This inverts the precession direction and the dephasing turns into a rephasing phenomenon. During this time, another gradient is applied, exactly equal to the first one, with the aim of recreating the conditions of the first τ

period and, consequently, generating an echo at $t = 2\tau$. The net result is the conventional spectrum, distorted for the *J*-coupling and weighted for the T_2. But the nuclei that in the first τ period were in the position z', were in the second τ period in the position z" because of the Brownian motion. Consequently, they experience two different magnetic fields during the dephasing and the rephasing periods, which causes an incomplete rephasing and, therefore, an attenuation of the signal intensities. Such attenuation depends on the difference between z' and z" and, since small molecules diffuse faster than large ones, the attenuation for the former will be more severe than for the latter, leading to a discrimination of the species in solution depending on their D_t (Figure 4).

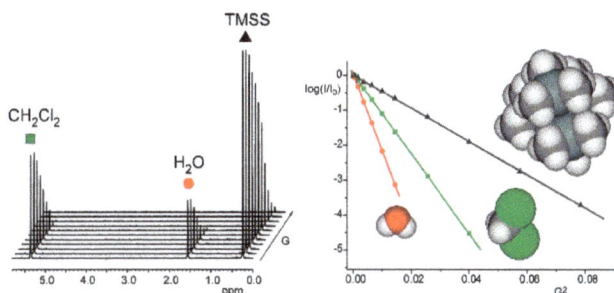

Figure 4. Reproduced from ref. [73]. Left: A series of 1D nuclear magnetic resonance (NMR) spectra recorded at different *G* values. Right: Plot of the $\log(I/I_0)$, where *I* and I_0 are the intensities of the signals in the presence and in the absence of *G*, respectively, vs. G^2. Note the inverse proportionality between the slope of the fitting line and the molecular size.

Equation (9) describes the relationship between the intensity of a signal (*I*) and D_t.

$$I(2\tau) = I(2\tau)_{G=0} \cdot e^{\left(-\gamma^2 G^2 \delta^2 D_t (\Delta - \frac{\delta}{3})\right)} \tag{9}$$

where *I* and I_0 are the intensities of a signal at time 2τ in the presence and in the absence of *G* [74]. By performing a series of spectra with different values of *G*, the experimental data can be fitted and D_t can be evaluated. This technique is generally called pulsed-field gradient spin echo (PGSE). Also, in this case many other more complex pulse sequences are available. Processing the data as a 2D spectrum with chemical shifts on the F2 axis and diffusion constants on the F1 axis, a diffusion-ordered NMR spectroscopy (DOSY) plot is obtained.

The subsequent passage from D_t to the hydrodynamic volume is possible through the Stokes–Einstein equation [75] successively modified by Chen for medium-size molecules [76]. The experimental conditions and data-processing are particularly important to obtain accurate values of D_t, and useful instructions can be found in reference [73], which illustrates, among other things, the importance of the internal standard and what to do in the presence of non-spherical species.

Taking advantage of this, diffusion NMR techniques, can be used in different ways: for example, if the NMR sample is contaminated with one or more solvents whose signals overlap with the compound of interest, the application of a small gradient will eliminate, partially at least, such signals [77]. More quantitative information can be useful in determining the molecular weight distribution for polymers [78], the purity and composition of functionalized carbon nanotubes [79], the formation of ion pairs and quadruples [80–83], or, in the case of neutral species, of supramolecular adducts [84,85].

4. Advanced NMR Techniques: Applications

4.1. Nuclear Overhauser Effect

The first paper employing a NOE analysis for the structural characterization of a XB system in solution, to the best of my knowledge, was published in 2004 by Tatko and Waters [86]. In this paper, the authors synthesized a model β-hairpin peptide and demonstrated that, in water, the substitution of a hydrogen with a halogen has a stabilizing effect on the folded conformation ($\Delta\Delta G = -0.12, -0.34, -0.47$ and -0.54 for F, Cl, Br and I, respectively; $\Delta\Delta G = -1.01$ kcal/mol in the case of substitution of two hydrogens with iodine atoms). The NOE analysis was important to ensure that (i) the conformation was folded; and (ii) the iodine was actually facing the aromatic ring of the N-terminal phenylalanine, allowing the authors to demonstrate the presence of the halogen-π interaction. They also studied the impact of the substitution on the thermodynamic parameters of the folding, finding out that the presence of iodine provides an enthalpic driving force but also an additional entropic cost. Combining these results with other thermodynamic data, the authors concluded that dispersion forces are responsible for the improved stability.

In 2010, Beer and coworkers synthesized some interlocked host systems in which a chloride is held into the macrocycle **1** (Figure 5) by two HBs and, at the same time, a functionalized imidazolium **2a-e** interacts with the chlorine by XB or HB [87].

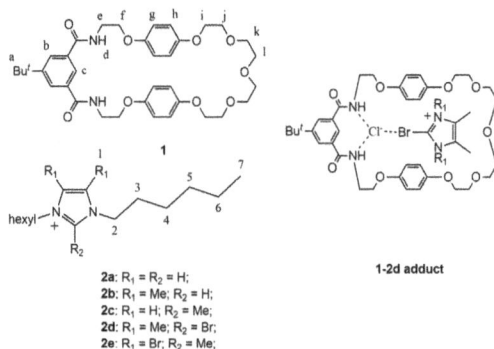

Figure 5. Numbering of the compounds studied in reference [87] and structure of the **1-2d** adduct.

The whole system is held together by a mixture of weak interactions (ion pairing, HB, π-π stacking, XB), but the orientation of the imidazolium moiety and the strength of the K_a is dictated by the functional groups. In the case of **1-2a** or **1-2b**, the hydrogen in the 2-position of the imidazolium interacts with the chloride, in both cases with a K_a around 95 M^{-1}. In the case of **1-2c**, the 2-position is occupied by a methyl group and the HB is not possible any more. Despite this, K_a is much higher than before (245 M^{-1}). Analyzing the pattern of chemical shifts, the authors conclude that for **1-2c**, the orientation of **2c** is different than before and now the hydrogens in position 4 and 5 interact with the chloride; while the methyl in position 2 establishes an additional weak HB with the oxygens, explaining the increased value of K_a. The authors employed ^{1}H ROESY for the characterization of **1-2d**, demonstrating not only that the adduct was formed, but also that the bromine in position 2 is facing the chloride (Figure 6). Indeed, the two methyl moieties in position 4 and 5 of **2d** (number 1 in Figure 6) show NOE intermolecular contacts with the protons g/h/k/l/j, which are far from the amine moieties, whereas there are no NOE contacts with the protons e/f/d, which are close to the amine moieties.

Notably, K_a(**1-2d**) = 254 M^{-1}, demonstrating that in some cases XB can be more efficient than HB in the construction of supramolecular adducts. The insertion of bromine in positions 4 and 5 or the substitution of Cl$^-$ with PF$_6$$^-$ did not lead to the formation of the adduct.

The same research group recently used systems similar to **1** to create catenane systems [88], also in this case using multiple weak interactions together to form a supramolecular adduct and, in a second step, closing the second cycle through a Grubbs-catalyzed ring closing metathesis. The final catenane system selectively binds iodide and bromide anions, whereas there is no evidence of binding in the presence of acetate anions.

Inspired by the Wang's paper [49], Ciancaleoni and others recently used the ^{19}F, ^{1}H HOESY technique to study in detail the structure in solution of small and well-known XB adducts and find other evidences of the contemporary presence of XB and non-XB adducts [61]. Given the excellent electron-withdrawing properties of fluorine, this nucleus is present in many XB systems, making ^{19}F, ^{1}H HOESY a technique with a great potential. In fact, given the high directionality of XB [89,90], the geometry of a XB adduct and, therefore, the NOE intermolecular pattern, can be accurately predicted. For this, any deviation of the experimental pattern from the predicted one can be due to the presence of other weak interactions, and, consequently, of adducts with a different geometry than that of the XB one.

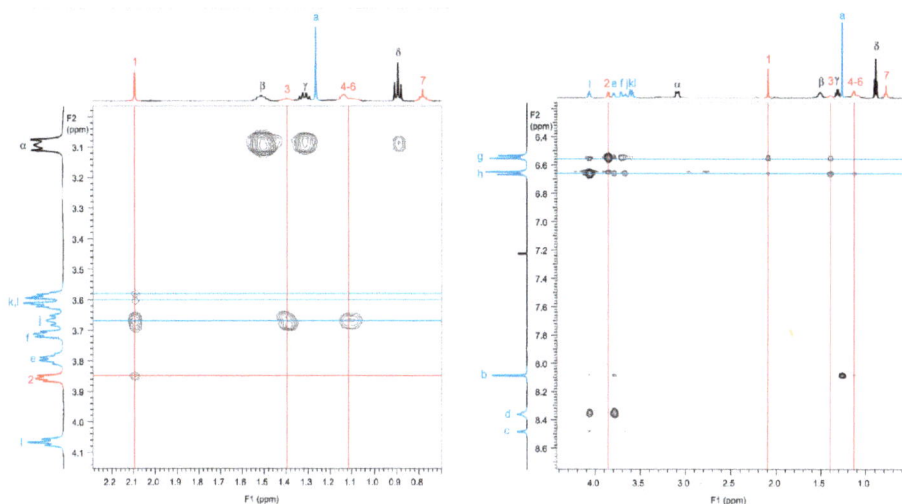

Figure 6. Reproduced from reference [87]. Through-space coupling between protons on imidazolium chloride **2d** and the posterior polyether protons of **1** (left) and between protons on imidazolium chloride **2d** and the hydroquinone protons of **1** (right). Blue and red peaks refer to compound **1** and **2d**, respectively. Labels in greek refer to the tetrabutylammonium cation. Copyright 2010 Wiley-VCH Verlag GmbH & Co. KGaA.

For example, considering **DABCO** and perfluorohexyl iodide (**I1**), the α-F/-CH$_2$- (whereas α refers to the fluorine atoms germinal to the iodine) heteronuclear NOE contact is a good indicator for the presence of the XB adduct. Experimentally, the α-F/-CH$_2$- contact is clearly visible (solvent = benzene-d$_6$), but also the γ-F/-CH$_2$- contact is strong, even stronger than the α-F/-CH$_2$- one, whereas the β-F/-CH$_2$- is very weak and all the others are almost undetectable. The reason why the γ-F/-CH$_2$- contact is stronger than the α-F/-CH$_2$- one can be rationalized by using density functional theory (DFT) calculations: the structure in which the chain is folded is almost isoenergetic to that with the unfolded chain (ΔE = 0.6 kcal/mol at B3LYP-D3/TZVP level of theory). Because of the folding, the γ-F results to be closer to -CH$_2$- than α-F (4.3 and 5.4 Å, respectively, according to DFT-optimized geometries). Therefore, the two conformers of the XB adduct can explain all the experimental NOE contacts. On the other hand, other non-XB adducts can be modeled by DFT, but they are higher in energy (ΔE = 5.6 kcal/mol) and can likely be neglected.

The same technique was also applied to the pair **DABCO**/pentafluoroiodobenzene (**I2**): given the structure of the XB adduct, only the *ortho*-F/-CH$_2$- NOE contact should be visible, because it is the only one for which the predicted internuclear distance is reasonable for NOE (5.1, 7.8 and 9.3 Å for *ortho*-, *meta*- and *para*-CH$_2$ distances, respectively). But experimentally, *meta*-F and *para*-F also give measurable contacts (Figures 1 and 7), the presence of which demonstrates that a portion of the adducts has a different structure and, therefore, is held by other interactions than XB. DFT calculations showed that the lone pair/π adduct is less stable than the XB one (ΔE = 3.0 kcal/mol at B3LYP-D3/TZVP level of theory), but the internuclear H/F distances are extremely short: 3.3–3.5 Å. Indeed, we need to remember that the intensity of an intermolecular contact depends, among other things, on two parameters: the average H/F distance in that adduct and the concentration of the structure in solution. The former can be derived by theoretical geometry optimizations, which are quite accurate [91], whereas the latter can be approximately evaluated by combining the intensity of NOE contacts, or better, their σ_{HF} constant evaluated by the NOE build-up (Figure 1), and the DFT-derived internuclear distances. According to this strategy, in the **DABCO**/**I2** mixture (solvent = benzene-d$_6$) 4% of the adducts are held by lone pair/π interactions and the remaining 96% by XB.

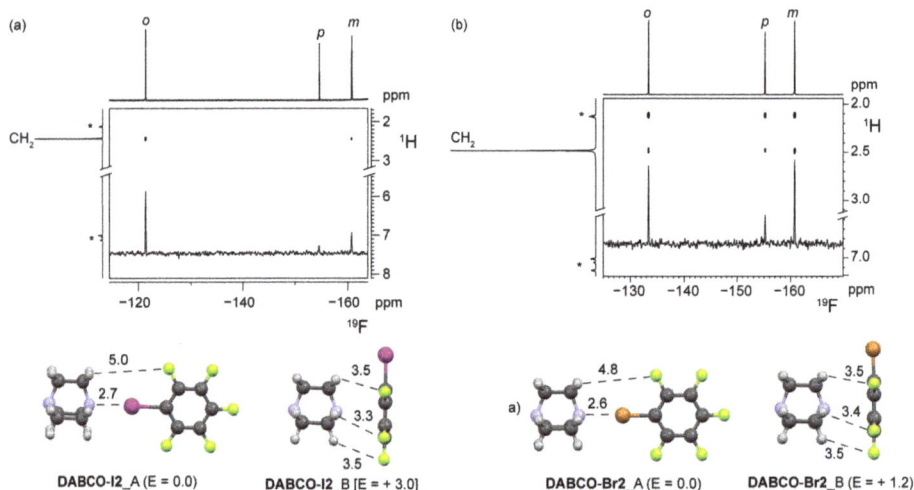

Figure 7. Adapted with permission from reference [61]. ^{19}F, ^{1}H HOESY NMR spectrum of a mixture of (a) **DABCO** and **I2** and (b) **DABCO** and **Br2**. The trace is relative to the frequency of the -CH$_2$-. Asterisks denote residual solvent peaks. Bottom: DFT-optimized geometries for **DABCO**/**I2** and **DABCO**/**Br2** adducts with relevant distances and relative energies [kcal mol^{-1}, B3LYP-D3/TZVP level of theory]. Copyright 2015 Wiley-VCH Verlag GmbH & Co. KGaA.

By using pentafluorobromobenzene (**Br2**), the *meta*-F/-CH$_2$- NOE contact is as intense as the *ortho*-F/-CH$_2$- one (Figure 7). The relative concentration of XB and non-XB adducts can be evaluated as 56:44. In this case, it is clear that a titration is not enough to accurately establish the strength of the XB, since the contribution of the two structures to the physical property (a NMR chemical shift, a UV–Vis absorbance peak ...) should firstly be disentangled before the fitting [92].

In 2017, Jiang and co-workers synthesized a series of alanine-based halogen-substituted bilateral N-amidothioureas containing two β-turns structural motifs. An X-ray of the crystal structure demonstrated that the monomer self-organizes in supramolecular helices held together by iodine ...

π interactions, while other inter-helices XBs create a complex network [93]. Substituting iodine with chloride, helical structures do not form in the solid state. Comparing X-ray structures with NOESY spectra, the authors deduced that when X = Cl (**LL-ACl**), only intramolecular NOE contacts are visible and, in particular, the e–f contact is clearly visible (internuclear e–f distance in the solid-state structure: 3.189 Å) and the e–g is not (internuclear e–g distance in the solid-state structure: 4.348 Å, Figure 8).

On the contrary, when X = I (**LL-AI**), also the e–g contact is visible and, therefore, it should be due to an intermolecular contact rather than intramolecular (inter- and intramolecular internuclear e–g distances in the solid-state structure: 2.368 and 4.550 Å, respectively, Figure 8). The authors underline but do not comment upon the absence of the e–f contact for which both inter- and intramolecular internuclear e–g distances in the solid-state structure (3.686 and 3.106 Å, respectively) should be short enough for a NOE contact. The existence of supramolecular helices is also supported by the peak intensities of **LL-AI**, which are lower, at the same concentration, than those of **LL-ACl**. This has been explained by the fact that the formation of the supramolecular helix leads to peak broadening and, therefore, invisibility in the NMR spectrum [94].

Extending the same topic, a double helix held by XBs has also been characterized in the solid state [95], whereas Berryman and co-workers succeeded in synthesizing a triple helix stabilized by XB. For the latter, a complete characterization in solution is also available, including the presence of inter-strand NOE contacts, to verify the helicity and measure the diffusional coefficient in order to further characterize the adduct.

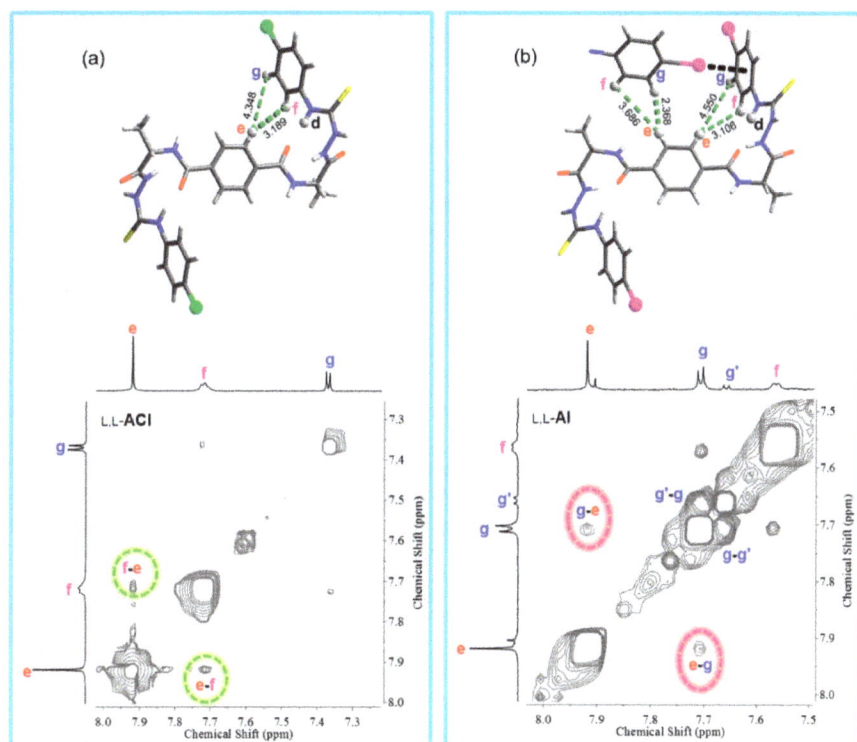

Figure 8. Reproduced from ref. [93]. Sections of ^{1}H-NOESY spectra showing couplings between protons in phenyl rings in **LL-ACl** (a) and **LL-AI** (b). On top, crystal structures are also shown. Copyright 2017 American Chemical Society.

In a very recent work by Erdélyi and co-workers, the authors took advantage of XB to stabilize the β-hairpin conformation in an artificial peptide model system [96]. After optimization of the amino acid sequence and the synthesis of the actual system, all the signals of the NMR spectrum were assigned by a total correlation spectroscopy (TOCSY) NOESY strategy, while the presence of two β-turns were confirmed by the measurement of the $\Delta\delta_{NH}/\Delta T$ coefficients. Structural information was obtained by measuring the NOE buildup for all the correlations, in order to ascertain the average distance ratios, and the $^3J_{C\alpha H,NH}$ values, which were converted in average dihedral angles through a version of the Karplus equation. The experimental data were coupled with restraint-free Monte Carlo calculations, and all the collected data were used as input for the NAMFIS analysis. The latter consists of the generation of a "complete" set of conformation that can potentially contribute to the experimental ensemble by theoretical methods and, in turn, the evaluation of their relative population by comparing experimental and computational data. As a result, the probability of the β-hairpin conformation existing in solution was 74%. Substituting the chlorine atom with a methyl group, the probability for the corresponding HB-stabilized conformer was only 29%. The substitution of chlorine with bromine did not increase the probability of the β-hairpin conformation, as expected, likely because of the higher steric demand of the bromine, which would cause a deformation of the backbone.

These examples demonstrated how NOE-based techniques can be of fundamental importance for the structural identification of simple and complex, inter- and intramolecular XB interactions.

4.2. Diffusion NMR

For diffusion NMR techniques (DOSY or PGSE), the first application, to the best of my knowledge, was in a study published in 2012 by Erdélyi and co-workers, dealing with the characterization of the $[N-X-N]^+$ bond [97]. The authors started from the fact that the analogous $[N-H-N]^+$ is generally asymmetric in solution (the central hydrogen is located closer to one basic center, $[N-H \ldots N]^+$) but symmetric in crystals (the hydrogen is located exactly in the midway between the two basic centers), [98] and synthesized molecular systems in which the N–X distances were free to adjust and others in which they were not. By means of the isotopic perturbation of equilibrium technique through the ^{13}C NMR detection, the authors suggested that, in contrast with HB, XB always prefers to be symmetrical in solution. In addition, the measurement of the diffusion coefficient of the cation (D_t^+) and the anion (D_t^-) allowed them to ascertain that all the systems form tight ion pairs [99] in CD_2Cl_2. However, even the proximity of the triflate is not enough to perturb the symmetry of the two XBs. It must be said that the simple presence of tight ion pairs could be not enough to perturb a similar system, since the anion should also be located close to the X^+ moiety to exert its effect, which is not obvious [100–103]. In the author's opinion, since some of the employed anions contains fluorine nuclei, a ^{19}F, ^1H HOESY experiment between the anion and the cation (see above) could have further clarified this point. The $[N-X-N]^+$ system remains symmetrical, also varying the nature of the anion [104]. A notable exception has been found when X = F, for which the asymmetric conformation is preferred [105]. Interestingly, in this case the authors did not observe any NOE contact between the fluorine and the pyridine moieties.

In 2012, Beer and co-workers synthesized a macrocyclic halo-imidazolium receptor able to recognize iodide and bromide in water and produce a fluorescence signal as a response [106]. Interestingly, the crystal structure critically depends on the anions: in the presence of two PF_6^-, no XB can be observed between the iodine of the cation and the anion; on the contrary, in the presence of one PF_6^- and an iodide, or bromide, a dimeric structure is found in the solid-state, with two cations interacting with two different halide in a pincer-like arrangement (Figure 9).

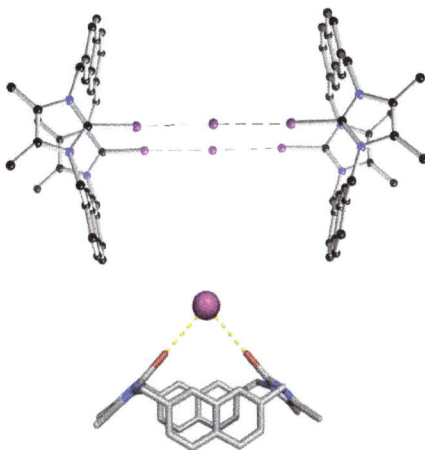

Figure 9. Adapted with permission from reference [106]. (Top) Dimeric crystal structure of iodo-imidazolium (PF$_6^-$ and hydrogens are omitted for clarity) and (bottom) density functional theory (DFT) structure of the monomeric structure present in solution. Copyright 2012 American Chemical Society.

The titration technique produced a very high value of K_a (>104 M^{-1}) and the Job plot suggested a cation:anion = 1:1 stoichiometry. But the latter cannot exclude the 2:2 structure found in the solid-state, therefore a technique able to determine the absolute hydrodynamic volume of the adduct in solution, as DOSY, was necessary. The authors measured the D_t of the cation with 0, 6 and 10 equivalents of NBu$_4$I. The presence of a dimeric structure would have led to a decrease of D_t as [I$^-$] increased, but the three values of D_t resulted in being very similar, once corrected for the viscosity of the solution. Consequently, only monomers were present in solution. DFT calculations provided a plausible, alternative structure (Figure 9), in which the two iodine atoms form two XBs with the same iodide anion. This structure has also a lower entropic cost. The message of this example is clear and of primary importance: never accept a priori the solid-state structure as a good model for the solution structure. Often they will be similar, but looking for experimental proofs confirming this is always a good idea.

Philp and co-workers synthesized an iodotriazole with a pending pentafluorophenyl moiety, demonstrating that it is a XB acceptor as good as the prototypical I2 [47]. As already noted, the K_a between the iodotriazole and 4-methylpyridine cannot be estimated through a ^{19}F NMR titration, as the fluorine atoms are too far from the XB acceptor group, but the use of ^{1}H, ^{15}N HMBC NMR spectroscopy led to a value of 1.67 M^{-1} (K_a = 2.67 M^{-1} for the I2/4-methylpyridine). The substitution of the *para*-F with a 3-hydroxyl-pyridine group led to a molecule having both XB acceptor and donor groups on the same side, leading, according to DFT calculations, to a stable doubly-halogen bonded homodimer. Interestingly, the solid state structural characterization showed that homodimers were not present and the iodine and the pyridine were connected by XB to different molecules, with the formation of supramolecular chains. In solution, the analysis of the ^{1}H NMR spectrum at different concentrations led to a self-aggregation constant of 3.4 M^{-1}, indicating that the two XBs behave almost independently from each other. Measuring the D_t of the molecule at different concentrations, the data could be satisfactorily fitted with a dimerization model and the value of D_t remained practically constant between 150 and 200 mM, an indication that a *plateau* had been reached. Combining the value of the self-aggregation constant, which is double with respect to the iodotriazole/4-methylpyridine case, and the trend of D_t vs. the concentration, the homodimer model indeed seems to be more accurate.

Finally, Ciancaleoni and co-workers published a paper completely focused on the NMR diffusion technique as a tool for characterizing single and multi-site XB adducts [107]. Firstly, they show how converting the values of D_t in volumes, using the corrected version of the Stokes–Einstein equation [73], can make the technique more useful and intuitive. The main advantage is that we can easily predict the volume of the adducts (Equation (10)), since generally the latter is the simple sum of the volumes of its components (this is not true when we deal with D_t or hydrodynamic radius values) [108].

$$V_H^{agg}(n\mathbf{D}, m\mathbf{A}) = n * V_H^0(\mathbf{D}) + m * V_H^0(\mathbf{A})E \qquad (10)$$

where n and m are the stoichiometric coefficient in the case of n:m adducts.

The first consequence is that a rough estimation of the K_a can be obtained from a single measurement, as done for the weak interaction between 2,4,6-trimethylpyridine and **I1** (1.6 M^{-1}) by using Equation (11) [109].

$$V_H^{exp}(\mathbf{D}) = \alpha\, V_H^{agg}(n\mathbf{D}, m\mathbf{A}) + (1-\alpha)\, V_H^0(\mathbf{D}) \qquad (11)$$

where α is the association coefficient, from which the calculation of K_a is possible. This can be useful, for example, when the low solubility of the adduct or the single component does not allow for a standard titration. But importantly, the value of $1.6 \pm 0.5\,\mathrm{M}^{-1}$ cannot be assigned exclusively to XB, since as ^{19}F, ^{1}H HOESY data demonstrated [61], in solution the adducts held from a XB between the nitrogen and the iodine coexist with others held from dispersion forces between the fluorinated chain and the hydrogens of the pyridine. As diffusion NMR results depend only on the total presence of the adducts and not on the kind of interaction that lead to the formation of the adduct, $1.6\,\mathrm{M}^{-1}$ has to be the sum of all the possible equilibrium constants leading to 1:1 adducts. Interestingly, the K_a measured by the ^{19}F NMR titration is much lower, $0.85 \pm 0.01\,\mathrm{M}^{-1}$. Under the hypothesis that ^{19}F NMR titration results refer exclusively to XB adducts, which is questionable but a likely first approximation, the K_a for non-XB adducts can be estimated as $0.75\,\mathrm{M}^{-1}$.

In the same paper, it is also shown that mixing hexamethylentetramine (**HMTA**), a base with four equivalent XB-acceptor groups, and N-bromosuccinimide (**NBS**), 1:3 and 1:4 adducts are not present in solution. Also in this case, the use of hydrodynamic volumes made the analysis much easier: since the hydrodynamic volumes of isolated **HMTA** and **NBS** can be easily measured (137 and 189 Å^3, respectively), the V_H values of the different adducts can be calculated, and result in being 323, 457, 591 and 729 Å^3 for 1:1, 1:2, 1:3 and 1:4 adducts, respectively. Experimental values for V_H(**HMTA**) go from 189 to 438 Å^3 (in the presence of 0 and 44 equivalents, respectively) confirming the presence of just 1:1 and 1:2 adducts. Actually, 438 Å^3 is an average value and does not exclude a priori the presence of larger adducts, but analyzing the trend of the data with the concentration it can be seen that V_H(**HMTA**) reaches a *plateau* in correspondence with the 1:2 adduct (Figure 10), indicating that larger aggregates are absent. Having the stoichiometry of the adduct as an experimental result and not as a hypothesis, the experimental data could be fitted without ambiguity with the correct model. The authors used the theoretical charge displacement function analysis method [110–113] to demonstrate the anti-cooperative nature of XB interactions in this system: when one nitrogen donates electronic density to a **NBS** moiety, the other nitrogens become less basic and, therefore, less able to establish a new XB.

It is also interesting to see that, indeed, in the solid-state structure of **HMTA/NBS** only 1:2 adducts can be detected, whereas substituting **NBS** with N-iodosuccinimide (**NIS**), 1:4 adducts are clearly visible [114]. Unfortunately, the **HMTA/NIS** adduct was too insoluble to allow any PGSE measurement.

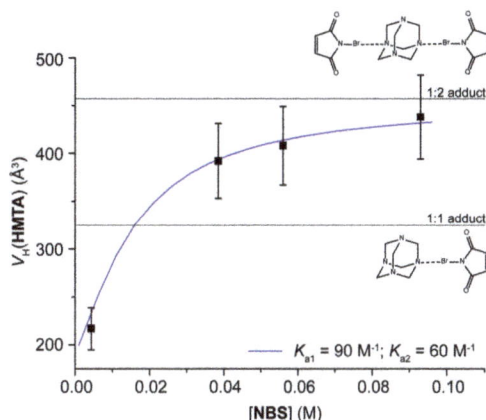

Figure 10. Adapted with permission from reference [107]. Experimental hydrodynamic volume of **HMTA** (C = 2.1 mM) at different concentrations of **NBS**. The solid line represents the best fitting equation for the 1:2 model. Copyright 2016 The Royal Society of Chemistry.

5. Conclusions

The present review demonstrates that XB adducts in solution should not be characterized only by titration and Job's plot techniques, and the latter should always be coupled with other, more sophisticated techniques enabled by modern NMR spectrometers. The most important information that can be derived concerns the internal structure (NOE) and the size (diffusion) of the adduct; but, depending on the system studied, the presence of competing interactions beyond the XB and the value(s) of equilibrium constants and thermodynamic parameters can also be derived. In some cases, the combination of experimental data with theoretical results is beneficial for providing a thorough description of the system.

For these reasons, it is expected that advanced NMR techniques will be used increasingly in the near future in the consolidated, but still fruitful and rapidly evolving, field of halogen bonding in solution.

Conflicts of Interest: The author declares no conflict of interest.

References and Notes

1. Hassel, O.; Hvoslef, J. The Structure of Bromine 1,4-Dioxanate. *Acta Chem. Scand.* **1954**, *8*, 873. [CrossRef]
2. Hassel, O. Structural aspects of interatomic charge-transfer bonding. *Science* **1970**, *170*, 497–502. [CrossRef] [PubMed]
3. Cavallo, G.; Metrangolo, P.; Milani, R.; Pilati, T.; Priimagi, A.; Resnati, G.; Terraneo, G. The Halogen Bond. *Chem. Rev.* **2016**, *116*, 2478–2601. [CrossRef] [PubMed]
4. Beale, T.M.; Chudzinski, M.G.; Sarwar, M.G.; Taylor, M.S. Halogen bonding in solution: Thermodynamics and applications. *Chem. Soc. Rev.* **2013**, *42*, 1667–1680. [CrossRef] [PubMed]
5. Bulfield, D.; Huber, S.M. Halogen Bonding in Organic Synthesis and Organocatalysis. *Chem. Eur. J.* **2016**, *22*, 14434–14450. [CrossRef] [PubMed]
6. Kolár, M.H.; Hobza, P. Computer Modeling of Halogen Bonds and Other σ-Hole Interactions. *Chem. Rev.* **2016**, *116*, 5155–5187. [CrossRef] [PubMed]
7. Wang, H.; Wang, W.; Jin, W.J. σ-Hole Bond vs π-Hole Bond: A Comparison Based on Halogen Bond. *Chem. Rev.* **2016**, *116*, 5072–5104. [CrossRef] [PubMed]

8. Desiraju, G.R.; Ho, P.S.; Kloo, L.; Legon, A.C.; Marquardt, R.; Metrangolo, P.; Politzer, P.; Resnati, G.; Rissanen, K. Definition of the Halogen Bond (IUPAC Recommendations 2013). *Pure Appl. Chem.* **2013**, *85*, 1711–1713. [CrossRef]

9. Zhang, Q.; Xu, Z.; Shi, J.; Zhu, W. The Underestimated Halogen Bonds Forming with Protein Backbone in Protein Data Bank. *J. Chem. Inf. Model.* **2017**, *57*, 1529–1534. [CrossRef] [PubMed]

10. Aragoni, M.C.; Arca, M.; Devillanova, F.A.; Isaia, F.; Lippolis, V. Adducts of S/Se Donors with Dihalogens as a Source of Information for Categorizing the Halogen Bonding. *Cryst. Growth Des.* **2012**, *12*, 2769–2779. [CrossRef]

11. Poznański, J.; Poznańska, A.; Shugar, D. A Protein Data Bank Survey Reveals Shortening of Intermolecular Hydrogen Bonds in Ligand-Protein Complexes When a Halogenated Ligand Is an H-Bond Donor. *PLoS ONE* **2014**, *6*, e99984. [CrossRef] [PubMed]

12. Mooibroek, T.J.; Gamez, P. Halogen bonding versus hydrogen bonding: What does the Cambridge Database reveal? *Cryst. Eng. Comm.* **2013**, *15*, 4565–4570. [CrossRef]

13. Wang, Y.; Wu, W.; Liu, Y.; Lu, Y. Influence of transition metal coordination on halogen bonding: CSD survey and theoretical study. *Chem. Phys. Lett.* **2013**, *578*, 38–42. [CrossRef]

14. Farina, A.; Meille, S.V.; Messina, M.T.; Metrangolo, P.; Resnati, G.; Vecchio, G. Resolution of Racemic 1,2-Dibromohexa-fluoropropane through Halogen-Bonded Supramolecular Helices. *Angew. Chem. Int. Ed. Engl.* **1999**, *38*, 2433–2436. [CrossRef]

15. Takeuchi, T.; Minato, Y.; Takase, M.; Shinmori, H. Molecularly Imprinted Polymers with Halogen Bonding-Based Molecular Recognition Sites. *Tetrahedron Lett.* **2005**, *46*, 9025–9027. [CrossRef]

16. Martí-Rujas, J.; Meazza, L.; Keat Lim, G.; Terraneo, G.; Pilati, T.; Harris, K.D.M.; Metrangolo, P.; Resnati, G. An Adaptable and Dynamically Porous Organic Salt Traps Unique Tetrahalide Dianions. *Angew. Chem. Int. Ed.* **2013**, *52*, 13444–13448. [CrossRef] [PubMed]

17. Nguyen, H.L.; Horton, P.N.; Hursthouse, M.B.; Legon, A.C.; Bruce, D.W. Halogen Bonding: A New Interaction for Liquid Crystal Formation. *J. Am. Chem. Soc.* **2004**, *126*, 16–17. [CrossRef] [PubMed]

18. Bolton, O.; Lee, K.; Kim, H.-J.; Lin, K.Y.; Kim, J. Activating Efficient Phosphorescence from Purely Organic Materials by Crystal Design. *Nat. Chem.* **2011**, *3*, 205–210. [CrossRef] [PubMed]

19. Yan, D.; Delori, A.; Lloyd, G.O.; Friščić, T.; Day, G.M.; Jones, W.; Lu, J.; Wei, M.; Evans, D.G.; Duan, X. A Cocrystal Strategy to Tune the Luminescent Properties of Stilbene-Type Organic Solid-State Materials. *Angew. Chem. Int. Ed.* **2011**, *50*, 12483–12486. [CrossRef] [PubMed]

20. Atzori, M.; Serpe, A.; Deplano, P.; Schlueter, J.A.; Laura Mercuri, M. Tailoring Magnetic Properties of Molecular Materials through Non-Covalent Interactions. *Inorg. Chem. Front.* **2015**, *2*, 108–115. [CrossRef]

21. Fourmigué, M.; Batail, P. Activation of Hydrogen- and Halogen-Bonding Interactions in Tetrathiafulvalene-Based Crystalline Molecular Conductors. *Chem. Rev.* **2004**, *104*, 5379–5418. [CrossRef] [PubMed]

22. Sarwar, M.G.; Dragisić, B.; Dimitrijević, E.; Taylor, M.S. Halogen Bonding between Anions and Iodoperfluoroorganics: Solution-Phase Thermodynamics and Multidentate-Receptor Design. *Chem. Eur. J.* **2013**, *19*, 2050–2058. [CrossRef] [PubMed]

23. Sarwar, M.G.; Dragisic, B.; Sagoo, S.; Taylor, M.S. A Tridentate Halogen-Bonding Receptor for Tight Binding of Halide Anions. *Angew. Chem. Int. Ed.* **2010**, *49*, 1674–1677. [CrossRef] [PubMed]

24. Gilday, L.C.; White, N.G.; Beer, P.D. Halogen- and hydrogen-bonding triazole-functionalised porphyrin-based receptors for anion recognition. *Dalton Trans.* **2013**, *42*, 15766–15773. [CrossRef] [PubMed]

25. Walter, S.M.; Kniep, F.; Herdtweck, E.; Huber, S.M. Halogen-Bond-Induced Activation of a Carbon-Heteroatom Bond. *Angew. Chem. Int. Ed.* **2011**, *50*, 7187–7191. [CrossRef] [PubMed]

26. Coulembier, O.; Meyer, F.; Dubois, P. Controlled Room Temperature ROP of L-Lactide by ICl3: A Simple Halogen-Bonding Catalyst. *Polym. Chem.* **2010**, *1*, 434–437. [CrossRef]

27. He, W.; Ge, Y.C.; Tan, C.H. Halogen-Bonding-Induced Hydrogen Transfer to C=N Bond with Hantzsch Ester. *Org. Lett.* **2014**, *16*, 3244–3247. [CrossRef] [PubMed]

28. Thordarson, P. Determining association constants from titration experiments in supramolecular chemistry. *Chem. Soc. Rev.* **2011**, *40*, 1305–1323. [CrossRef] [PubMed]

29. Laurence, C.; Queignec-Cabanetos, M.; Dziembowska, T.; Queignec, R.; Wojtkowiak, B. 1-Iodoacetylenes. 1. Spectroscopic evidence of their complexes with Lewis bases. A spectroscopic scale of soft basicity. *J. Am. Chem. Soc.* **1981**, *103*, 2567–2573. [CrossRef]

30. Walker, O.J. Absorption spectra of iodine solutions and the influence of the solvent. *Trans. Faraday Soc.* **1935**, *31*, 1432–1438. [CrossRef]

31. Erdélyi, M. Halogen bonding in solution. *Chem. Soc. Rev.* **2012**, *41*, 3547–3557. [CrossRef] [PubMed]

32. Webb, J.E.A.; Crossley, M.J.; Turner, P.; Thordarson, P. Pyromellitamide Aggregates and Their Response to Anion Stimuli. *J. Am. Chem. Soc.* **2007**, *129*, 7155–7162. [CrossRef] [PubMed]

33. Pastor, A.; Martınez-Viviente, E. NMR spectroscopy in coordination supramolecular chemistry: A unique and powerful methodology. *Coord. Chem. Rev.* **2008**, *252*, 2314–2345. [CrossRef]

34. Ciancaleoni, G.; Zuccaccia, C.; Zuccaccia, D.; Macchioni, A. *Techniques in Inorganic Chemistry*; Fackler, J.P., Jr., Falvello, L., Eds.; CRC: Boca Raton, FL, USA, 2011; pp. 129–180, ISBN 978-1-4398-1514-4.

35. Macchioni, A.; Ciancaleoni, G.; Zuccaccia, C.; Zuccaccia, D. Diffusion Ordered NMR Spectroscopy (DOSY). In *Supramolecular Chemistry: From Molecules to Nanomaterial*; Gale, P.A., Steed, J.W., Eds.; John Wiley & Sons, Ltd.: New York, NY, USA, 2012; Volume 2, Chapter 4; ISBN 978-0-470-74640-0.

36. Bellachioma, G.; Ciancaleoni, G.; Zuccaccia, C.; Zuccaccia, D.; Macchioni, A. NMR investigation of non-covalent aggregation of coordination compounds ranging from dimers and ion pairs up to nano-aggregates. *Coord. Chem. Rev.* **2008**, *252*, 2224–2238. [CrossRef]

37. Rocchigiani, L.; Macchioni, A. Disclosing the multi-faceted world of weakly interacting inorganic systems by means of NMR spectroscopy. *Dalton Trans.* **2016**, *45*, 2785–2790. [CrossRef] [PubMed]

38. Neuhaus, D.; Williamson, M. *The Nuclear Overhauser Effect in Structural and Conformational Analysis*, 2nd ed.; WILEY-VCH: New York, NY, USA, 2000; ISBN 978-0-471-24675-6.

39. Stilbs, P. Fourier transform pulsed-gradient spin-echo studies of molecular diffusion. *Prog. Nucl. Magn. Reson. Spectrosc.* **1987**, *19*, 1–45. [CrossRef]

40. Price, W.S. Pulsed-field gradient nuclear magnetic resonance as a tool for studying translational diffusion. Part I. Basic theory. *Concepts Magn. Res.* **1997**, *9*, 299–336. [CrossRef]

41. Price, W.S. Pulsed-field gradient nuclear magnetic resonance as a tool for studying translational diffusion. Part II. Experimental aspects. *Concepts Magn. Res.* **1998**, *10*, 197–237. [CrossRef]

42. Bertrán, J.F.; Rodríguez, M. Detection of halogen bond formation by correlation of proton solvent shifts. 1. Haloforms in n-electron donor solvents. *Org. Magn. Reson.* **1979**, *12*, 92–94. [CrossRef]

43. Metrangolo, P.; Resnati, G. Halogen Bonding: A Paradigm in Supramolecular Chemistry. *Chem. Eur. J.* **2001**, *7*, 2511–2519. [CrossRef]

44. Carlsson, A.-C.C.; Veiga, A.X.; Erdelyi, M. Halogen Bonding in Solution. *Top. Curr. Chem.* **2015**, *359*, 49–76. [CrossRef] [PubMed]

45. Thorson, R.A.; Woller, G.R.; Driscoll, Z.L.; Geiger, B.E.; Moss, C.A.; Schlapper, A.L.; Speetzen, E.D.; Bosch, E.; Erdélyi, M.; Bowling, N.P. Intramolecular Halogen Bonding in Solution: ^{15}N, ^{13}C, and ^{19}F NMR Studies of Temperature and Solvent Effects. *Eur. J. Org. Chem.* **2015**, *2015*, 1685–1695. [CrossRef]

46. Hakkert, S.B.; Gräfenstein, J.; Erdelyi, M. The ^{15}N NMR chemical shift in the characterization of weak halogen bonding in solution. *Faraday Discuss.* **2017**. [CrossRef] [PubMed]

47. Maugeri, L.; Asencio-Hernández, J.; Lébl, T.; Cordes, D.B.; Slawin, A.M.Z.; Delsuc, M.-A.; Philp, D. Neutral iodotriazoles as scaffolds for stable halogen-bonded assemblies in solution. *Chem. Sci.* **2016**, *7*, 6422–6428. [CrossRef] [PubMed]

48. Webb, J.A.; Klijn, J.E.; Hill, P.A.; Bennett, J.L.; Goroff, N.S. Experimental Studies of the ^{13}C NMR of Iodoalkynes in Lewis-Basic Solvents. *J. Org. Chem.* **2004**, *69*, 660–664. [CrossRef] [PubMed]

49. Ma, N.; Zhang, Y.; Ji, B.; Tian, A.; Wang, W. Structural Competition between Halogen Bonds and Lone-Pair$\cdots\pi$ Interactions in Solution. *Chem. Phys. Chem.* **2012**, *13*, 1411–1414. [CrossRef] [PubMed]

50. Rocchigiani, L.; Ciancaleoni, G.; Zuccaccia, C.; Macchioni, A. Probing the Association of Frustrated Phosphine—Borane Lewis Pairs in Solution by NMR Spectroscopy. *J. Am. Chem. Soc.* **2014**, *136*, 112–115. [CrossRef] [PubMed]

51. Korenaga, T.; Shoji, T.; Onoue, K.; Sakai, T. Demonstration of the existence of intermolecular lone pair$\cdots\pi$ interaction between alcoholic oxygen and the C_6F_5 group in organic solvent. *Chem. Commun.* **2009**, 4678–4680. [CrossRef] [PubMed]

52. Viger-Gravel, J.; Leclerc, S.; Korobkov, I.; Bryce, D.L. Correlation between ^{13}C chemical shifts and the halogen bonding environment in a series of solid para-diiodotetrafluorobenzene complexes. *CrystEngComm* **2013**, *15*, 3168–3177. [CrossRef]

53. Vioglio, P.C.; Chierotti, M.R.; Gobetto, R. Solid-state nuclear magnetic resonance as a tool for investigating the halogen bond. *Cryst. Eng. Comm.* **2016**, *18*, 9173–9184. [CrossRef]

54. Vioglio, P.C.; Catalano, L.; Vasylyeva, V.; Nervi, C.; Chierotti, M.R.; Resnati, G.; Gobetto, R.; Metrangolo, P. Natural Abundance 15N and 13C Solid-State NMR Chemical Shifts: High Sensitivity Probes of the Halogen Bond Geometry. *Chem. Eur. J.* **2016**, *22*, 16819–16828. [CrossRef] [PubMed]

55. Cabot, R.; Hunter, C.A. Non-covalent interactions between iodo-perfluorocarbons and hydrogen bond acceptors. *Chem. Commun.* **2009**, 2005–2007. [CrossRef] [PubMed]

56. Shen, Q.J.; Jin, W.J. Strong halogen bonding of 1,2-diiodoperfluoroethane and 1,6-diiodoperfluorohexane with halide anions revealed by UV-Vis, FT-IR, NMR spectroscopes and crystallography. *Phys. Chem. Chem. Phys.* **2011**, *13*, 13721–13729. [CrossRef] [PubMed]

57. Liu, Z.-X.; Sun, Y.; Feng, Y.; Chen, H.; He, Y.-M.; Fan, Q.-H. Halogen-Bonding for Visual Chloride Ion Sensing: A Case Study Using Supramolecular Poly(aryl ether) Dendritic Organogel System. *Chem. Commun.* **2016**, *52*, 2269–2272. [CrossRef] [PubMed]

58. Mungalpara, D.; Stegmüller, S.; Kubik, S. A neutral halogen bonding macrocyclic anion receptor based on a pseudocyclopentapeptide with three 5-iodo-1,2,3-triazole subunits. *Chem. Commun.* **2017**, *53*, 5095–5098. [CrossRef] [PubMed]

59. Noggle, J.H.; Schirmer, R.E. *The Nuclear Overhauser Effect*; Academic Press: New York, NY, USA, 1971; ISBN 9780323141390.

60. Keeler, J. Understanding NMR Spectroscopy. 2002. Available online: http://www-keeler.ch.cam.ac.uk/lectures/Irvine/ (accessed on 28 August 2017).

61. Ciancaleoni, G.; Bertani, R.; Rocchigiani, L.; Sgarbossa, P.; Zuccaccia, C.; Macchioni, A. Discriminating Halogen-Bonding from Other Noncovalent Interactions by a Combined NOE NMR/DFT Approach. *Chem. Eur. J.* **2015**, *21*, 440–447. [CrossRef] [PubMed]

62. Zuccaccia, C.; Bellachioma, G.; Cardaci, G.; Macchioni, A. Solution structure investigation of Ru(II) complex ion pairs: Quantitative NOE measurements and determination of average interionic distances. *J. Am. Chem. Soc.* **2001**, *123*, 11020–11028. [CrossRef] [PubMed]

63. Zuccaccia, C.; Bellachioma, G.; Cardaci, G.; Macchioni, A. Specificity of interionic contacts and estimation of average interionic distances by NOE NMR measurements in solution of cationic Ru(II) organometallic complexes bearing unsymmetrical counterions. *Organometallics* **1999**, *18*, 1–3. [CrossRef]

64. Ciancaleoni, G.; Zuccaccia, C.; Zuccaccia, D.; Macchioni, A. Diffusion and NOE NMR studies on the interactions of neutral amino-acidate arene ruthenium(II) supramolecular aggregates with ions and ion pairs. *Magn. Reson. Chem.* **2008**, *46*, S72–S79. [CrossRef] [PubMed]

65. Geldbach, T. J.; Ruegger, H.; Pregosin, P. S. NOESY, HOESY, T1 and solid-state NMR studies on [RuH(h6-toluene)(Binap)](CF3SO3): A molecule with a strongly distorted piano-stool structure. *Magn. Reson. Chem.* **2003**, *41*, 703–708. [CrossRef]

66. Lingscheid, Y.; Arenz, S.; Giernoth, R. Heteronuclear NOE Spectroscopy of Ionic Liquids. *Chem. Phys. Chem.* **2012**, *13*, 261–266. [CrossRef] [PubMed]

67. Braun, D.; Steinhauser, O. The intermolecular NOE is strongly influenced by dynamics. *Phys. Chem. Chem. Phys.* **2015**, *17*, 8509–8517. [CrossRef] [PubMed]

68. Gabl, S.; Schröder, C.; Braun, D.; Weingärtner, H.; Steinhauser, O. Pair dynamics and the intermolecular nuclear Overhauser effect (NOE) in liquids analysed by simulation and model theories: Application to an ionic liquid. *J. Chem. Phys.* **2012**, *140*, 184503. [CrossRef] [PubMed]

69. Crank, J. *The Mathematics of Diffusion*, 2nd ed.; Clarendon Press: Oxford, UK, 1975; ISBN 978-0198534112.

70. Einstein, A. *Investigations in the Theory of Brownian Movements*; Dover: New York, NY, USA, 1956; ISBN 9781607962854.

71. Hahn, E.L. Spin echoes. *Phys. Rev.* **1950**, *80*, 580–594. [CrossRef]

72. Stejskal, E.O.; Tanner, J.E. Spin diffusion measurements: Spin echoes in the presence of a time-dependent field gradient. *J. Chem. Phys.* **1965**, *42*, 288–292. [CrossRef]

73. Macchioni, A.; Ciancaleoni, G.; Zuccaccia, C.; Zuccaccia, D. Determining accurate molecular sizes in solution through NMR diffusion spectroscopy. *Chem. Soc. Rev.* **2008**, *37*, 479–489. [CrossRef] [PubMed]

74. By using this formulation, the effect of T_2 can be neglected, since it is the same with and without G.

75. Edward, J.T. Molecular volumes and the Stokes-Einstein equation. *J. Chem. Educ.* **1970**, *47*, 261–270. [CrossRef]

76. Chen, H.-C.; Chen, S.-H. Diffusion of crown ethers in alcohols. *J. Phys. Chem.* **1984**, *88*, 5118–5121. [CrossRef]
77. Sinnaeve, D. Simultaneous solvent and J-modulation suppression in PGSTE-based diffusion experiments. *J. Magn. Res.* **2014**, *245*, 24–30. [CrossRef] [PubMed]
78. Chen, A.; Wu, D.; Johnson, C.S., Jr. Determination of Molecular Weight Distributions for Polymers by Diffusion-Ordered NMR. *J. Am. Chem. Soc.* **1995**, *117*, 7965–7970. [CrossRef]
79. Marega, R.; Aroulmoji, V.; Bergamin, M.; Feruglio, L.; Dinon, F.; Bianco, A.; Murano, E.; Prato, M. Two-Dimensional Diffusion-Ordered NMR Spectroscopy as a Tool for Monitoring Functionalized Carbon Nanotube Purification and Composition. *ACS Nano* **2010**, *4*, 2051–2058. [CrossRef] [PubMed]
80. Rocchigiani, L.; Bellachioma, G.; Ciancaleoni, G.; Crocchianti, S.; Lagana, A.; Zuccaccia, C.; Zuccaccia, D.; Macchioni, A. Anion-Dependent Tendency of Di-Long-Chain Quaternary Ammonium Salts to Form Ion Quadruples and Higher Aggregates in Benzene. *Chem. Phys. Chem.* **2010**, *11*, 3243–3254. [CrossRef] [PubMed]
81. Pettirossi, S.; Bellachioma, G.; Ciancaleoni, G.; Zuccaccia, C.; Zuccaccia, D.; Macchioni, A. Diffusion and NOE NMR Studies on Multicationic DAB-Organoruthenium Dendrimers: Size-Dependent Noncovalent Self-Assembly to Megamers and Ion Pairing. *Chem. Eur. J.* **2009**, *15*, 5337–5347. [CrossRef] [PubMed]
82. Pregosin, P.S. NMR diffusion methods in inorganic and organometallic chemistry. *Spectrosc. Prop. Inorg. Organomet. Compd.* **2011**, *42*, 248–268. [CrossRef]
83. Rocchigiani, L.; Busico, V.; Pastore, A.; Macchioni, A. Probing the interactions between all components of the catalytic pool for homogeneous olefin polymerisation by diffusion NMR spectroscopy. *Dalton Trans.* **2013**, *42*, 9104–9111. [CrossRef] [PubMed]
84. Allouche, L.; Marquis, A.; Lehn, J.M. Discrimination of Metallosupramolecular Architectures in Solution by Using Diffusion Ordered Spectroscopy (DOSY) Experiments: Double-Stranded Helicates of Different Lengths. *Chem. Eur. J.* **2006**, *12*, 7520–7525. [CrossRef] [PubMed]
85. Ciancaleoni, G.; Zuccaccia, C.; Zuccaccia, D.; Clot, E.; Macchioni, A. Self-Aggregation Tendency of All Species Involved in the Catalytic Cycle of Bifunctional Transfer Hydrogenation. *Organometallics* **2009**, *28*, 960–967. [CrossRef]
86. Tatko, C.D.; Waters, M.L. Effect of Halogenation on Edge—Face Aromatic Interactions in a β-Hairpin Peptide: Enhanced Affinity with Iodo-Substituents. *Org. Lett.* **2006**, *6*, 3969–3972. [CrossRef] [PubMed]
87. Serpell, C.J.; Kila, N.L.; Costa, P.J.; Félix, V.; Beer, P.D. Halogen Bond Anion Templated Assembly of an Imidazolium Pseudorotaxane. *Angew. Chem. Int. Ed.* **2010**, *49*, 5322–5326. [CrossRef] [PubMed]
88. Gilday, L.C.; Beer, P.D. Halogen- and Hydrogen-Bonding catenanes for Halide-Anion Recognition. *Chem. Eur. J.* **2014**, *20*, 8379–8385. [CrossRef] [PubMed]
89. Huber, S.M.; Scanlon, J.D.; Jimenez-Izal, E.; Ugalde, J.M.; Infante, I. On the directionality of halogen bonding. *Phys. Chem. Chem. Phys.* **2013**, *15*, 10350–10357. [CrossRef] [PubMed]
90. Politzer, P.; Murray, J.S.; Clark, T. Halogen Bonding: An Electrostatically-Driven Highly Directional Noncovalent Interaction. *Phys. Chem. Chem. Phys.* **2010**, *12*, 7748–7757. [CrossRef] [PubMed]
91. Kozuch, S.; Martin, J.M.L. Halogen Bonds: Benchmarks and Theoretical Analysis. *J. Chem. Theory Comput.* **2013**, *9*, 1918–1931. [CrossRef] [PubMed]
92. Yan, X.Q.; Zhao, X.R.; Wang, H.; Jin, W.J. The Competition of σ-Hole···Cl− and π-Hole···Cl− Bonds between C_6F_5X (X = F, Cl, Br, I) and the Chloride Anion and Its Potential Application in Separation Science. *J. Phys. Chem. B* **2014**, *118*, 1080–1087. [CrossRef] [PubMed]
93. Cao, J.; Yan, X.; He, W.; Li, X.; Li, Z.; Mo, Y.; Liu, M.; Jiang, Y.-B. C−I···π Halogen Bonding Driven Supramolecular Helix of Bilateral N-Amidothioureas Bearing β-Turns. *J. Am. Chem. Soc.* **2017**, *139*, 6605–6610. [CrossRef] [PubMed]
94. In the author's opinion, NOESY and peak intensities should have been coupled with a simple DOSY experiment (see below) that would have proved the existence of supramolecular oligomers more effectively.
95. Casnati, A.; Liantonio, R.; Metrangolo, P.; Resnati, G.; Ungaro, R.; Ugozzoli, F. Molecular and Supramolecular Homochirality: Enantiopure Perfluorocarbon Rotamers and Halogen-Bonded Fluorous Double Helices. *Angew. Chem. Int. Ed.* **2006**, *45*, 1915–1918. [CrossRef] [PubMed]
96. Danelius, E.; Andersson, H.; Jarvoll, P.; Lood, K.; Gräfenstein, J.; Erdélyi, M. Halogen Bonding: A Powerful Tool for Modulation of Peptide Conformation. *Biochemistry* **2017**, *56*, 3265–3272. [CrossRef] [PubMed]
97. Carlsson, A.-C.C.; Gräfenstein, J.; Budnjo, A.; Laurila, J.L.; Bergquist, J.; Karim, A.; Kleinmaier, R.; Brath, U.; Erdélyi, M. Symmetric Halogen Bonding Is Preferred in Solution. *J. Am. Chem. Soc.* **2012**, *134*, 5706–5715. [CrossRef] [PubMed]

98. Perrin, C.L. Are Short, Low-Barrier Hydrogen Bonds Unusually Strong? *Acc. Chem. Res.* **2010**, *43*, 1550–1557. [CrossRef] [PubMed]

99. Macchioni, A. Ion Pairing in Transition-Metal Organometallic Chemistry. *Chem. Rev.* **2005**, *105*, 2039–2074. [CrossRef] [PubMed]

100. For example, in [L^1AuL^2]X ion pairs, the positive charge is formally located on the gold, but the anion position depends on the nature of the ligands. See references 101–103.

101. Zuccaccia, D.; Belpassi, L.; Tarantelli, F.; Macchioni, A. Ion Pairing in Cationic Olefin–Gold(I) Complexes. *J. Am. Chem. Soc.* **2009**, *131*, 3170–3171. [CrossRef] [PubMed]

102. Ciancaleoni, G.; Belpassi, L.; Tarantelli, F.; Zuccaccia, D.; Macchioni, A. A combined NMR/DFT study on the ion pair structure of [($PR^1_2R^2$)Au($\eta^{2–3}$-hexyne)]BF_4 complexes. *Dalton Trans.* **2013**, *42*, 4122–4131. [CrossRef] [PubMed]

103. Biasiolo, L.; Ciancaleoni, G.; Belpassi, L.; Bistoni, G.; Macchioni, A.; Tarantelli, F.; Zuccaccia, D. Relationship between the anion/cation relative orientation and the catalytic activity of nitrogen acyclic carbene–gold catalysts. *Catal. Sci. Technol.* **2015**, *5*, 1558–1567. [CrossRef]

104. Bedin, M.; Karim, A.; Reitti, M.; Carlsson, A.-C.C.; Topić, F.; Cetina, M.; Pan, F.; Havel, V.; Al-Ameri, F.; Sindelar, V.; et al. Counterion influence on the N–I–N halogen bond. *Chem. Sci.* **2015**, *6*, 3746–3756. [CrossRef]

105. Karim, A.; Reitti, M.; Carlsson, A.-C.C.; Gräfenstein, J.; Erdélyi, M. The nature of [N–Cl–N]$^+$ and [N–F–N]$^+$ halogen bonds in solution. *Chem. Sci.* **2014**, *5*, 3226–3233. [CrossRef]

106. Zapata, F.; Caballero, A.; White, N.G.; Claridge, T.D.W.; Costa, P.J.; Félix, V.; Beer, P.D. Fluorescent Charge-Assisted Halogen-Bonding Macrocyclic Halo-Imidazolium Receptors for Anion Recognition and Sensing in Aqueous Media. *J. Am. Chem. Soc.* **2012**, *134*, 11533–11541. [CrossRef] [PubMed]

107. Ciancaleoni, G.; Macchioni, A.; Rocchigiani, L.; Zuccaccia, C. A PGSE NMR approach to the characterization of single and multi-site halogen-bonded adducts in solution. *RSC Adv.* **2016**, *6*, 80604–80612. [CrossRef]

108. If inter-penetrating adducts are involved, as in the case of helices or cages, this could be not true.

109. If the adduct and the components are not in rapid exchange, the Equation (11) is not valid anymore. But the kinetics of XB formation is generally very fast.

110. Belpassi, L.; Infante, I.; Tarantelli, F.; Visscher, L. The Chemical Bond between Au(I) and the Noble Gases. Comparative Study of NgAuF and NgAu$^+$ (Ng = Ar, Kr, Xe) by Density Functional and Coupled Cluster Methods. *J. Am. Chem. Soc.* **2008**, *130*, 1048–1060. [CrossRef] [PubMed]

111. Ciancaleoni, G.; Santi, C.; Ragni, M.; Braga, A.M. Charge–displacement analysis as a tool to study chalcogen bonded adducts and predict their association constants in solution. *Dalton Trans.* **2015**, *44*, 20168–20175. [CrossRef] [PubMed]

112. Ciancaleoni, G.; Belpassi, L.; Marchetti, F. Back-Donation in High-Valent d^0 Metal Complexes: Does It Exist? The Case of Nb^V. *Inorg. Chem.* **2017**, *56*, 11266–11274. [CrossRef] [PubMed]

113. Bistoni, G.; Rampino, S.; Scafuri, N.; Ciancaleoni, G.; Zuccaccia, D.; Belpassi, L.; Tarantelli, F. How π back-donation quantitatively controls the CO stretching response in classical and non-classical metal carbonyl complexes. *Chem. Sci.* **2016**, *7*, 1174–1184. [CrossRef]

114. Raatikainen, K.; Rissanen, K. Interaction between amines and N-haloimides: A new motif for unprecedentedly short Br$\cdots$N and I$\cdots$N halogen bonds. *CrystEngComm* **2011**, *13*, 6972–6977. [CrossRef]

MDPI

St. Alban-Anlage 66

4052 Basel

Switzerland

Tel. +41 61 683 77 34

Fax +41 61 302 89 18

www.mdpi.com